Improving the thermal processing of foods

Related titles from Woodhead's food science, technology and nutrition list:

Thermal technologies in food processing (ISBN 1 85573 558 X)

Thermal technologies have long been at the heart of food processing. The application of heat is both an important method of preserving foods and a means of developing texture, flavour and colour. An essential issue for food manufacturers is the effective application of thermal technologies to achieve these objectives without damaging other desirable sensory and nutritional qualities in a food product. Edited by a leading authority in the field, and with a distinguished international team of contributors, *Thermal technologies in food processing* addresses this major issue. It provides food manufacturers and researchers with an authoritative review of thermal processing and food quality.

Novel food packaging techniques (ISBN 1 85573 675 6)

Packaging continues to be one of the most innovative areas in food processing. With its distinguished international team of contributors, *Novel food packaging techniques* summarises the key developments in the field. The first part of the book discusses general issues such as packaging design, consumer attitudes to novel packaging and the legislative context. Part II looks at new techniques such as the use of oxygen and other scavengers, freshness indicators and antimicrobial packaging. The final part of the book discusses packaging materials and considers how packaging can be used with other preservation techniques to improve the quality of particular foods.

Food preservation techniques (ISBN 1 85573 530 X)

Extending the shelf-life of foods whilst maintaining safety and quality is a critical issue for the food industry. As a result there have been major developments in food preservation techniques, which are summarised in this authoritative collection. The first part of the book examines the key issue of maintaining safety as preservation methods become more varied and complex. The rest of the book looks both at individual technologies and how they are combined to achieve the right balance of safety, quality and shelf-life for particular products.

Details of these books and a complete list of Woodhead's food science, technology and nutrition titles can be obtained by:

- visiting our web site at www.woodhead-publishing.com
- contacting Customer Services (e-mail: sales@woodhead-publishing.com; fax: +44 (0) 1223 893694; tel.: +44 (0) 1223 891358 ext. 30; address: Woodhead Publishing Limited, Abington Hall, Abington, Cambridge CB1 6AH, England)

Selected food science and technology titles are also available in electronic form. Visit our web site (www.woodhead-publishing.com) to find out more.

If you would like to receive information on forthcoming titles in this area, please send your address details to: Francis Dodds (address, tel. and fax as above; e-mail: francisd@woodhead-publishing.com). Please confirm which subject areas you are interested in.

Improving the thermal processing of foods

Edited by
Philip Richardson

CRC Press
Boca Raton Boston New York Washington, DC

WOODHEAD PUBLISHING LIMITED
Cambridge England

Published by Woodhead Publishing Limited
Abington Hall, Abington
Cambridge CB1 6AH
England
www.woodhead-publishing.com

Published in North America by CRC Press LLC
2000 Corporate Blvd, NW
Boca Raton FL 33431
USA

First published 2004, Woodhead Publishing Limited and CRC Press LLC

British Library Cataloguing in Publication Data
A catalogue record for this book is available from the British Library.

Library of Congress Cataloging-in-Publication Data
A catalog record for this book is available from the Library of Congress.

Woodhead Publishing Limited ISBN 1 85573 730 2 (book); 1 85573 907 0 (e-book)
CRC Press ISBN 0-8493-2549-6
CRC Press order number: WP2549

The publisher's policy is to use permanent paper from mills that operate a sustainable forestry policy, and which has been manufactured from pulp which is processed using acid-free and elementary chlorine-free practices. Furthermore, the publisher ensures that the text paper and cover board used have met acceptable environmental accreditation standards.

Project managed by Macfarlane Production Services, Markyate, Hertfordshire
(e-mail: macfarl@aol.com)
Typeset by MHL Typesetting Limited, Coventry, Warwickshire
Printed by TJ International Limited, Padstow, Cornwall, England

Contents

Contributor contact details

Chapter 1

Mr S. D. Holdsworth
Withens
Stretton-on-Fosse
Moreton-in-Marsh
Glos GL56 9SG
UK

Tel: +44(0)1608 661782
E-mail:
sdholdsworth@ukonline.co.uk

Chapter 2

Dr J. C. Oliveira
Department of Process Engineering
University College Cork
Cork
Ireland

Tel: 353 21 4902006
Fax: 353 21 4270249
E-mail: j.oliveira@ucc.ie

Chapter 3

Dr R. Simpson
Departamento de Procesos Químicos, Biotecnológicos, y Ambientales
Universidad Técnica Federico Santa María
Casilla 110-V
Valparaíso
Chile

Tel: 56 32 654302
Fax: 56 32 654478
E-mail: ricardo.simpson@usm.cl

Chapter 4

Dr P. Verboven, Dr J. De Baerdemaeker and Dr B. M. Nicolaï
Laboratory/Flemish Centre of Postharvest Technology
K U Leuven
W de Croylaan 42
B-3001 Leuven
Belgium

Tel: +32 (0) 16 32 14 53
Fax: +32 (0) 16 32 29 55
E-mail:
pieter.verboven@agr.kuleuven.ac.be

Chapter 5

Dr G. Bown

E-mail:
graham.bown@btinternet.com

Chapter 6

Mr G. S. Tucker
Campden and Chorleywood Food Research Association
Chipping Campden
Gloucestershire
GL55 6LD
UK

Tel: +44 (0) 1386 842000
Fax: +44 (0) 1386 842100
E-mail: g.tucker@campden.co.uk

Chapter 7

Mr N. May
Campden and Chorleywood Food Research Association
Chipping Campden
Gloucestershire
GL55 6LD
UK

Tel: +44 (0) 1386 842000
Fax: +44 (0) 1386 842100
E-mail: n.may@campden.co.uk

Chapter 8

Dr S. Ghazala
Memorial University of Newfoundland
St John's
Newfoundland
Canada A1B 3X9

Tel: 709 737 2085
Fax: 709 737 8542
E-mail: ghazala@mun.ca

Chapter 9

Dr K. P. Sandeep, Dr J. Simunovic and Professor K. R. Swartzel
129 Schaub Hall
Box 7624
North Carolina State University
Raleigh
NC 27695 7624
USA

Tel: 919 515 2957
Fax: 919 515 7124
E-mail: kp_sandeep@ncsu.edu

Chapter 10

Mr G. S. Tucker
Campden and Chorleywood Food Research Association
Chipping Campden
Gloucestershire
GL55 6LD
UK

Tel: +44 (0) 1386 842000
Fax: +44 (0) 1386 842100
E-mail: g.tucker@campden.co.uk

Dr U. Bolmstedt
Tetra Pak Processing Components AB
Sweden

Chapter 11

Dr L. Wang and Dr B. Sundén
Division of Heat Transfer
Lund Institute of Technology
Box 118
221 00 Lund
Sweden

Tel: +46 46 222 86 05
Fax: +46 46 222 86 12
E-mail: Lieke.Wang@vok.lth.se
Bengt.Sunden@vok.lth.se

Chapter 12

Dr R. Ruan, Dr X. Ye and Dr P. Chen
University of Minnesota
USA

Tel: 612 625 1710
Fax: 612 624 3005
E-mail: ruanx001@tc.umn.edu

Dr C. Doona and Dr T. Yang
Combat Feeding Directorate
US Army Natick Soldier System Center
USA

Chapter 13

A. Sarkar and Professor R. P. Singh
Department of Biological and Agricultural Engineering
University of California, Davis
CA-95616
USA

Tel: 530 752 0811
Fax: 530 752 5293
E-mail: rpsingh@ucdavis.edu

Chapter 14

Dr K. Warriner
Department of Food Science
University of Guelph
Guelph
Ontario
Canada N1G 2W1

Tel: 519 824 4120 extn 56072
Fax: 519 824 6631
E-mail: kwarrine@uoguelph.ca

Dr S. Movahedi and Dr W. M. Waites
Division of Food Sciences
School of Biosciences
University of Nottingham
Loughborough
Leics
LE12 5RD
UK

Chapter 15

Dr F. Eszes and Dr R. Rajkó
College Faculty of Food Engineering
University of Szeged
H-6701 Szeged POB 433
Hungary

Tel: +36 62 546030
Fax: +36 62 546034
E-mail: rajko@sol.cc.u-szeged.hu

Chapter 16

Mr G. S. Tucker
Campden and Chorleywood Food Research Association
Chipping Campden
Gloucestershire
GL55 6LD
UK

Tel: +44 (0) 1386 842000
Fax: +44 (0) 1386 842100
E-mail: g.tucker@campden.co.uk

Chapter 17

Mr G. H. Shaw
Campden and Chorleywood Food Research Association
Chipping Campden
Gloucestershire
GL55 6LD
UK

Tel: +44 (0) 1386 842000
Fax: +44 (0) 1386 842100
E-mail: g.shaw@campden.co.uk

Chapter 18

Professor A. Van Loey, Dr Y. Guiavarc'h, Dr W. Claeys and Professor M. Hendrickx
Laboratory of Food Technology
Kasteelpark Arenberg 23
B-3001 Leuven
Belgium

Tel: +32 16 32 15 67
Fax: +32 16 32 19 60
E-mail: ann.vanloey@agr.kuleuven.ac.be

Chapter 19

Dr K. P. Nott and Professor L. D. Hall
Herchel Smith Laboratory for Medicinal Chemistry,
University of Cambridge School of Clinical Medicine
University Forvie Site, Robinson Way,
Cambridge
CB2 2PZ
UK

Tel: +44 (0) 1223 336805
Fax: +44 (0) 1223 336748
E-mail: kpn20@hslmc.cam.ac.uk

Chapter 20

Dr M. Peleg
Department of Food Science
Chenoweth Laboratory
University of Massachusetts at Amherst
MA 01003-1410
USA

Tel: 413 545 5842
Fax: 413 545 1262
E-mail: micha.peleg@foodsci.umass.edu

Chapter 21

Dr A. H. Geeraerd, V. P. Valdramidis, Dr K. Bernaerts and Professor J. F. Van Impe
BioTeC-Bioprocess Technology and Control
Department of Chemical Engineering
Katholieke Universiteit Leuven
W. de Croylaan 46
B-3001 Leuven
Belgium

Tel: +32 16 322675
Fax: +32 16 322991
E-mail: jan.vanimpe@cit.kuleuven.ac.be

Chapter 22

Dr J. T. Rosnes
Norconserv AS
Seafood Processing Research
Niels Juelsgt. 50
PO Box 327
4002 Stavanger
Norway

Tel: +47 51844600
Fax: +47 51844651
E-mail: jtr@norconserv.no

Chapter 23

Dr N. Heppell
Oxford Brookes University
Oxford
UK

Tel: +44 (0) 1865 483956
Fax: +44 (0) 1865 484017
E-mail: njheppell@brookes.ac.uk

The Editor

Professor P. S. Richardson
Campden and Chorleywood Food Research Association
Chipping Campden
Gloucestershire
GL55 6LD
UK

Tel: +44 (0) 1386 842000
Fax: +44 (0) 1386 842100
E-mail: p.richardson@campden.co.uk

Part I

Opimising thermal processes

1

Optimising the safety and quality of thermally processed packaged foods

S. D. Holdsworth, formerly Campden and Chorleywood Food Research Association, UK

1.1 Introduction: reconciling safety and quality

Ever since the invention of thermal processing as a method of preserving packaged foods by the Frenchman Nicolas Appert in the early 19th century, there has been a relentless search to reduce the amount of thermal damage to the quality of food products. Today there is a large range of packaging materials available, metallic cans, glass containers and plastic, which may be presented in a variety of different geometrical shapes. Whatever the material and its shape or the food product it is necessary to apply a suitable *process*, i.e., a given time at a specified temperature, to ensure that the products do not pose a public health problem, e.g., food poisoning. Equally it is necessary to ensure that the product has received sufficient heat to cook it and to maintain the highest possible quality. Thus the art of thermal processing of foods is to select suitable time/temperature combinations and cooling regimes which will ensure the above criteria. The term *process* should not be confused with the more conventional meaning of the word as a sequence of engineering operations.

For the adequate destruction of the spores of pathogenic microorganisms, whose toxins may cause food poisoning, practical processing temperatures of 110–130°C are required for times depending on the nature of the food product. The higher the temperature the shorter the time required. One of the most heat resistant spores is *Clostridium botulinum*, which unless inactivated will produce the lethal botulin toxin under the anaerobic conditions in the container. Consequently most *processes* are chosen on the basis of the destruction of this microorganism; the argument being that any products containing less resistant microorganisms will thereby be inactivated by a 'botulinum process'. It is important to realise that there are more heat-resistant organisms, i.e., thermophiles, present in thermally processed foods which do not present a threat to human health. Under normal storage conditions these will be innocuous; however, should the products be stored at tropical temperatures, e.g., 35°C and above, thermophiles will grow and ultimately swell or burst the container. Processed foods intended for such climatic conditions require a more severe process to stabilise them. Products, which receive a safe 'botulinum process' but still contain thermophiles, are described as 'commercially sterile'. The meaning of sterile as used in these processes differs from the absolute definition of sterility used by the medical profession, indicating free from living organisms.

An important source of post-process contamination is 'leaker spoilage' which, as the name implies, indicates that microorganisms have penetrated the container after processing and usually during cooling with water. Metallic containers are particularly prone to this problem, since during the cooling of the container a vacuum develops and microorganisms may enter through imperfections in the sealing. The problem is resolved by chlorinating or decontaminating the cooling water.

When food products are heated the components are generally affected by the length of the heating process and the level of the temperature. Some of the desirable effects are enzyme destruction (usually in the case of vegetables, this is achieved by pre-process blanching and cooking); undesirable effects include loss of vitamin potency, flavour changes, and texture and structure changes. Each product behaves differently and it is necessary to know the principal components which affect the quality, especially in the processing of formulated food, where the textural attributes are important in the finished product. Whilst this subject is of considerable commercial significance, relatively little is known about the kinetics of these complex processes compared with microbial destruction. In general, biochemical processes are much slower than microbial destruction processes, which is helpful in preserving the quality attributes of the products.

1.2 The kinetics of microbial inactivation during heat treatment

1.2.1 Heat resistance of microorganisms

The amount of heat required to inactivate microorganisms is an important property, which must be known or determined in order to specify a suitable process for a product, usually known as the specified process. Some typical data for some types of organism initially present in foods is given in Table 1.1. Traditional canning technology makes use of two factors to determine the time/temperature process required to produce a heat-stable food. The first of these is the decimal reduction time or D-value, which is defined as the time in minutes at any given temperature to destroy 90% of the spores or vegetative cells in a given

Table 1.1 Some inactivation data for microorganisms

Organism	Time/temperature
Vegetative cells	10 min/80°C
Yeast ascopspores	5 min/60°C
Fungi	30–60 min/88°C
Thermophiles:	
Clostridium thermosccharolyticum	3–4 min/121°C
Bacillus stearothermophilius	4 min/121°C
Mesophiles	
Clostridium botulinum	3 min/121°C
Botulinum A &B toxins	0.1–1 min/121°C
Clostridium sporogenes	1.5 min/121°C
Bacillus subtillis	0.6 min/121°C

organism. It may be obtained from heat-resistance studies by determining the number of survivors resulting from a given process. The plot of logarithm of the number of survivors versus temperature for a given organism versus time, see Fig. 1.1, is used to determine the D-value. This is known as the semi-logarithmic survivor curve, which has a slope of $-1/D$, the equation of the curve being given in equation 1.1.

$$\log N = \log N_o - t/D \tag{1.1}$$

where N is the number of surviving microorganisms, N_o is the initial number of microorganisms, t is the time in minutes and D is the decimal reduction time in minutes. Logarithms to the base 10 are indicated by log. Figure 1.2 shows various types of survivor curves encountered in canning microbiology.

The second factor is the thermal death constant z. This is the change of D-value with temperature and is obtained from a plot of log *D* versus temperature (see Fig 1.3). The equation for the D/z plot is given in equation 1.2:

$$\log D_T = \log D_{ref} - (T - T_{ref})/z \tag{1.2}$$

where D_T is the D-value in minutes at any temperature T and D_{ref} is the corresponding value at the reference temperature T_{ref}, The usual temperature in Celsius is 121.1° (this is the equivalent of 250°F, previously widely used in the canning industry). The z-value has units of Celsius degrees C° (for conversion purposes 1°C ≡ 1.8°F). Some typical values of D and z values are given in Table 1.2. For extensive tabulated data see Holdsworth (1997).

An alternative method of expressing the kinetics of microbial destruction is to assume first-order kinetics and express equation (1.1) as

$$N = N_0 e^{-kt} \tag{1.3}$$

where k is the specific reaction rate in reciprocal seconds s^{-1} and $D = 2.3/60k$.

Using the Arrhenius kinetic theory $k = Ae^{-E/RT}$, where A is the pre-exponential factor (s^{-1}), R is the molar gas constant (8.135 J /molK) and E is the activation energy kJ/mol and is equivalent 2.303 RT T_{ref}/z.

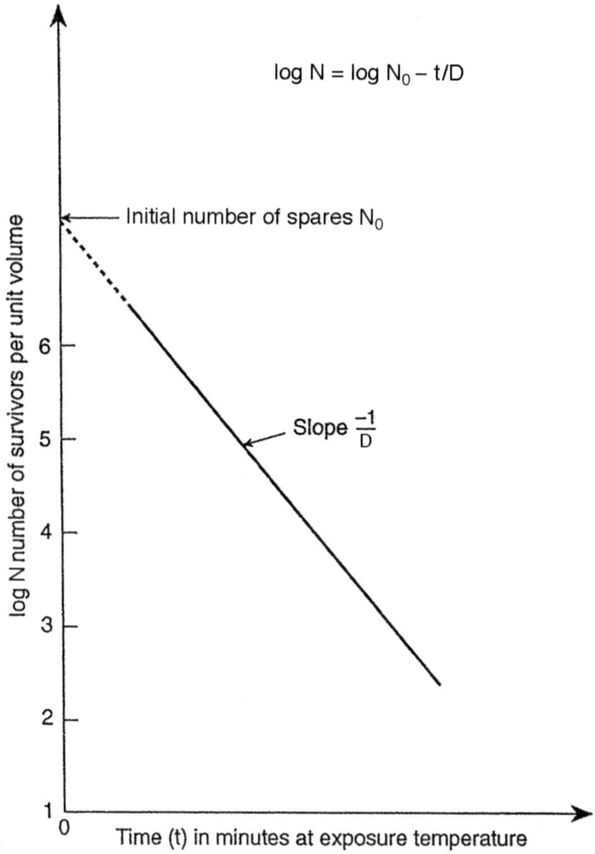

Fig. 1.1 Logarithm of the number of microbial survivors versus time for a given organism showing D-value determination.

Equation 1.2 may be expressed as follows:

$$\ln k = \ln k_{ref} - (E/R[1/T - 1/T_{ref}] \qquad (1.4)$$

where ln is the natural logarithm (base 2.303) and k_{ref} is rate constant at the reference temperature T_{ref}.

At normal canning temperatures 120–125°C, the two approaches give sufficiently similar results for either to be used (Nunes *et al.*, 1993). However, with the use of higher temperatures and shorter times more accuracy will be required for kinetic factors and the k/E approach may be more adventitious. Datta (1993) has made a full analysis of the two approaches and shown that under normal canning conditions relatively low errors are incurred. A new equation is proposed for modifying the D-z approach to take into account the variation of reaction rate constant with temperature. The D-z approach is widely used and a well-proven practical system in the traditional canning industry.

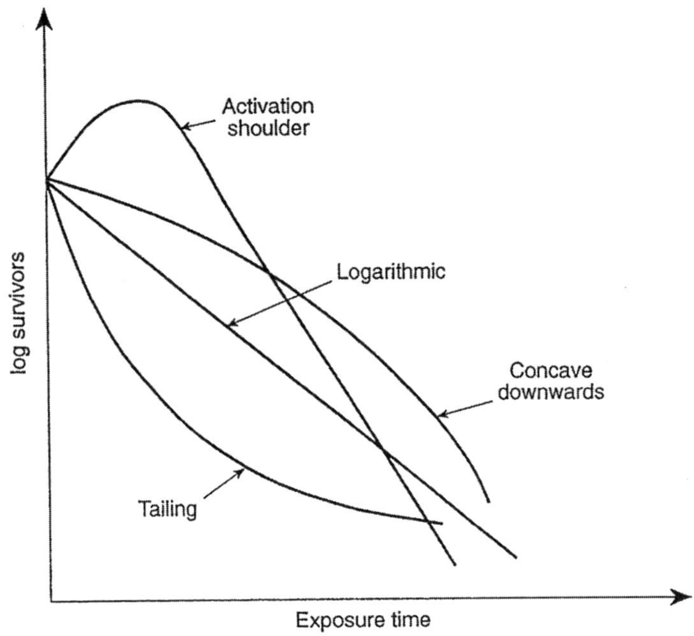

Fig. 1.2 Various types of microbial spore survivor curves encountered in canning microbiology.

1.2.2 Factors affecting heat resistance

A number of factors influence the heat resistance of microorganisms, i.e., water activity, pH, and composition and consistency of the food.

Water activity

The water activity of most food products is sufficiently high for this not to affect the heat resistance. However, in circumstances where dry powders can exist in formulated products or the substrate is oily or fatty then the heat resistance is marked higher. This also applies to direct heating with steam; dry superheated steam will be less effective for inactivation.

pH

pH has a marked effect on microbial inactivation. In general for acidic products, pH < 4.5, e.g., a wide range of fruits and their juices, pathogenic organisms do not cause a problem, hence only a mild heat treatment, usually referred to as pasteurisation, is required for stabilising the product. For pH > 4.5 e.g., most vegetables, fish and meat products, the scheduled process must be sufficient to inactivate *Clostridium botulinum* spores. For products which fall close to the dividing line 4.4–4.7 special care must be taken, e.g., tomato products and pears, depending on the variety and maturity. In some cases it is possible to acidify the product to ensure that a pasteurisation process is adequate. For products for

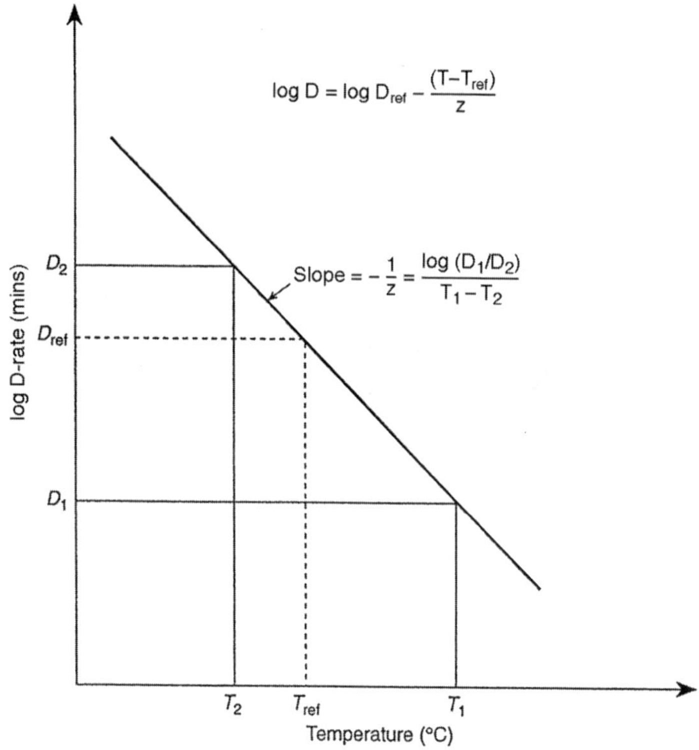

Fig. 1.3 Determination of the z-value from log D versus temperature.

which there is no scheduled process it is necessary to determine the pH beforehand and if this falls into a borderline case to do inoculated or other microbiological tests. It is usual to identify four categories of products as shown in Table 1.3.

Table 1.2 Some typical D values for spores

Organism	Temperature °C	D-value (min)
Bacillus coagulans	121	3
Bacillus coagulans var. thermoacidurans	96	8
Bacillus licheniformis	100	13
Bacillus stearothermophilus	121	3–5
Bacillus subtilis	121	0.3–0.7
Clostridium botulinum	121	0.2
Clostridium butyricum	85	8
Clostridium sporogenes	121	0.2–1.5
Clostridium thermosaccharolyticum	121	3–5
Desulfotomaculum nigrificans	121	3–5

Source: CCFRA Database.

Table 1.3 pH values for some food products

Category	Designation	pH range	Products
Group 1	low-acid	$\geq$5.0	meat, fish, milk, some soups and most vegetables
Group 2	medium-acid	5.0–4.5	meat and vegetable mixtures, pasta, soups and pears
Group 3	acid	4.5–3.7	tomatoes, figs, pineapple and other fruits
Group 4	high-acid	$\leq$3.7	citrus juices, pickles, grapefruit and rhubarb

Other factors
These include presence of oily or fatty constituents, dielectric constant, ionic species, e.g., salt or nitrite, ionic species, oxygen content, organic acids and antibiotics (Gould, 1995). Some of these materials are used to enhance preservation processes by reducing the scheduled heat process required.

1.3 Setting the limits for sterilisation and pasteurisation processes

1.3.1 F-values and the lethal rate concept

A measure of the lethal effect of a process can be obtained using the decimal reduction time ratio D/D_T and this is known as the lethal rate in minutes defined by equation 1.5

$$L = 10^{(T-T_{ref}/z)} \tag{1.5}$$

The lethal rate for $T = 111.1°C$ and $T_{ref} = 121.1°C$ will be $10^{-(10/10)} = 0.1$ min. Thus one minute at 111.1°C is worth 0.1 min at 121.1°C.

Since the temperature at the slowest-heating point in the food in a container changes with time it is necessary to determine the lethal contribution of each temperature. The summation of the lethal rate or lethality obtained at each temperature for unit time is known as the F-value.

$$F = \int L \, dt \tag{1.6}$$

In using these two formulae it is necessary to know the z-value of the target organism. For low-acid foods $z = 10°C$ corresponding to the generally used value for *Clostridium botulinum*. The most usual form for F-value equation is given in equation 1.7:

$$F = \int 10^{(T-T_{ref})/z} dt \tag{1.7}$$

or for a low-acid food cook the F-value at the reference temperature of 121.1°C, known universally as the F_0-value and referred to as F-nought or F-zero:

$$F = \int 10^{(T-121.1)/10} dt \tag{1.8}$$

A typical temperature-time profile obtained for a canned food undergoing processing is shown in Fig 1.4. This has been obtained by placing a thermocouple at the point of slowest heating in the food product and it includes the effect of the coming-up to retort temperature time and also the cooling time (see Section 1.3.5). The argument is that if the F-value at this point is greater than the minimum specified for the process then all other points in the container will have received at least the minimum required. On the plot the lethal rate is also shown and from the area under the curve the total integrated lethality, i.e., the F-value, can be obtained.

F_c is sometimes used to indicate the F-value at the centre, i.e., the slowest heating point, and F_s for the total integrated lethality. The standard F-value is also written using adscripts $F^z{}_T$ or $F^{10}{}_{121.1°C}$.

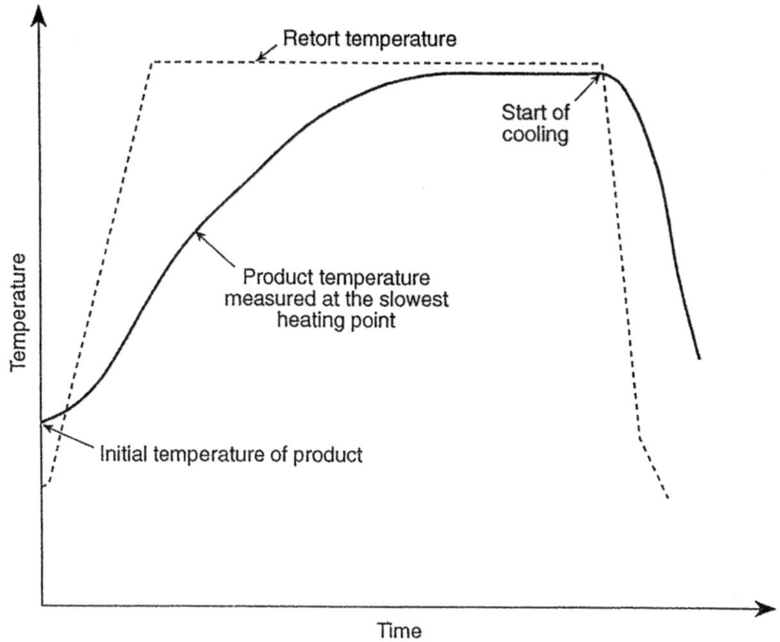

Fig. 1.4 A typical temperature-time profile obtained from a canned food undergoing processing.

1.3.2 The botulinum cook

Based on a z-value of 10°C for spores of *Clostridium botulinum* a practical concept for determining a satisfactory process for a low-acid food has evolved. This is that any process which achieves an F_0 of 3 minutes is considered to render the product free from pathogenic organisms which could create a public health danger. Another concept, which again is traditional in the canning industry, is the concept of a 12–D, i.e., 12× decimal reduction time or 12 log reductions in the bacterial population. A botulinum cook is often referred to this when a D-value of 0.25 is used; however, the $F_0 3$ is obtained completely independently of this concept. The 12-D concept can be explained on a statistical basis. A survival rate of 1 spore in a population of 10^{12} will be achieved by applying the process. Whilst the assumptions underlying this concept are not entirely satisfactory, it should be noted that the botulinum cook concept has a long record of proven safety.

1.3.3 Pasteurisation processes

For high acid foods it is only necessary to inactivate food spoilage organisms, e.g., moulds and yeasts, since spores of *Clostridium botulinum* will not germinate under acid-conditions. However, if the conventional F-value concept were used very low and impractical values would be obtained. A pasteurising unit PU based on z = 10 and a reference temperature of 65°C is therefore defined as in equation 1.9

$$P_{65}^{10} = \int 10^{(T-65)/10} dt \tag{1.9}$$

A P* value has also been used in relation to milk processing; this is based on a reference temperature of 72°C and z-value of 8°C. The criterion for a satisfactory milk pasteurisation process is P* = 1 minute.

1.3.4 Recommended commercial processes

The actual processes used by the canning industry are usually much greater than the minimum processes defined by the foregoing considerations. It is necessary to take into account any additional cooking requirements, excessively high initial spore loads, product formulation variations, filled weight variation or the presence of highly heat-resistant pathogenic organisms. Some examples of typical processes used in commercial practice are given in Table 1.4. Where additives are used to alter the pH, e.g., acidification or antibiotics or similar materials used to control bacterial behaviour, then lower processes may be used.

Guidelines for the safe processing of canned foods are available in many countries of the world. The most extensively documented information comes from the laboratories of the National Food Processors Association (formerly National Canners Association) in the USA. Bulletin 26L deals with processes for low-acid foods in metal containers (NFPA, 1983) and Bulletin 30L deals with

Table 1.4 Some commercial F_o values[a]

Product	Can size (mm) d × h	F_0–value	Source
Fish products:			
Mackerel in brine	301 × 411	3–4	d
Herrings in tomato sauce	oval	6–8	b
Meat products:			
Chilli con carne	all	6	d
Curried meats and vegetables	up to 16Z (73 × 117)	8–12	b
Hams 3.3% brine	1 lb	0.3–0.5	c
4.0% brine	1 lb	0.1–0.2	c
Ham 'sterile'	1 and 2lb	3–4	b
Luncheon meat 3–4% brine	1 lb	1.0–1.5	
Luncheon meat 5–5.5% brine	1 lb	0.5	c
Meat in gravy	all	12–15	b
Meat in gravy sliced	oval	10	b
Meat pies tapered	flat	10	b
Pet food	A2 (83 × 114)	12	d
Pet food	up to 16Z (73 × 117)	15–18	b
Pet food	A10 (153 × 178)	6	d
Sausages 'Frankfurters'	up to 16Z (73 × 178)	3–4	b
Vegetables:			
Asparagus	all	2–4	d
Beans in tomato sauce	all	4–6	b
Carrots	all	3–4	b
Celery	A2 (83 × 114)	3–4	b
Corn whole kernel in brine	A2 (83 × 114)	9	d
Corn whole kernel in brine	A10 (153 × 178)	15	d
Corn cream–style	A2 (83 × 114)	5–6	d
Green beans in brine	up to A2 (83 × 114)	4–6	b,d
Mushrooms in brine	A1 (65 × 101)	8–10	b
Peas in brine	up to A2 (83 × 114)	7	b,d
Peas in brine	A10 (153 × 178)	10	d
Poultry:			
Chicken boned	small sizes	6–8	d
Chicken breast in jelly	up to 16oz (73 × 117)	6–10	b
Poultry/game, whole in brine	A2½ (99 × 119)		
	A10 (153 × 178)	15–18	b
Formulated and other products:			
Baby foods	52 × 72	3–5	b
Soups			
– meat	up to 16Z (73 × 117)	10	b
– non–cream tomato	all	3	b
– cream soups	A1(65 × 101)–16Z (73 × 117)	4–5	b
– cream soups	up to A10 (153–178)	6–10	b
Milk puddings	up to 16Z (73 × 117)	4–10	b
Cream	up to 6oz (65 × 58)	3–4	b
Cream	16Z (73 × 117)	6	b
Evaporated milk	up to 16Z (73 × 117)	5	b

Sources:
[a] Adapted from Holdsworth (1997); for extensive American and European data see Eisner (1988).
[b] Collected UK data held by CCFRA, Chipping Campden.
[c] FAO (1986).
[d] Alstrand and Eklund (1952).

similar requirements for glass containers (NFPA, 1971). The Food and Drugs Administration has also produced important regulations (FDA, 1983), as well as the Food Processors Institute (FPI, 1988). In the United Kingdom the government has produced guidelines for the safe production of heat preserved foods (UK-DH 1994), which supersedes Food Hygiene Code of Practice No.10 (UK-DHSS 1981). More specific commercial guidelines are available from CCFRA, Chipping Campden, in particular a guide to the processing of canned fruits and vegetables (CCFRA, 1980). In France the Institut Appert also produces advice on the canning processes (Institut Appert, 1979). Codex Alimentarius has also produced an international Code of Practice for low-acid foods (FAO, 1983).

1.3.5 Heat penetration determination

In order to evaluate the lethal rates (equation 1.6) and total integrated F value (equations 1.7 and 1.8) it is necessary to have the time-temperature profile at the slowest point of heating. This is determined experimentally by placing a temperature-measuring device at the appropriate point in the container. These are available in sizes such that when fitted to a particular container the temperature-sensing element is at the appropriate position. The most commonly used thermocouple is made of copper and constantan (a copper alloy with 45% nickel) wires and known as a type T thermocouple. Two commercial systems are widely used the (a) Eklund system of thermocouples (Eklund Custom Thermocouples, Cape Coral, Florida, USA) and (b) the Ellab system (Ellab A/S, Copenhagen, Denmark). Both these systems have to be fitted into the containers by mechanical means and are particularly well suited to static processes. They may however, be adapted for use in systems where cans are rotated to increase the heat penetration by using a system of slip rings. For processes which make transmission of the signal using wires impracticable, i.e., continuous processes, a device which records the temperatures during the processes, is available, namely, the Ball Datatrace®, (Datatrace Division Mesa Medical Inc., Wheat Ridge, CO, USA). The device is fitted inside the container such that the temperature measuring sensor, a thermistor with a range 10–150°C, is at the appropriate point. The unit may also be used to record the temperature profile in a steriliser by allowing the sensor to be in contact with the sterilising atmosphere. The Micropack™ version has a cylindrical body 35 mm in diameter and length varying from 55 to 156 mm, with a probe 25–125 mm long. The unit can be programmed to read from 1 per second to 1 per day, depending on requirements. The performance of the unit has been thoroughly tested for a variety of canned foods (May and Cossey, 1989; May, 1991, 1992).

For retortable plastic pouches a variety of different methods of arranging a thermocouple in the pouch have been reported. In one method (Bhowmik and Tandon, 1987; Spinak and Wiley, 1982) a folded strip welded to the pouch sides which straightens up when the pouch is filled and located in the thermocouple at the appropriate point, is used. Rigid plastic inserts, e.g., PTFE or nylon, have also been used for supporting the thermocouple.

For some applications the conduction errors inherent in commercial thermocouples are too great for accuracy and consequently thin thermocouple wires of thickness of the order of 0.1 mm are used. It is essential for temperature measurement work that the sensor is calibrated to an appropriate standard using a constant temperature source and a thermometer calibrated to a national standard, (ASTM, 1988; Cossey and Richardson, 1991; Dobie, 1993).

Thermocouple location should be made at the point of slowest heating, often known as the critical point. This point varies, depending on the nature of the product and the type of cooker, e.g., stationary or rotating. One method of determining this point is to place a number of thermocouples in the container at differing positions and to observe which is the slowest heating. For small sized cans of conduction-heating food the critical point will be near the geometrical centre of the food mass. For large sized cans, e.g., A10 and larger, this is not necessarily so because the centre of the can will continue to heat until the cooling effect is felt. There will therefore be an additional contribution to the lethality. Flambert and Deltour (1972) showed that the critical point location depended on the h/d ratio for the can. For the particular conditions of their experimental work they showed that the critical point would be at the geometrical centre of the food mass for $h/d < 0.3$ and greater than 0.95. For $0.3<h/d<0.95$ the critical point was located symmetrically along the vertical axis with respect to the central plane. For values $0.95<h/d< 1.9$ the critical point lies in a ring-shaped space across the can. For convection heating products the critical point is located on the central axis of the container but at points lower than the geometrical centre. The UK recommendation is that the thermocouple should be placed at a height from the base of 20% of the total height (CCFRA 1977).

For products that show a change of heating from convection to conduction during the processing, i.e., broken-heating curves, the convection heating position should be used. Complex heating products should always be studied with multiple thermocouple positions initially. For experimental purposes involving the study of steriliser performance it is convenient to use simulant materials. The most common of these is a carefully prepared suspension of a clay mineral, bentonite in water. Dilute suspensions 1% may be used to simulate convection heating packs and more concentrated ones for broken-heating 3.5% and conduction heating packs 5%. A useful simulant for conduction-heating packs is the silicone elastomer – Sylgard.

1.3.6 Factors affecting heat penetration

(a) Process-related factors. These include retort temperature and process time, the nature of the heat transfer medium and container agitation. Saturated steam is the most effective heat transfer medium and provides an effective pressure to balance the internal pressure developed in the container. With water and steam air mixtures the heat transfer rate depends on the velocity of the heating medium. In batch retorts, there is an initial period, known as the come-up time before the retort reaches processing temperature and this must be considered in process determination. Conversely, continuous retorts are at a steady-state before the cans are introduced.

(b) Product-related factors. These include product consistency, initial temperature. Initial spore load, thermal properties, pH, additives. It is possible to categorise various types of behaviour, namely, most rapid convection heating – thin liquids, juices, broths and milk, less rapid convection heating – fruits/syrup, vegetables/brine, low-starch purées, some vegetable soups, slower convection/conduction heating products – cream soups, noodle soups, tomato juice; conduction-heating products containing water – cream-style corn, thick purées, solid pack products, pet foods, rice, spaghetti and conduction-heating products not water based – high fat or oil meat and marine products, high sugar products and low-moisture puddings.

(c) Packaging-related factors. These include container materials and shape. The thermal properties of the material determine the rate of heat penetration; metallic materials have a low resistance whereas glass and plastic materials have a higher resistance.

1.3.7 Analysis of heat penetration data

The typical temperature time profile for a canned product heated in a batch steam retort is shown in Fig. 1.5. This may be converted to a linear curve using a

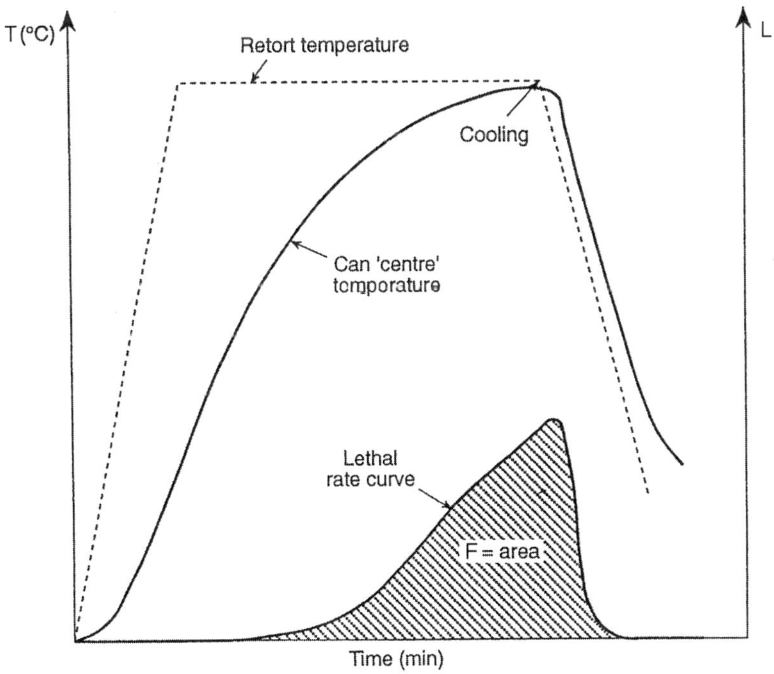

Fig. 1.5 Determination of F-value from lethal rate versus temperature.

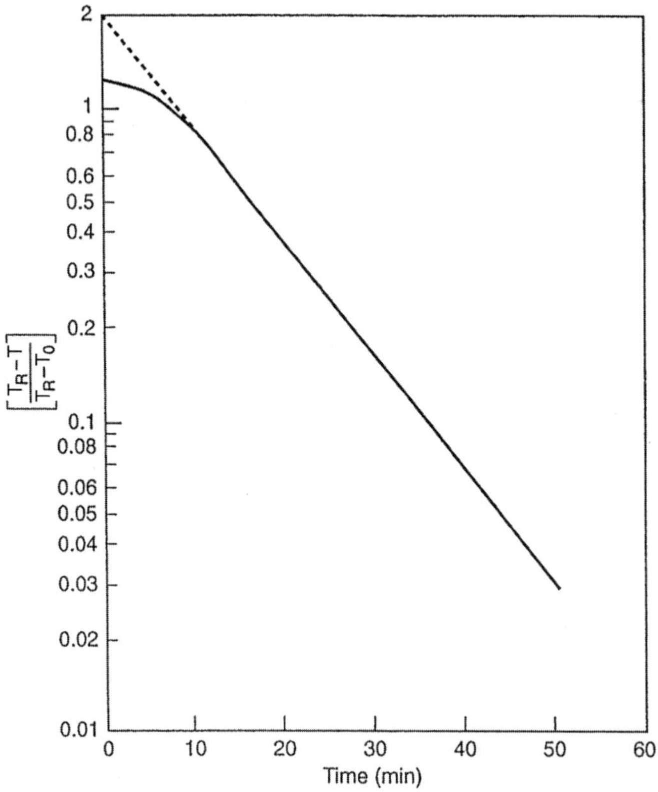

Fig. 1.6 Heat penetration curve for the heating process.

logarithm temperature scale (Fig 1.6). From this, two important parameters may be obtained – the lag of the heating curve j and the f value f_h for heating and f_c for cooling. The relationship for the logT/t curve is given by equation (1.10)

$$(T_R - T_t)/(T_R - T_0) = j\ 10^{-t/f}\ j\ e^{-2.303t/f} \tag{1.10}$$

where T_R is the retorting temperature, T_t is the measured temperature in the can at any time t, T_0 is the initial temperature of the food in the can at time $t = 0$, j is the lag factor in mins and f is the heating time for one log cycle of temperature.

For convection-heating foods $j = 1$ and for conduction-heating packs j is usually about 2. Some typical f_h-values are given in Table 1.5. The importance of the f_h-value is that all aspects of the heat penetration are contained in the one parameter. The f_h-value is inversely related to the thermal diffusivity α of the product. Equation (1.11), which is derived from heat transfer considerations (Holdsworth, 1997), may be used to determine the f_h-value for differing sizes of cylindrical container.

$$f_h = 0.398/[\alpha(1/a^2 + 0.427/4b^2)] \tag{1.11}$$

where 2a is the diameter and 2b the height of the can respectively.

Table 1.5 Some typical values for f_h for canned products processed in steam-heated static retorts.

Can size D × h mm	Conduction-heating food min	Convection-heating food min
66 × 54 (5oz)	25	4.0
66 × 78 (picnic)	34	4.5
66 × 102 (A1)	39	5.0
73 × 62 (8Z)	34	4.5
73 × 115 (UT)	47	4.5
74 × 116 (16Z)	52	5.5
84 × 114 (A2)	62	6.0
99 × 119 (A2½)	83	7.0
154 × 235 (A10)	198	11.0

Source: Collected data CCFRA, Chipping Campden.

1.4 Setting thermal process parameters to maximise product quality: C-values

1.4.1 Problems with thermal processing

The thermal process delivered to a packaged food not only inactivates potential spoilage organisms, but it also cooks the food, in many cases to produce a food with an acceptable texture, in accordance with producer's brand image. Many canned foods are essentially pre-cooked so that, as convenience foods, they only require the minimum of reheating before being eaten. The amount of cooking depends very much on the nature of the produce. For convection-heating foods such as vegetables in brine heat penetration is fairly rapid and uniform across the contents of the can. With conduction-heating products, where the contents of the can are not mobile then the product nearest the outside of the container receives far more heat than that delivered at the point-of-slowest heating near the centre of the product mass. Whilst can foods have played a very important part in feeding people, under varying circumstances, they have generally been perceived as being of a lower quality then their chilled or frozen counterparts. However, because heat-preserved foods may be stored at ambient temperatures, they are extremely useful for a variety of purposes. Consequently, there has been considerable effort to reduce the thermal processes for canned foods so that less heat damage is done to the food components, e.g., vitamins, colour and other thermo-labile components. Some of the methods, which have been used to reduce thermal processes, have included: (a) alteration of container geometry, e.g., using thin layers of product in flexible plastic packages and trays, (b) higher temperatures and shorter times, which require special processing techniques to counteract the internal pressures developed in the containers, (c) reducing the processes and storing under chilled conditions, (d) acidification of the products followed by pasteurisation, and (e) the use of microwave or ohmic heating.

Despite the importance of this subject there are only a limited number of studies, which show that long processes at lower temperature adversely affect the product quality compared with shorter processes at a higher temperature. Brown (1950) published work on canned vegetables and showed that the texture of carrots, but not of peas was influenced markedly by the duration of the process for temperatures between 100 and 132°C. Process conditions did not significantly affect the levels of carotene or ascorbic acid, however, vitamin B_1 retention was favoured by the 'high-short' process. Tischer *et al.* (1953) in a very extensive coverage of the processing conditions for canned beef, especially in terms of tenderness and drained juice levels, showed that longer processes favoured the tenderness of meat. Apart from a number of general attributes, this text is not concerned with the actual effects of heating on specific components of food products; this is treated in texts on food science and in more detail by Hoyem and Kvale (1977) and Priestley (1979).

1.4.2 Cooking versus microbial inactivation

Commercially it has always been recognised that there is a judicious balance between the requirements of microbial inactivation, which have to be achieved, and the quality attributes which may be achieved. The art of canning is therefore to optimise these opposing requirements in the most practical way possible. This fact is recognised in the way in which tables of process values give different requirements for cooking and sterilisation at different temperature levels (CCFRA 1977, 1980). Table 1.6 shows the choices that the canner has in producing canned beans-in-tomato sauce in a rotary cooker. All these processes lead to a product which is essentially organoleptically the same in relation to flavour and texture. The main advantage of the higher temperature and shorter time is that a greater throughput can be realised.

The kinetic factors for the effect of heat on quality parameters, which are essentially chemical in nature are of the order of about four times those for microbial destruction. The z values for the effect of heat on cooking and nutrients vary from approximately 25–45°C compared with microbial destruction of 7–12°C. In very general terms, for every 10°C rise of temperature the cooking effect is doubled, whereas the microbial inactivation increases tenfold.

1.4.3 Determining the cook-value

A quantitative measure of the effect of heat on quality factors is the C-value or cooking value, originally proposed by Tom Mansfield of FMC, St Jose (Mansfield 1962; 1974). This is defined in its simplest form by the equation

$$C = \int 10^{(T - T_{ref})/z_c} dt \qquad (1.12)$$

where z_c is the thermal destruction rate for a specified component. C is usually meant to imply $C^{z_c}_{T_{ref}}$. The reference temperature for cooking is usually taken as 100°C, whereas for microbial inactivation T_{ref} is always 121.1°C for a

Table 1.6 Minimum equivalent times (min) for sterilisation and cooking of beans in tomato sauce in a rotary cooker

		Process temperature (C)		
Can size		115.5	121.1	126.6
A1	Sterilisation	22	13½	10
	Cooking	38	26	19
A2	Sterilisation	25	16½	12½
	Cooking	40	28	21
A2½	Sterilisation	27	18	14½
	Cooking	41	29	22
A10	Sterilisation	35	26	21
	Cooking	46	34	47

Source: CCFRA Tech. Bulletin No.4.

'botulinum process.' C-values for $T_{ref} = 100°C$ are usually of the order of 5–30 min., but for $T_{ref} = 121°C$ rather lower at 1–7 min. Some typical z_c-values for heat vulnerable components are given in Table 1.7

Sterilising values are usually evaluated at the point of slowest heating, but C-values are for the whole of the contents and are often designated C_s-values as defined by the equation

$$C_s = D_{ref} \log(c/c_o) \tag{1.13}$$

where c_0 and c are the concentrations of the heat-labile component at time 0 and t.

The volume average C-value for a container is given by the equation

$$C_{ave} = (1/V) \int \int 10^{(T-T_{ref})/z_c} dtdV \tag{1.14}$$

For applications of this concept see Ohlsson (1980a,b,c). This was also used indirectly by Tucker and Holdsworth (1990, 1991). However, Silva *et al.* (1992b)

Table 1.7 Some typical z-values for heat-vulnerable components

Component	z-value range °C
Bacterial spores	7–12
Vegetative cells	4–8
Enzymes	10–50
Vitamins	25–30
Proteins	15–37
Sensory factors	
Overall	25–47
Texture-softening	25–47
Colour	24–50

Source: Holdsworth (1992).

pointed out that the C_s-value depends on the D_{ref} for the specified component and this should be taken into account by using equations (1.15) and (1.16):

$$c/c_0 = (1/V) \int 10^{C_c/D_{ref}}.dV \qquad (1.15)$$

$$\text{where } C_c = (1/V) \int 10^{(T-T_{ref})/z_c}.dt \qquad (1.16)$$

This allows for the use of different heating profiles. Equations (1.14) and (1.15) give essentially the same results for high D_{ref}-values found for vitamin destruction; however for low values of, e.g. colour degradation or texture softening then the latter is superior. McKenna and Holdsworth (1990) have reviewed the published models for determining both F_s and C_s. Using the simple relation equation (1.12) Figure 1.7 compares the rate of establishment of the C-value with that of the F-value for the same process.

Relatively little data has been obtained comparing C_0-values for food products; however, Preussker (1970) produced some for static processes, and Eisner (1988) for rotary processes. More recently Tucker and Holdsworth (1991) have determined $C_{121.1}$-values for a number of ready meals (see Table 1.8.). These data help to determine the magnitude of the effect of the process and to give some practical processes for sterilising ready meals.

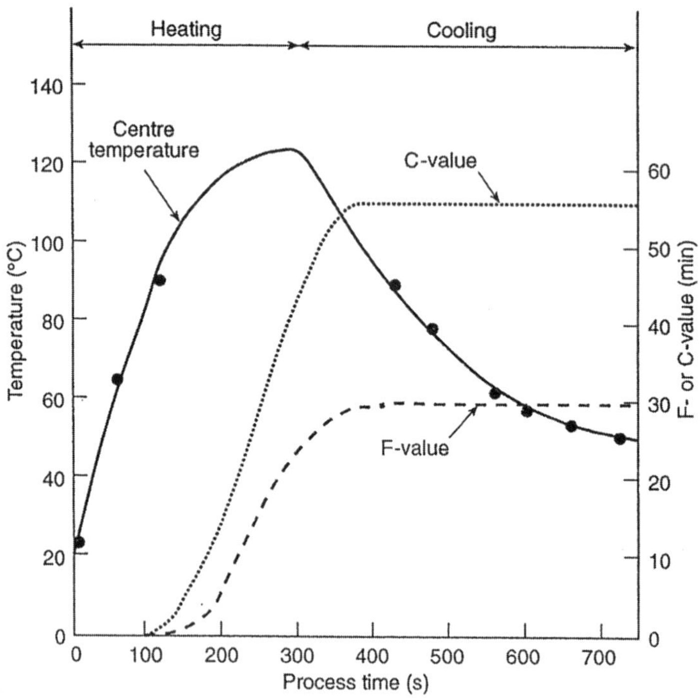

Fig. 1.7 C-value variation with heating profile.

Table 1.8 Some $C_{121.1}$-values for some prepared canned foods

Product	Container size (°C)	Process temperature (min)	Process time (min)	$F_{121.1}$ total (min)	$C_{121.1}$ centre (min)	$C_{121.1}$ total
Beans in tomato sauce	A2	121.1	121	11.6	196.5	331.6
Beans in tomato sauce	UT	121.1	94	8.3	155.0	267.2
Carrot purée	A1	121.1	74	8.0	128.5	210.3
Celeraic purée	A1	121.1	72	6.0	117.2	199.8
Chicken supreme sauce	UT	121.1	86	6.6	138.3	242.1
Chilli con carne	UT	121.1	91	6.6	144.0	255.3
Mackerel in tomato sauce	UT	121.1	97	10.9	168.0	280.5
Pet food	UT	125.8	84	20.0	180.4	323.6
Stewed steak	UT	121.1	105	12.0	188.3	308.2
Spaghetti in Tomato sauce	A2½	121.1	83	9.0	148.5	245.1

Note: for can sizes see Table 1.5.
Source: Tucker and Holdsworth (1991).

1.5 Optimising thermal process conditions for product safety and quality

1.5.1 Graphical methods

A simple graphical method of comparing the effects of sterilisation and cooking is to plot the log time versus temperature for processes with differing z-values. In its simplest form the graph is represented by Fig. 1.8 (Mansfield 1962, 1974). This shows that any combinations of time-temperature to the right of the F_1–F_2 line result in a sterilised product; however, those in the sector F_1-O-C_2 are cooked in relation to the z-value chosen, but those in the sector F_2-O-C_2 are uncooked. Combinations of time and temperature to the left of F_1-O-F_2 are of no commercial interest, being insufficiently sterilised. These lines represent conditions of instantaneous heating of the product. If the centre of a product is considered then the lines become progressively concave (Holdsworth, 1997). If C-O-C lines representing differing levels of destruction of a heat-vulnerable component are imposed on the sterilisation time-temperature graph, then it is possible to choose an appropriate process to minimise the loss. A typical idealistic representation of this is shown in Fig. 1.9 for the enzyme thiamin, which shows that to minimise the loss of thiamin shorter times at higher temperatures are required. Table 1.9 summarises some of the graphical optimisation procedures used by various workers (Holdsworth, 1985, 1997).

1.5.2 Analytical methods of optimisation

There have been a number of optimisation procedures used to determine the effect of heating on heat-vulnerable components in canned foods. One of the

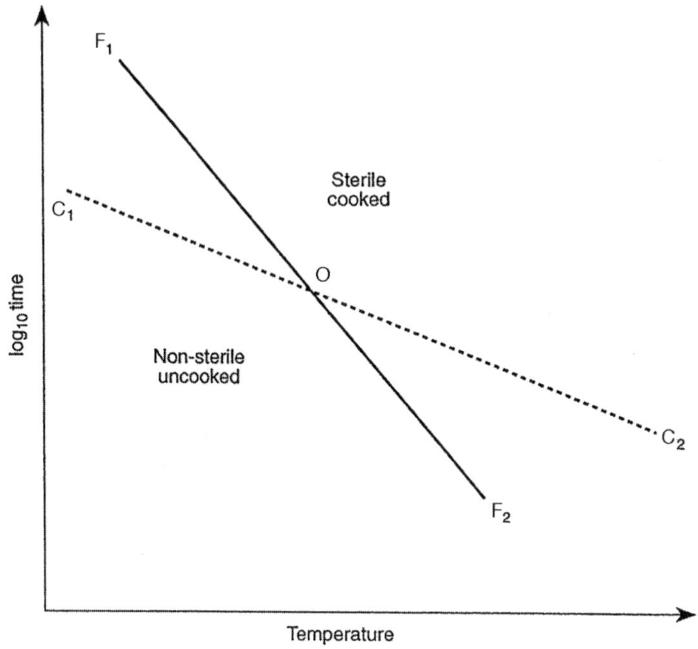

Fig. 1.8 Log time/temperature showing competing effect of sterilisation and cooking.

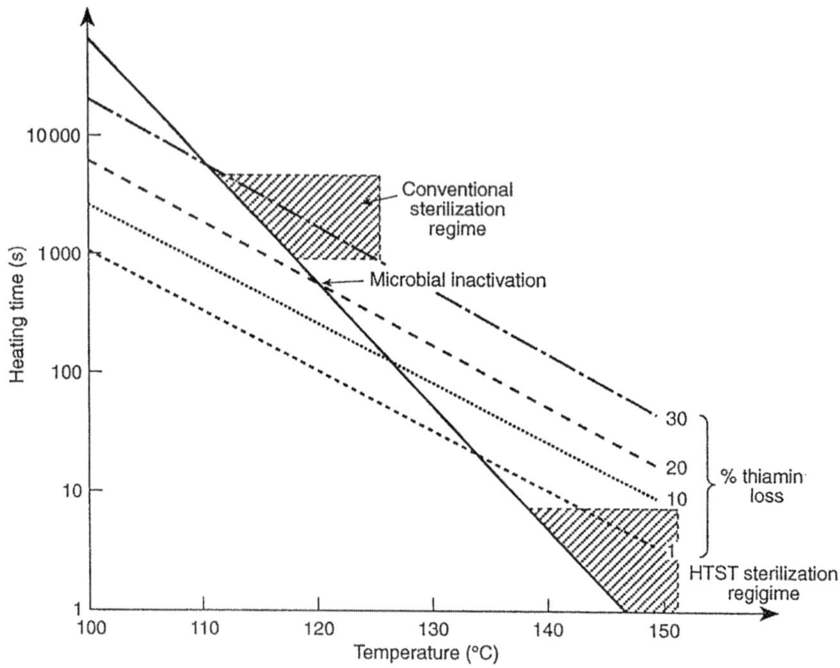

Fig. 1.9 Degradation of thiamin for differing time'temperature combinations.

Table 1.9 Optimisation of heat–vulnerable components in canned foods using graphical procedures in chronological order

Heat-vulnerable components			Sterilisation conditions		Reference
Description	z_c (°C)	C_{100} (min)	z (°C)	$F_{121.1}$ (min)	
Thiamin/cured meat	–	–	10.0	0.25	Greenwood *et al* (1944)
Thiamin/cured meat					Jackson *et al.* (1945)
Thiamin/cured meat					Ball and Olson (1957)
Cooking	33	5–30	10.0	2–30	Mansfield (1962)
Betanin	–	–	10.0	1.0	Hermann (1969)
Cooking/linear heating	33	0.1–50	10.0	0.1–50	Preussker (1970)
Enzymes/green beans	48.9	–	8.9	0.9	Reichert (1977)
Enzymes/potatoes	10.3	–	10.0	2.5	Reichert (1977)
Enzymes	17.5	–	8.9	0.9	Reichert (1977)
Vitamin C	23.2	–	8.9	0.9	Reichert (1977)
Vitamin B_1	26.1	–	8.9	0.9	Reichert (1977)
Cooking	25–40	–	8.9	0.9	Reichert (1977)
Sensory	26.5	–	8.9	0.9	Reichert (1977)
Cholorphyll/green beans	87.8	–	8.9	0.9	Reichert (1977)
Cooking	33.0	10,36,52	10.0	1.0	Reichert (1974, 1977)
Cooking/peas	29.0	42,45,62	10.0	76.0	Reichert (1974, 1977)
Vitamin B_1/liver	26.1	–	10.0	5–10	Bauder and Heiss (1975)
Lipase (microbial)	3.1	–	10.0	2.7	Svensson (1977)
Peroxidase	35.0	–	10.0	10.0	Svensson (1977)
Thiamin	–	–	10.0	6.0	Lund (1977)
Anthocyanin/grapes	23.0	18.0	10.0	24.0	Newman and Steele (1978)
Thiamin	–	–	10.0	5.0	Ohlsson (1980b)
Thiamin/milk	–	–	10.5	2.0	Kessler (1981)
Lysine/milk	–	–	10.5	2.0	Kessler (1981)
Protease	–	–	10.5	2.0	Kessler (1981)
Lipase	–	–	10.5	2.0	Kessler (1981)
Colour	–	–	10.5	2.0	Kessler (1981)
Enzymes/particulates	27.0	–	10.0	3.0	Brown and Ayres (1982)
Browning/protease	25.0	–	10.0	4.0	Jelen (1983)
Quality/tomato purée	–	–	10.0	various	Zanoni *et al.* (2003)

Source: Holdsworth (1985).

earliest was due to Teixeira *et al.* (1969), who used a finite-difference method to solve the sterilisation and cooking equations involving the heat transfer into the cans and the process conditions. Using the models and experimental data for the retention of thiamin the effect of various process times/temperatures was studied. Figure 1.10 shows that percentage of thiamin retained reached an optimum value for a process of 90 min/120°C. The effect of container size was also studied and it was shown that for equal volumes the thiamin retention decreased from 68 to 41% for values of L/D increasing from 0.96 to 1.270 and

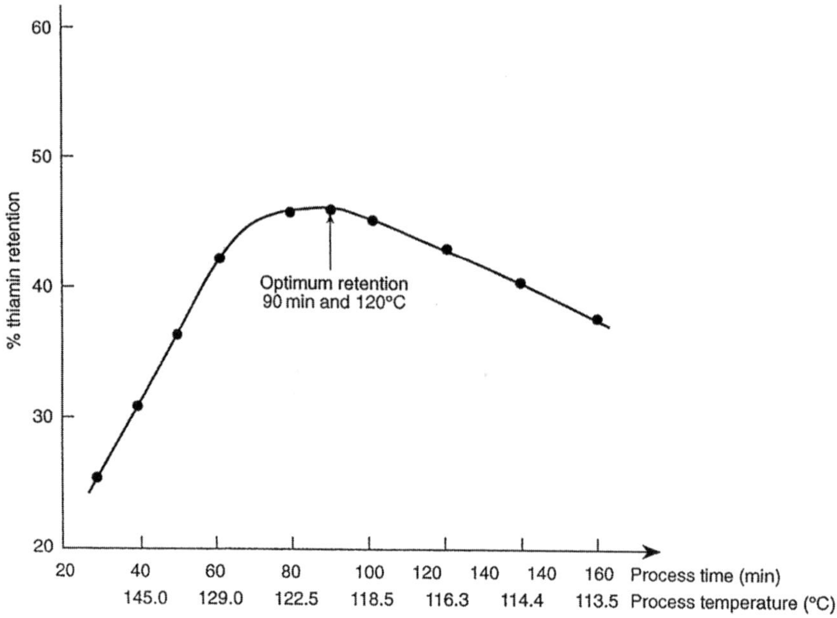

Fig. 1.10 Optimisation curve for the percentage thiamin retention for various time/temperature combinations.

then increased from 43 to 63% for values of L/D increasing from 1.710. Using time varying surface temperature profiles had little effect on thiamin retention.

Lenz and Lund (1977) studied the statistical distribution of C_s-values and found this to be normal for processing times of less than 20 min; however, for longer times an increase in the standard deviation was observed with pronounced skewness of the distribution.

Sjöström and Dagerskog (1977) reported an important study of the browning of canned chopped fish ($z_c = 33$°C) for processing at temperatures between 110 and 145°C. For a range of times and temperatures equivalent to $F_0 = 7.5$ min they showed that for a t/T combination of 60 min at 127°C the C-value was a minimum for a position intermediate between the surface and the centre of the food. Again variable temperature profiles had little effect on colour retention.

Ohlsson (1980a,b,c) made an extensive study of the C-values for a range of products, including fish paste, liver paste, strained beef, strained vegetables, tomato sauce and vanilla sauce and a range of sensory factors, odour, appearance, taste, consistency, hardness, coarseness and lightness. This work showed that the volume average cook value C_{av} for a given F-value, showed minimum values, which decreased with increasing temperatures and decreasing can sizes. The optimal processing temperature was found to be between 117 and 199°C for a 73 × 99 mm can, which was found to be consistent with the earlier work of Teixeira *et al.* (1969). Richardson *et al.* (1988) used a finite-difference model to determine nutrient retention in conduction-heating packs. This work showed that the experimental results correlated better with the theoretical results when the temperatures were measured at the surface of the containers, rather than using the retort temperature.

A number of workers have applied formal optimisation techniques (e.g. the continuous maximum principle theory of Pontryagin *et al.* (1962)). Saguy and Karel (1979) optimised the retention of thiamin in pea purée in 401 × 411 cans. A constant heating temperature was shown to be almost as good as the theoretically derived profiles. A similar result was obtained by Nadkarni and Hatton (1985) for the retention of thiamin in canned pork purée.

Banga *et al.* (1991) developed an optimisation algorithm called ICRS (Integrated Control Random Search). The results of percentage retention versus equivalent t-T processes showed a maximum (similar to Fig. 1.10) for both the overall nutrition retention and the quality factor surface retention. This work also showed that a significant increase of quality at the surface is achieved with a variable retort temperature profile as against the optimum-temperature profile. Tucker and Holdsworth (1990, 1991) reported on the optimisation of quality factors for foods thermally processed in three sizes of rectangular container 150 × 100 × 30, 40 and 50.

Silva *et al.* (1992a) critically reviewed the two objective functions volume-average retention (equation 1.14) and the volume average cook value (equation 1.15) and studied the effect of D-value on the two functions. For values higher than 150 min, the two objective functions yield the same results; however, for values below 150 min there is a divergence in the results. They concluded that the volume-average cook value could severely underestimate optimal processing temperatures. Silva *et al.* (1992b) also showed that the effect of finite heat transfer coefficient at the surface had an effect on the optimal sterilisation temperature see also Tucker and Holdsworth (1990, 1991). However, the initial temperature of the pack and the coming-up time to processing temperature had little effect.

Nasari *et al.* (1993) showed that between 29 and 70% thiamin retention was obtained in a thiamin-enriched pea puree in 303 × 406 cans. The retort temperature for optimum retention was found to be about 120°C for an $F_0 = 10$ min, which was in agreement with previous work. The experimental method produced values of $D_{121.1}$ of 304±32 min and z_c-values of 30±3°C.

Hendrickx *et al.* (1989, 1992, 1993) developed a semi-empirical model for determining the optimal sterilisation temperatures, which included the effect of cooling period and come-up-time. The latter factors were found to have only a minor influence on the optimal processing temperature. Silva *et al.* (1994a,b,c) also studied the maximisation of the surface quality retention of conduction heated foods in pouches using a one-dimensional heat transfer model. Optimal temperatures were obtained for a variety of processing conditions. The semi-empirical method was shown to be a very effective way of studying the optimal problems of thermal processing compared with the complexities of the analytical approach.

Various workers have demonstrated the importance of variable temperature profiles for improving the surface quality of canned foods, e.g., Banga *et al.*

(1991), Almonacid-Merino *et al.* (1993), Durance *et al.* (1996), Durance (1997) and the authors whose work is discussed below.

Norohna *et al.* (1993) studied the effects of variable temperature profiles on maximising the surface quality retention and found that the surface quality was improved by up to 20% compared with a constant temperature profile. Terajima and Nonaka (1996) have made a study of retort temperature profiles required to obtain optimum quality retention in retortable pouches. These containers having a thin profile have much to offer in retaining quality factors because of the shorter processes required. Balsa-Canto *et al.* (2000, 2002a,b) have employed novel methods for thermal process design and optimisation using the derivation of reduced-order models which allows very fast and accurate solutions for optimisation problems. The study used data for the retention of thiamin in a canned pork purée. This work showed that the optimum retention was highest at 53% for the lowest F_{centre}- value = 8 min.

Chen and Ramaswamy (2002) have also developed a sophisticated model for optimising variable retort temperature using neural networks and genetic algorithms. The workers conclude that variable retort temperature profiles are effective in improving the quality of canned foods and reducing the processing time. The retort temperature profiles studied included sine-wave and exponential functions. These were shown to reduce the process time by more than 20% and the surface cook value by about 7–10% compared to the best constant retort temperature processes. It was considered that by combining functions further improvements might be possible. A method of reducing the dispersion in product quality induced by a known variability in product thermal behaviour due to the distribution in thermal diffusivity has been reported by Baucour *et al.* (2003). The selection of the optimum processing conditions to minimise quality variability is analysed using the selected simple models for microbial destruction and quality degradation.

1.6 Future trends

The subject of optimising processing conditions is particularly relevant at the present time. There is a need to conserve energy as well as a demand for products with better nutritive value. Although a variety of methods of optimising processing conditions have been reported here, there are many areas that require attention. The most important is the need for more experimental work on consumer products and the application of the theoretical models to determine how suitable and satisfactory they are for predicting the fate of heat vulnerable components in the food. There is still a demand for heat processed foods which are stable at ambient temperature and much more work is required to meet the objectives discussed in this chapter.

1.7 Sources of further information and advice

BALL C O and OLSON F C W (1957), *Sterilization in Food Technology – Theory, Practice and Calculations*, New York, McGraw-Hill.

FOOTITT R J and LEWIS A A (EDS) (1995), *The Canning of Fish and Meat*, Glasgow, Blackie Academic and Professional.

HERSOM A C and HULLAND E D (1980), *Canned Foods, Thermal Processing and Microbiology*, 7th Edition, Edinburgh, Churchill-Livingstone,.

HOLDSWORTH S D (1997), *Thermal Processing of Packaged Foods*, Glasgow, Blackie Academic and Professional.

REES J A G and BETTISON J (EDS) (1991), *The Processing and Packaging of Heat Preserved Foods*, Glasgow, Blackie and Sons Ltd.

1.8 References

ALMONACID-MERINO, S F, SIMPSON R and TORRES J A (1993), 'Time-variable retort temperature profiles for cylindrical cans: Batch process time, energy consumption, and quality retention model', *J Food Process Eng*, 16, 271–287.

ALSTRAND D V and EKLUND O F (1952), 'The mechanics and interpretation of heat penetration tests in canned foods', *Food Technol*, 6(6), 185–189.

ASTM (1988), *Standard Guide for the Use in the Establishment of Thermal Processes for Foods Packaged in Flexible Containers*, Philadelphia, PA, American Society for the Testing of Materials.

BALL C O and OLSON F C W (1957), *Sterilization in Food Technology – Theory, Practice and Calculations*, New York, McGraw-Hill.

BALSA-CANTO E, BANGA J R, ALONSON A A and VASSILIADIS V S (2000), 'Optimal control of distributed processes using restrictive second order information', in Biegler L T, Brambilla A and Scali C (eds), *Int. Symp. Advanced Control of Chemical Processes*, ADCHEM, 905–910.

BALSA-CANTO E, BANGA J and ALONSO A A (2002a), 'A novel, efficient and reliable method for thermal process design and optimization. Part I: theory.' *J Food Eng*, 52(3), 227–234.

BALSA-CANTO E, BANGA J and ALONSO A A (2002b), 'A novel, efficient and reliable method for thermal process design and optimization. Part II: applications', *J Food Eng*, 52(3), 235–247.

BANGA J R, PEREZ-MARTIN R I, GALLARDO J M and CASARES J J (1991), 'Optimization of the thermal processing of conduction-heated canned foods: study of several objective functions', *J Food Eng* 14, 25–51.

BAUCOUR P, CRONIN K and STYNES M (2003), 'Process optimization strategies to diminish variability in the quality of discrete packaged foods during thermal processing', *J Food Eng* 60(2), 147–154.

BAUDER U and HEISS R (1975), 'Heat sterilization and quality of food', *Verfahrenstechnik*, 9, 566–573.

BHOWMIK S R and TANDON S (1987), 'A method for thermal process evaluation of conduction heated foods in retortable pouches', *J Food Eng*, 52(1), 202–209.

BROWN F (1950), 'Effect of time and temperature of processing on the quality and nutritive value of canned vegetables'. Progress Report, Chipping Campden, CCFRA.

BROWN K L and AYRES C (1982), 'Thermobacteriology of UHT processed foods', in Davies R, *Developments in Food Microbiology 1,* Barking, Elsevier Applied Science, 129–152.

CCFRA (1977), *Guidelines for the Establishment of Scheduled Heat Processes for Low Acid Canned Foods*, Technical Manual No. 3, Chipping Campden, CCFRA.

CCFRA (1980), *The Processing of Canned Fruits and Vegetables*, Technical Manual No. 29, Chipping Campden, CCFRA.

CHEN C R and RAMASWAMY H S (2002),'Modeling and optimization of variable retort temperature (VRT) thermal processing using coupled neural networks and genetic algorithms', *J Food Eng*, 53(3), 209–220.

COSSEY R and RICHARDSON P S (1991), *The Use of Constant Temperature Blocks for the Calibration of Master Temperature Indicators*, Technical Memorandum No. 613, Chipping Campden, CCFRA.

DATTA A K (1993), 'Error estimates for approximate kinetic parameters used in food literature', *J Food Eng*, 18, 181–199.

DOBIE P A P (1993), 'Calibration: DIY or ISO?', *Food Manufacture*, 68(4), 28–29.

DURANCE T (1997), 'Improving canned food quality with variable retort temperature processes', *Trends in Food Sci and Technol*, 8, 113–118.

DURANCE T, DOU J and MAZZA J (1996), 'Selection of variable retort temperature processes for canned salmon', *J Food Process Eng,* 20, 65–76.

EISNER M (1988), *Introduction to the Technique of Rotary Sterilization*, privately published, Milwaukee, WI.

FAO (1983), *Recommended International Code of Practice for Low-acid and Acidified Low-acid Canned Foods*, Codex Alimentarius, Vol. G, CAC/RCP23–1978, Rome, Food and Agricultural Organisation, World Health Organization.

FAO (1986), *Codex Elimentarius*, ALINORM 86/16, Appendix VI, Rome, Food and Agricultural Organisation, World Health Organization.

FDA (1983), *Thermally-processed Low-acid Foods Packaged in Hermetically Sealed Containers*, Regulations 21, Part 113, Washington, Food and Drugs Administration.

FLAMBERT F and DELTOUR J (1972), 'Localization of the critical area in thermally-processed conduction heated food', *Lebensm Wiss u Technol*, 5(1), 7, 1972.

FPI (1988), *Canned Foods, Principles of Thermal Process Control, Acidification and Container Closure Evaluation*, Washington, Food Processors Institute.

GOULD G W (ed.) (1995), *New Methods of Food Preservation*, London, Blackie.

GREENWOOD D A, KRAYBILL H R, FEASTER J F and JACKSON J M (1944), 'Vitamin retention in processed meats', *Ind Eng Chem,* 36, 922–28.

HENDRICKX M, VAN GENTCHEN K and TOBBACK P (1989), 'Optimizing quality attributes of conduction heated foods, a simulation approach', *5th Int. Congress Engineering and Food*, Cologne, Germany.

HENDRICKX M, SILVA C, OLIVEIRA F and TOBBACK, P (1992), 'Optimization of heat transfer in thermal processing of conduction heated foods', in Singh R P and Wirakartakusumah M A, *Advances in Food Engineering* Boca Raton, FL, CRC Press, 221–235.

HENDRICKX M, SILVA C, OLIVEIRA F, and TOBBACK, P. (1993), 'Generalized (semi)-empirical formulae for optimal sterilization temperatures of conduction-heated foods with infinite surface heat transfer coefficients', *J Food Eng*, 19(2), 141–158.

HERMANN J (1969), 'Optimization of sterilization processes with regard to the destruction of bacteria and chemical changes', *Die Nahrung*, 13, 639–661.

HOLDSWORTH S D (1985), 'Optimisation of thermal processing review', *J Food Eng,* 4, 89–116.

HOLDSWORTH S D (1992), *Aseptic Processing and Packaging of Foods*, London, Elsevier Applied Science.

HOLDSWORTH S D (1997), *Thermal Processing of Packaged Foods*, Glasgow, Blackie Academic and Professional.

HOYEM T and KVALE O (1977), *Physical, Chemical and Biological Changes in Food Caused by Thermal Processing*, London, Applied Science Publishers.

INSTITUT APPERT (1979), *Barèmes de Sterlisation pour Aliments Appertisés*, Paris, Institut Appert.

JACKSON J M, FEASTER J F and PILCHER R W (1945), 'The effect of canning procedures on vitamins in food'. *Proc Inst Food Technologists*, 81–90.

JELEN P (1983), 'Review of basic technical principles and current research in UHT processing of foods', *J Food Sci*, 36, 692–698.

KESSLER H G (1981), *Food Engineering and Dairy Technology*, Chapter 6, Freissing, Germany, Verlag A. Kesler, 186–207.

LENZ M K and LUND D B (1977), 'The lethality Fourier number method: experimental verification of a model for calculating average quality value retention in conductive-heating canned foods', *J Food Sci*, 42(4), 997–1001.

LUND D B (1977), 'Design of thermal processes for maximizing nutrient retention', *Food Technology* 31(2), 71–75.

MANSFIELD T (1962), 'High-temperature short-time sterilization', in *Proc 1st Int Congress Food Sci Technol* 4, London, Gordon & Breach, 311–316.

MANSFIELD T (1974), *A Brief Study of Cooking*, San José, CA, FMC.

MAY N S (1991), *Evaluation of the Datatrace® Micropack system. Part II*, Technical Memorandum No.620, Chipping Campden, CCFRA.

MAY N S (1992), *Evaluation of the Datatrace® Micropack system. Part III*, Technical Memorandum No.661, Chipping Campden, CCFRA.

MAY N S and COSSEY R (1989), *Review of data acquisition units for in-container temperature measurement and initial evaluation of the Ball Datatrace® system Part I,* Technical Memorandum No. 567 Chipping Campden, CCFRA.

MCKENNA A B and HOLDSWORTH S D (1990), *Sterilization and Cooking of Food Particulates. A Review*, Technical Bulletin No. 75, Chipping Campden, CCFRA.

NARDKARNI M M and HATTON T A (1985), 'Optimal nutrient retention during the thermal processing of conduction-heated canned foods: application of the distributed minimum principle', *J Food Sci*, 50, 1312–1321.

NASARI H, SIMPSON R, BOUZAS J and TORRES J A (1993), 'An unsteady-state method to determine kinetic parameters for heat inactivation of quality factors: conduction-heated foods', *J Food Eng*, 19(3), 291–301.

NEWMAN M and STEELE D A (1978), *Methods of Thermal Process Calculations for Food Particles*, Technical Memorandum No. 210, Chipping Campden, Glos., CCFRA.

NFPA (1971), *Processes for Low-acid Canned Foods in Glass Containers*, Bulletin 30–L, National Food Processors Association, Washington.

NFPA (1983), *Thermal Processes for Low-acid Canned Foods in Metal Containers*, Bulletin 26–L, National Food Processors Association, Washington.

NOROHNA J, HENDRICKX M, SUYS J and TOBBACK P (1993), 'Optimization of surface quality retention during thermal processing of conduction heated foods using variable temperature retort profiles', *J Food Proc Preserv*, 17, 75–91.

NUNES R V, SWARTZEL K R and OLLIS D F (1993), 'Thermal evaluation of food processes: the role of a reference temperature', *J Food Eng* , 20(1), 1–15.

OHLSSOHN T (1980a), 'Optimal sterilization temperatures for flat containers', *J Food Sci*, 45, 848–853.

OHLSSON T (1980b),'Optimal sterilization temperatures for sensory quality in cylindrical containers', *J Food Sci* 45, 1517–1521.

OHLSSON (1980c), 'Optimization of heat sterilization using C-values', in Linko P *et al., Food Process Engineering*, London, Elsevier Applied Science, 137–145.

PONTRYAGIN L S, BOLTANSKII V G, GAMKRELIDZE R V and MISHCHENKO E F (1962), *The Mathematical Theory of Optimal Processes*, New York, Wiley-Interscience.

PREUSSKER H (1970), 'Preservation techniques: selection of temperature and processing times for sterilization and cooking of food', *Die Ernährungwirtschaft/ Lebensmittel*, 17(10), 770–788.

PRIESTLEY R G (ed.) (1979), *Effects of Heating on Foodstuffs*, London, Applied Science Publishers.

REICHERT J E (1974), 'Optimum sterilization temperatures for prepared foods', *Fleisch-Wirtschaft,* 1305–1313.

REICHERT J E (1977), 'The C-value as an aid to optimization', *Z Lebensm-Technol u-Verfahr*, 28(1), 1–7.

RICHARDSON P S, KELLY P T and HOLDSWORTH S D (1988), 'Optimization of in-container sterilization processes', in *Proc Int. Symp. Progress in Food Preservation Processes,* Vol. 2, Brussels, CERIA, 1–11.

SAGUY I and KAREL M (1979), 'Optimal retort temperature profile for optimizing thiamin retention in conduction-type heating canned foods', *J Food Sci*, 44, 1485–1490.

SILVA C, HENDRICKS M, OLIVEIRA F and TOBACK P (1992a), 'Critical evaluation of commonly used objective functions to optimize overall quality and nutrient retention of heat-preserved foods', *J. Food Eng*, 17(4), 241–258.

SILVA C, HENDRICKS M, OLIVEIRA F and TOBACK P (1992b), Optimal sterilization temperatures for conduction heating foods considering finite surface heat transfer coefficients. *J Food Sci,* 57(3), 743–748.

SILVA C, OLIVEIRA F A R and HENDRICKX M (1994a), 'Optimum sterilization: A comparative study between average and surface quality', *J Food Process Eng*, 17, 155–176.

SILVA C L M, OLIVEIRA F A R, LAMB J, TORRES A P and HENDRICKX M (1994b), 'Experimental validation of models predicting optimal surface quality sterilization temperatures', *Int J Food Sci Technol,* 29, 227–241.

SILVA C L M, OLIVEIRA F A R and HENDRICKX M (1994c), 'Quality optimization of conduction heating foods sterilized in different packages', *Int J Food Sci Technol.* 29, 515–530.

SJÖSTRÖM, C and DAGERSKOG, M. (1977), 'Optimization of time/temperature relationship for heat serilization of canned food by computer simulation'. In *Proc. Mini-Symposium Mathematical Modelling in Food Processing,* Orenas, Sweden, 227–234.

SPINAK S H and WILEY R C (1982), 'Comparison of the general and Ball formula methods for retort pouch process calculations,' *J Food Sci*, 56(1), 27–30.

SVENSSON S (1977), 'Inactivation of enzymes during thermal processing' in Hoyem T and Kvale O, *Physical, Chemical and Biological Changes in Food Caused by Thermal Processing* , London, Applied Science Publishers, 202–208.

TEIXEIRA A A, DIXON J R, ZAHRADNIK J W and ZINSMEISTER G E (1969), 'Computer simulation of variable retort control and container geometry as a possible means of improving thiamine retention in thermally processed foods', *J Food Sci,* 40, 656–659.

TERAJIMA Y and NONAKA Y (1996), 'Retort temperature profile for optimum quality during conduction-heating of foods in retortable pouches', *J. Food Sci,* 61(4), 673–678.

TISCHER R G, HURWICZ H and ZOELLNER J A (1953), 'Heat processing of beef. III. Objective measurement of changes in tenderness, drained juice and sterilization value during heat processing', *Food Research*, 18, 539–554.

TUCKER G S and HOLDSWORTH S D (1990), 'Optimisation of quality factors for foods thermally processed in rectangular containers', in Field W and Howell J, *Process Engineering in the Food Industries-1 Convenience Foods and Quality Assurance*, London, Elsevier Applied Science, 59–74.

TUCKER G S and HOLDSWORTH S D (1991), 'Mathematical modelling of cooking and sterilization processes for heat preserved foods: applications of a new heat transfer model', *Food and Bioproducts Processing, Trans I Chem E,* 69(C1), 5–12.

UK-DHSS (1981), *The Canning of Low Acid Foods*, Food Hygiene Code of Practice No. 10, Department of Health and Social Security, London, HMSO.

UK-DH (1994), *Guidelines for the safe production of heat preserved foods*, Department of Health, London, HMSO.

ZANONI B, PAGLIARINI E, GIOVANELLI G and LAVELLI V (2003), 'Modelling the effects of thermal sterilization on the quality of tomato puree', *J Food Eng*, 56(2/3), 201–206.

2

Optimising the efficiency and productivity of thermal processing

J. C. Oliveira, University College Cork, Ireland

2.1 Introduction: the role of thermal processing in extending shelf-life

Subjecting food products to heat for a specified period of time may be done with one of two objectives:

1. Develop the nutritional and/or sensory quality of the product so that it is edible, digestible and tasteful (e.g. baking bread, grilling steaks).
2. Extend the shelf-life of the product.

This chapter is concerned with the second objective only. What limits the shelf-life of a food product?

- *Quality factors*. This is the most important issue for vegetable-based products, for instance. A process of senescence (that in some products sets up right after harvest) will eventually make the product inedible. Even before that, the food may dry, shrivel, soften, change colour (mostly browning), that is, lose market value and eventually become unsalable altogether.
- *Safety factors*. Human foods are foods for micro-organisms too. As they feed and grow they release products of their own metabolism onto the food. This may have a simple quality impact, such as a strong acid flavour (e.g. lactic acid bacteria synthesise the sugars of the food and release lactic acid). However, some micro-organisms (pathogens) release chemicals that are highly toxic to humans: for instance, *Clostridium botulinum* releases one of the most powerful toxins known to Man. Every year, hundreds of people all over the world (including the affluent societies) are severely affected by food poisoning due to the activity of pathogenic micro-organisms in some food product they consumed, and some die (Bryan *et al.*, 1997).

How can shelf-life be extended? The quality-deteriorating metabolisms in food tissues involve biochemical reactions. Most, though not all, are catalysed by enzymes. Microbial metabolism itself is also strongly dependent on the enzymic profiles of the microbial cells. Enzymes are proteins, sensitive to denature when subjected to heat. Hence, by heating food for a sufficiently long time so that all enzymes are denatured, we will prevent the enzymic-mediated quality deterioration pathways and the food will maintain its quality (except for the processes that do not require enzymes, such as non-enzymatic browning). Similarly, if all microbial cells are killed by a sufficiently high thermal processing, and the product is hermetically sealed later to prevent subsequent contamination, the food will be microbially stable indefinitely. Thermal processing can therefore extend the shelf-life of food products for very many years. It is one of the oldest forms of preserving food products.

Some microbial cells are, however, rather difficult to kill. Spore-forming bacteria (which includes some of the most dangerous pathogens) protect the genetic material of the cell before dying: when subjected to heat stress, they form spores, which are like capsules that protect cellular DNA and are much more resistant to heat, as they are not live cells, and hence have no enzymatic-mediated metabolism to be destroyed by thermal processing. Once the thermal stress is gone, spores may germinate and create new cell colonies. Hence, a food will only be microbially stable if heat processing destroys the spores themselves. An exception is foods with a pH below 4.5, as spores do not germinate in such low pH; in that case destroying the live (vegetative) cells would suffice for long-term storage. Therefore, there may be three objectives for thermal processing as a means of extending shelf-life:

1. denaturing enzymes;
2. killing vegetative microbial cells;
3. destroying bacterial spores.

These three biological entities have different resistances to thermal treatment and different sensitivity to temperature changes, so we might need to assess each of them independently. We can aim at all three objectives, in which case the product will be stable for a very long time. In broad terms, the thermal processes aimed at each of those objectives are:

- blanching (inactivate extracellular enzymes);
- pasteurisation (kill vegetative cells);
- sterilisation (destroy bacterial spores in addition to vegetative cells) – more precisely, should be 'commercial sterilisation', as absolute sterilisation is not needed for long-term food storage.

In some products, such as milk, sterilisation is very difficult with acceptable sensory results, and the product is not really sterilised. However, spore destruction is very extensive and the product has a much extended shelf-life compared to pasteurisation. That is the case of UHT (Ultra High Temperature) processes, also known as ultra-pasteurisation. The severity of these thermal

Table 2.1 Advantages and disadvantages of thermal processing compared to other long-term methods of food preservation

	Thermal processing (sterilisation)	Freezing	Drying
Advantages	Unit cost of production lower than freezing Lower capital investment than freezing No special need for storage (other than food in hermetically sealed container)	Lowest impact on sensory qualities	Lowest capital investment (depends on technology) No special need for storage (other than impermeable container)
Disadvantages	Loss of texture, particularly detrimental for vegetables Loss of nutritional quality Loss of colour Possible formation of off-flavours	Higher unit cost of production High capital investment Special storage requirements (freezing temperatures) implies high storage and distribution costs	Quality-technologies with high capital investment Resulting product extremely different from original raw material Loss of aroma and flavour Loss of nutritional quality (depends on technology)

treatments is, however, more comparable to general food commercial sterilisation than to pasteurisation (Simpson *et al.*, 2000).

The main 'competitors' of thermal processing as a means for long-term preservation of foods are processes that totally inhibit microbial growth and tissue metabolism. The microbial cells may well be present in the food, but they cannot grow, and the enzymatic-mediated reactions cannot take place either. That can be done by removing liquid water, because all these processes require an aqueous environment. This may be achieved by removing the water from the matrix (drying) or by making it solid (freezing). Table 2.1 compares these methods with thermal processing. Another alternative would be to make the food environment inhospitable to microbial growth by lowering the pH significantly, below 3 (pickling), or by adding salt (salting). However, the acid or salt required to achieve microbial stability imparts very strong acidic or salty flavour to the food, which makes it suitable only for some speciality products with decreasing market interest.

2.1.1 Advantages and disadvantages of thermal processing

If all that heat would do to food was to inactivate enzymes and kill microbial cells, thermal processing (sterilisation) would be the paramount process of food preservation. Unfortunately, subjecting foods to heat induces several deteriorative processes, such as (Ryley and Kajda, 1994; Fellows, 1997):

1. Loss of nutritional quality. Vitamins are also sensitive to heat, as there are various enzymic and non-enzymic metabolic pathways for their degradation to other (nutritionally uninteresting) components which are accelerated or triggered at high temperatures (Hardy *et al.*, 1999). Maillard browning reactions are the major route of nutrient loss in thermally processed products.
2. Loss of texture, specially in vegetable tissues (Moreira *et al.*, 1994).
3. Formation of off-flavours. Some of the metabolisms triggered at high temperatures may create components with unpleasant aroma or taste (e.g. so-called cooked flavour in heated milk).
4. Strong modifications of colour. Chlorophyll is sensitive to high temperature and green colour fades with thermal treatments; the opposite problem, particularly strong browning, may also occur, specially if there are sugars or fats in the matrix.

Sterilised and UHT products currently have a poor market image. Considering all products that are available in supermarkets these days, that is unfair to these products. However, the perception of poorer sensory quality prevails in consumers, and that is what controls buying behaviour.

2.1.2 Minimising the detrimental effects of thermal processing

In very broad terms, the less severe the exposure to heat, the better the product quality. Hence, if we can accept a more limited shelf-life (two weeks to six months, depending on the product), we could settle for killing the vegetative cells only. Even though spores remain, will germinate and new colonies will appear, this gives us a lag time up to spoilage (Breand *et al.*, 1999), and that window is the shelf-life. We can extend it by keeping the product at low (chilled) temperatures, to retard spore germination and microbial growth even more. Consequently, pasteurised products have a better sensory quality than sterilised ones, but generally require a chilled storage and distribution chain, and therefore, although the thermal process itself is less expensive (less heat), the product will end up being more expensive to the consumer.

It should also be noted that while the consumer will find higher prices in pasteurised products *versus* their commercially sterile and UHT counterparts, the industrial product (prior to distribution) may actually have a low margin for the manufacturing company in spite of the higher quality, as the higher costs of storage and sales shift the percentual value added distribution downstream from manufacturing.

2.2 Setting commercial objectives for thermal processes: process optimisation

Whether we are considering pasteurisation, UHT or sterilisation of a food product, there is a target defined in terms of the extent of microbial death, and there is a detrimental impact on quality. Hence, there is a typical optimisation problem to solve: achieve the target of thermal processing with minimum detrimental impact on the product quality (Ávila and Silva, 1999). Another optimisation objective would obviously be to achieve the thermal treatment target with minimum production costs, an equation that may involve both energy costs (Barreiro *et al.*, 1984) and productivity value (Banga *et al.*, 1991). The precise conditions that optimise one do not necessarily optimise the other. Most scientific research has focused on the former (quality optimisation).

This is unfortunate, as the most important optimisation to achieve is evidently to maximise profits. That objective would need to consider not really the maximum intensity of quality factors achievable, but the actual increased product value, while integrating also the production cost factors. That optimisation function would be the really winning proposition, but has rarely been addressed in literature. The main reason is likely that while the intensity of a quality factor (e.g. vitamin C content, sensory rating) or the microbial death achieved have very objective quantitative figures, product value has both a quantitative and an intangible element, which would make the overall model rather fuzzy.

Many products state on the label the composition of nutritional components present in the original product mix, as legislation in many countries does not force to state the value actually present at the end of shelf-life. Hence, product labels may ignore the loss of nutritional value in processing and storage – in that case, optimising nutritional quality factors has actually no monetary value *per se*. At the other extreme, bioactive components that have a noticeable impact on human metabolism (for instance, caffeine) are detected by the consumer, and in that case what actually matters is the intensity of them at the time of consumption. Maximising the intensity of these quality factors therefore has an obvious quantitative value: to ensure the end result we need to use more ingredients in the mix, so the savings due to optimisation times the unit cost of these ingredients is the monetary value of optimisation. However, the relationship between the actual intensity of these factors and market value is blurred – the same can be said for sensory quality.

For instance, let us consider a process with optimised production cost that therefore puts on the market a product slightly cheaper than another which optimised for sensory factors. Which is the proposition that maximises profit? That depends on how the products perform (higher sales of a cheaper product may result in higher profits), and therefore on the value given by the consumers at large to the quality advantage *versus* the price. In conventional economics, this means that it depends on the level of elasticity of demand. Figure 2.1 illustrates some demand curves (sales *versus* price) typical of supermarket food

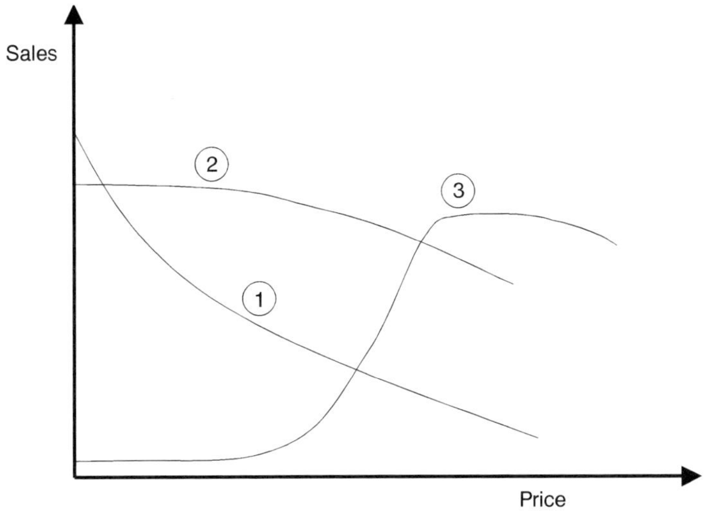

Fig. 2.1 Typical demand curves. 1: elastic demand, stronger for low prices (typical of commodities); 2: inelastic demand in the lower price range (may be found in strong brand products); 3: so-called 'snob' product demand, with reverse elasticity below a threshold and fairly inelastic around it (sometimes found in high quality niche products).

products. The elasticity of demand will vary substantially with the product category, but will always be a somewhat fuzzy question. Furthermore, the consequences of thermal process optimisation may be strongly related to the business strategic objectives of the company. How has the senior management decided to compete on the market for the product category in question? A company that set business targets of cutting down on production costs by x% in the year may be uninterested in optimising for quality, specially if the market gains are not clear.

The following expressions are simplified, but suffice for a quantitative view of this discussion:

$$Profit = Sales\_value - Costs \qquad [2.1]$$

(*Profit* is the operating profit, removing fixed costs)

$$Sales\_value = Sales \times Price \qquad [2.2]$$

$$Costs = Production \times Unit\_cost \qquad [2.3]$$

(*Sales* is the amount sold; *Price* is the average unit price obtained by the company; *Unit_cost* is the production cost allocated to this product divided by the units produced, which may include amortising capital investment)

$$Sales = Production - Waste \qquad [2.4]$$

(*Waste* is the amount of product not sold)

The overall maximisation function should be:

$$Profit = Sales \times Price - Sales \times Unit\_cost - Waste \times Unit\_Cost \quad [2.5]$$

These factors are a function of:

- Sales = $\mathbf{f_1}$ (Price, Sensory_rating, Bioactive_ingredients) [2.6]
(*Sensory_rating* is the perceived quality rated by consumers in terms of flavour, taste, appearance, etc.; *Bioactive_ingredients* is the content of specific nutrients or stimulating components to which the consumer gives a value)
- Unit_cost = $\mathbf{f_2}$ (Energy_costs, Productivity, Raw_materials) [2.7]
(*Raw-materials* is the cost of raw materials, which includes ingredients used in the product formulation that may be sensitive to thermal processing)
- Waste = $\mathbf{f_3}$ (Sensory_rating, Price) [2.8]

$\mathbf{f_1}$, $\mathbf{f_2}$ and $\mathbf{f_3}$ are functions that may be very different. In fact, $\mathbf{f_2}$ is a very objective and deterministic function, while $\mathbf{f_1}$ and $\mathbf{f_3}$ are market-type functions, that is, they depend on assumptions that historical market data may be used to infer future product performance (if no historical data are available, they can also be estimated from consumer/market studies). $\mathbf{f_1}$ shows how elastic is the demand (similar to depictions in Fig. 2.1). As these functions depend also on competition, not only between similar products from rival companies, but also between different product categories as market preferences may shift, they are not more than forecasts, highly product- and market-specific.

Optimising for quality may maximise *Sensory_rating* and *Bioactive_ingredients* and minimise *Raw_materials*. Optimising for energy costs minimises *Energy_costs*. These specific elements of the overall picture are those that have received most attention in scientific literature (Ávila and Silva, 1999). Optimising for productivity, although a simple proposition, has received little attention, and there is no paper really looking at equation 2.5. A good example of a comprehensive business analysis involving thermal process optimisation is given by Simpson *et al.*, 2003.

2.3 Assessing the potential of in-container, aseptic and HTST processing

The significant detrimental impact of thermal processing on product quality (Hardy *et al.*, 1999) means that there are huge shifts in market performance between different process categories of thermal treatments that have been developed to obtain optimum quality products. We will briefly discuss three categories:

1. In-container processing.
2. Aseptic processing.
3. HTST (High Temperature Short Time).

The most conventional concept for thermal processing is to seal the product in a hermetic container and then thermally process the container and product. The thermal process may be (Fellows, 1997):

- batch, using a retort to provide the heating, holding and cooling phases of the thermal treatment;
- continuous, with products rolling on conveyors into a tunnel with the three sections (heating, holding, cooling).

The latter can only be used for process temperatures below boiling point, as above it the container will increase internal pressure significantly and it is necessary that the heating medium provides counter-pressure. Only retorts (closed environments) are suitable for this (in-container sterilisation is known as Apertisation, specially in francophone regions, in memory of Nicolas Apert).

In aseptic processing the product is packed only after processing. It must therefore be transported through equipment where it will be heated, held at the required temperature for the required time, and then cooled, and it must then be packed in an aseptic environment, into sterilised packages. The advantage stems from the resistance to heat transfer that food products themselves exhibit. In a container, internal layers of the product are thermally insulated by the outer layers. If there is some mixing inside the container the difference is minimised (e.g. rotating retorts and particulate products, with a substantial liquid phase), while if the product is solid, there is little that can be done to mitigate this insulating effect. The most difficult zone of product to be heated is termed the 'cold spot', and obviously, it is its temperature history that will dictate whether the microbial safety target was reached. The concept of aseptic processing comes from the obvious fact that we minimise this problem if we have thin layers of product. Aseptic processing makes sense with continuous processes, although this is not a conceptual requirement.

As aseptic processing minimises the resistance to heat transfer in the system, it minimises also the energy needs and the processing times for heating and cooling. Hence, aseptic processing achieves higher quality, lower energy costs and higher productivity compared to in-container processing (Ramaswamy *et al.*, 1997).

In general, aseptic processing is very easy to implement with liquids and so there is almost no more in-container batch processing of liquids. On the other hand, solids are not easily transported; while a liquid may be pumped, a solid needs to be mechanically or pneumatically conveyed. In-container thermal processing is therefore the choice for solids. For particulate products (liquids containing solid particles, like soups) aseptic processing is now proving more popular.

Whether the food is processed in-container or aseptically, a HTST (High Temperature Short Time) process would result in significant quality gains and also minimise energy costs and maximise productivity (Simpson *et al.*, 2000). HTST therefore has everything going for it in equation 2.5. If we assume a constant temperature for treatment, we could use any temperature high enough to kill microbes. The higher the temperature, the shorter the time required. Therefore, if we use a higher temperature we can achieve the microbial safety target in a shorter time – the higher the better. Productivity is thus increased, and due to energy efficiency, unit costs are lower for producing higher heating energy for shorter periods of time. However, the drive for HTST has not been the cost factors, but actually the quality gains, as these can be quite substantial.

At a constant temperature, microbes will be killed at a given rate. This rate increases with temperature. The same occurs for the rate at which quality factors are degrading. However, the sensitivity of the microbial death rate to temperature is much higher (two to five times) than that of quality factor degradation (Martinez *et al.*, 1999). Therefore, even though both the microbial death rate and the quality loss rate increase with temperature, the former increases much more significantly. As processing time is shortened by the higher temperatures, we may eventually end up with a much lower loss of quality for the same microbial death target. This concept is depicted in Fig. 2.2. Assuming

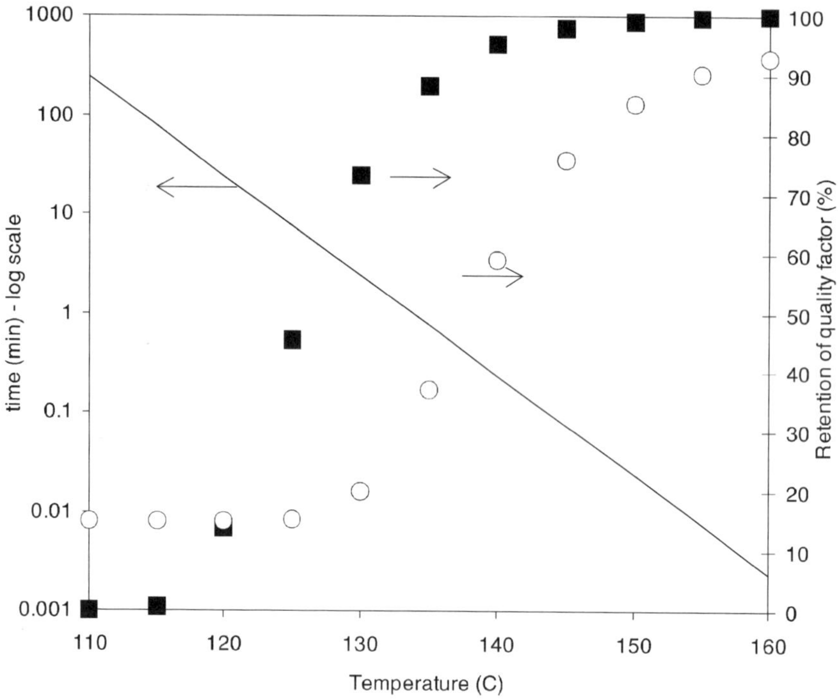

Fig. 2.2 Examples of retention of quality factors in samples treated at a constant temperature (conceptual graph, valid for instantaneous heating only). The process temperature is indicated in the abcissa. The solid line gives the process times required to achieve the target lethality (12 log reduction), to be read in the left y-axis. The points indicate the percentual retention of the quality factor after the corresponding time of treatment at that temperature, to be read in the right y-axis. Open circles represent texture (maximum load force in a compression test) compared to that of the raw material and closed squares indicate the vitamin C content relative to that of the raw product.

the simplest and most widely used models for these kinetics (first-order kinetics with Arrhenius or Bigelow temperature dependence – Ávila and Silva, 1999), the graph shows for a typical (vegetable) product the time required for achieving a given microbial death target and the loss of two quality factors (vitamin C content and texture). It is evident that the higher the temperature and the shorter the time, the better the quality.

The limiting factor in HTST is therefore how fast can we heat and cool a product. For solids, with their in-container processing, the insulating effect is particularly problematic. The temperature of surface layers and that of the centre will have very different histories, and in order to process the centre adequately, the surface may be severely overprocessed. The higher the process temperature, the greater this differential, and the product surface may suffer unacceptable cooking if we were to try too high temperatures. There are therefore various optimisation constraints that can be considered for defining what the optimum processing conditions will be in a case like this, and it may be necessary to consider various quality factors in different parts of the product. An example would be the following constrained optimisation target: 'Maximise average vitamin content with microbial death equal or higher to the target and surface browning equal or lower to quality threshold criteria.' In this type of problem, variable retort temperatures have proved beneficial (Durance, 1997).

Microwave and radio-frequency heating systems are very advantageous in this respect for the heating phase in solid products, as they have better volumetric heating characteristics (Ohlsson, 1999). For continuous (aseptic) processing, Ohmic heating offers the same advantages (Eliot-Godereaux *et al.*, 2001). Furthermore, their energy efficiency is good. If quality gains of HTST heating are important in the product, the capital investment would be justified. The fact that these heating technologies are not really widespread in industry reflects the small interest that quality optimisation actually has in food business strategies for products that are considered by the market as little more than commodities; the margin is not there to recover the capital investment of the better heating systems.

2.4 Techniques for optimising the efficiency of thermal processes

We could conceive two types of optimisation strategy:

- *Deterministic*. We make some assumptions which enable the application of mathematically exact models. The targets are quantitative (for instance, reduce the number of colony forming units of a specific pathogen by six orders of magnitude). Recent examples of this strategy are given by Pornchaloempong *et al.* (2003) and Balsa-Canto *et al.* (2002b).
- *Stochastic*. We consider variability and given distributions in the relevant variables and factors. The targets are probabilistic (for instance, the probability of survival of a spore of a specific pathogen is 1 in 10 million).

It is also possible to consider a mixed strategy, using deterministic models where some factors have a given stochastic distribution (Smout *et al.*, 2000).

Both food legislation and scientific research have preferred the deterministic approach. While in engineering terms this is a perfectly good choice, in safety terms it has created the wrong impression. The fact that the safety target and the optimisation designs are exact values does not mean that the result is a precise and exact figure. The mathematical model is indeed exact, but only once the assumptions are valid. Nature, however, is not deterministic, and many factors have a significant variability. This includes the kinetic parameters of the micro-organisms, the operating variables themselves, etc. We could say that each of the assumptions has a given error, or a given probability of being right/wrong, so in actual fact both techniques should be seen as providing a stochastic result.

The most widely considered problem of thermal processing optimisation is the constrained optimisation target expressed as: 'find the combination of operating variables that minimises the loss of a specific quality factor while delivering the quantitative safety target' (Jung and Fryer, 1999). Variations on the theme include the quality factor specified, its location (e.g. average *versus* surface quality – Silva *et al.*, 1994b; cold-spot *versus* hot-spot), and the technology itself. This is a very deterministic scenario:

- the quality target to optimise is the intensity of a factor that can be measured analytically (e.g. vitamin C content, texture, colour) in the location to be considered;
- the safety constraint is the microbial lethality delivered in the cold-spot (or portion of product that received the least lethality), quantified in terms of one microbial species.

The question of 'location' makes obvious sense in the case of in-container processing of solids, while it is less well defined in aseptic processing. For turbulent flow, we cannot talk about a precise flow pathway where thermal treatment is lower, and would always need to consider the problem in a stochastic way, for instance, analysing the residence time distribution of the fluid and the average thermal treatment for the minimum residence time (Torres and Oliveira, 1998). The case of rotational retorts is somewhere in between these two extremes (Ghani *et al.*, 2001)

The quality and the safety factors can be measured experimentally, and from there it is necessary to obtain a mathematical model that quantifies these factors as a function of the operating variables of the process (these are the process design factors, that is, those parameters that can be changed by the operator in the piece of equipment, such as temperature of the heating medium, process time, flow rates, etc.). Then, a deterministic optimisation routine can be applied (e.g. Simplex method). We need two types of model:

1. Heat transfer models, which predict the temperature at any location and any time as a function of the operating variables (Balsa-Canto *et al.*, 2002a). For aseptic processing (Jung and Fryer, 1999) and for containers with liquids

(Ghani *et al.*, 2001), the flow model needs to be integrated with heat transfer.

2. Kinetic models, which predict the intensity of a quality or safety factor as a function of temperature and time.

In the processing of particulate foods (solids immersed in a liquid phase), and particularly in blanching, it may also be necessary to consider losses of water-soluble nutrients by leaching, which requires mass transfer models (Ryley *et al.*, 1990).

We need two kinetic models, which must be valid in non-isothermal conditions (for the heating and cooling phases):

- intensity of the quality factor;
- microbial death (number of colony forming units remaining after recovery from thermal stress, which includes spores).

Figure 2.3 depicts this general procedure. Microbial kinetics have been widely simplified to a first-order model, and process assessment and validation has been defined following the so-called TDT (thermal death time) or Bigelow model (Guiavarc'h *et al.*, 2002). We assume that at constant temperature the death rate is proportional to the number of colony forming units remaining. The reciprocal of this proportionality is the thermal death time, or D value, which is then assumed to decrease exponentially with temperature. The lethality achieved in a given thermal treatment is therefore given by:

$$\log(N_o) - \log(N) = \frac{1}{D_r}\int_0^t 10^{(T-T_r)/z}.dt \qquad [2.9]$$

where log is a decimal logarithm, N_o is the initial number of colony forming units, N their remaining number after the thermal treatment – hence (log N_o – log N) is the reduction of the order of magnitude of the colony forming units, which is the microbial safety target (e.g. 12 reduction is the target for sterilisation, for pasteurisation it varies, but a value of 6 is common), T_r is a reference temperature (can be anything, in sterilisation 250°F/121°C, is the common standard), T is the temperature at a given time t, D_r is the thermal death time of the selected microbial species in the food matrix in question at the reference temperature T_r, and z is the sensitivity to temperature of the D value of this species in this matrix. Process targets are often expressed as the so-called F-value, which is the product of the microbial lethality achieved by the reference D value: $F = D_r \times (\log N_o - \log N)$. Analysis of whether the microbial safety target was achieved therefore depends on the selection of pathogen and the product itself, as the food matrix may affect the microbial death kinetic parameters significantly. For sterilisation of low-acid foods non-proteolytic *Clostridium botulinum* is considered, as it is the most heat resistant pathogen, with a z value of 10°C, which is the most pessimistic (higher) value for the range of z-values of *Clostridia* species (Pflug and Odlaug, 1978). For pasteurisation the target micro-organism and hence the kinetic parameters to use for safety assessment depend on the product (Gaze, 1992).

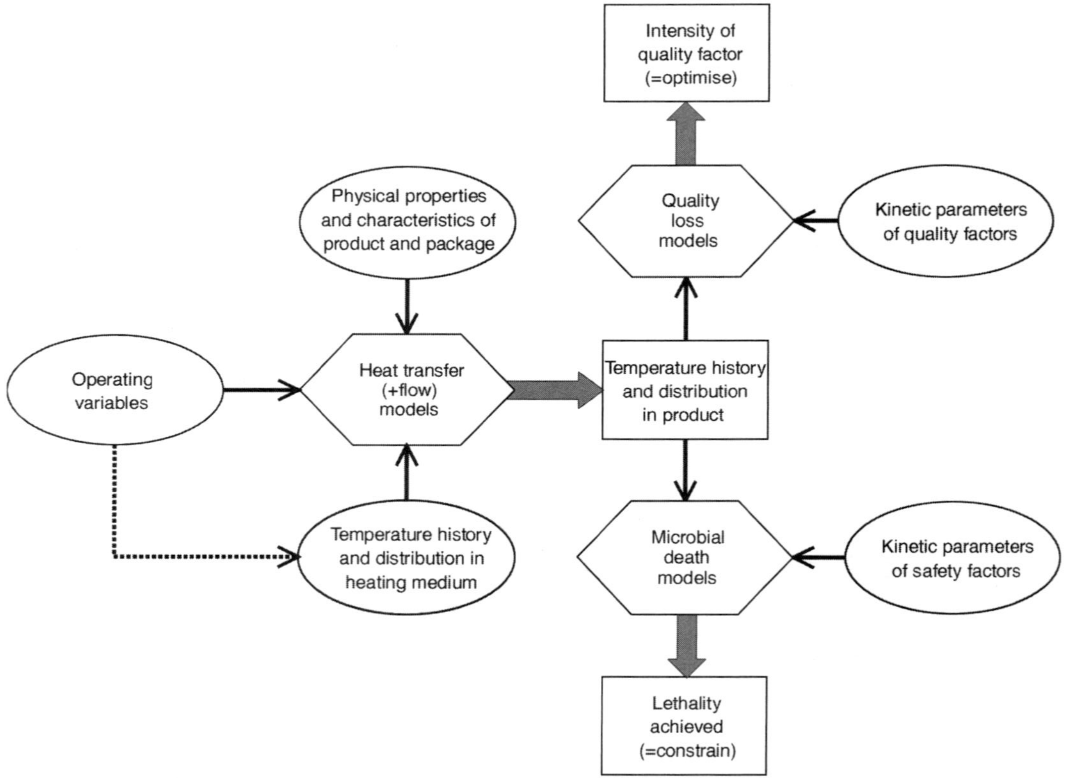

Fig. 2.3 Scheme of information flow for quality optimisation in thermal processing. Ellipsoids represent inputs, hexagons indicate models (information processing elements), and rectangles depict the model predictions.

This first-order model is fairly suitable for high temperatures and relatively short times. Microbial death tends to tail off, that is, the D value does not really decrease exponentially, it starts to level off at lower temperatures, though various other patterns have been reported (Peleg, 2000, Peleg and Penchina, 2000). The kinetic pattern also depends on the particular micro-organism and the food media itself. For lower processing temperatures (70–90°C) it is highly advisable to check the microbial death pattern and adjust the model accordingly.

Quality factors have been described with similar models (in which case it is usual to define a C-value, named 'cook value', similar to the F-value of microbial death – Silva *et al.*, 1994a). Alternatively, an Arrhenius-temperature dependency model of the rate constant is also used, as this model is more common for chemical and biochemical reactions. For the range of temperatures of interest to food processing, the difference is not really important. Kinetic patterns that deviate significantly from a first order are more common than in microbial death, however. This means that the degradation or loss rate is often not constant and proportional to the intensity of the quality factor – usually, the rate decreases with the intensity. There are two main models used in this case (Martinez *et al.*, 1999): (i) a n-order model, where it is assumed that the loss rate is proportional not to the intensity but to a power n of it; (ii) a two-fraction model, where it is assumed that there are actually two populations (iso-forms) of the factor, each with a first-order decay, but with different rate constants.

There are few studies on the application of stochastic models, and particularly of optimisation using this approach. Variability of operating variables has been described with normal distributions, and with time-series (Varga and Oliveira, 2000). Some systems have shown a good correlation with the Weibull distribution, as opposed to normal distribution (Peleg and Penchina, 2000). The Weibull distribution is a time-to-failure function, which exhibits a tailing compared to the symmetry of normality and is therefore more universal (Cunha *et al.*, 1998).

2.5 Future trends

As companies modernise their manufacturing facilities there will be a wider application of volumetric heating (microwave, radio-frequency and Ohmic heating systems). If a manufacturing equipment investment is going to be made, it is easier to conceive that within it, it will make sense to consider a somewhat higher investment that shows a better return. The return has a tangible and quantitative element: the higher energy efficiency in heating and the higher productivity; and also another aspect that may be more intangible, or actually concede a valuable market edge: higher product quality. Due to the limitations of these heating technologies (namely heterogeneity) it is likely that mixed heating systems (conventional heating enhanced with microwave/radio-frequency/Ohmic heating) will prove more popular than volumetric heating systems on their own.

Similarly, HTST aseptic processing systems for particulate products will become more popular and progressively replace in-container sterilisation for these products, and this will offer more innovation possibilities for packaging. However, rotating retort systems improve somewhat heat resistance problems in in-container processing, and have a much better control of processing conditions over continuous flow, specially in particulate products. This may weigh more in some business strategies, and is likely to be a product-specific issue. Which is more important – lower costs and higher quality, or better control (minimum product variability)?

The greater interest for new heating technologies and aseptic processing, and particularly for particulate products, raises questions about process assessment and validation, as it is not possible to measure physically the temperature at the cold spot with inexpensive thermocouples. The development of Time-Temperature Integrators (TTI's) is very promising for these situations. These are systems with a kinetic response that mimics that of the target micro-organisms, and therefore the lethality achieved can be inferred from the TTI reading (Guiavarc'h *et al.*, 2002).

In general, the business objectives and market strategies of companies regarding their specific products should preside to the establishment of the optimisation functions. While research has done well in proposing various minimal and combined processing systems towards optimum quality products, it has faltered in the commercial exploration of these findings due to lack of information on actual cost and profit-related implications.

A very important field of development of thermal processing in the modern food industry is the application of mild heat treatments. Some form of thermally induced stress is an essential element in virtually all minimum processing systems for convenience products, and specially those that attempt to be as close to fresh or just-cooked as possible (Breand *et al.*, 1999). Minimum processing in this context means that the preservation techniques used modify sensory properties minimally. These systems usually take advantage of more than one source of microbial stability or hindrance, combining the effect of various protective factors: for instance, heat, low pH, higher water activity, modified atmosphere packaging, etc. Each factor on its own would be unable to achieve a significant shelf-life extension, but all put together, plus chilled storage and distribution, result in suitable shelf-life while modifying sensory properties minimally (Gorris and Tauscher, 1999).

The most crucial factor for this type of thermal processing is the variability of the thermal treatments. As they are light, we cannot really have either over-processing or underprocessing without jeopardising the product concept itself. Temperature and heat penetration distribution in ovens and retorts have therefore received due attention, and they need to be assessed regularly in the industrial equipment itself. The application of solutions designed with the assistance of Computational Fluid Dynamics (CFD) is likely to become a best practice (Verboven *et al.*, 1999).

Conceiving and designing minimal processing systems with equation 2.5 in mind rather than the simple bipolar quality-safety target would be a very interesting proposition. There are several unit operations, and therefore various cost and quality-related factors to analyse in terms of what is the combination that maximises profit.

One of the most exciting possibilities recently proposed is the production of long shelf-life (commercially sterile or UHT equivalent) ready-meals obtained by combined thermal and high-pressure treatment (Meyer *et al.*, 2000). This would result in ready-meals with the quality of frozen products without the need for a frozen distribution chain. It is an ideal system for production-on-demand or mass customisation in the food supply chain, as it is amenable for division of the production system into intermediate components plus final customised assembly close to the demand (geographically or in time).

2.6 References

ÁVILA I M and SILVA C L M (1999), 'Methodologies to optimize thermal processing conditions: an overview', in Oliveira F A R and Oliveira J C, *Processing Foods – Quality Optimization and Process Assessment*, Boca Raton, FL, USA, CRC Press, 67–82.

BARREIRO J A, PEREZ C R and GUARIGUATA C (1984), 'Optimization of energy consumption during the heat processing of canned foods', *Journal of Food Engineering*, 3 (1), 27–36.

BALSA-CANTO E, BANGA J R and ALONSO A A (2002a), 'A novel, efficient and reliable method for thermal process design and optimization. Part I: theory', *Journal of Food Engineering*, 52 (3), 227–234.

BALSA-CANTO E, BANGA J R and ALONSO A A (2002b), 'A novel, efficient and reliable method for thermal process design and optimization. Part II: applications', *Journal of Food Engineering*, 52 (3), 235–247.

BANGA J R, PEREZ-MARTIN R I, GALLARDO J M and CASARES J J (1991), 'Optimisation of the thermal processing of conduction-heated canned foods: study of several objective functions', *Journal of Food Engineering*, 14 (1), 25–37.

BREAND S, FARDEL G, FLANDROIS J P, ROSSO L and TOMASSONE R (1999), 'A model describing the relationship between regrowth lag time and mild temperature increase for *Listeria monocytogenes*', *International Journal of Food Microbiology*, 46 (3), 251–261.

BRYAN F L, GUZEWICH J J and TODD E C D (1997), 'Surveillance of foodborne disease. 2. Summary and presentation of descriptive data and epidemiologic patterns, their value and limitations', *Journal of Food Protection*, 60 (5), 567–578.

CUNHA L M, OLIVEIRA F A R, BRANDÃO T S and OLIVEIRA J C (1998), 'Optimal experimental design for estimating the kinetic parameters of the Bigelow model', *Journal of Food Engineering*, 33, 111–128.

DURANCE T D (1997), 'Improving canned food quality with variable retort temperature processes', *Trends in Food Science and Technology*, 8 (4), 113–118.

ELIOT-GODEREAUX S C, FAIRHURST P G, GOULLIEUX A and PAIN J P (2001), 'Passage time distributions of cubes and spherical particles in an ohmic heating pilot plant',

Journal of Food Engineering, 47 (1), 11–22.

FELLOWS P (1997), *Food Processing Technology: Principle and Practice*, Cambridge, England, Woodhead.

GAZE J E (1992), 'Food Pasteurization Treatments', Technical Manual No. 27, Chipping Campden, England, Campden and Chorleywood Food Research Association.

GHANI AGA, FARID M M, CHEN X D and RICHARDS P (2001), 'A computational fluid dynamics study on the effect of sterilization temperatures on bacteria deactivation and vitamin destruction', *Proceedings of the Institution of Mechanical Engineers Part E – Journal of Process Mechanical Engineering*, 215 (E1), 9–17.

GUIAVARC'H Y P, DINTWA E, VAN LOEY A M, ZUBER F T and HENDRICKX M E (2002), 'Validation and use of an enzymic time-temperature integrator to monitor thermal impacts inside a solid/liquid model food', *Biotechnology Progress,* 18 (5), 1087–1094.

GORRIS L and TAUSCHER B (1999), 'Quality and safety aspects of novel minimal processing technologies', in Oliveira F A R and Oliveira J C, *Processing Foods – Quality Optimization and Process Assessment*, Boca Raton, FL, USA, CRC Press, 325–340.

HARDY J, PARMENTIER M and FANNI J (1999), 'Functionality of nutrients and thermal treatments of food', *Proceedings of the Nutrition Society,* 58 (3), 579–585.

JUNG A AND FRYER P J (1999), 'Optimising the quality of safe food: Computational modelling of a continuous sterilisation process', *Chemical Engineering Science*, 54 (6), 717–730.

MARTÍNEZ A, RODRIGO C, FERNÁNDEZ P S, OCIO M J, RODRIGO F and RODRIGO M (1999), 'Microbial and quality assessment in thermal processing', in Oliveira F A R and Oliveira J C, *Processing Foods – Quality Optimization and Process Assessment*, Boca Raton, FL, USA, CRC Press, 83–96.

MEYER R S, COOPER K L, KNORR D and LELIEVELD H L M (2000), 'High-pressure sterilization of foods', *Food Technology*, 54 (11), 67–74.

MOREIRA L A, OLIVEIRA F A R, OLIVEIRA J C and SINGH R P (1994), 'Textural changes in vegetables during processing. Part I – A descriptive method to segregate effects of process treatments', *Journal of Food Processing and Preservation*, 18, 483–496.

OHLSSON T (1999), 'Minimal processing of foods with electric heating methods', in Oliveira F A R and Oliveira J C, *Processing Foods – Quality Optimization and Process Assessment*, Boca Raton, FL, USA, CRC Press, 97–106.

PELEG M (2000), 'Microbial survival curves – the reality of flat "shoulders" and absolute thermal death times', *Food Research International,* 33 (7), 531–538.

PELEG M and PENCHINA C M (2000), 'Modelling microbial survival during exposure to a lethal agent with varying intensity', *Critical Reviews in Food Science and Nutrition,* 40 (2), 159–172.

PFLUG I J and ODLAUG T E (1978), 'A review of z and F values used to ensure the safety of low-acid canned foods', *Food Technology*, 6 (1), 63–74.

PORNCHALOEMPONG P, BALABAN M O, TEIXEIRA A and CHAU K V (2003), 'Optimization of quality retention in conduction-heating foods of conical shape', *Journal of Food Process Engineering*, 25 (6), 557–570.

RAMASWAMY H S, AWUAH G B and SIMPSON B K (1997), 'Heat transfer and lethality considerations in aseptic processing of liquid/particle mixtures: A review', *Critical Reviews in Food Science and Nutrition*, 37 (3), 253–286.

RYLEY J, ABDELKADER Z M, GUEVARA L V, LAPPO B P, LAMB J, MANAN F, POVEY M I W and RACE P (1990), 'The prediction of vitamin retention in food-processing', *Journal of*

Micronutrient Analysis, 7 (3–4), 329–347.

RYLEY J and KAJDA P (1994), 'Vitamins in thermal-processing', *Food Chemistry*, 49 (2), 119–129.

SANDEEP K P, ZURITZ C A and PURI V M (1999), 'Determination of lethality during aseptic processing of particulate foods', *Food and Bioproducts Processing*, 77 (C1), 11–17.

SILVA C L M, OLIVEIRA F A R and HENDRICKX M (1994a), 'Quality optimization of conduction heating foods sterilized in different packages', *International Journal of Food Science and Technology*, 29 (5), 515–530.

SILVA C L M, OLIVEIRA F A R, LAMB J, TORRES A P and HENDRICKX M (1994b), 'Experimental validation of models for predicting optimal surface quality sterilization temperatures', *International Journal of Food Science and Technology*, 29 (3), 227–241.

SIMPSON R, JIMENEZ M, VEGA M, ROMERO A and COSTA M (2000), 'Evaluation of commercial UHT milk and optimization of industrial process', *Archivos Latinoamericanos de Nutricion*, 50 (4), 353–360.

SIMPSON R, ALMONACID S and TEIXEIRA A (2003), 'Optimization criteria for batch retort battery design and operation in food canning-plants', *Journal of Food Process Engineering*, 25 (6), 515–538.

SMOUT C, VAN LOEY A M L and HENDRICKX M E G (2000), 'Non-uniformity of lethality in retort processes based on heat distribution and heat penetration data', *Journal of Food Engineering*, 45 (2), 103–110.

TORRES A P and OLIVEIRA F A R (1998), 'Residence time distribution studies in continuous thermal processing of liquid foods: a review', *Journal of Food Engineering*, 36 (1), 1–30.

VARGA S, OLIVEIRA J C, SMOUT C and HENDRICKX M E (2000), 'Modelling temperature variability in batch retorts and its impact on lethality distribution', *Journal of Food Engineering*, 44 (3), 163–174.

VERBOVEN P, SCHEERLINCK N, DE BAERDEMAEKER J and NICOLAI B (1999), 'Possibilities and limitations of computational fluid dynamics for thermal process optimization', in Oliveira F A R and Oliveira J C, *Processing Foods – Quality Optimization and Process Assessment*, Boca Raton, FL, USA, CRC Press, 41–66.

3

Optimising the efficiency of batch processing with retort systems in thermal processing

R. Simpson, Universidad Técnica Federico Santa María, Chile

3.1 Introduction: batch processing in food canning plants

Batch processing has been widely practised but little analysed in the context of canned food plants. Although high speed processing with continuous rotary or hydrostatic retort systems can be found in very large canning factories, such systems are not economically feasible in the majority of small to medium-sized canneries (Simpson *et al.*, 2003a).

In this chapter, we will analyse batch processing in a retrospective and a prospective view. Firstly, batch processing problem structure will be defined in relation to canned food plants. Then, a batch system optimisation will be discussed. To bridge the gap between thermal processing and industrial engineering in optimising design and operation of food canning plants we will discuss and present specific procedures to determine: (a) optimum number of retorts to maximise net present value, and (b) optimum process conditions for simultaneous processing of different product lots in the same retort.

Finally, we will try to discuss and analyse this large and diverse field where should be plenty of room for surprises, perhaps particularly for those who take time to look closely enough with an open and speculative mind.

3.1.1 Problems in maximising the efficiency of batch processing

Batch processing with a battery of individual retorts is a common mode of operation in many food-canning plants (canneries). Although high speed processing with continuous rotary or hydrostatic retort systems can be found in very large canning factories (where they are cost-justified by high volume throughput), such systems are not economically feasible in the majority of small to medium-sized canneries (Norback and Rattunde, 1991). In such smaller canneries, retort operations are carried out as batch processes in a cook room in which the battery of retorts is located. Although the unloading and reloading operations for each retort are labour intensive, a well designed and managed cook room can operate with surprising efficiency if it has the optimum number of retorts and the optimum schedule of retort operation.

This type of optimisation in the use of scheduling to maximise efficiency of batch processing plants has become well known, and is commonly practised in many process industries. Several models, methods and implementation issues related to this topic have been published in the process engineering literature (Rippin, 1993; Kondili *et al.*, 1993; Reklaitis, 1996; Barbosa and Macchietto, 1993; Lee and Reklaitis, 1995a, 1995b). However, specific application to retort batteries in food canning plants has not been addressed in the food process engineering literature. Food canneries with batch retort operations are somewhat unique in that the cannery process line as a whole is usually a continuous process in that unit operations both upstream and downstream from the retort cook room are normally continuous (product preparation, filling, closing, labelling, case packing, etc.). Although retorting is carried out as a batch process within the cook room, unprocessed cans enter and processed cans exit the cook room continuously at the same rate (see Fig. 3.1). Since the entire process line operates continuously, food canneries are often overlooked as batch process industries. The focus of this work was to apply these batch process optimisation techniques only to the retort operations within the cook room, and not the entire process line of the cannery.

Food processing, and thermal processing in particular, is an industry confronted with strong global competitiveness. Continuous innovation and improvement of processing procedures and facilities is needed. Although the literature in food science and thermal processing is very extensive, most of the references deal with the microbiological and biochemical aspects of the process or with engineering analysis of a single unit process operation, and rarely analyse the processing operations in the context of manufacturing efficiency. The early stages of a project usually involve studies of alternative processes, plant configurations and type of equipment. Among problems confronted by canned food plants with batch retort operations are peak energy/labour demand, underutilisation of plant capacity and underutilisation of individual retorts.

3.1.2 Issues in optimising batch processing

In batch retort operations, maximum energy demand occurs only during the first few minutes of the process cycle to accommodate the venting step, while very little is needed thereafter in maintaining process temperature. Likewise, peak labour demand occurs only during loading and unloading operations, and is not required during the holding time at processing temperature. In order to minimise

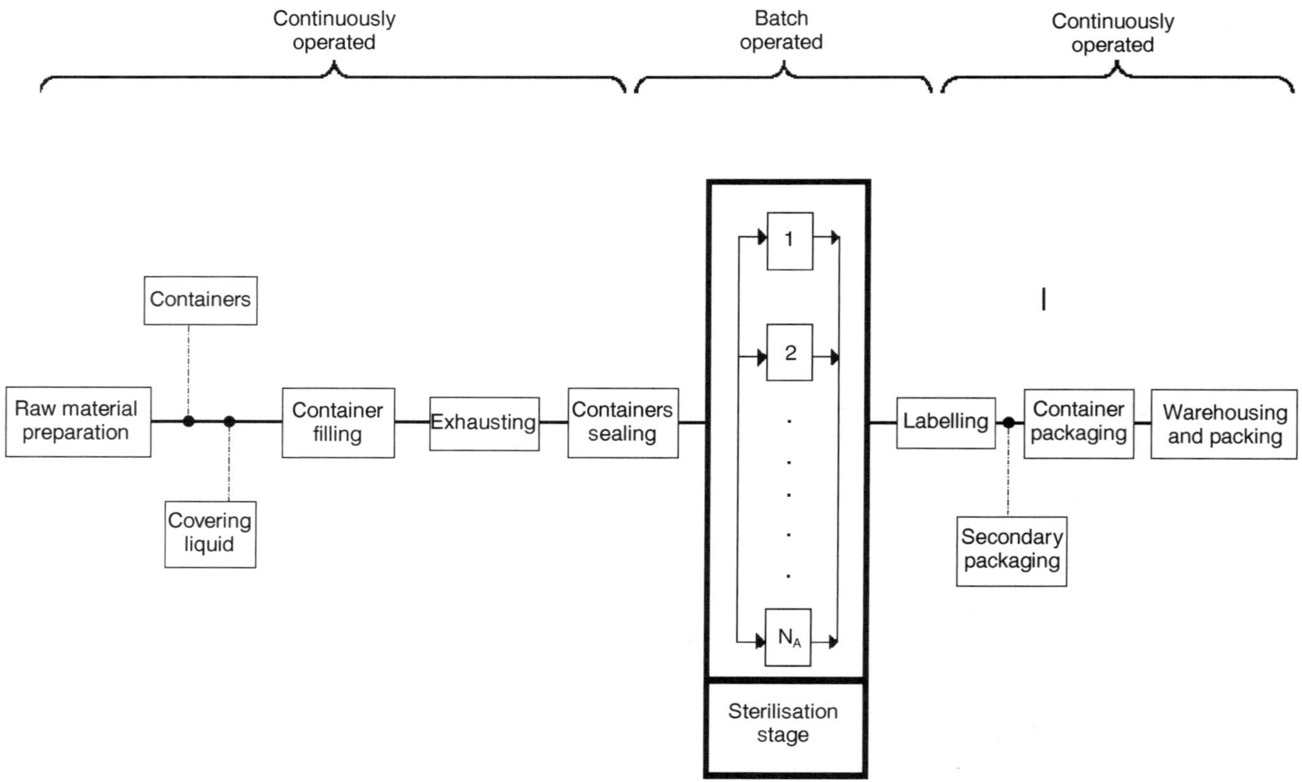

Fig. 3.1 General simplified flow diagram for a canning plant.

peak energy demand it is customary to operate the retorts in a staggered schedule, so that no more than one retort is venting at any one time. Similar rationale applies to labour demand, so that no more than one retort is being loaded or unloaded at any one time. Too few retorts in a battery can leave labour unutilised, while too many will leave retorts unutilised. The optimum number will maximise utilisation of labour and equipment, thus minimising ongoing processing costs. Alternatively, the optimum number of retorts may be based upon maximising the economic rate of return on the capital investment in the project measured in terms of net present value, which takes many additional factors beyond processing costs into account. In the case of maximising output from a fixed number of retorts for different products and container sizes, iso-lethal processes can be identified for each of the various products (alternative combinations of retort temperature and process time that deliver the same lethality), from which a common set of process conditions can be chosen for simultaneous processing of different product lots in the same retort.

3.2 Criteria for optimal design and operation of batch processing

The hierarchical approach consists of successive refinements and the design procedure is similar to the hierarchical planning strategy discussed in the artificial intelligence (AI) literature (Douglas, 1988). In contrast to normal true batch processes, canned food plants are operated with just one stage functioning in a batch mode. During normal operation of the sterilisation stage (Fig. 3.1) the various retort units are filled with cans, perform the retorting process for a specified period and then they shut down and the cycle is repeated. As previously mentioned, in canned food plants, all units, with the exception of retorts, operate continuously. The distinctions between batch and continuous processes are sometimes somewhat 'fuzzy' (Douglas, 1988). According to the literature, when a plant has one or two batch operations with large production rates that otherwise operate continuously they are normally referred to as a continuous process. Although most of the food science and food engineering literature refers to a canning plant as a batch plant, when the sterilisation stage is operated in batch mode, and the hierarchical approach is applied, it is assumed that it is better to classify it as a continuous process.

The design effort will be to decide whether a concept is sufficiently promising from an economic point of view that a more detailed study could be justified. In our specific case the flow scheme of the process is presented in Fig. 3.1. Although some exceptions to this flow scheme could be justified, the following analysis will consider it as a general flow scheme for canned food plants. The main target in the following sections will be to decide the optimum number of retorts that can be allocated in a canned food plant. The approach will be to identify a general procedure that can be applied to canned food plants.

To decide and optimise canned food plant design and operation the Net Present Value (or Net Present Worth) profitability evaluation method will be utilised.

3.3 Optimising energy consumption

Several models for the energy distribution in retorts have been developed, but without studying the dynamic response (Holdsworth, 1997). The transient energy balance for a system defined as the retort including cans without their contents, and the steam and condensate in the retort requires no work term (see Fig. 3.2). The heat transfer terms – between the system and its environment – include radiation and convection to the plant environment, and heat transfer to the food within the cans. Equations were solved simultaneously and the heat transfer model for the food material was solved using an explicit finite difference technique. Correlations valid in the range of interest (100°C through 140°C) were utilised to estimate the thermodynamics properties of steam, condensed water, and food material.

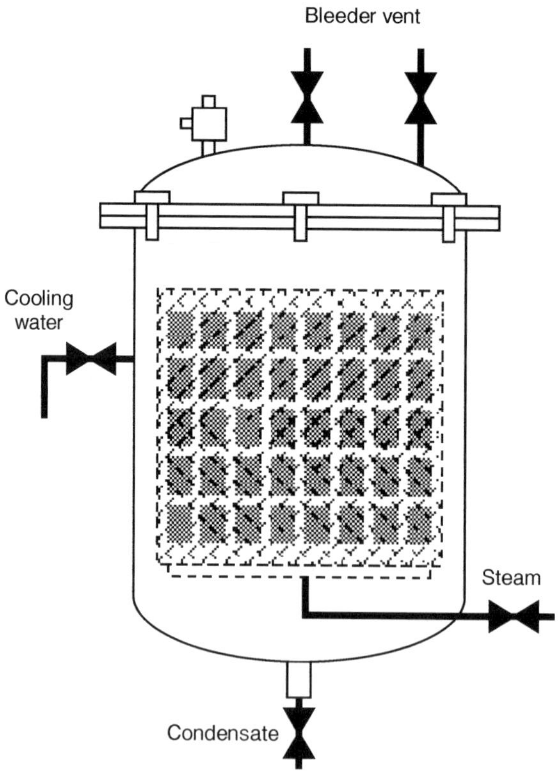

Fig. 3.2 Still vertical retort (cross-sectional view of vertical retort used for study).

Due to the particular characteristics of the process, to develop the model the process was divided into three steps, as indicated: (a) venting period, (b) period after venting to reach process temperature, and (c) holding time. Firstly, a brief mathematical model is presented for the food material and then a complete development of the energy model for the thermal process.

3.3.1 Mathematical model for food material

Food material was assumed to be homogeneous and isotropic, therefore the heat conduction equation could be expressed as:

$$\frac{1}{r}\frac{\partial T}{\partial r} + \frac{\partial^2 T}{\partial r^2} + \frac{\partial^2 T}{\partial z^2} = \frac{1}{\alpha}\frac{\partial T}{\partial t} \qquad [3.1]$$

Where T ($T(r,z,t)$) is a function of the position (r,z) and time (t). The respective boundary and initial conditions are as follows: $T(\textit{food material}, 0) = T_0$; where T_0 is a known and uniform value through the food material at time 0.

To estimate the temperature at food surface at any time t, a finite energy balance was developed at the surface:

$$-kA\frac{\partial T}{\partial r} + hA\partial T = MCp\frac{\partial T}{\partial t} \qquad [3.2]$$

In most practical cases, it can be assumed that Biot number is well over 40, meaning that the temperature of the surface of the food material could be equalised, at any time, with retort temperature (Teixeira *et al.*, 1969; Datta *et al.*, 1986; Simpson *et al.*, 1989; Almonacid-Merino *et al.*, 1993; Simpson *et al.*, 1993). The aforementioned statement is not necessarily applicable for retortable pouch processing (Simpson *et al.*, 2003b). The model (equations 3.1 and 3.2) considers the possibility of a Biot number less than 40, but is also suitable for a Biot number equal to or larger than 40.

3.3.2 Mass and energy balance during venting

Before expressing the energy balance, it is necessary to define the system to be analysed: Steam-air inside the retort at any time t $(0 \leq t \leq t^*)$, during venting – was considered as the system (see Fig. 3.2).

Global mass balance

$$\dot{m}_s - \dot{m}_{sv} - \dot{m}_a = \frac{dM}{dt} \qquad [3.3]$$

Mass balance by component

Air:

$$-\dot{m}_a = \frac{dM_a}{dt} \qquad [3.4]$$

Vapour:
$$\dot{m}_s - \dot{m}_{sv} - \dot{m}_w = \frac{dM_{sv}}{dt} \quad [3.5]$$

Condensed water:
$$\dot{m}_w = \frac{dM_w}{dt} \quad [3.6]$$

Where:
$$M = M_a + M_{sv} + M_w; m = \dot{m}_{sv} + \dot{m}_a \quad [3.7]$$

General energy balance

$$[\underline{H}_s\dot{m}_s]_{IN} - [\underline{H}_{sv}\dot{m}_{sv} + \underline{H}_a\dot{m}_a]_{OUT} + \delta\dot{Q} - \delta\dot{W} = \frac{dE_{system}}{dt} \quad [3.8]$$

Where
$$\delta\dot{Q} = \delta\dot{Q}_c + \delta\dot{Q}_r + \delta\dot{Q}_p + \delta\dot{Q}_e + \delta\dot{Q}_{rt} + \delta\dot{Q}_{in} \quad [3.9]$$

and
$$\delta\dot{W} = 0 \quad [3.10]$$

Replacing the respective terms into equation 3.9, the term δQ in equation 3.8 can quantified as:

$$\delta\dot{Q} = hA(T_{in} - T_{amb}) + \sigma\epsilon A(T_{in}^4 - T_{amb}^4) + M_pCp_p\frac{d\bar{T}_p}{dt} + M_{rt}Cp_{rt}\frac{d\bar{T}_{rt}}{dt} + M_eCp_e\frac{d\bar{T}_e}{dt} + M_{in}Cp_{in}\frac{d\bar{T}_{in}}{dt} \quad [3.11]$$

The following expression shows how the cumulative term of equation 3.8 was calculated. Because of the system definition, changes in potential energy as well as kinetic energy were considered negligible:

$$\frac{dE_{system}}{dt} = M_{sv}\frac{d\underline{H}_{sv}}{dt} + \underline{H}_{sv}\frac{dM_{sv}}{dt} - P_{sv}\frac{dV_{sv}}{dt} - V_{sv}\frac{dP_{sv}}{dt} + M_a\frac{d\underline{H}_a}{dt} + \underline{H}_a\frac{dM_a}{dt} - P_a\frac{dV_a}{dt} - V_a\frac{dP_a}{dt} + M_w\frac{d\underline{H}_w}{dt} + \underline{H}_w\frac{dM_w}{dt} - P_w\frac{dV_w}{dt} - V_w\frac{dP_w}{dt} \quad [3.12]$$

The mass flow of condensed water was estimated as:

$$\dot{m}_w(\underline{H}_{sv} - \underline{H}_{si}) = \delta\dot{Q} = hA(T_{in} - T_{amb}) + \sigma\epsilon A(T_{in}^4 - T_{amb}^4) + M_pCp_p\frac{d\bar{T}_p}{dt} + M_{rt}Cp_{rt}\frac{d\bar{T}_{rt}}{dt} + M_eCp_e\frac{d\bar{T}_e}{dt} + M_{in}Cp_{in}\frac{d\bar{T}_{in}}{dt} \quad [3.13]$$

Therefore:

$$\dot{m}_w = \left[hA(T_{in} - T_{amb}) + \sigma\epsilon A(T_{in}^4 - T_{amb}^4) + M_pCp_p\frac{d\bar{T}_p}{dt} + M_{rt}Cp_{rt}\frac{d\bar{T}_{rt}}{dt} + M_eCp_e\frac{d\bar{T}_e}{dt} + M_{in}Cp_{in}\frac{d\bar{T}_{in}}{dt}\right] / (\underline{H}_{sv} - \underline{H}_{sl}) \quad [3.14]$$

Therefore the mass flow demand during venting should be obtained replacing equations 3.5, 3.6, 3.7, 3.11, 3.12, and 3.14 into equation 3.8.

3.3.3 Mass and energy consumption between venting and holding time (to reach process temperature)

As was mentioned before, first, it is necessary to define the system to be analysed: steam and condensed water inside the retort were considered as the system (see Fig. 3.2).

Global mass balance: $$\dot{m}_s - \dot{m}_b = \frac{dM}{dt} \qquad [3.15]$$

Vapour: $$\dot{m}_s - \dot{m}_b - \dot{m}_w = \frac{dM_{sv}}{dt} \qquad [3.16]$$

Condensed water: $$\dot{m}_w = \frac{dM_w}{dt} \qquad [3.17]$$

Energy balance on the bleeder

System: Steam flow through the bleeder. Considering an adiabatic steam flow:

$$[\underline{H}_{sv}\dot{m}_b]_{IN} - \left[\left(\underline{H}_b + \frac{v^2}{2g_c}\right)\dot{m}_b\right]_{OUT} = 0 \qquad [3.18]$$

Where the bleeder is assumed to be operating in steady state condition, with no heat, no work, and negligible potential energy effects, the energy balance around the bleeder reduces to (Balzhiser *et al.*, 1972):

$$(\underline{H}_b - \underline{H}_{sv}) + \frac{v_b^2 - v_{sv}^2}{2g_c} = 0 \qquad [3.19]$$

For a gas that obeys the ideal gas equation (and has a Cp independent of T):

$$[\underline{H}_{sv} - \underline{H}_b] = Cp(T_{sv} - T_b) \qquad [3.20]$$

Neglecting v_{sv}^2 in relation to v_b^2, and replacing equation 3.20 into equation 3.19, it reduces to:

$$v_b^2 = 2g_c Cp T_{sv}\left(\frac{T_b}{T_{sv}} - 1\right) \qquad [3.21]$$

Considering an isentropic steam flow in the bleeder, which obeys the ideal gas equation, equation 3.21 could be re-written as:

$$v_b^2 = -\frac{2g_c P_{sv}}{\rho_{sv}}\left(\frac{\gamma}{\gamma - 1}\right)\left(\left(\frac{P_b}{P_{vs}}\right)^{(\gamma-1)/\gamma} - 1\right) \qquad [3.22]$$

Where the continuity equation is:

$$\dot{m}_b = \rho v_b A \qquad [3.23]$$

Therefore, combining equations 3.22 and 3.23:

$$\dot{m}_b = \frac{P_s A_b}{\sqrt{\frac{RT_s}{\gamma}}} \left(\frac{P_{amb}}{P_s}\right) \sqrt{\left(\frac{2}{\gamma - 1}\right)\left(1 - \left(\frac{P_{amb}}{P_s}\right)^{(\gamma-1)/\gamma}\right)} \qquad [3.24]$$

The maximum velocity of an ideal gas in the throat of a simple converging nozzle is identical to the speed of sound at the throat conditions. The critical pressure is P_c (Balzhiser *et al.*, 1972):

$$P_c = P_{amb}\left(\frac{2}{\gamma + 1}\right)^{\gamma/(\gamma-1)} \qquad [3.25]$$

Then equation 3.23 will be valid for P_s in the following range:

$$\left(\frac{2}{\gamma + 1}\right)^{\gamma/(\gamma-1)} \leq \frac{P_{amb}}{P_s} < 1 \qquad [3.26]$$

If P_s is bigger than P_c, substituting equation 3.25 into equation 3.24, the expression for the mass flow is as follows:

$$\dot{m}_b = \frac{P_s A_b}{\sqrt{\frac{RT_s}{\gamma}}}\left(\frac{2}{\gamma + 1}\right)^{(\gamma+1)/2(\gamma-1)}; \quad \text{for } \frac{P_{amb}}{P_s} < \left(\frac{2}{\gamma + 1}\right)^{\gamma/(\gamma-1)} \qquad [3.27]$$

3.3.4 Mass and energy balance during holding time

System: Steam inside the retort (see Fig. 3.2)

Global mass balance: $$\dot{m}_s - \dot{m}_b = \frac{dM}{dt} \qquad [3.28]$$

Vapour: $$\dot{m}_s - \dot{m}_b - \dot{m}_w = \frac{dM_{sv}}{dt} \qquad [3.29]$$

Condensed water: $$\dot{m}_w = \frac{dM_w}{dt} \qquad [3.30]$$

Energy balance on the bleeder

The steam flow through the bleeder was estimated as previously mentioned, therefore:

$$\dot{m}_b = \frac{P_s A_b}{\sqrt{\frac{RT_s}{\gamma}}} \left(\frac{P_{amb}}{P_s}\right) \sqrt{\left(\frac{2}{\gamma - 1}\right)\left(1 - \left(\frac{P_{amb}}{P_s}\right)^{(\gamma-1)/\gamma}\right)};$$

$$\text{if } \left(\frac{2}{\gamma + 1}\right)^{\frac{\gamma}{\gamma-1}} \leq \frac{P_{amb}}{P_s} < 1$$

$$\dot{m}_b = \frac{P_s A_b}{\sqrt{\frac{RT_s}{\gamma}}} \left(\frac{2}{\gamma + 1} \right)^{(\gamma+1)/2(\gamma-1)};$$

if $\frac{P_{amb}}{P_s} < \left(\frac{2}{\gamma + 1} \right)^{\gamma/(\gamma-1)}$

$$\dot{m}_w = \left[hA(T_{in} - T_{amb}) + \sigma\epsilon A(T_{in}^4 - T_{amb}^4) + M_p Cp_p \frac{d\bar{T}_p}{dt} + M_{rt} Cp_{rt} \frac{d\bar{T}_{rt}}{dt} + M_e Cp_e \frac{d\bar{T}_e}{dt} + M_{in} Cp_{in} \frac{dT_{in}}{dt} \right] / (\underline{H}_{sv} - \underline{H}_{sl})$$

Therefore the steam mass flow was estimated as:

$$\dot{m}_s = \dot{m}_w + \dot{m}_b \qquad [3.31]$$

An application example

Using bibliographic data (Barreiro *et al.*, 1984), the following data were utilised for calculations:

Retort: m_{rt}: 163.6 kg, c_{rt}: 500 J/kg°C, A_b: 7.94 × 10^{-6} m^2, A_{rt}: 2.97 m^2, V_{rt}: 0.356 m^3.

Containers: Can dimensions: 307 × 409; number of cans inside the retort: 180, m_e: 0.06 kg, c_e: 500 J/kg°C.

Product: Pea purée with a thermal diffusivity of 1.70 × 10^{-7} m^2/s.

Processing conditions: The time–temperature requirements were calculated for the thermal sterilisation of the pea purée in 307 × 409 cans reaching an integrated lethal value F_c of 2.52 min for *Clostridium botulinum* at 121.1°C (Barreiro *et al.*, 1984). T_{po}: 37.8°C, T_{cw}: 26.7°C.

Surface heat transfer coefficient and emissivity: h: 5.77 w/m^2°C, ϵ: 0.94

Retort insulation: Figures 3.3 and 3.4 compare the transient and total steam consumption for insulated (3.8 cm thick layer of asbestos cement) and non-insulated retort. Using the insulated retort the total steam consumption reduction was in the order of 19% when compared to non-insulated retort (see Fig. 3.3). Insulation did not play a significant role in terms of reducing peak energy consumption (see Fig. 3.4).

Initial food temperature (T_o): Figure 3.5 compares the transient steam consumption for different initial food temperatures (37.8 and 75°C). Figure 3.5 shows that for the high temperature the maximum peak and the total steam consumption are reduced 30% and 29% respectively. Clearly the most important variable to reduce peak energy consumption was the initial temperature of the food material.

Retort size: Figure 3.6 compares the steam consumption per processed can for different retort with the same ratio product / retort volume (0.47).

Case 1: Retort dimensions: m_{rt}: 163.6 kg, A_{rt}: 2.97 m^2, V_{rt}: 0.356 m^3. Containers: Can dimensions: 307 × 409; number of cans inside the retort: 180.

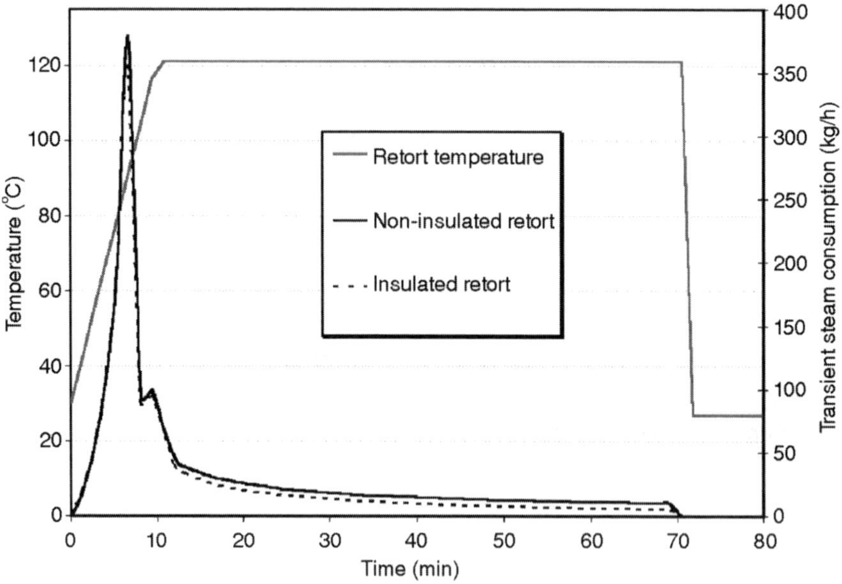

Fig. 3.3 Process temperatures and transient steam consumption profiles for insulated and non-insulated retort.

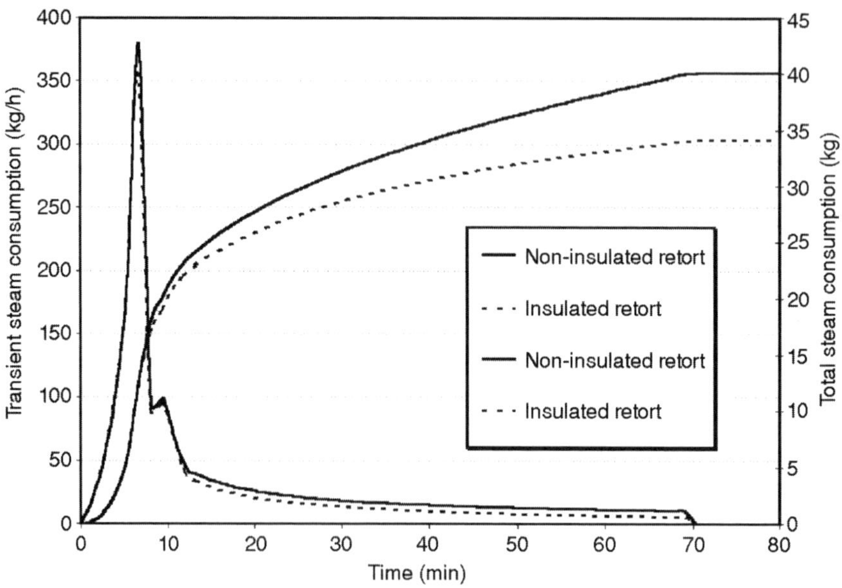

Fig. 3.4 Transient and total steam consumption profiles for insulated and non-insulated retort.

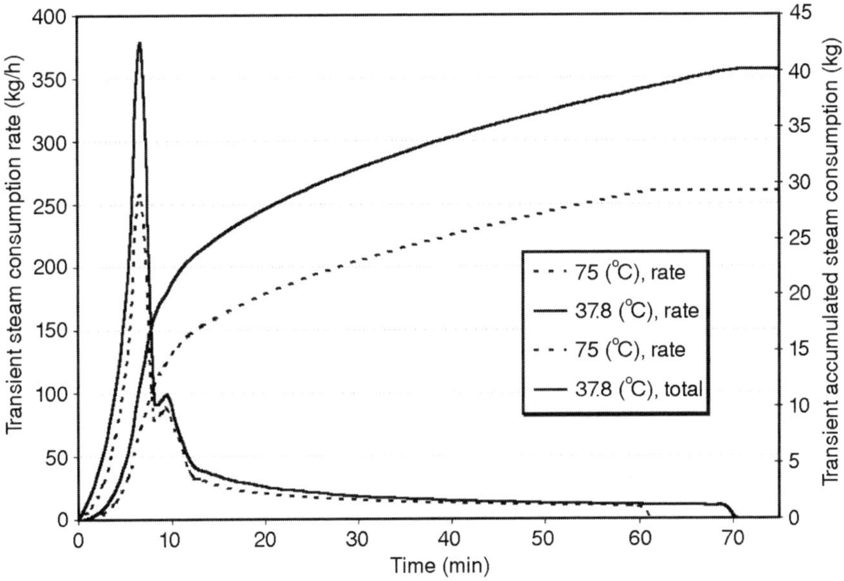

Fig. 3.5 Transient and total steam consumption profiles for different initial food temperatures.

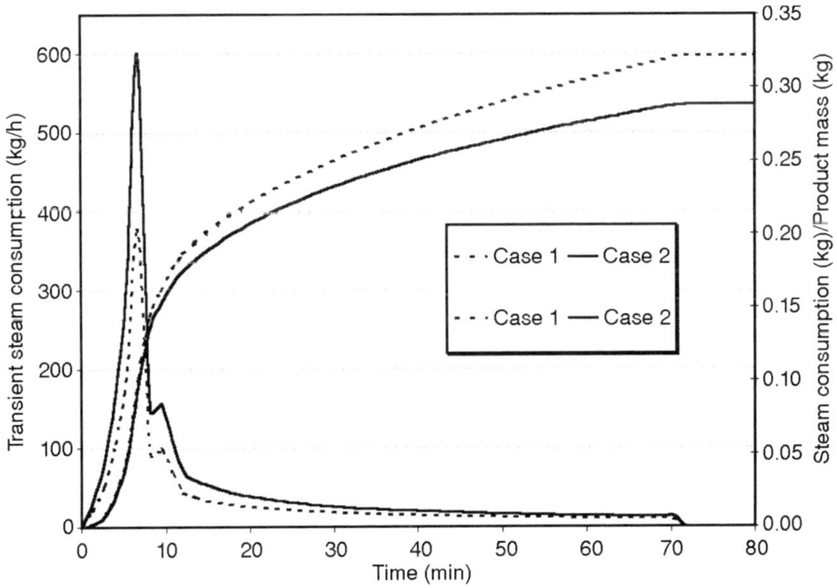

Fig. 3.6 Transient steam consumption profiles and ratio product mass/steam consumption mass for different retort sizes.

Case 2: Retort dimensions: m_{rt} : 230.4 kg, A_{rt} : 3.50 m^2, V_{rt} : 0.6 m^3 (Bhowmik *et al.*, 1985), Containers: Can dimensions: 307 × 409; number of cans inside the retort: 303.

Figure 3.6 shows that for the larger retort size the ratio steam consumption kg/mass product kg is reduced (13%) but the maximum steam consumption is increased (85%). Possibly the main reason behind this result is that the area over volume ratio decreases as the retort increases.

3.4 Optimising retort scheduling

Batch processing in food canneries consists of loading and unloading individual batch retorts with baskets or crates of food containers that have been filled and sealed just prior to the retorting operation. Each retort process cycle begins with purging of all the atmospheric air from the retort (venting) with inflow of steam at maximum flow rate, and then bringing the retort up to operating pressure/temperature, at which time the flow rate of steam falls off dramatically to the relatively low level required to maintain process temperature. The retort is then held at the process temperature for the length of time calculated to achieve the target lethality (F_o value) specified for the product. At the end of this process time, steam to the retort is shut off and cooling water is introduced to accomplish the cool down process, after which the retort can be opened and unloaded.

One of the factors that should be considered to decide retort scheduling is the energy demand profile during sterilisation processing (Almonacid-Merino *et al.*, 1993). In batch retort operations, maximum energy demand occurs only during the first few minutes of the process cycle to accomplish the high steam flow venting step. Very little steam is needed thereafter to compensate for the bleeder (and convection and radiation losses) in maintaining process temperature (Bhowmik *et al.*, 1985; Barreiro *et al.*, 1984). A typical representation of the energy demand profile during one cycle of a retort sterilisation process is shown in Fig. 3.3. As shown, at the initial stage of the process a high peak of energy consumption occurs (venting before reaching the retort temperature), later decreasing dramatically, and finally reaching a low and constant value (convection, radiation and bleeder). Thus, the energy demand for the whole plant will be conditioned upon this acute venting demand in the sterilisation process of each retort operating cycle. To minimise the boiler capacity and maximise energy utilisation, it is necessary to determine adequate scheduling for each individual retort.

Likewise, peak labour demand occurs only during loading and unloading operations, and is not required during the holding time at processing temperature. Therefore, a labour demand profile would have a similar pattern to the energy demand profile. In order to minimise these peak energy and labour demands the retort must operate in a staggered schedule so that no more than one retort is venting at any one time, nor being loaded or unloaded at any one time. When a battery consists of the optimum number of retorts for one labour

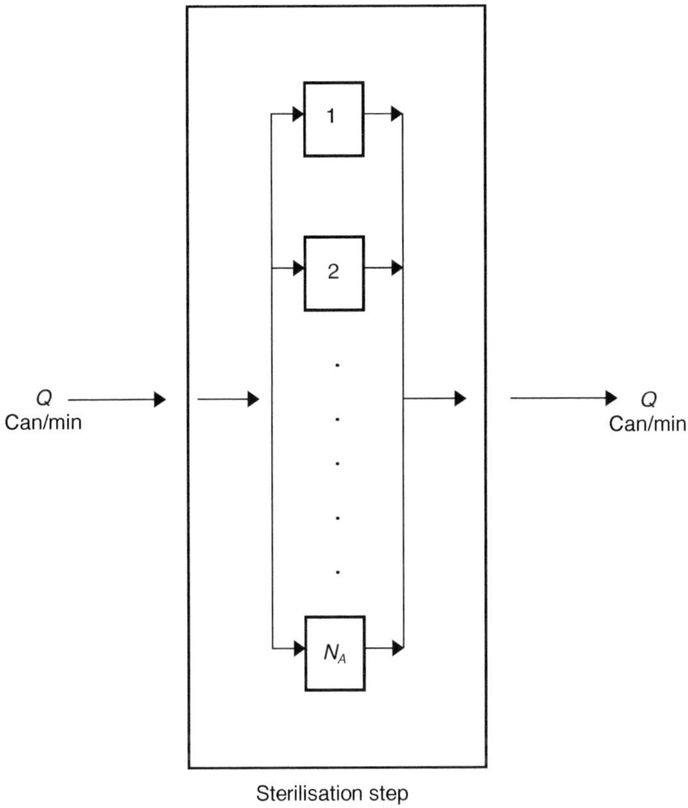

Fig. 3.7 Diagram for operation of a battery with optimum number (N_A) of retorts such that the cook room system operates with continuous inflow and outflow of product.

crew, the workers will be constantly loading and unloading a retort throughout the workday, and each retort will be venting in turn, one at a time. Under these optimum circumstances, unprocessed product will flow into and processed product will flow out of the retort battery system as though it were a continuous system as shown in Fig. 3.7, while the energy profile will appear as in Fig. 3.8.

The optimum number of retorts in the battery will maximise utilisation of labour and equipment, thus minimising unit-processing costs. Too few retorts in a battery can leave labour unutilised, while too many will leave retorts unutilised. A Gantt chart showing the temporal programming schedule of the battery retort system (see Fig. 3.9) can be used as a first step in determining the optimum number of retorts. Optimum operation of the retort battery can be achieved if the loading step of the last retort starts at the same time as the first retort finishes its cycle and is ready for unloading. This means that the loading time multiplied by the number of retorts must fit within the total time to load, process, and unload one retort. This relationship can be expressed mathematically:

$$t_c + t_p + t_d = t_c N_A \qquad [3.32]$$

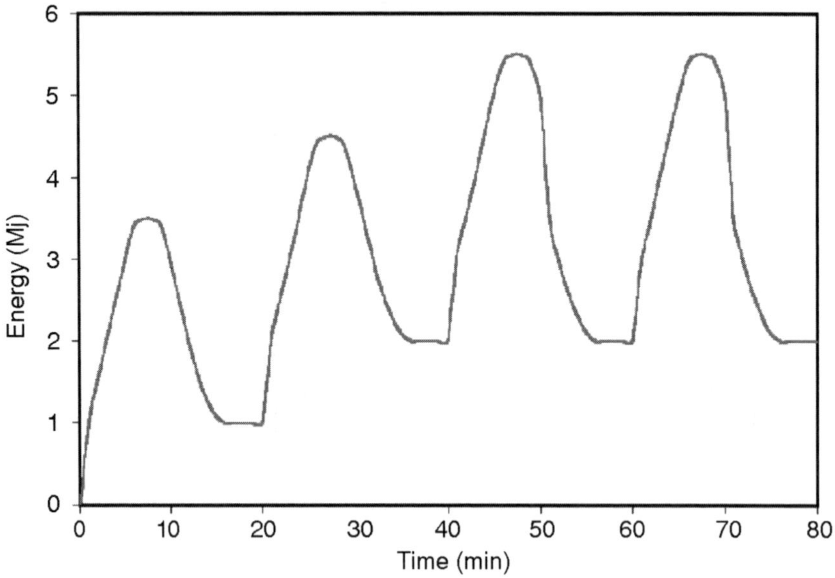

Fig. 3.8 Energy demand profile from retort battery operating with optimum number of retorts and venting scheduling.

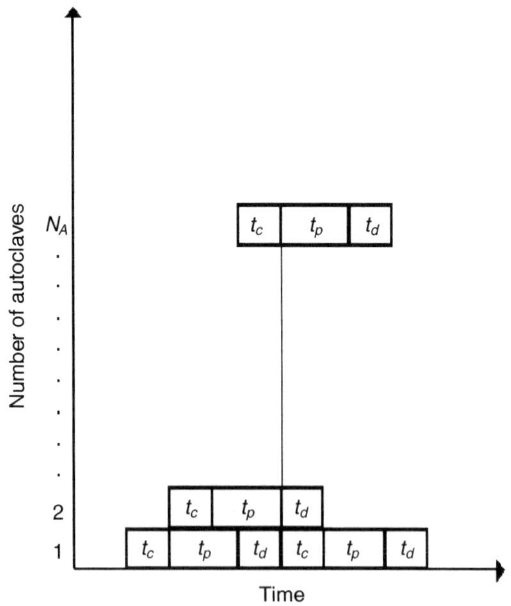

Fig. 3.9 Gantt chart showing temporal programming schedule of the battery retort system operation.

where N_A is number of retorts and t_c, t_p, and t_d are loading, process, and unloading times, respectively. Considering that loading and unloading times are equal ($t_c = t_d$), we get:

$$N_A = 2 + \frac{t_p}{t_c} \quad [3.33]$$

Therefore: $3 \leq N_A \leq \infty$,

and the minimum number of retorts for optimum operation under this criterion is three. The number of retorts for any given situation will depend upon the ratio of process time to loading/unloading time.

Moreover, according to the operation scheme presented in Fig. 3.7, the following mathematical relationships can relate the plant production capacity (Q) to loading time and retort size:

$$Qt_c = KV \quad [3.34]$$

Rearranging equation 3.34 and replacing t_c from equation 3.33 it is possible to obtain an expression for production capacity (Q) as a function of processing time (t_p) and retort number (N_A) as follows:

$$Q = \frac{KV(N_A - 2)}{t_p} \quad [3.35]$$

From equation 3.35 it is possible to infer that production capacity is directly influenced by process temperature because the higher the process temperature the shorter the process time, and so the higher the production capacity (more batches per day).

3.5 Maximising net present value of capital investment for batch processing

A criterion to optimise plant design and operation is to determine the number of retorts that will maximise the net present value (NPV) of the invested capital for the new process line. This can be approached on the basis of microeconomics. Equation 3.36 is the expression for Net Present Value (NPV):

$$NPV = -I + \sum_{j=1}^{n} \frac{\beta_j}{(1+i)^j} \quad [3.36]$$

where two main terms can be distinguished, total investment ($-I$) and annual benefits (β_j). The total investment for the project will be expressed as the cost requirement in retorts, fittings, boiler, general equipment, construction and engineering, working capital, etc. Expressing the total investment mathematically:

$$I = I_N + I_F + I_B + I_x \quad [3.37]$$

According to Guthrie (1969) the cost of equipment (retorts, etc.) could be expressed as being in proportion to its capacity. Therefore, in the specific case of retorts, a mathematical expression is:

$$C_R = k_R V^a, 0 < a < 1 \qquad [3.38]$$

Therefore the required investment in N_A retorts, could be expressed as:

$$I_N = N_A(k_R V^a) \qquad [3.39]$$

According to the schemes presented in Figs 3.7 and 3.9 and utilising equation 3.34, it is possible to obtain a mathematical relation between the retort size (V) and the number of retorts (N_A):

$$V = Q\frac{t_p}{(N_A - 2)K} \qquad [3.40]$$

Replacing equation 3.40 in equation 3.39 an expression for retort investment as a function of retort number is obtained:

$$I_N = N_A\left[k_R\left(\frac{Qt_p}{(N_A - 2)K}\right)^a\right] \qquad [3.41]$$

The fittings investment (I_F) will be considered as directly related to the retort investment. Therefore from equation 3.41:

$$I_F = \phi N_A\left[k_R\left(\frac{Qt_p}{(N_A - 2)K}\right)^a\right]; 0 < \phi < 1 \qquad [3.42]$$

The boiler size is related to the size of each retort but also to retort scheduling (total magnitude of peak energy consumption). In this analysis the retort scheduling was adopted from the Gantt chart presented in Fig. 3.9. Each retort has a time delay equivalent to the loading time (t_c) resulting in a boiler size mainly dependent on the retort size, then:

$$B = B(N_A, Scheduling) \qquad [3.43]$$

$$C_B = K_1[B]^b; 0 < b < 1 \qquad [3.44]$$

Arranging equation 3.44, the boiler investment (I_B) could be expressed as:

$$I_B = k_1(B)^b = k_1(B(N_A, Scheduling))^b; 0 < b < 1 \qquad [3.45]$$

Replacing equations 3.41, 3.42 and 3.45 into equation 3.37, the total investment could be expressed as:

$$I = (1 + \phi)N_A\left[k_R\left(\frac{Qt_p}{(N_A - 2)K}\right)^a\right] + k_1(B(N_A, Scheduling))^b + I_x \qquad [3.46]$$

As is shown in equation 3.46, investment (I) is a function of retort number (N_A), process time (t_p), and boiler size (B):

$$I = I(N_A, t_p B)$$

Annual incomes

Analysing the second term of equation 3.36, annual benefits (or debits) could be expressed as:

$$\sum_{j=1}^{n} \frac{\beta_j}{(I+i)^j} \qquad [3.47]$$

where $\beta_j = Q^*(P_u - C_u)y$; and $Q^* = QT$

$$\beta_j = QT_yP_u - QT_yC_u \qquad [3.48]$$

Replacing equation 3.35 into equation 3.48;

$$\beta_j = \frac{KV(N_A - 2)}{t_p} T_y(P_u - C_u) \qquad [3.49]$$

From equation 3.49 it is possible to infer:

$$\beta_1 = \beta_2 = \ldots = \beta_n = \beta = \frac{KV(N_A - 2)}{t_p} T_y(P_u - C_u) \qquad [3.50]$$

Expressing the annual benefits as a present value, the second term of equation 3.36 could be reduced to:

$$B_E = \sum_{j=1}^{n} \frac{B_j}{(1+i)^j} = K' \frac{KV(N_A - 2)}{t_p} T_y(P_u - C_u) \qquad [3.51]$$

where
$$K' = \frac{([1+i]^n - 1)}{i(1+i)^n}$$

and then
$$B_E = B_E(N_A, t_p)$$

Replacing equations 3.47 and 3.51 into equation 3.36:

$$NPV = -(1+\phi)N_A\left[k_R\left(\frac{Qt_p}{(N_A - 2)K}\right)^a\right] - k_1(B(N_A, Scheduling))^b - I_X + K' \frac{KV(N_A - 2)}{t_p} T_y(P_u - C_u) \qquad [3.52]$$

where
$$NPV = f(N_A, t_p, B) \qquad [3.53]$$

Therefore *NPV* is a function of process time, number of retorts (N_A), and boiler size (B). To maximise *NPV* it is necessary to get the partial derivatives of equation 3.52 and find the critical values for N_A, t_p, and boiler size.

By inspection, and analysing equation 3.54, it can be demonstrated that:

$$\left(\frac{\partial(V(N_A - 2))}{\partial N_A}\right)_{t_p} = 0 \qquad [3.54]$$

In addition, the *NPV* function (Equation 3.52) increases indefinitely as process time decreases (with corresponding increasing process temperature). Therefore, in terms of maximising *NPV* the process temperature must be as high as possible. The upper limit on process temperature will be restricted by adverse effects on product quality and upper operating limits of the retort. To find the critical value for number of retorts (N_A for maximum *NPV*), equation 3.52 will be treated as a variable function (N_A,*B*), then:

$$NPV = -I(N_A, B) + B_E(N_A) \qquad [3.55]$$

Taking derivatives:

$$\frac{d(NPV)}{dN_A} = -\frac{dI(N_A, B)}{dN_A} + \frac{dB_E(N_A)}{dN_A} \qquad [3.56]$$

But, according to equation 3.54:

$$\frac{dB_E(N_A)}{dN_A} = 0 \qquad [3.57]$$

Therefore:

$$\frac{d(NPV)}{dN_A} = -\frac{dI(N_A, B)}{dN_A} \qquad [3.58]$$

According to equation 3.58 the maximisation of *NPV* corresponds to the minimisation of investment ($I(N_A, B)$). To find the maximum *NPV* value, it is imperative to find a relation between boiler sizes (*B*) and retort number (N_A).

Case study 1

As a starting point, a simplified situation will be analysed in which the investment for a specified plant size is affected, only, by the number of retorts (N_A). The investment on fittings and boiler (I_F and I_B) would be independently considered to N_A. Therefore equation 3.52 can be re-written for this simplified case as:

$$NPV = -N_A\left[k_R\left(\frac{Qt_P}{(N_A - 2)K}\right)^a\right] - I_B - I_F - I_X + K'\frac{KV(N_A - 2)}{t_p}t_y(P_u - C_u) \qquad [3.59]$$

To maximise *NPV* it is necessary to take the derivative of equation 3.59, equalise it to zero and find the critical value for N_A. If the second derivative is less than zero then the critical value for N_A will represent a maximum value for *NPV*. Taking the derivative from equation 3.59, considering equation 3.54, and equalising to zero:

$$\frac{d(NPV)}{dN_A} = -k_R\left(\frac{Qt_p}{(N_A - 2)K}\right)^a + k_R a N_A\left(\frac{Qt_p}{(N_A - 2)^2K}\right)\left(\frac{Qt_p}{(M_A - 2)K}\right)^{a-1} = 0 \qquad [3.60]$$

Rearranging equation 3.60 the critical value for N_A is:

$$N_A^* = \frac{2}{1-a} \quad [3.61]$$

Given that the second derivative of equation 3.60 is less than zero, N_A* represents a maximum for *N.P.V.* According to Guthrie (1969) the exponent **a** has a value of 0.6 for horizontal retorts and 0.65 for vertical retorts. Therefore:

$$N_{AH}^* = \frac{2}{1-0.6} = 5 \quad [3.62]$$

$$N_{AV}^* = \frac{2}{1-0.65} \cong 6 \quad [3.63]$$

A general approach should analyse and consider, as a basic starting point, equation 3.52. Following the same procedure, it is necessary to find the first and second derivative of equation 3.52 to identify the critical value for N_A and scheduling to see if it represents a maximum or minimum for *NPV.* Therefore, taking derivatives of equation 3.52:

$$\frac{d(NPV)}{dN_A} = (1+\phi)N_A k_R a\left(\frac{Qt_p}{(N_A-2)K}\right)^{a-1}\frac{Qt_p}{K(N_A-2)^2} - (1+\phi)k_R\left(\frac{Qt_p}{(N_A-2)K}\right)^{a} - \frac{d[(k_1(B(N_A, Scheduling))^b]}{dN_A} = 0 \quad [3.64]$$

Arranging and simplifying equation 3.64:

$$\frac{(1+\phi)N_A k_R a}{(N_A-2)}\left(\frac{Qt_p}{(N_A-2)K}\right)^{a} - (1+\phi)k_R\left(\frac{Qt_p}{(N_A-2)K}\right)^{a} - \frac{d[(k_1(B(N_A, Scheduling))^b]}{dN_A} = 0 \quad [3.65]$$

Case study 2

In this situation, the concept is to analyse a more general case. At least, the boiler size is intended to be related with the number of retorts (N_A) but also with the scheduling. To investigate the relationship between boiler size and scheduling a specific plant was analysed (Table 3.1) and evaluated utilising the procedure described in Section 3.3. According to Fig. 3.10 boiler size is linearly related to retort sizes. Therefore:

$$C_B = k_1[B^b = k_1[k_B V]^b = k_2[V]^b; \quad 0 < b < 1 \quad [3.66]$$

Arranging equation 3.66, the boiler investment (I_B) could be expressed as:

$$I_B = k_2(V)^b = k_2\left(\frac{Qt_p}{(N_A-2)K}\right)^b; \quad 0 < b < 1 \quad [3.67]$$

Table 3.1 Simulation conditions in a specific processing plant to relate boiler size with retort volume

Property	Value
Thermal diffusivity (pea purée), m^2/s	1.70×10^{-7}
Process temperature, °C	121.1
Initial food temperature, °C	26.7
Lethal value F_o, min	6.0
CUT min	5

Reference microorganism: *Clostridium Botulinum* (Barreiro *et al.*, 1984).

where, replacing into equation 3.52 and taking derivatives, equation 3.68 is obtained:

$$\frac{(1+\phi)N_A k_R a}{(N_A-2)}\left(\frac{Qt_p}{(N_A-2)K}\right)^a - (1+\phi)k_R\left(\frac{Qt_p}{(N_A-2)K}\right)^a + \frac{k_2 b}{(N_A-2)}\left(\frac{Qt_p}{(N_A-2)K}\right)^b = 0 \qquad [3.68]$$

Equation 3.68 should be solved numerically by using a procedure like Newton-Raphson, regula falsi, etc. If the exponent ***a*** and ***b*** are similar ($\boldsymbol{a \cong b}$), equation 3.68 has an analytical and interesting solution (Equation 3.69). Reported values

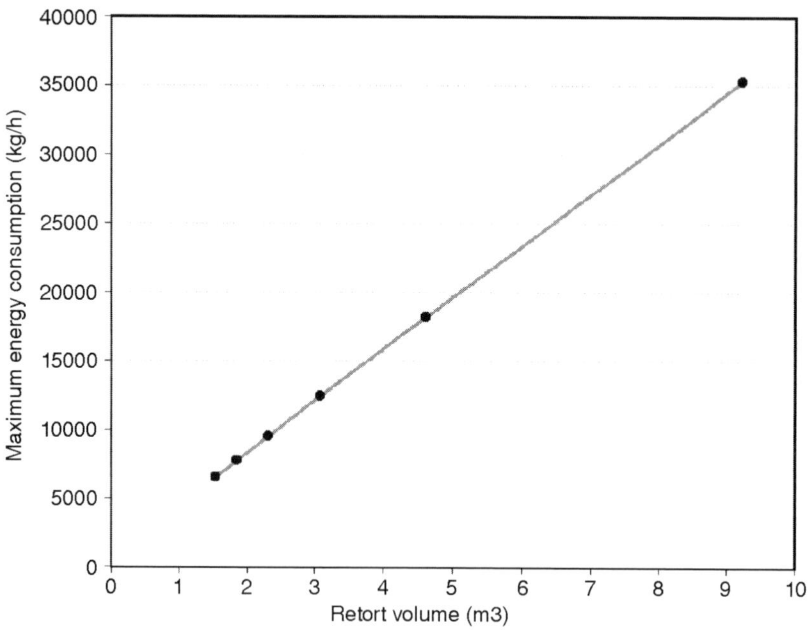

Fig. 3.10 Comparison between retort volume and maximum steam consumption.

for ***a*** and ***b*** exponents are in the range of 0.6 to 0.65 for **a** and 0.55 to 0.65 for **b** (Guthrie, 1969; Peters and Timmerhaus, 1991; Vilbrandt and Dryden, 1959).

$$N_A^* = \frac{k_2 a + 2k_R(1+\phi)}{k_R(1-a)(1+\phi)} \quad [3.69]$$

Rearranging:

$$N_A^* = \frac{k_2 a}{k_R(1-a)(1+\phi)} + \frac{2}{1-a} \quad [3.70]$$

as was stated before, the exponent ***a*** could be 0.6 or 0.65 depending on retort type (horizontal or vertical); then for horizontal retorts, equation 3.70 could be expressed as

$$N_A^* = \frac{1.5k_2}{k_R(1+\phi)} + 5 \quad [3.71]$$

To evaluate N_A it is necessary to quantify k_2, k_R and ϕ. Values for these constants can be found in the literature (Guthrie, 1969; Peters and Timmerhaus, 1991; Vilbrandt and Dryden, 1959) or directly from the manufacturers of equipment (retorts and boilers). It is interesting to note that independent of its respective values (k_2, k_R and and ϕ) in this general analysis the optimum number of retorts is always larger than five for horizontal retorts and larger than six for vertical retorts.

3.6 Simultaneous processing of different product lots in the same retort

This optimisation criterion applies to the case of small canneries with few retorts that are frequently required to process small lots of different products in various container sizes that normally require different process times and retort temperatures. In these situations, retorts often operate with only partial loads because of the small lot sizes, and are underutilised. The proposed approach to this optimisation problem is to take advantage of the fact that, for any given product and container size, there exists any number of alternative combinations of retort temperature (above the lethal range) and corresponding process time that will deliver the same lethality (F_o value). These can be called iso-lethal processes. They were first described by Teixeira *et al.* (1969) to find optimum iso-lethal process conditions that would maximise nutrient retention (thiamin) for a given canned food product, and later confirmed by others (Lund, 1977; Ohlsson, 1980). Barreiro *et al.* (1984) used a similar approach to find optimum iso-lethal process conditions that would minimise energy consumption. Results from both studies are as superimposed in Fig. 3.11, and show that conditions optimum for thiamine retention are not necessarily the same as those optimum for energy consumption.

Important to this study is the fact that the differences found in the absolute level of quality retention were relatively small over a practical range of iso-

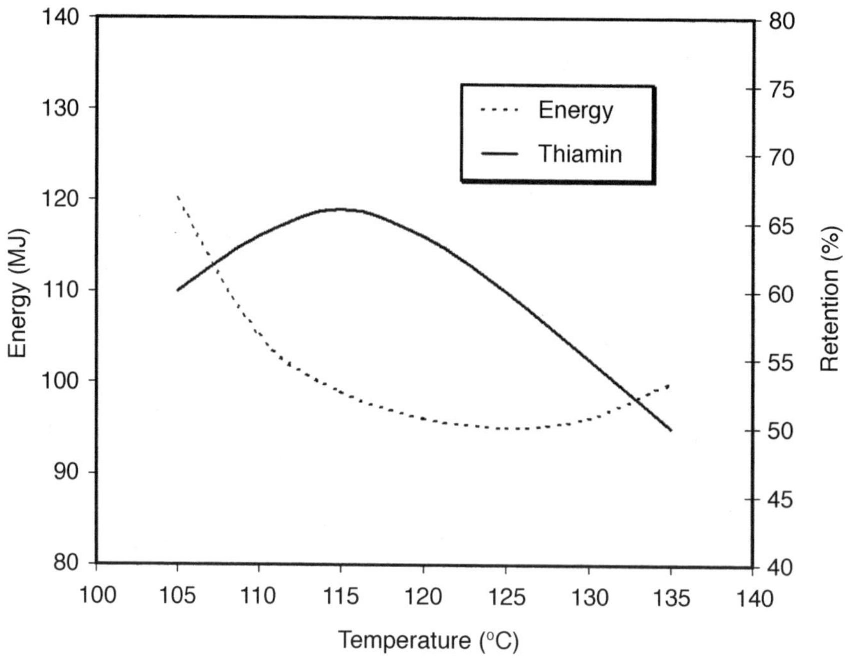

Fig. 3.11 Effect of process temperature on quality retention and energy consumption over a range of iso-lethal processes.

lethal process conditions. This relative insensitivity of quality over a range of different iso-lethal process conditions opens the door to maximising output from a fixed number of retorts for different products and container sizes. Iso-lethal processes can be identified for each of the various products from which a common set of process conditions can be chosen for simultaneous processing of different product lots in the same retort.

3.6.1 Simultaneous processing characterisation

In terms of analysis, a range of iso-lethal processes for selected products and container sizes should be obtained from experimental work. Heat penetration tests should be conducted for each product in order to establish process time at a reference retort temperature to achieve target lethality (F_o values). A computer program can be utilised to obtain the equivalent lethality processes according to the following specifications:

- Two F_o values should be considered for each product ($F_{o\ min}$ and $F_{o\ max}$).
- For each F_o value ($F_{o\,min\,j}$ and $F_{o\,max\,j}$) iso-lethal processes at retort temperatures of TRT_1, TRT_2, TRT_3, …, TRT_N should be obtained for each product.
- The discrete values that define each process per product at different temperatures will be transformed as a continuous function through the cubic

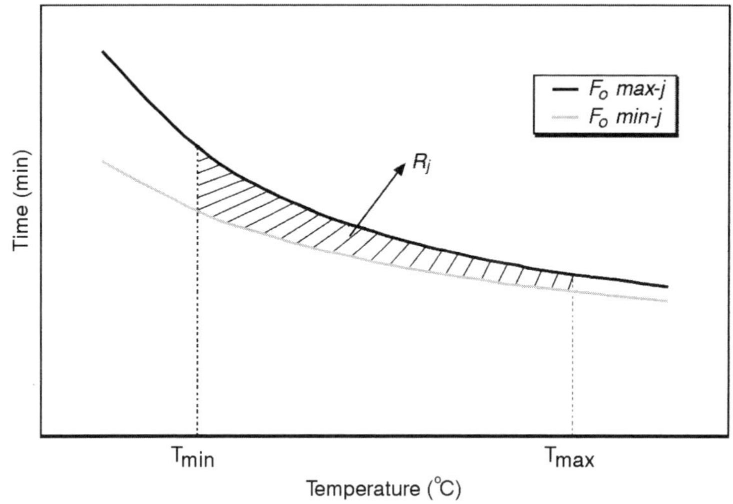

Fig. 3.12 Region restricted for the maximum and minimum iso-lethal curves ($F_{o\ min\text{-}j}$ and $F_{o\ max\text{-}j}$) for the j-nth product.

spline procedure (for both $F_{o\,min}$ and $F_{o\,max}$, per product), obtaining a set of two continuous curves per product (Fig. 3.12).

In addition, the following criteria should be established for choosing the optimum set of process conditions for simultaneous sterilisation of more than one product:

- The total lethality achieved for each product must be equal or greater than the pre-established $F_{o\ min}$ value for that specific product.
- The total lethality for each product must not exceed a pre-established maximum value ($F_{o\ max}$) to avoid excessive overprocessing.

3.6.2 Mathematical formulation for simultaneous processing

Let us assume we have $n = 2$ products, say $P_1, \ldots P_n$, which are processed at the plant location. Let us consider the index set:

$$X = \{1, 2, \ldots, n\} \qquad [3.72]$$

considering the temperature interval $[T_{\min}, T_{\max}$, which denotes the temperature capabilities of the process. Each product P_j, for j $\in X$, has attached two strictly decreasing continuous functions, say

$$m_j, M_j : [T_{\min}, T_{\max}] \rightarrow (0, +\infty) \qquad [3.73]$$

where $m_j(T) \leq M_j(T)$, for each $T \in [T_{\min}, T_{\max]}$. The meaning of $m_j(T)$ (respectively, $M_j(T)$) is the minimum time (respectively, maximum time) needed to process the product P_j at temperature T. Defining the region R_j:

$$R_j = \{(T, t) : T_{\min} \leq T \leq T_{\max}, m_j(T) \leq t \leq M_j(T)\} \qquad [3.74]$$

the interpretation of R_j is that the product P_j can be processed at temperature T with time t if and only *if* $(T, t) \in R_j$ (Fig. 3.12). It is clear that a sub-collection of products will be:

$$P_{j_1}, \ldots, P_{j_r} \qquad [3.75]$$

where $$1 \leq j_1 \leq j_2 \leq \ldots \leq j_r \leq n$$

can be simultaneously processed at temperature T and time t if and only if

$$(T, t) \in R_{j_1} \cap R_{j_2} \cap \ldots \cap R_{j_r}$$

It follows that in order to obtain all possible sub-collection of products, which can be simultaneously processed, is equivalent to find all possible subsets

$$Q = \{j_1, \ldots, j_r\} \subset X$$

$$r > 0$$

$$1 \leq j_1 \leq j_2 \leq \ldots \leq j_r \leq n$$

For which it holds that:

$$I_Q = R_{j_1} \cap R_{j_2} \cap \ldots \cap R_{j_r} \neq \oslash$$

Computational procedure
In the practical sense, we have the products $P_1 \ldots, P_n$ and the temperature interval $[T_{\min}, T_{\max}]$.

(1) We choose a positive integer $k \in \{1, 2, 3, \ldots\}$ and a partition $P = \{T_0 = T_{\min}, T_1, \ldots, T_k = T_{\max}\}$ where $T_{\mathrm{m}} < T_{m+1}$ for m $= 0, \ldots, k - 1$.
(2) For each product P_j we compute the values $m_j(T_m), M_j(T_m), m = 0, 1, \ldots, k$.
(3) For each $m \in \{0, 1, \ldots, k\}$, we define the values $m_p(T_m) =$ Maximum $\{m_p(T_m) : j \in P\}$; $M_p(T_m) =$ Minimum $\{M_p(T_m) : j \in P\}$.
(4) If for some $m \in \{0.1, \ldots, k\}$ we have that $m_p(T_m) \leq M_p(T_m)$, then we observe that products $P_1, \ldots, P_n$ can be simultaneously processed at temperature T_m with time $t \in \lfloor m_p(T_m), M_p(T_m) \rfloor$.

3.6.3 An application example of simultaneous processing

The following canned seafood food products were selected for heat penetration tests and development of iso-lethal process conditions: Mussels (*Mytilus chilensis*); Salmon (*Salmo salar*); Crab (*Cancer adwardsii*); Clams (*Protothaca thaca*); Razor clams (*Mesodesma donacium*).

Experimental procedure
The experimental procedure was divided into four steps, as follows:

1. Can sizes for each of the selected products were chosen according to NFPA (1982). The can sizes chosen for seafood products are shown in Table 3.2.

Table 3.2 Canned seafood products showing combinations of product and can sizes selected for study

	Can size			
Product	RO-200	RO-150	RR-125	Oval 1 lb
Salmon (natural)	✓			
Salmon w/corn and peas (oil)	✓			
Salmon w/beans and corn (oil)	✓			
Smoked salmon (oil)			✓	
Chunk salmon (oil)	✓			
Crab (natural)	✓			
Mussels (natural}		✓		
Seafood mixture (tomato sauce)	✓			
Seafood mixture (natural)	✓			
Clams (natural)				✓
Razor clams (natural)		✓		

2. Each canned seafood product was processed at 114°C with the exception of razor clams and clams, which were processed at 117°C and 110°C respectively.
3. Each process was calculated to achieve F_o value of 6 minutes, and then repeated for F_o value of 10 minutes (Pflug and Odlaug, 1978).
4. Equivalent lethality processes (six per product) were obtained by computer program for retort temperatures in the range of 105°C to 125°C. Data management, to adjust each new process to a F_o of 6 and 10 minutes was carried out according to the procedure developed by Simpson *et al.* (2003c).

Heat penetration tests for products under study

Heat-penetration experiments at different temperatures and processing times were conducted to examine the nature of the heat-penetration curves (centre temperature histories). The retort heating profile used consisted of an initial equilibrium phase at 20°C, followed by a linear coming-up-time (CUT) of 7–9 min to accomplish venting, and holding phase over the calculated process time required to achieve the target lethality for the specific product under thermal processing.

Results and analysis

Table 3.3 show the feasibility of simultaneous sterilisation of various canned seafood products, respectively. Table 3.3 shows the opportunities available at widely different retort temperatures with each column representing a different process time at that temperature. Therefore, all the cells in one column experience the same retort process. Although the F_o value received by each product will differ, it will lie in the acceptable range (between 6 and 10 min) if the cell is marked with the numeral 1, which indicates the process is feasible for that product. When a zero appears in a cell it means the F_o value received by the

Table 3.3 Feasibility matrix for simultaneous sterilisation of canned seafood products

Product	Retort temperature (°C)					
	112	114*			125*	
Chunk salmon	1	0	0	1	0	0
Mussels (natural)	1	1	1	0	1	0
Smoked salmon	1	1	1	1	1	0
Crab chunk	1	1	1	1	0	1
Seafood mixture (sauce)	1	0	0	1	0	0
Salmon w/beans and corn	1	1	1	1	0	0
Salmon w/corn and peas	1	0	1	1	0	1
Seafood mixture (natural)	1	1	1	1	1	0
Salmon (natural	1	0	0	1	0	0
Clams	1	1	1	1	1	0
Razor clams	1	1	1	0	0	0

1 indicates that products are feasible for simultaneous sterilisation.
* Columns represent different process times for simultaneous sterilisation under each retort temperature.

product under that process falls outside the acceptable range, and the process is not feasible for that product. From this table it is possible to infer that the lower the process temperature the higher the possibility to attain simultaneous sterilisation. For all of the products considered in this study, it was possible to sterilise simultaneously at a retort temperature of 112°C. Computer software was developed to generate all combinations of practical iso-lethal processes at any process temperature. As an example, according to Table 3.3 at 114°C the plant manager will be able to utilise three different alternative process times for simultaneous sterilisation.

The opportunity to carry out simultaneous sterilisation and the possibility to employ alternative processes (same F value) provides flexibility to optimise retort utilisation. Within a pre-established range of F values, it was possible to obtain all the combinations for simultaneous sterilisation. This procedure is of special relevance for small companies that normally work with many different products at the same time. Practical implementation of this flexibility will require close attention to batch record-keeping requirements of the FDA Low Acid Canned Food regulations, and the need to file with the FDA each of the alternative processes as an acceptable scheduled process for each product.

3.7 Conclusion

Batch processing has been extensively practised since the development of the canning industry but barely analysed. The batch process implies a lack of accuracy in production planning. As discussed and analysed in this chapter, food-canning plants are not a true batch process. As mentioned, if one or two stages are batch operated, the whole plant will be better classified as a continuous process.

The transient energy balance (dynamic response), for the sterilisation process, is an essential tool to optimise quantitatively batch retort battery design and operation in food-canning plants. Considering a hierarchical approach, the cook room system (retorts battery) operates with continuous inflow and outflow of product.

When the number of retorts was considered as the only variable affecting plant investment for a fixed capacity, the optimum number of autoclaves – to maximise NPV – was five for horizontal retorts and six for vertical retorts. In a more general situation (considering boiler capacity as a function of retort number (N_A) and scheduling) the optimum was displaced to a larger number of autoclaves.

The opportunity to carry out simultaneous sterilisation and the possibility to employ alternative processes (iso-lethality) provides flexibility to optimise retort utilisation. As discussed, within a pre-established range of F values, it was possible to obtain all the combinations for simultaneous sterilisation. Practical implementation of this proposed procedure (simultaneous processing) will require close attentions to batch record-keeping requirements of the FDA, or the correspondent low acid food regulator.

As has been shown in the chemical industry, the manner in which the products will be delivered to the customers in the future will further favour batch processing. Customers' requirements will be more specific and more demanding with respect to specification, quality and delivery.

Several challenges are ahead. In the near future we should see much research in batch design and operation related to canning food plants. Hopefully we will be able to look for really big surprises as Japanese researchers have proposed a multi-purpose pipeless batch plant in which the materials are contained in movable vessels and guided automatically within the plant locations.

3.8 List of symbols

A: area (m^2)
a: constant (exponent); $0 < a < 1$
b: constant (exponent); $0 < b < 1$
B: boiler size (Kg. steam/h)
B_E: annual benefits as a present value (US$)
$B_E(N_A)$: as B_E but as a function of N_A (US$)
C_R: retort cost (US$/unit)
C_B: boiler cost (US$/unit)
C_u: cost per processed can (US$/can)
Cp: specific heat (J/kg K)
E: energy (J)
F_o : process lethality at $T_R = 121°C$ (min)
F_{max}: pre-established maximum value for F_o (min)

g_c: universal conversion factor; 1 (Kg m / N s^2)
$\underline{H}$: enthalpy (J / Kg)
h: heat convection coefficient (W/m^2 K)
i: interest rate
I: total investment (US$)
I(N_A): total investment as a function of number of retorts (US$)
I_N: investment in N_A retorts (US$)
I_x: investment in equipment (other than retorts), construction and engineering, working capital cost and cost of land (US$)
I_F: fittings investment (US$)
I_B: boiler investment (US$)
j: period j
k_R: constant (US$/m^3)
k_B: constant
k_1: constant (US$/kg/h)
k_2: constant, $= k_1 {*} k_B{}^b$
K: constant (cans/ m^3 of retort)
K′: constant
$\dot{m}$: mass flow rate (kg / s)
M: mass (kg)
n: number of period for NPV evaluation
N_A: number of retorts
$N*_A$: critical number of retorts
$N*_{AH}$: critical number of retorts (Horizontal)
$N*_{AV}$: critical number of retorts (Vertical)
NVP: net Present Value (US$)
P: pressure (Pa)
Pm: molecular weight (kg / kmol)
P_u: can price (US$/can)
P_t: operator process time (min)
$\dot{Q}$: thermal energy flow (W)
Q: production capacity (cans/min)
Q*: annual production (cans/year)
R: ideal gas constant 8.315 (Pa m^3/kmol K); (J/kmol K)
T: temperature (K)
T_0: initial temperature
$\bar{T}$: average product temperature (K)
t: time (s)
t^*: time required to eliminate air from retort
t_y: annual operation time of the plant per year (h)
t_c: loading time (min)
t_d: unloading time (min)
t_p: processing time (including CUT and cooling) (min)
V: retort volume (m^3)
v : velocity (m / s)

Sub indices

a:	air
b:	bleeder
amb:	ambient
c:	convection
cv:	condensed vapour
cr:	critical value
cw:	cooling water
e:	metal container
in:	insulation
p:	food product
r:	radiation
rt:	retort
s:	steam
sl:	saturated liquid
sv:	saturated vapour
rs:	retort surface
t:	time
v:	vapour
w:	condensed water

Greek symbols

ρ:	density (kg/m^3)
ϵ:	surface emissivity of retort shell at an average of emitting and receiving temperatures (dimensionless)
γ:	ratio of specific heat at constant pressure to specific heat at constant volume (dimensionless)
σ:	Stefan-Boltzmann constant, 5.676×10^{-8} (W/m^2 K^4)
β_j:	benefits in period j
ϕ:	proportionality constant; $0 < \phi < 1$

3.9 References

ALMONACID-MERINO, S., SIMPSON, R. and TORRES, J.A. 1993. 'Time variable retort temperature for cylindrical cans: batch process time, energy consumption, and quality retention model', *J. Food Proc. Eng.*, 16, 271–287.

BALZHISER, R.E., SAMUELS, M.R. and ELIASSEN, J.D. 1972. *Chemical Engineering Thermodynamics*, Prentice-Hall, Englewood Cliffs, NJ, USA.

BARBOSA, A.P. and MACCHIETTO, S. 1993. 'Optimal design of multipurpose batch plants 1. Problem formulation', *Computers and Chemical Engineering*, 17, S33–S38.

BARREIRO, J., PEREZ, C. and GUARIGUATA, C. 1984. 'Optimization of energy consumption during the heat processing of canned foods', *J. of Food Eng.* 3, 27–37.

BHOWMIK, S.R., VISCHENEVETSKY, R. and HAYAKAWA, K. 1985. 'Mathematical model to estimate steam consumption in vertical still retort for thermal processing of canned foods', *Lebensmittel Wissenschaft und Technologie*, 18, 15–23.

DATTA, A. K., TEIXEIRA, A. A. and MANSON, J. E. 1986. 'Computer based retort control logic for on-line correction of process deviations', *J. of Food Sci.*, 51, 480–483, 507.

DOUGLAS, J. M. 1988. *Conceptual Design of Chemical Processes*, McGraw-Hill International Edition. Chemical Engineering Series.

GUTHRIE, K. 1969. 'Data and techniques for preliminary capital cost estimating', *Chemical Engineering*, 3, 114–142.

HOLDSWORTH, S.D. 1997. *Thermal Processing of Packaged Foods*. Blackie Academic & Professional. London, UK.

KONDILI, C., PANTELIDES, R. and SARGENT, H. 1993. 'A general algorithm for short-term scheduling of batch operations – I. MILP formulation', *Computers and Chemical Engineering*, 17, 211–227.

LEE, B. and REKLAITIS, G.V. 1995a. 'Optimal scheduling of cyclic batch processes for heat integration – I. Basic formulation', *Computers and Chemical Engineering*, 19 (8), 883–905.

LEE, B. and REKLAITIS, G.V. 1995b. 'Optimal scheduling of cyclic batch processes for heat integration-II. Extended problems', *Computers and Chemical Engineering*, 19 (8), 907–931.

LUND, D.B. 1977. 'Design of thermal processes for maximizing nutrient retention', *Food Technol*, 31, 71–78.

NATIONAL FOOD PROCESSORS ASSOCIATION (NFPA) 1982. *Thermal Processes for Low-Acid Foods in Metal Containers*. Bulletin 26–L. Twelfth Edition. Washington, DC, USA.

NORBACK, J and RATTUNDE, M. 1991. 'Production planning when batching is part of the manufacturing sequence', *J. of Food Proc. Eng.*, 14, 107–123.

OHLSSON, T. 1980. 'Temperature dependence of sensory quality changes during thermal processing', *J. of Food Sci.*, 45, 836–839.

PETERS, M.S. and TIMMERHAUS, K.D. 1991. *Plant Design and Economics for Chemical Engineers*, Chemical Engineering Series, McGraw-Hill, New York. USA.

PFLUG, I.J. and ODLAUG, T.E. 1978. 'A review of z and f values used to ensure the safety of low-acid canned foods', *Food Technol.*, 32 (2), 63–70.

REKLAITIS, G.V. 1996. 'Overview of scheduling and planning of batch process operations'. In *Batch Processing System Engineering*. (G.V. Reklaitis, A.K. Sunol, D.W. Rippin and O. Hortacsu, eds) pp. 660–705, Springer, Berlin.

RIPPIN, D.W. 1993. 'Batch process system engineering: A retrospective and prospective review', *Computer and Chemical Engineering*, 17, S1–S13.

SIMPSON, R., ARIS, I.and TORRES, J.A. 1989. 'Sterilization of conduction-heated foods in oval-shaped containers', *J of Food Sci.*, 54, 1327–1331, 1363.

SIMPSON, R., ALMONACID, S. and TORRES, J.A. 1993. 'Mathematical models and logic for the computer control of batch retorts: Conduction-heated foods', *J. Food Eng.*, 20, 283–295.

SIMPSON, R., ALMONACID, S. and TEIXEIRA, A.A. 2003a. 'Optimization criteria for batch retort battery design and operation in food canning-plants', *J. of Food Proc. Eng.*, 25, 515–538.

SIMPSON, R., MITCHELL, M. and ALMONACID, S. 2003b. 'Mathematical model development, experimental validation and process optimisation: retortable pouches packed with seafood in cone frustum shape', *J of Food Eng.*, 63, 153–162.

SIMPSON, R., ALMONACID, S. and TEIXEIRA, A. 2003c. 'Bigelow's general method revisited:

development of a new calculation technique', *J. of Food Science*, 68 (4), 1324–1333.

TEIXEIRA, A., DIXON, J., ZAHRADNIK, J. and ZINSMEITER, G. 1969. 'Computer optimization of nutrient retention in the thermal processing of conduction-heated foods', *Food Technol.*, 23: 845–850.

VILBRANDT, F.C. and DRYDEN, C.E. 1959. *Chemical Engineering Plant Design*, McGraw-Hill, Tokyo. Japan.

4

Using computational fluid dynamics to optimise thermal processes

P. Verboven, J. de Baerdemaeker and B. M. Nicolaï, Katholieke Universiteit, Leuven, Belgium

4.1 Introduction: computational fluid dynamics and the importance of fluid flow in thermal processes

Computational fluid dynamics (CFD) is the simulation of fluid flow by means of a mathematical model solved on a computer. It serves as a basis to simulate heat and mass transfer in flow systems, such as liquid foods in cans, continuous sterilisation of liquid foods and heating of foods in ovens by hot air. Since the early development in the 1950s of programs on supercomputers to calculate particular aspects of aerodynamics, CFD has evolved to a user-friendly, versatile and interactive software environment that can be run on most computer platforms. CFD has now become the engineering tool for the design of fluid flow processes, reducing costs and design cycles. Many commercial CFD codes have become available over the years, which, if not the basis, is the proof of its success in a wide range of fields, from the aerospace to the automobile industry, from meteorology to biomedical engineering, from the chemical to the food industry. The latter first picked up CFD in the late 1980s as a basic research tool to investigate the, often complex, transport phenomena related to thermal processing of foods. It has gradually evolved since then, until now it has become a tool to design thermal food processes.

This chapter will outline the principles of CFD, its possibilities and current limitations. We will start with an overview of fluid flow processes in food applications in the next section. Section 4.2 will deal with the available tools for the study of fluid flow, both by measurement and simulation. The equations used in CFD are discussed in Section 4.3, with an emphasis on the required approximations in order to solve them for thermal food processes. In Section 4.4, some recent examples of CFD analysis in thermal food processes are presented, and some points for further development are outlined in Section 4.5. Some additional resources of CFD related information are given in Section 4.6.

4.1.1 Fluid flow in food processes

Many thermal food processes involve fluid flow. Flow of gases and liquids increases heat exchange rates and improves mixing, which should allow faster and more uniform processing. The direct effect of flow velocity on heat and mass transfer rates is quantified by a heat and a mass transfer coefficient, a single linear parameter that can be used for mathematical analysis and design of processes and equipment. The heat transfer coefficient, therefore, has been recognised as the most important characteristic of fluid flow in thermal food processes.

Sterilisation of canned liquid foods is an example of a thermal food process involving fluid flow. Doubling the rotation speed of cans with liquid-particle foods increases the overall wall-to-fluid heat transfer coefficient by approximately 30% and the fluid-to-particle heat transfer coefficient by about 50% (Sablani and Ramaswamy, 1998). Fluids mixed with particles represent challenging fluid flow problems, which have to be resolved for optimal processing, food safety and quality.

A further example of this is aseptic processing. Aseptic processes are continuous processes that are designed to meet criteria of commercial sterility, maximal quality and nutritional value by a proper residence time and temperature of liquid-particle foods in heat exchangers and holding tubes, which are based on the values of the wall-to-fluid and fluid-to-particle heat transfer coefficient. Fluid properties and flow rate, and particle shape, size and concentration determine these values. While flows of liquid foods in tubes have been investigated widely over the last decades, fluid flows laden with particles have received attention only in the last 10 to 15 years (Alhamdan and Sastry, 1997). In heat exchangers, fouling deposits increase the resistance to fluid flow and heat transfer and serve as places for microbial growth (Changani *et al.*, 1997). The investigation of fouling requires insight in chemical kinetics and mass transfer in addition to fluid flow and heat transfer.

In ovens, heat transfer rates are increased by circulation of air around foods (Sato *et al.*, 1987). Air impingement ovens use high velocity air jets perpendicular to the food surface to rapidly transfer heat to the food, even at low temperatures (Xue and Walker, 2003). Furthermore, air supply/circulation/extraction schemes are used to optimise heating (Broyart and Trystram, 2002) and to improve crispiness of microwaved foods (Datta and Ni, 2002). Air flow in ovens is generally a complicated process, strongly affected by buoyancy and turbulence and it, besides fluid-to-food heat transfer, also affects water transport (Verboven *et al.*, 2003).

In general, representing fluid flow by a single parameter, such as the heat transfer coefficient, is a simplification of the process, but it hardly serves as a basis for process improvement when:

- large variations of the heat transfer coefficient exist in space and/or time, which indicates variations in local fluid velocity;
- complex phenomena such as mass transfer and chemical kinetics in the fluid are involved, which could affect magnitude and direction of local velocity;
- particle flow and fluid flow strongly interact, resulting in complex flow patterns and important variations of the heat transfer coefficient;
- multiple fluids mix on a macroscopic, visible scale with consequent spatial variations in volume fraction of each fluid that may change with time;
- the process involves complex geometrical designs with multiple or complex supplies of the processing fluid.

In these cases, a good process design can only be achieved by means of a thorough understanding of the flow patterns involved.

4.2 Measurement and simulation of fluid flow in thermal processes

Even the most fundamental measurements of fluid velocity and temperature are difficult in thermal food processes, due to the invasive nature of most measurement techniques, the sometimes extreme processing conditions and the inaccessibility of equipment (Broyart and Trystram, 2002). Changani *et al.* (1997) state that experiments on plant where it is difficult to determine flows and temperature are of little fundamental value. A few techniques are available for in-depth measurement of fluid and particle flow in food processes, but they are usually expensive and require considerable skill and expertise. Non-invasive techniques to study air and solid-liquid flow systems are not always directly applicable to real food processing situations, but, in cases where they are, fundamental understanding of the fluid flow process is achieved (McCarthy *et al.*, 1992; Ramaswamy *et al.*, 1995; Chandrasekaran, 1997; Fairhurst *et al.*, 2001).

Computer simulations based on mathematical models offer a different route to gain understanding of thermal food processes. Based on fundamental governing equations of physics, they offer a versatile tool for process design and optimisation. The more the models contain empirical features, the more the simulations lose their versatility and cannot be directly transferred from one application to another. The more parameters in the models have physical meaning (such as thermo-physical fluid properties), the more the model becomes universally applicable over a wide range of processes. An example of such a model is the set of Navier-Stokes equations for fluid flow, presented in the nineteenth century by both Navier and Stokes. Developed from conservation principles on an infinitesimal scale, the equations cover all single fluid flows: liquids as well as gases, compressible as well as incompressible flows, inviscid as well as creeping flows, laminar as well as turbulent flows, Newtonian as well as non-Newtonian flows. Depending on the purpose of the simulation, the fluid and the flow type, the Navier-Stokes model may be simplified, empirical features (such as turbulence models) must be introduced or additional physics needs to be modelled (e.g., heat transfer, chemical reaction, mass transfer, particulate flows). Depending on the complexity of the resulting model, additional approximations need to be introduced to solve the model equations for the system under consideration. Such approximations include geometrical simplifications and simplifications of the boundary conditions, the numerical discretisation (the discrete points in space and time where a solution is required) and the convergence tolerance (the error that is allowed on the iterative solution of the discretised model, if required). The more severe these approximations, the less accurate the model will become. Thus, although a fundamentally exact description of the macroscopic continuous flow phenomenon, the Navier-Stokes model in practice is an approximation of reality. To solve fluid flow problems based on the Navier-Stokes equations by numerical, and thus approximating means on a computer, is known as Computational Fluid Dynamics (CFD).

Computational power of computers and fast, accurate numerical solution algorithms lay at the basis of the success of CFD in a wide variety of industries, including food manufacturing and processing (Scott and Richardson, 1997; Xia and Sun, 2002). Although CFD computer codes offer user-friendly desktop interfaces that allow a step-by-step definition and analysis of the problem, the user should have sufficient skill to judge the accuracy of the approximations and of available models, which may be strongly empirical.

4.3 Using computational fluid dynamics (CFD) to analyse thermal processes

4.3.1 The foundation: a closed system of equations with physical constants

The basic equations of CFD are the Navier-Stokes equations, written here for incompressible flow:

$$\nabla \cdot \mathbf{u} = 0 \tag{4.1}$$

$$\frac{\partial(\rho u_i)}{\partial t} + \nabla \cdot (\mathbf{u}(\rho u_i)) = \frac{\partial}{\partial x_i} p + \nabla \cdot (\tau_i) + \rho g_i \tag{4.2}$$

Equation [4.1] is the incompressible continuity equation (stating conservation of mass). Equation [4.2] states the conservation of momentum (ρu_i) in direction i ($i = 1, 2$ for 2D and i = 1, 2, 3 for 3D). The velocity vector **u** consists of components u_i, each of which is the solution of a separate equation. In the above equations, p is pressure, ρ is density, g_i the gravity vector component in direction i and τ_i is the stress vector, which for incompressible flow of a Newtonian fluid is simplified to:

$$\tau_i = \mu \nabla u_i \tag{4.3}$$

with μ the dynamic viscosity of the Newtonian fluid, independent of shear rate. The Newtonian model describes well ideal gases and most dilute water solutions. In the case of non-Newtonian fluids, viscosity becomes an apparent 'modelled' parameter μ_a, which is defined as a function of shear rate γ to express the non-linear relationship between stress and shear. An example of a non-Newtonian model is the Power Law model:

$$\mu_a = k_a \gamma^{n-1} \tag{4.4}$$

with k_a and n product-dependent Power Law constants. Equation [4.4] was used, e.g., by Jung and Fryer (1999) to model a continuous sterilisation process of different non-Newtonian foods. A more complex temperature-dependent model was used by Yang and Rao (1998) and Tattiyakul *et al.* (2001) to describe the flow behaviour of a corn starch dispersion, including the gelatinisation process:

$$\mu_a = \left[C\mu^*(T) \left(\frac{\gamma_r}{\gamma} \right) \right]^{1/n} \tag{4.5}$$

where μ^* is the complex viscosity and γ_r a reference shear rate, C and n are phenomenological constants.

In thermal processes, the energy equation needs to be solved to obtain the temperature profile:

$$\frac{\partial(\rho c T)}{\partial t} + \nabla \cdot (\mathbf{u}(\rho c T)) = \nabla \cdot (\lambda \nabla T) \tag{4.6}$$

with c the specific heat, λ the thermal conductivity of the fluid and T the temperature.

Boundary conditions to the above equations need to be supplied for the velocity components and/or pressure and temperature. Initial conditions should be supplied for these variables at all positions in the solution domain. At inflow boundaries, velocity components or pressure can be specified together with temperature. At outflow boundaries, pressure or an outgoing mass flux can be set. Overall mass conservation must be met at outlets. At walls, no-slip conditions (zero velocity) are usually applied. Alternatively, wall velocity, free slip (zero shear stress) or a specified wall shear stress can be set. With regard to heat transfer, wall temperatures or wall heat fluxes can be defined. If solids are part of the solution domain, continuity of temperature and heat flux is required at the interface between solid and fluid.

4.3.2 The implementation: need for approximation

The transport phenomena at the interface between solids and fluids are of prime importance for thermal food processes. In its purest form, CFD is able to resolve the boundary layers and the corresponding temperature gradients at the surface of solid objects, and therefore does not require the use of empirical heat transfer coefficients or boundary layer models. In an optimal world, therefore, the above

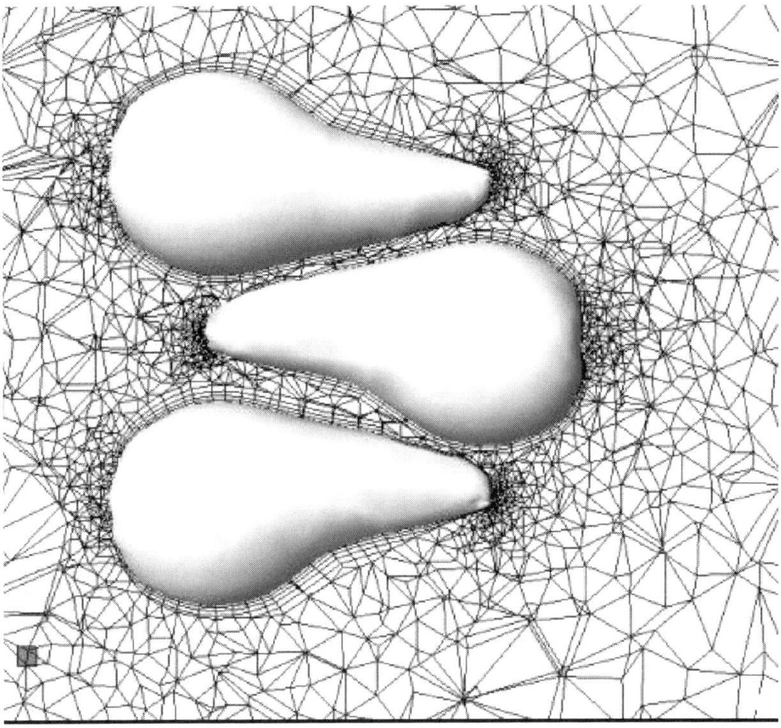

Fig. 4.1 CFD geometry of three pears and computational mesh of the air space.

set of equations should render an exact empirical-free simulation of a thermal food process, if the following criteria are met:

- The geometry is simulated exactly.

With the last generation of computer codes, the geometry of objects can be simulated with high resolution by means of CAD import facilities. Figure 4.1 shows a computer image of pears that was reproduced from real pears by means of a computer vision system (Jancsok *et al.*, 2001) and then imported into a CFD program.

- The equations are solved on a sufficiently small discrete scale so that the analytical solution is approximated with disappearing error.

The meshing tools available in the last generation of CFD codes allow meshing of complex geometries, with curved and intersecting objects (see Figs 4.1 and 4.2). Automatic meshing tools are now available that refine meshes in regions that are identified by the user to contain large gradients, or without manual intervention in regions of high curvature, near surfaces, proximate edges, etc. Mesh elements (that define the discrete points where a solution is required) can be any combination of tetrahedrals, hexahedrals, pyramids and prisms, with possibilities to allocate certain mesh shapes to certain regions (e.g., prisms on

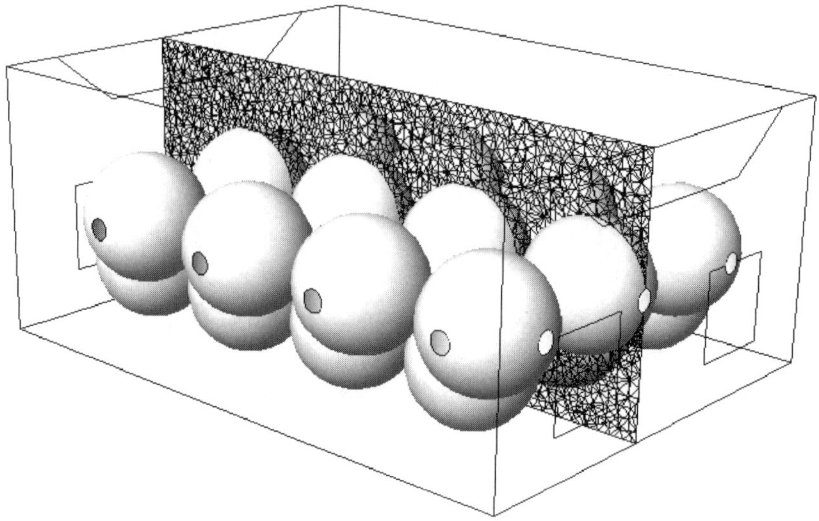

Fig. 4.2 Unstructured mesh of the air voids in a bulk of 24 spheres in a bin.

surfaces). In Fig. 4.2, an unstructured mesh is displayed of a bulk of spherical particles in a bin. Figure 4.1 shows a prismatic mesh near the surface of the objects, which is beneficial to resolve the boundary layers.

Although fine complex meshes can be defined, it should be realised that mesh independence of the solution is rarely achieved. It is then required to perform a sensitivity study with respect to mesh density. It is said that the solution converges towards the true solution of the model equations if the solution monotonically approaches a constant value if the mesh density is increased. Figure 4.3 displays the mesh convergence error of the CFD prediction of the average surface heat transfer coefficient to a food in an oven. In the limit of a very fine mesh, the solution becomes constant, which can be assumed to be the mesh-independent solution of the model (Verboven *et al.*, 2003). By means of plots like Fig. 4.3, it can be decided what mesh density is required for a specific application to achieve sufficient accuracy of the CFD prediction. Adaptive meshing is now being introduced in CFD codes to refine meshes in regions of high gradients during the solution process, to achieve this goal automatically.

- All model parameters are thermo-physical properties of the fluid that are exactly known.

In the case of Newtonian fluids, the only parameters required are density, viscosity, specific heat and thermal conductivity. These physical properties are well documented for most ideal gases and most common liquids in textbooks and on the web. Some references are listed in Section 4.6. In the case of non-

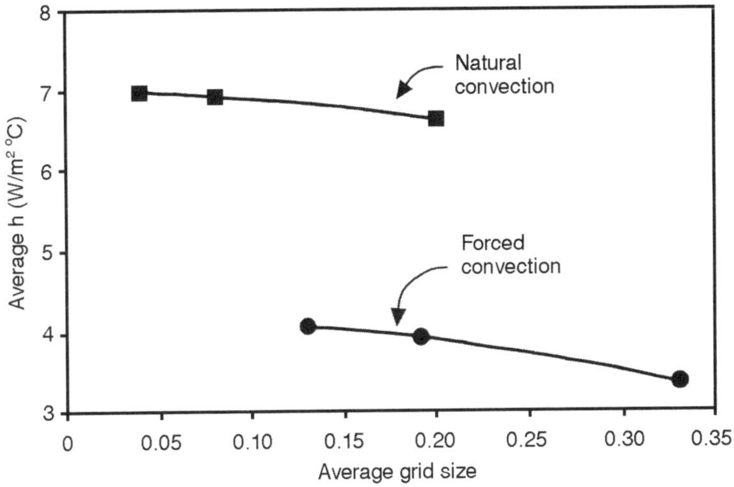

Fig. 4.3 Mesh refinement study of a CFD model of air flow and heat transfer to a food in an oven under different model assumptions. The solution approaches a constant value when the mesh volume size is reduced to a small value (reprinted from *Journal of Food Engineering*, Vol. 59, P. Verboven *et al.*, Computation of airflow affects on heat and mass transfer in a microwave oven, pp. 181–190, © copyright 2003, with permission from Elsevier).

Newtonian fluids, the quality of the simulation strongly depends on the quality of the available thermo-physical properties. Considerable experimental efforts are required for less well-documented foods.

- Boundary and initial conditions are known exactly as a function of space and time.

Uniform initial conditions may be a crude, but sufficient assumption. Inflow and outflow conditions should at least represent accurately the total supply of mass, momentum and energy to the domain under consideration. In the event that exact profiles are not known it is better to choose inflow and outflow boundaries far from the region of interest. In well-controlled continuous food processing plants, the definition of boundary and initial conditions should not be a major problem. In large-scale facilities such as ovens or chillers, however, initial temperature variations can severely affect the solution. Furthermore, air flow from a fan is often very complex and the detailed modelling of the fan is in many cases not feasible (Verboven *et al.*, 2000). A sensitivity analysis can then help to shine a light on the extent of the influence of boundary and initial conditions on the process. In the end, CFD simulations often reveal that uncontrolled initial or boundary conditions are the main cause of the poor operation of a certain process.

Even if the above criteria are met in practice, the complexity of the flow (e.g., due to turbulence) and the geometry (e.g., flow through a bulk of stacked foods) may lead to computational requirements that far exceed the resources and time allowed for solving a specific thermal food processing problem. Herein lies a major limitation of CFD for a breakthrough as the engineering design tool of thermal food processing, regardless of the additional features (mass transfer, multi-phase flows, chemical kinetics) that one tries to solve. What should become clear from the above discussion is that sensitivity studies are a vital part of any CFD analysis.

4.3.3 Additional features

Provided the above aspects are well considered, CFD is a powerful tool to simulate any physical phenomenon coupled to fluid flow. It should be realised that the nature of the equations that describe these additional features is much less versatile than the basic fluid flow equations. In addition to being described by semi- or fully empirical equations, additional features also often imply empirical modifications to the basic fluid flow and heat transfer equations.

Mass transfer

In many thermal food processes some form of mass transfer takes place. An additional component may be transported by the fluid (e.g., water vapour in air or micro-organisms in a liquid food). If one assumes that such substances do not affect the fluid flow (e.g., when their concentration is low) a convection-diffusion equation describes the mass transfer of the additional component with mass fraction X:

$$\frac{\partial(X)}{\partial t} + \nabla \cdot (\mathbf{u}X) = \nabla \cdot (D\nabla X) \qquad [4.7]$$

In this equation, the right-hand side is a modelled diffusion expression of the true mass flux of the component relative to the mean flow; D is the apparent diffusivity of the considered component in the fluid. In multi-component mass transfer, interaction effects may need to be taken into account (Bird *et al.*, 2000). Accurate determination of D is important for laminar flows. For gases in air and chemicals in water, diffusivity values can be found in literature (see, e.g., Lide, 2000; Anon., 1993). One can also rely on the kinetic theory of gas diffusion in dilute mixtures (Anon., 1993). In binary liquids the theory is less developed and depends on hydrodynamic and activated state models (Bird *et al.*, 2000), and the Stokes-Einstein equation is a popular means to estimate mass diffusivity of large spherical molecules in solvents of low molecular weight, and of small suspended particles. Experimental data on mass diffusivity of components in liquid foods can be found in only a few bibliographic references (Rao and Rizvi, 1994). The key to a good model then is a sensitivity analysis around an intelligent guess.

Turbulence

Although turbulence can be predicted by means of direct simulation of the given governing equations, this is currently not possible for complex flows due to computational restrictions. Instead, turbulence models need to be applied. A wide range of models is available, each with their merits and limitations. Eddy-viscosity models rely on the assumption that turbulent mixing can be viewed as a mechanism similar to viscous action: turbulent eddies break through laminar profiles, resulting in sharper gradients of the average flow properties. This assumption therefore modifies the momentum equation by adding a turbulence component to the laminar viscosity. The effective viscosity becomes:

$$\mu_{eff} = \mu + \mu_{TUR} \tag{4.8}$$

where μ_{TUR} is a modelled parameter that requires additional semi-empirical model equations of turbulence quantities such as the k-ϵ and k-ω models (Wilcox, 2000), where k is the turbulence kinetic energy, ϵ the turbulence energy dissipation rate and ω the turbulence frequency (Wilcox, 2000). Turbulence also enhances mixing of energy and mass. Effective parameters are therefore also introduced in the energy and mass transfer equations:

$$\lambda_{eff} = \lambda + \frac{\mu_{TUR}}{\sigma_e}(\rho c) \tag{4.9}$$

$$D_{eff} = D + \frac{\mu_{TUR}}{\sigma_m}\rho \tag{4.10}$$

with σ_e and σ_m empirical turbulent Prandtl numbers. Alternatively, more versatile but more complex turbulence closures include Reynolds stress models and Large Eddy Simulations (Wilcox, 2000). In many turbulent flows, the turbulence contributions dominate equations [4.8–4.10]. It is therefore of the utmost importance that turbulent flows are modelled adequately.

In addition, turbulent boundary layers are very complex and cannot be modelled by means of the standard turbulence models. Separate models are therefore required to capture the profiles and gradients near surfaces. A simple algebraic equation is the logarithmic wall function, that sufficiently describes a part of the boundary layer to calculate the gradients at a flat surface in a fully turbulent flow. Effects of boundary layer separation cannot be described by means of these functions. Furthermore, in case wall functions are used, the solution becomes strongly mesh size-dependent. Mesh size near the wall should be chosen such that the dimensionless distance to the wall, expressed by the variable y^+, is in the range of 30 to 100 corresponding to the logarithmic region of the velocity profile. Figure 4.4 gives the wall Nusselt number in a pipe for turbulent flow, calculated by different turbulence models and different mesh sizes. It is shown that models that use wall functions (standard and RNG k-ϵ) can render a variety of values depending on mesh size, with larger errors if the mesh is further refined below $y^+ = 30$. Instead of using wall functions, the turbulence model equations themselves can be modified to be valid in the turbulent wall profile. These so-called low-Reynolds-number models are much more effective in the near wall region, but require much finer meshes near the surface (Fig. 4.4).

The use of standard turbulence models to predict surface heat transfer coefficients to food products in turbulent flows is in most cases not possible

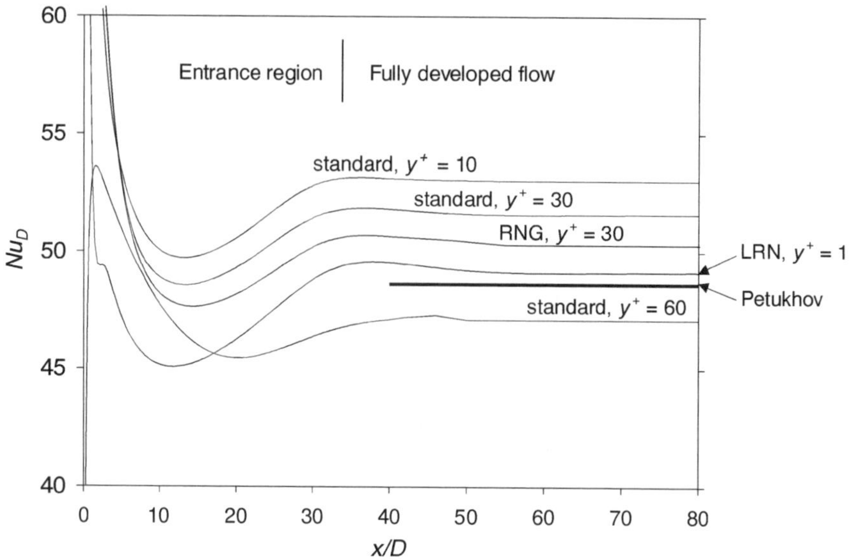

Fig. 4.4 Nusselt number in a pipe, $Re_D = 20{,}000$ and $D = 0.1$m, the reference is calculated with the correlation given by Petukhov (1970). Comparison of different turbulence models and mesh sizes (standard: k-ϵ model, RNG: Renormalisation Group k-ϵ model, LRN: low-Reynolds-number k-ϵ model).

(Verboven *et al.*, 1997; Kondjoyan and Boisson, 1997), because boundary layers are far more complex than those described by the logarithmic wall function. One then has to rely on low-Reynolds-number models or higher order models such as Reynolds stress models.

Porous media fluid flow

In porous media, a detailed geometrical model of air or fluid voids in between particles is often not feasible and the mesh volumes will be larger than the smallest void size. The complex flow in the porous medium then needs to be represented on a superficial scale by a semi-empirical model that relates superficial fluid velocity to the pressure gradient across the bulk of material. This model replaces the original Navier-Stokes equations in the region of the porous medium. A full development of the porous medium model can be found in Lage (1997). It has been applied to bulks of foods by e.g., Alvarez *et al.* (2003), Hoang *et al.* (2003) and van der Sman (2002).

Particle transport

If particles in a fluid significantly change the fluid flow and have distinct velocities from the fluid, then the above mass transfer equation cannot be used. This is the case for most liquid-solid foods (Lareo *et al.*, 1997). The particles then become a separate phase, which can be described using continuum ('Eulerian') or discrete models ('Lagrangian'). In Eulerian models the particle concentration is modelled by means of a convection-diffusion equation in which the parameters depend on particle properties and dimensions and dispersion needs to be modelled empirically as a diffusion-like process:

$$\nabla \cdot (r_a \mathbf{u}_a) = m_a \qquad [4.11]$$

$$\frac{\partial (r_a \rho_a u_{ai})}{\partial t} + \nabla \cdot (r_a \mathbf{u}_a (\rho_a u_{ai})) = -\frac{\partial}{\partial x_i} p + \nabla \cdot (r_a \mu_a \nabla u_{ai}) + r_a \rho_a g_i + c_d u_{di} + \mathbf{F}_i \qquad [4.12]$$

with r_a the volume fraction of phase a of the fluid-particle mixture. One thus needs two sets of full equations to describe the process; one for the fluid and one for the dispersed phase. It may be quite difficult to model the 'apparent' viscosity μ_a of the dispersed phase. There are additional terms that couple the transfer processes in the two phases: m_a is mass transfer into phase a from the other phase, and $c_d u_{di}$ is a drag force, with u_{di} the relative drag velocity between the phases. There may be additional forces **F** acting on the phases (due to virtual mass, turbulent dispersion, etc.), as well as an added mass term. In addition to the above equations, two energy equations must be solved, one for each phase:

$$\frac{\partial (r_a \rho_a c_a T_a)}{\partial t} + \nabla \cdot (r_a \mathbf{u}_a (\rho_a c_a T_a)) = \nabla \cdot (r_a \lambda_a \nabla T_a) + h A_s (T_b - T_a) \qquad [4.13]$$

with possibly an added mass contribution as well. In equation [4.13], h is the heat transfer coefficient, modelling empirically the heat transfer at the interphase boundary of the two phases. It may also be difficult to obtain a good model for the 'apparent' conductivity λ_a of the dispersed phase.

Lagrangian models calculate the trajectories of particles with a representative dimension through the fluid. This approach has the advantage that individual particle motions can be predicted and is therefore less empirical than the Eulerian model:

$$m_p \frac{d\mathbf{u}_p}{dt} = \mathbf{F} \qquad [4.14]$$

with m_p the mass of the particle and $\mathbf{u}_p$ its velocity, **F** are the forces working on the particle. Mass and heat transfer can be modelled by means of simple formulations based on Newton's law of cooling. In the Lagrangian approach, the fluid flow still needs to be modelled by an Eulerian formulation, basically the Navier-Stokes equations, extended with the additional terms given in equations [4.11–4.13].

In the above approaches, turbulent dispersion needs be accounted for in the forces acting on the particles. Dispersion occurs under the influence of the random velocity fluctuations of the turbulent flow. However, as explained above, these fluctuations are normally not calculated; instead the turbulence properties such as the turbulence kinetic energy and dissipation rate are modelled. Stochastic models are thus required to calculate correctly fluctuating fluid velocities from the distribution of the turbulence properties (Burfoot *et al.*,

1999). Particle interaction and wall effects need to be accurately modelled. Dispersion of fluid droplets in air has been modelled by Reynolds (1997), where particles were smaller than the mesh size. In case particle dimensions exceed mesh size, with order of magnitudes comparable to that of the domain of interest (such as in liquid-solid food flows in heat exchangers) other procedures need to be implemented, on fixed meshes (Duchanoy and Jongen, 2003) or deforming meshes (Hu *et al.*, 1996).

Chemical/microbial kinetics
Kinetics appear as source terms in the mass transfer equations (Ghani *et al.*, 2002). Sometimes kinetics contain interaction effects between different species and then require careful numerical implementation to correctly treat the coupling between equations. Numerical instability is not a rarity in coupled mass transfer problems, and a thorough numerical convergence study is then a must. Solutions may overshoot the limits of applicability and unrealistic results may be obtained, such as negative concentrations. Adequate solution algorithms should be implemented to prevent non-physical solutions.

Evaporation/condensation
At food surfaces there may be moisture loss by evaporation and moisture deposit by condensation, which is the case, e.g., in hot air ovens or during steaming. Few efforts have been made so far to describe these processes by CFD models in food applications and usually it requires considerable semi-empirical modelling (Hoang *et al.*, 2003; Xu and Burfoot, 1999). In case of phase change, mass and energy equations become strongly coupled. Latent heat has to be entered into the energy equation, which depends on the mass exchange. Most computer codes do not provide this coupling term, which has to be programmed by the user, if possible. More fundamental approaches to condensate formation and flows are referred to and used by, e.g., Panday (2003).

4.4 Improving thermal food processes by CFD: packaged foods, heat exchangers and ovens

4.4.1 Canned and packaged foods

Canning was recognised as one of the first thermal food processes that could benefit from CFD calculations. Indeed, heating of liquid foods creates natural convection currents in the can that affect the temperature distribution. However, the experimental observation of the temperature distribution in a can filled with liquid is almost impossible, because inserted thermometers will change the flow patterns. The first efforts to model natural convection heating in cans were done by Engelman and Sani (1983) and Datta and Teixeira (1988), who successfully modelled flow and temperature in a beer bottle and a cylindrical can, respectively.

More recently, the approach was applied to study flow and heat transfer in non-Newtonian foods in containers (Kumar and Bhattacharya, 1991; Yang and Rao,

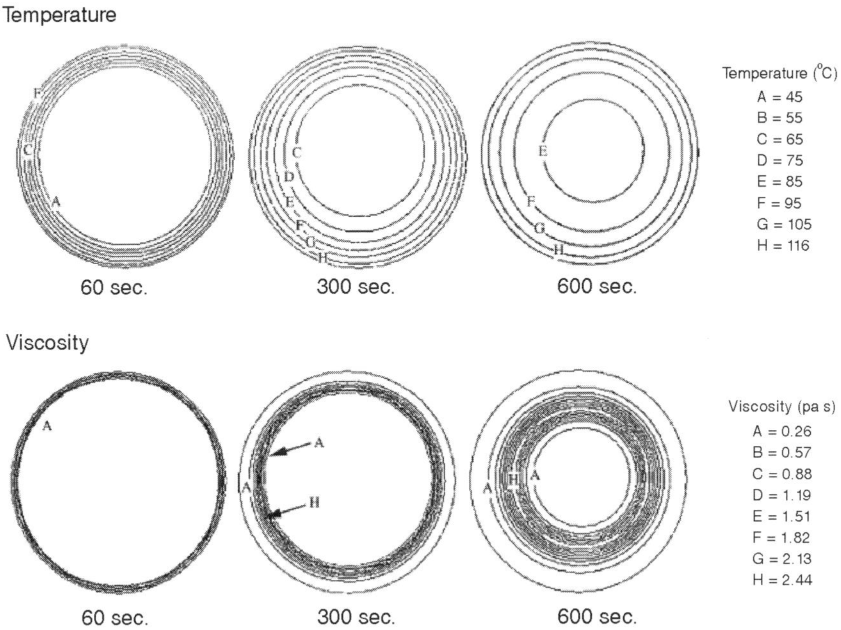

Fig. 4.5 Temperature and viscosity in corn starch in a can during axial rotation at 146 rpm, processing temperature 121°C (reprinted from *Chemical Engineering and Processing*, Vol. 40, J. Tattiyakul *et al.*, Simulation of heat transfer to a canned corn starch dispersion subjected to axial rotation, pp. 391–399, © copyright 2001, with permission from Elsevier).

1998; Tattiyakul *et al.*, 2001; Ghani *et al.*, 2002). Applied to food pouches, Ghani *et al.* (2002), coupled the CFD calculations to a convection-diffusion-reaction model of the concentration distribution and destruction of bacteria. The bacterial inactivation kinetics were simple first-order. The predicted values of average bacterial concentration were compared to experimental counts, which showed a good comparison for the number of survivors as a function of time of heating.

Tattiyakul *et al.* (2001) modelled non-Newtonian flow and heat transfer in an axially rotating can. A different slowest heating point was detected for rotating and stationary cans, with a strong effect of rotation speed: at high speeds a viscous layer of gelatinised starch formed at the can wall, hindering heat transfer in the radial direction, resulting in slower heat penetration (Fig. 4.5). Intermittent rotation helps to dislocate the gelatinised starch from the boundary region resulting in faster heating and more uniform temperature profiles in the can (Tattiyakul *et al.*, 2002).

4.4.2 Heat exchangers

Continuous sterilisation of liquids is a typical modelling problem that can be dealt with by means of CFD. The prediction of flow patterns, temperature distribution and bacterial concentration in heat exchangers and holding tubes facilitates the design and optimisation of the process. Jung and Fryer (1999) used CFD to investigate the effect of rheological properties of the fluid on sterilisation process during laminar flow in a heated tube. The bacterial and quality destruction was modelled in post processing, by means of formulas for the sterility and quality value based on the calculated temperature history. First-order destruction and quality kinetics were used. It was found that in some conditions the temperature gradients are such that the near-wall regions are overprocessed. The authors emphasise that natural convection and the presence of particulates need to be taken into account.

Solid-liquid foods were modelled by Hu (1996) and Duchanoy and Jongen (2003), the latter of which emphasised food flows in heat exchangers. Hu (1996) used moving unstructured meshes to track large particles through the fluid with the Lagrangian model to predict the pathway of each particle. At each time step, the particle boundaries were then updated and the domain was re-meshed. Up to 400 particles were simulated in a vertical channel and the effect of gravity, applied pressure gradient and solid loading was studied.

Duchanoy and Jongen (2003) applied a slightly different approach, which is less computational intensive, as no re-meshing is required. The Lagrangian tracks of particles were followed by assigning at each time certain mesh elements to a particle according to its position. The drawback of this approach is that the fluid-particle interface is not smooth because of the Cartesian element shapes, but the authors found that for particle diameters larger than 10 elements this did not present any significant errors in the flow regime with very low particle Reynolds numbers (< 3) studied. Liquid-solid flow in a 2D bend with

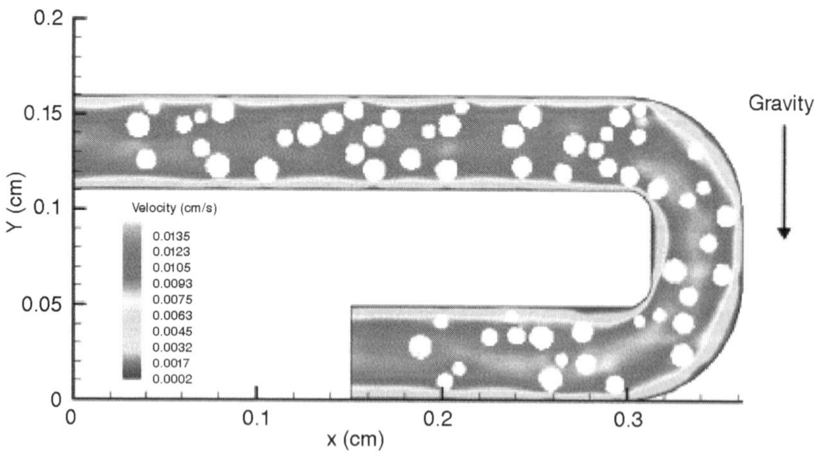

Fig. 4.6 Simulated particle and fluid flow in a bend of a heat exchanger (reprinted from *Computers and Fluids*, Vol. 32, C. Duchanoy *et al.*, Efficient simulation of liquid-solid flows with high solids fraction in complex geometries, pp. 1453–1471, © copyright 2003, with permission from Elsevier).

periodic boundaries, representative of consecutive bends in a tubular heat exchanger, was simulated (Fig. 4.6).

4.4.3 Ovens

Air flow is much more prone to the development of turbulence than liquid food flows, because of the higher Reynolds numbers that are achieved. This explains partly the success of hot air convection and air impingement ovens, as turbulence increases mixing and surface heat transfer rates. The CFD activity on these devices is mainly directed to modelling the air flow in the cavity and around the food. Air flow around foods results in a certain distribution of the surface heat transfer coefficient along the perimeter of the food. Turbulent air flows around foods were evaluated by Verboven *et al.* (1997) and Kondjoyan and Boisson (1997). Standard k-ϵ based turbulence models require modifications to deal with the complex boundary layers around foods. These restrictions can be partly overcome by using k-ω models, that are more versatile in the boundary layer region, but which are more sensitive to free stream conditions of the turbulence properties. Impingement of the food by means of jets has not received that much attention yet as far as CFD modelling is concerned, although CFD has been used to study surface jets in other fields (e.g., Roy *et al.*, 2002).

Air flow and heat transfer in the cavity of a forced convection oven was modelled by Verboven *et al.* (2000). Predicted differences in flow patterns and temperature distribution were related to slow food heating zones. Verboven *et al.* (2003) modelled the effect of cavity dimensions, inlet and outlet position and shape, air flow rate and turntable rotation speed of a microwave oven on the surface heat transfer coefficient to the food and the moisture accumulation in the cavity (Fig. 4.7). The supply of small amounts fresh air enhanced uniform food heating and avoided moisture accumulation. If air flow rates were too large, fresh air was shortcut from inlet to outlet without benefits for the food.

4.5 Future trends

The three application fields discussed in the previous section are subject to further improvement, based on simulation by means of CFD. Some examples are given.

Packaged foods often contain a considerable headspace region. To date no attempts have been made to model the changing liquid/gas interface as a consequence of rotation of the container during processing in a retort and the effect on the heat transfer and lethality. The combination of two-phase free surface models with enhanced tools to update the geometry and mesh during the simulation could provide a means to achieve a simulation model of this important process.

Further developments to models are expected to simulate more accurately solid-liquid food flows during continuous sterilisation. Adaptive meshing

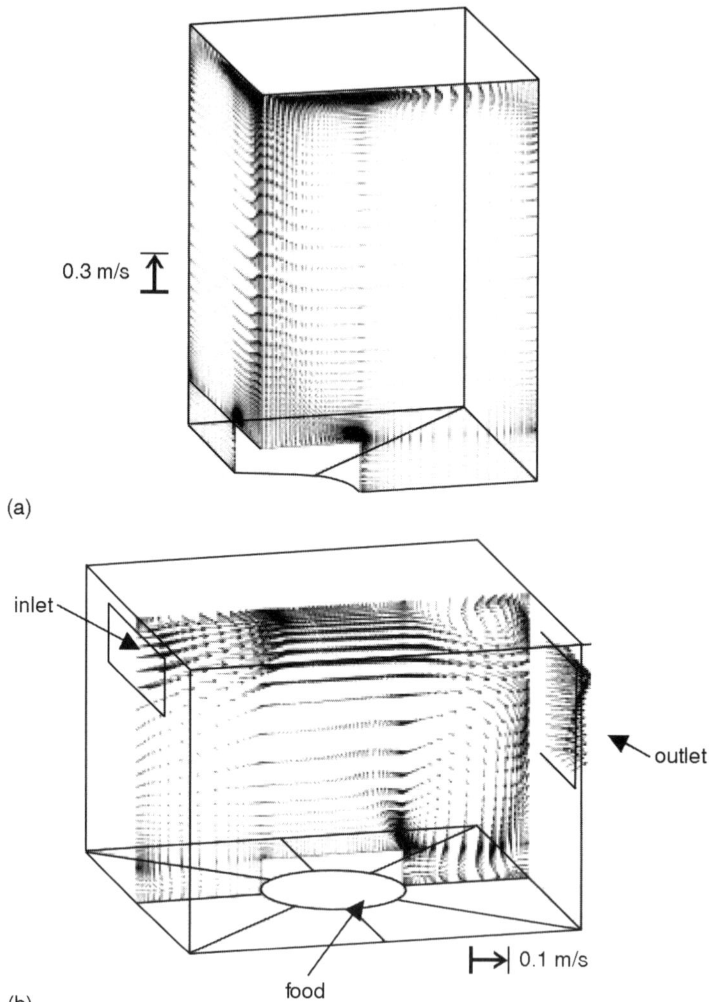

Fig. 4.7 Air flow patterns in a microwave oven under natural (a) and forced (b) convection (reprinted from *Journal of Food Engineering*, Vol. 59, P. Verboven *et al.*, Computation of airflow effects on heat and mass transfer in a microwave oven, pp. 181–190, © copyright 2003, with permission from Elsevier).

combined with Lagrangian models already have been implemented, but particle-particle and particle-wall interaction still requires additional modelling. A combination of CFD with the Discrete Element Method (DEM) may provide a means to achieve this goal. The discrete element procedure is used to determine the dynamic contact topology of moving bodies. It accounts for complex non-linear interaction phenomena between bodies and numerically solves the equations of motion. Since the DEM is a very computationally intensive procedure, many existing computer codes are limited to modelling either two-dimensional or small three-dimensional problems that employ simple body geometries.

Thermal processing by means of forced hot air can further benefit from CFD studies through improved geometry modelling and enhanced unstructured meshing to model accurately complex shapes often encountered with foods. Furthermore, recent developments in turbulence models towards Large Eddy Simulation (LES) may provide a means to improve the accuracy of the prediction of turbulent flow. This method uses basic properties of turbulence in its application. First, the larger scales carry the majority of the energy, and hence are more important. Second, the smaller scales have been found to be more universal, and hence are more easily modelled. The resulting methodology is a hybrid between these two methods, which involves the filtering of the Navier-Stokes equations to separate those scales which will be modelled from those which will be solved for directly. As a result, with LES, boundary layers can be modelled more universally than with traditional eddy-viscosity turbulence models, but this technique is of course much more computational expensive.

4.6 Sources of further information and advice

Thermo-physical properties of foods:

- Database on the web: http://www.nelfood.com/
- Software database: *Food Properties Database*, by R.P. Singh, 1995, ISBN: 0-8493-0765-1
- Handbooks: Anon. (1993); Rao and Rizvi (1994); Lide (2000)

Basics of CFD:

- Handbooks: Anderson (1995); Versteeg and Malalasekera (1996); Ferziger and Peric (2001)
- Reviews of books: http://www.cfd-online.com/Books/

Guidelines to CFD analysis:

- Book: Best Practice Guidelines of ERCOFTAC Special Interest Group on 'Quality and Trust in Industrial CFD', eds. M Casey and T Wintergerste, 2000, ERCOFTAC
- On the web: https://pronet.wsatkins.co.uk/marnet/guidelines/guide.html

Validation of CFD:

- Overview of available databases: http://www.cfd-online.com/Resources/refs.html
- NEXUS Database Resources of ERCOFTAC: http://ercoftac.mech.surrey.ac.uk/
- CFD Verification and Validation Web Site of the NPARC Alliance CFD community: http://www.grc.nasa.gov/WWW/wind/valid/
- Flownet: http://dataserv.inria.fr/flownet/

Discussion forums:

- http://www.cfd-online.com/Forum/

4.7 References

ALHAMDAN A and SASTRY S (1997), 'Residence time distribution of food and simulated particles in a holding tube', *Journal of Food Engineering*, 34, 271–292.

ALVAREZ G, BOURNET P-E and FLICK D (2003), 'Two-dimensional simulation of turbulent flow and transfer through stacked spheres', *International Journal of Heat and Mass Transfer*, 46(15), 2459–2469.

BIRD R B, STEWART W E and LIGHTFOOT E N (2000), *Transport Phenomena*, Second edition, New York, Wiley.

ANDERSON J D (1995), *Computational Fluid Dynamics*, New York, McGraw-Hill Education.

ANON. (1993), ASHRAE Handbook, Fundamentals, Atlanta, American Society of Heating, Refrigeration and Air-Conditioning Engineers.

BROYART B and TRYSTRAM G (2002), 'Modelling heat and mass transfer during continuous baking of biscuits', *Journal of Food Engineering*, 51, 47–57.

BURFOOT D, HALL K, BROWN K and XU Y (1999), 'Fogging for the disinfection of food processing factories and equipment', *Trends in Food Science & Technology*, 10, 205–210.

CHANDRASEKARAN M, MARCROFT H, BAKALIS S and KARWE M V (1997), 'Application of laser Doppler anemometry in understanding food processing operations', *Trends in Food Science & Technology*, 8, 369–375.

CHANGANI S D, BELMAR-BEINY M T and FRYER P J (1997), 'Engineering and chemical factors associated with fouling and cleaning in milk processing', *Experimental Thermal and Fluid Science*, 14, 392–406.

DATTA A K and TEIXEIRA A A (1988), 'Numerically predicted transient temperature and velocity profiles during natural convection heating of canned liquid foods', *Journal of Food Science*, 53, 191–195.

DATTA A K and NI H (2002), 'Infrared and hot-air-assisted microwave heating of foods for control of surface moisture', *Journal of Food Engineering*, 51(4), 355–364.

DUCHANOY C and JONGEN T R G (2003), 'Efficient simulation of liquid-solid flows with high solids fraction in complex geometries', *Computers and Fluids*, 32, 1453–1471.

ENGELMAN, M S and SANI R L (1983), 'Finite-element simulation of an in-package pasteurization process', *Numerical Heat Transfer*, 6, 41–54.

FERZIGER J H, and PERIC M (2001), *Computational Methods for Fluid Dynamics*, Berlin, Springer-Verlag.

FAIRHURST P G, BARIGOU M, FRYER P J, PAIN J-P and PARKER D J (2001), 'Using positron emission particle tracking (PEPT) to study nearly neutrally buoyant particles in high solid fraction pipe flow', *International Journal of Multiphase Flow*, 27, 1881–1901.

GHANI A G A, FARID M M and CHEN X D (2002), 'Theoretical and experimental investigation of the thermal inactivation of *Bacillus stearothermophilus* in food pouches', *Journal of Food Engineering*, 51, 221–228.

HOANG M L, VERBOVEN P, BAELMANS M and NICOLAÏ B M (2003), 'A continuum model for air flow, heat and mass transfer in bulk of chicory roots,' submitted to *Transactions of the ASAE.*

HU H H (1996), 'Direct simulation of flows of solid-liquid mixtures', *International Journal of Multiphase Flow*, 22(2), 335–352.

JANCSÓK P T, CLIJMANS L, NICOLAÏ B M and DE BAERDEMAEKER J (2001), 'Investigation of the effect of shape on the acoustic response of "conference" pears by finite element modelling', *Postharvest Biology and Technology*, 23(1), 1–12.

JUNG A and FRYER P J (1999), 'Optimising the quality of safe food: Computational modelling of a continuous sterilisation process', *Chemical Engineering Science*, 54, 717–730.

KONDJOYAN A and BOISSON H C (1997), 'Comparison of calculated and experimental heat transfer coefficients at the surface of circular cylinders placed in a turbulent cross-flow of air', *Journal of Food Engineering*, 34(2), 123–143.

KUMAR A and BHATTACHARYA M (1991), 'Transient temperature and velocity profiles in a canned non-Newtonian liquid food in a still-cook retort', *International Journal of Heat and Mass Transfer*, 34, 1083–1096.

LAGE J L (1997), 'The fundamental theory of flow through permeable media from Darcy to turbulence, in Ingham D B and Pop I, *Transport Phenomena in Porous Media*, Oxford, Pergamon, 1–30.

LAREO C, FREYER P J and BARIGOU M (1997), 'The fluid mechanics of two-phase soilid-liquid food flows: a review', *Transactions of the Institute of Chemical Engineers* (Part C), 75(2), 73–105.

LIDE D R (2000), *CRC Handbook of Chemistry and Physics: a ready-reference book of chemical and physical data*, 81st edition, Boca Raton, CRC Press.

MCCARTHY K L, KAUTEN R J, MCCARTHY M J and STEFFE J F (1992), 'Flow profiles in a tube rheometer using magnetic-resonance-imaging', *Journal of Food Engineering*, 16(1–2), 109–125.

PANDAY P K (2003), 'Two-dimensional turbulent film condensation of vapours flowing inside a vertical tube and between parallel plates: a numerical approach', *International Journal of Refrigeration*, 26(4), 492–503.

PETUKHOV B S, IRVINE T F and HARTNETT J P (1970), *Advances in Heat Transfer, 6*, New York, Academic Press.

RAMASWAMY H S, ABDELRAHIM K A, SIMPSON B K and SMITH J P (1995), 'Residence time distribution (RTD) in aseptic processing of particulate foods: a review', *Food Research International*, 28(3), 291–310.

RAO M A and RIZVI S S H (1994), *Engineering Properties of Foods*, New York, Marcel Dekker.

REYNOLDS A M (1997), 'Modelling particle dispersion within a ventilated airspace', *Fluid Dynamics Research*, 22(3), 139–152.

ROY S, NASR K, PATEL P and ABDULNOUR B (2002), 'An experimental and numerical study of heat transfer off an inclined surface subject to an impinging airflow', *International Journal of Heat and Mass Transfer*, 45(8), 1615–1629.

SABLANI S S and RAMASWAMY H S (1998), 'Multi-particle mixing behavior and its role in heat transfer during end-over-end agitation of cans', *Journal of Food Engineering*, 38, 141–152.

SATO H, MATSUMURA T and SHIBUKAWA S (1987), 'Apparant heat transfer in a forced convection oven and properties of baked foods', *Journal of Food Science*, 52(1), 185–193.

SCOTT G and RICHARDSON P. (1997), 'The application of computational fluid dynamics in the food industry', *Trends in Food Science & Technology*, 8, 119–124.

TATTIYAKUL J, RAO M A and DATTA A K (2001), 'Simulation of heat transfer to a canned corn starch dispersion subjected to axial rotation', *Chemical Engineering and Processing*, 40, 391–399.

TATTIYAKUL J, RAO M A and DATTA A K (2002), 'Heat transfer to a canned corn starch dispersion under intermittent agitation', *Journal of Food Engineering*, 54(4), 321–329.

VAN DER SMAN R G M (2002), 'Prediction of airflow through a vented box by the DarcyForchheimer equation', *Journal of Food Engineering*, 55(1), 49–57.

VERBOVEN P, SCHEERLINCK N, DE BAERDEMAEKER J and NICOLAÏ B M (1997), 'Performance of different turbulence models for the calculation of the surface heat transfer coefficient', in Proceedings of the Workshop on Modelling of Thermal Properties and Behaviour of Foods during Production, Storage and Distribution, Prague, *Food Research Institute*, 242–248.

VERBOVEN P, SCHEERLINCK N, DE BAERDEMAEKER J and NICOLAÏ B M (2000), 'Computational fluid dynamics modelling and validation of the temperature distribution in a forced convection oven', *Journal of Food Engineering*, 43(2), 61–73.

VERBOVEN P, DATTA AK, NGUYEN TA, SCHEERLINCK N and NICOLAÏ B M (2003), 'Computation of airflow effects on heat and mass transfer in a microwave oven', *Journal of Food Engineering*, 59(2–3), 181–190.

VERSTEEG H K and MALALASEKERA W (1996), *An Introduction to Computational Fluid Dynamics: The Finite Volume Method*, Harlow, Longman Scientific & Technical.

WILCOX D C (2000), *Turbulence Modeling for CFD*, 2nd edn, La Canada CA, DCW Industries.

XIA B and SUN D-W (2002), 'Applications of computational fluid dynamics (CFD) in the food industry: a review', *Computers and Electronics in Agriculture*, 34, 5–24.

XU Y and BURFOOT D (1999), 'Predicting condensation in bulks of foodstuffs', *Journal of Food Engineering*, 40(1– 2), 121–127.

XUE J and WALKER C E (2003), 'Humidity and its effects on baking in an electrically heated air jet impingement oven', *Food Research International*, 36, 561–569.

YANG W H and RAO M A (1998), 'Transient natural convection heat transfer to starch dispersion in a cylindrical container: numerical solution and experiment', *Journal of Food Engineering*, 36, 395–415.

Part II

Developments in technologies for sterilisation and pasteurisation

5

Modelling and optimising retort temperature control

G. Bown

5.1 Introduction

This short chapter addresses the process optimisation of batch retorts used to pasteurise and sterilise food products. It discusses the basic needs for process control, outlines model systems used to simulate the delivery of thermal lethality in real time, offers an optimisation process and opens a debate reviewing definitions for 'Achieved Lethality'.

The risks to public health and the threats to the commercial viability of the processor from inappropriately controlled retort processes, are well known and obvious; those processes that do not succeed in delivering a sufficiently lethal thermal process risk causing sickness and death amongst consumers and risk significant damage to the food processor's reputation.

It is therefore vitally important to ensure that every thermal process achieves an appropriate lethality, sufficient to achieve 'Commercial Sterility' in all the containers inside the retort. It is necessary to make certain that the point of lowest lethal heat in a container spends sufficient time at an appropriate temperature so that the target organism population at that point is reduced to a level that will not present a hazard to public health and will not cause product degradation.

It would be all too easy to administer an overwhelming lethality by performing thermal processes far in excess of those required to achieve Commercial Sterility, but this would be to the detriment of product quality and the expenses associated with excess energy consumption and lost production capacity.

Hence, there is a need to optimise the thermal process; to balance the damage done to the product by the thermal process with the need to deliver sufficient lethal heat to achieve Commercial Sterility reliably and consistently, in the face of product, package and process line variables and given the constraints of the process engineering reflected in the choice and operation of retort and associated equipment.

It is therefore necessary for the processor to '*prevent pre-processing product spoilage or contamination, maintain container integrity, establish and consistently apply the Scheduled Process by properly trained, experienced and supervised personnel and prevent post-processing contamination*' DHSS (1994).

The food processor is supported in his quest for safe and reliable thermal processes by several sources of direct aid and reference including DHSS (1994) and CCFRA (1997a,b,c). These works, along with those of Bown (2003), May (2001), NFPA (1982) and others provide details of batch retort types, their safe and reliable operation and their interaction with the product and packaging format.

5.2 Factors affecting thermal process control

If a thermal process is to be successful, it must achieve Commercial Sterility in order to remove risks to public health (and company fortunes) and must produce a product with sufficient quality attributes to ensure an appropriate shelf life. It is therefore necessary to confirm the factors that influence the 'post-process' state of the product are in control, including:

- factors connected with the initial state of the product, prior to any factory processes;
- factors connected with factory process that may influence the microbial state of the product or that may influence quality attributes, including of course, the thermal process;
- factors connected with product packaging that influence filling and sealing operations, the thermal process itself and those factors that influence the rate of product degradation during its shelf life.

The popular maxim 'what is not measured cannot be managed' applies equally to food process control, and successful process control relies on appropriate measurements just as much as it relies on the means of process control.

The first step is to identify the influencing factors. Those factors connected with the canning process have been studied by many and good sources of help are the CCFRA Guidelines (CCFRA (1977a,b,c)), they link influencing factors with methods to measure Achieved Lethality in practical situations and provide methods to ensure the Scheduled Process is correctly defined and validated, particularly regarding the conditions that should be considered when defining the Scheduled Process. Bown (2003) provides a short review of influencing factors connected with semi-rigid and flexible packaging, particularly related to Achieved Lethality.

The most significant factors influencing Achieved Lethality include the following:

- *Container geometry.* The distance that heat needs to travel from the outside of the container to the point of lowest lethal heat, influences the time lethal heat requires to reach that point and so the value of Achieved Lethality at the end of the thermal process. The container shape during the thermal process may not be constant with time, particularly if the container is semi-rigid or flexible (e.g. thermoformed tray with heat sealed flexible laminate lid or a pouch). In these situations, container geometry change is induced by the differential pressure that develops across the container wall during a retort cycle.
- *Product thermal diffusivity.* The rate at which heat may be transferred from the container surface to the point of lowest lethal heat, influences the rate at which lethal heat arrives at that point. The complexity of heat transfer in multi-phase food products is further complicated since headspace or entrapped gases may collect, reducing heat transfer rates from some container surfaces into some parts of the container.
- *Temperature and heat transfer media.* The temperature at which the cooker is held during the 'cook' phase of the process influences the amount of heat available at the container surface. DHSS (1994) suggests the temperature controller should be capable of controlling the sterilising environment at −0.5°C to +1°C and that the environment should be the same for all the containers within the process vessel to within 0.5°C, i.e. no two points should be more than 1°C apart. Usually the period during which heat arriving at the point of lowest heat contributes to Achieved Lethality is limited to the 'cook' phase of the cycle, but 'high performance' retorts routinely achieve these temperature control constraints and a debate has arisen to discuss the potential to extend the 'cook' phase to include some parts of the pre-heat and cooling phases. This is discussed further, later in this chapter.

  The heat transfer medium within the cooker determines the rate at which that heat may be transferred to the container. Four styles of heating in batch retorts are amongst those in common use:

  1. *Pure steam*: the process vessel is filled with pure wet, saturated steam and so heat transfer to the container surface occurs as the steam condenses and gives up its heat. It is important that the process vessel contains no air during the cook phase since there is little forced agitation within the vessel to avoid that air from cooling on contact with cooler container surfaces and so producing a local cold spot with attendant risk of local sub-lethal processes.
  2. *Steam/air*: a mixture of steam and air which is blown past the container surfaces at high velocity (providing high heat transfer rates and ensuring good steam/air mixing). The turbulence created by the usually very large circulating fan ensures the steam and air are well mixed so that the air is at steam temperatures and does not 'collect' at container surfaces. The air is used to apply a pressure to the containers above that applied by the

steam and so allows the retort to operate with an 'overpressure'. Note that saturated steam pressure is a function of steam temperature and so in order to maintain a fixed overpressure, air may be added or released from the retort during normal operation.

3. *Water spray/raining*: re-circulating water is heated in an external heat exchanger (usually by steam) and is directed at the containers as a series of high velocity sprays. The spray design should ensure all containers receive similar volumes of water to ensure similar heating conditions. Heat transfer to the containers is a function of water temperature and water velocity. Higher water velocity reduces the temperature difference between water at the top of the retort and that at the bottom of the retort that has just travelled in contact with the containers and so reduces the time between 'visits' to the heat exchanger to recover set point temperature. The turbulence caused by so many high velocity water sprays ensures air introduced into the retort to achieve independent pressure (and overpressure) control is heated to water temperature. Note that it is usual to heat water to temperatures well above 100°C. A pressure slightly above that needed to maintain steam in a saturated state at the requisite temperature, is called for and is provided by adding compressed air to the process vessel; for instance to achieve a water temperature of 121.1°C, a pressure in excess of 1 bar (gauge pressure) is required. Note also that it is normal to operate with an overpressure during high temperature sections of the thermal process and so process vessel pressure is usually far higher than that needed to maintain water in its superheated state and so the thermodynamic system is generally stable.
4. *Water immersion*: all the containers are completely immersed in water that is circulated through the retort at high velocity ensuring high heat transfer rates at the container surface. Steam is usually injected directly to the re-circulating water to achieve set point temperature, and the engineering is usually arranged so that the water is forced to travel along a prescribed route so that all containers receive similar amounts of heat. Overpressure air is usually contained in a 'ballast pipe' or secondary vessel located above the process chamber and is usually kept away from the food containers to avoid slowing heat transfer to the containers.

- *Process vessel pressure.* The use of steam or superheated water requires a pressure in the process vessel to maintain the required steam or water phase. It is usual to apply an additional pressure within the process vessel to maintain container integrity (by minimising the differential pressure that develops across the container wall). This is especially important if the container is semi-rigid or flexible since this 'overpressure' also ensures container geometry is in control and behaves consistently from process to process. Some pasteurising systems operating at temperatures below 100°C may not require process vessel pressure to maintain the water phase but may use an overpressure to maintain container integrity/geometry.

- *Cook time.* The time during which heat arrives at the point of lowest lethal heat and is active on the target organism.

Given that a practical system may be engineered to provide appropriate measurement and process control during the thermal process, the most significant factors that influence Commercial Sterility include:

- *Initial microbial population size.* The thermal process reduces microbial population by a given amount proportional to the death kinetics of the target microorganism and the lethality of the Scheduled Process. Given that the Scheduled Process is defined prior to the start of a thermal process, the final microbial population within a batch of containers is directly proportional to the initial population, so if the final population is to be sufficiently small not to present a risk to health, the initial population must be known and 'in control'.
- *Selection of target 'viable micro-organism'.* In general sterilising situations, when pH is above about 4.3, *Clostridium botulinum* is regarded as the organism most likely to survive a thermal process and pose a risk to public health. In other situations where, for instance pH is around 3.4, then less heat resistant *butyric anaerobes*, yeasts and moulds may be the organisms of concern. The decision regarding target organism is made prior to establishing the Scheduled Process, and so assumptions made at that time regarding pH, water activity, death kinetics etc. must be maintained during commercial process activities in order to ensure the lethal effect of the Scheduled Process on the target '*viable micro-organism*' population was actually achieved.

5.3 Modelling techniques for predicting lethal heat

An objective of this chapter is to propose an approach to numerical modelling that may be used in real-time, to compute lethal heat and so be useful in process control. It is first necessary to link the death kinetics of a target organism with the thermodynamics of heat transfer. A model may then be developed to simulate heat transfer in real-time and thus allow the effect of that heat to be expressed in terms of lethal heat with respect to the target organism. In this short chapter a series of assumptions will be made to allow the model development process to be completed. The case studied here is intended to be in itself a model so that the approach may be used in different situations. In other situations, alternate container geometries, product heating characteristics or other influential variables may need to be considered.

The principal assumptions made here include:

- The container is a cylindrical can of fixed geometry such that the height of the can is in the order of three times its diameter and thus may be deemed a finite cylinder. This will allow the use of partial differential equations for finite cylinders as a start point for the model.

- The can will be placed in a sterilising situation where it is supposed that the product has a pH greater than 4.3 and that there is a risk from *cl. Botulinum*.
- Heat transfer within the can will be by conduction in a simple, single-phase homogeneous product.

Other assumptions will be given in the context of the modelling.

5.3.1 Empirical and theoretical equations

Kinetics of thermal destruction

The preservation of food by the application of heat is achieved by the large-scale destruction of food spoiling organisms and their spores. Food is spoiled when organisms that survive the thermal process produce spores that later germinate. Food becomes a risk to public health when these germinating spores produce toxins. The risk to consumers arises when these toxins are ingested, particularly those known as botulin produced in the anaerobic conditions within a container by *Clostridium botulinum* (one of the more heat resistant organisms) that, if ingested, often proves fatal.

Elsewhere in this book (Chapter 1 by S D Holdsworth), the reader can find a detailed review of the kinetics of microbial destruction, an authoritative discussion on the concept of Commercial Sterility and the development of empirical models describing the relationship between the thermal inactivation of microbial spores and the thermodynamics of heat transfer in real food systems that culminates in the following expression:

$$F^{z}_{Tref} = \log 10\left(\frac{N_0}{N_F}\right)D_{Tref} = \int_{t0}^{tF} 10^{-(T_{ref}-T)/z}dt \tag{5.1}$$

where:

- F^{z}_{Tref} describes the severity of the thermal process (the F value for a given z referenced to a given temperature);
- the ratio N_0/N_F is the probability of survival of the target organism expressed as the ratio of initial population size and final population size;
- D_{Tref} describes the rate of destruction of a given organism at a reference temperature T_{ref};
- T is the temperature at the point of concern at time t;
- z is the numerical value obtained by measuring the number of degrees Celsius required for the thermal death curve of the test organism in a specific substrate to traverse one log cycle, i.e. the temperature change required to effect a tenfold change in the rate of microbial destruction (DHSS, 1994).

Equation 5.1 neatly divides the two aspects of thermal process evaluation; the left side of the expression relates a given thermal process lethality to its effect on the population of a given organism whilst the right side relates thermal process lethality to the time/temperature profile applied to the given organism, by the retort.

5.3.2 Numerical approximations

The classical equation describing two-dimensional transient heat conduction in a finite cylinder (i.e. the can described earlier) is given by Carslaw and Jaeger (1959):

$$\frac{\delta T}{\delta t} = \alpha\left\{\frac{\delta^2 T}{\delta r^2} + \frac{1}{r}\frac{\delta T}{\delta t} + \frac{\delta^2 T}{\delta h^2}\right\} \tag{5.2}$$

where α is thermal diffusivity (cm^2/s), r is radial distance from the can vertical axis, h is the vertical distance from the horizontal axis, T is the temperature at the point of interest and t is the time at the moment of interest, elapsed since the previous moment of interest.

Several techniques to solve equation 5.2 have been proposed; those employing finite difference integration lend themselves to numerical solution by computer and have been described by many including Teixeria *et al.* (1969a,b) who use the method to determine the distribution of spore survival and nutrient retention throughout the container by considering each volume element separately.

Arpaci (1966) and Carnaham *et al.* (1969) consider the explicit form of the finite difference solution. They go into great detail and develop finite difference equations as approximations to the partial derivatives in equation 5.2 via Taylor's series expansion.

In this solution, it is assumed that the temperature surrounding the can is constant over the whole surface at any time and that the resistance to heat transfer at the surface of the can, across the wall of the can and into the product is not significant at any surface of the can (thus ignoring any headspace gas influences). It is also assumed that the thermal diffusivity of the product is constant throughout the can. Thus, it follows that symmetry exists either side of the vertical axis and on either side of the horizontal axis, of a two-dimensional section through the can.

The finite difference technique establishes a two-dimensional network of node points throughout the can. Given the symmetry noted above, it is only necessary to establish a nodal grid for a single quadrant of a section through the can as shown in Fig. 5.1. Three incremental variables also need to be defined: change in radial distance from the vertical axis (Δr); change in vertical distance from the horizontal axis (Δh); and the change in time from the start of the thermal process to the moment under consideration (Δt).

The temperature at a node point located in say, column i and row j of the nodal grid at time t, is described as $T_{i,j}^{(t)}$, at time $t + \Delta t$ as $T_{i,j}^{(t+\Delta t)}$ at one radial increment further away from the vertical axis as $T_{i+1,j}^{(t)}$ and at one vertical increment away from the horizontal axis as $T_{i,j+1}^{(t)}$. Similarly the subscripts of T are expressed as $i - 1$ or $t - 1$ if the nodal position is nearer to the vertical or horizontal axis.

Carnaham *et al.* (1969) show that an expansion in Taylor's series for $T_{i-1,j}^{(t)}$ and $T_{i+1,j}^{(t)}$ about $T_{i,j}^{(t)}$ and for $T_{i,j-1}^{(t)}$ and $T_{i,j+1}^{(t)}$ about $T_{i,j}^{(t)}$ provide a set of

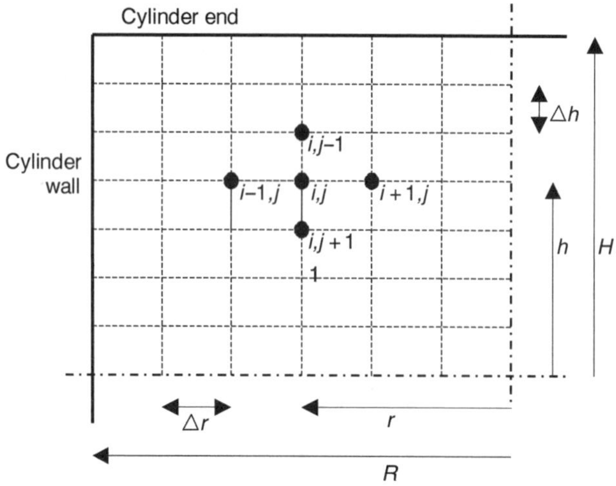

Fig. 5.1 Nodal grid system for a finite cylinder.

difference equations describing forwards and central differences as follows:

$$\frac{\delta T}{\delta t} = \frac{T_{i,j}^{(t+\Delta t)} - T_{i,j}^{(t)}}{\Delta t} \tag{5.3}$$

$$\frac{\delta^2 T}{\delta r^2} = \frac{[T_{i-1,j} - 2T_{ij} + T_{i+1,j}]^{(t)}}{(\Delta r)^2} \tag{5.4}$$

$$\frac{1}{r}\frac{\delta T}{\delta r} = \frac{1}{r} \cdot \frac{[T_{i-1,j} - T_{i+1,j}]^{(t)}}{2(\Delta r)} \tag{5.5}$$

$$\frac{\delta^2 T}{\delta h^2} = \frac{[T_{i-1,j} - 2T_{i,j} + T_{i,j+1}]^{(t)}}{(\Delta h)^2} \tag{5.6}$$

where r is the radial distance of the node under consideration from the vertical axis and h is the vertical distance of the node under consideration from the horizontal axis.

A general solution emerges if the explicit form is considered to allow direct calculation of $T_{i,j}^{(t+\Delta t)}$ as follows:

$$T_{i,j}^{(t+\Delta t)} = T_{i,j}^{(t)} + \frac{\alpha\Delta t}{(\Delta r)^2}\{T_{i-1,j} - 2T_{i,j} + T_{i+1,j}\}^{(t)} + \frac{\alpha\Delta t}{2r(\Delta r)}\{T_{i-1,j} - T_{i+1,j}\}^{(t)}$$
$$+ \frac{\alpha\Delta t}{(\Delta h)^2}\{T_{i,j-1} - 2T_{i,j} + T_{i,j+1}\}^{(t)} \tag{5.7}$$

Carnaham *et al.* (1969) show that the finite difference solution provides an acceptable approximation to equation 5.2 at the nodes but they also point out that computing values for $T_{i,j}^{(t)}$ involves rounding errors. Thus, if this numerical solution is used to compute internal can temperatures and thence lethality, the output can only be an approximation.

Croft and Lilley (1977) examine convergence, stability and accuracy from these rounding errors and give guidelines for the number of radial and vertical nodes required in a grid to achieve mathematical stability with reasonable computation efficiency. Teixeria *et al.* (1969a,b) and Bown (1985) show that larger numbers of nodal points do not add significantly to the sensitivity of the model when the model is used to compute Achieved Lethality. Bown (1989) provides a table of optimum numbers of node points for a range of can sizes.

The rules of symmetry discussed earlier may be used to derive four specific cases of equation 5.7:

1. The general case, when r lies between the side surface of the can and the vertical axis – but not on the vertical axis and when h lies between the end of the can and the horizontal axis – but not on the horizontal axis. That is given in equation 5.7.
2. The vertical centre axis case, when r lies on the vertical axis and h lies between the can end and the horizontal axis – but not on the horizontal axis, that is:

$$T_{i,j}^{(t+\Delta t)} = T_{i,j}^{(t)} + \frac{4\alpha\Delta t}{(\Delta r)^2}\{T_{i-1,j} - T_{i,j}\}^{(t)} + \frac{\alpha\Delta t}{(\Delta h)^2}\{T_{i,j-1} - 2T_{i,j} + T_{i,j+1}\}^{(t)} \tag{5.8}$$

3. The horizontal centre axis case, when h lies on the horizontal axis and r lies between the side surface of the can and the vertical axis – but not on the vertical axis, that is:

$$T_{i\,j}^{(t+\Delta t)} = T_{i\,j}^{(t)} + \frac{\alpha\Delta t}{(\Delta r)^2}\{T_{i-1,j} - 2T_{i,j} + T_{i+1,j}\}^{(t)} + \frac{\alpha\Delta t}{2r(\Delta r)}\{T_{i-1,j} - T_{i+1,j}\}^{(t)} + \frac{2\alpha\Delta t}{(\Delta h)^2}\{T_{i,j-1} - T_{i,j}\}^{(t)} \tag{5.9}$$

4. The geometric centre case, when both r and h are coincident at the geometric centre, that is:

$$T_{i,j}^{(t+\Delta t)} = T_{i,j}^{(t)} + \frac{4\alpha\Delta t}{(\Delta r)^2}\{T_{i-1,j} - T_{i,j}\}^{(t)} + \frac{2\alpha\Delta t}{(\Delta h)^2}\{T_{i,j-1} - T_{i,j}\}^{(t)} \tag{5.10}$$

This set of four equations may be used to define a series of temperature arrays with a progression in time (Δt). The temperature of each node in an array at a given time is related to the boundary conditions (can temperature) and adjacent nodes. The temperature array of a set of nodes at a time, t is related to that an interval later, at time $= t + \Delta t$. Thus it is possible to compute the temperature at any nodal point within a can at any time from the beginning of the thermal process according to the changing temperature at the can surface and within the assumptions made here.

If equation 5.1 is used to compute an F_{value} for this point (and any or all other points), it can be seen that the numerical model may be used to estimate the value of lethal heat according to the time/temperature history experienced by the can. If the model is calculated in real-time, at intervals of Δt, then a real-time estimate of lethal heat may be calculated as the thermal process progresses. Bown (1989) goes further into the detail of real-time calculations and offers spreadsheet and Basic program solutions. A number of calculation aids are also discussed to ease and speed the calculation process. The model is applied in a number of areas to show its value in predictive modelling to achieve a specific F_{value}, for process design and for on-line correction of thermal process deviations.

The modelling example used above describes the effect of the thermal process on temperature within a simple product, heating by conduction within a typical can. A similar approach may be used to model other situations by adjusting the boundary conditions of the cylinder model, including for example:

- variable heat transfer via the headspace within a can;
- variations in thermal diffusivity within a product either with time or space, to simulate real multi-component foods;
- the influence of package material on heat transfer (for instance glass).

A wider range of situations may be modelled by employing other partial differential equations, including for example:

- the effects of convection heat transfer;
- alternate package geometries including flat pouches, taller and much shorter cans.

5.4 On-line process control of retort temperature

The concept of 'derived-value' control is discussed by Bown (1985) and Bown (1989) in which real-time data are collected and used within a process control environment to adjust the process according to the computed effects of the measurements on a process variable that cannot be directly measured, thus controlling a 'derived' value rather than a measured variable. In the context of the thermal process, derived-value control allows a thermal process to be controlled according to its lethal effect on the point of lowest lethal heat. The control system computes the state of lethality at that point in real-time and may adjust cooker temperature or cook time to ensure, for instance, that sufficient Achieved Lethality is applied to that point before the cooling phase of the process is started.

This approach may be used to estimate the effect of the thermal process on the assumed microbial population at the point of lowest lethal heat. It may also be used to compute the effect of heat on other 'product attributes', some of which may be associated with product quality. Bown (1989) takes the opportunity to describe a number of situations where modelling can have positive impacts on thermal process control. Four such applications are discussed here.

5.4.1 On-line process deviation correction

Given that the model describes the actual state of lethality within the can accurately enough and may be computed whilst the process is under active control, it is possible to compare the actual progress towards Achieved Lethality and Delivered Lethality with that intended to occur when the thermal process was first validated. Thus it is possible to identify process deviations as they may relate to lethality. These deviations may be due to many reasons including:

- Those related to the food; for example: initial temperature, thermal diffusivity – assuming on-line measurement or data input from experience is possible prior to the start of the process.
- Those related to the initial conditions of the retort; for example: initial water temperature, available steam pressure and hence steam flow rate.
- Those that may occur during the thermal process; for example: system temperature response according to product load (i.e. smaller-than-normal product load accelerates rates of temperature rise and fall), interruptions in the steam supply or real time variations in steam flow (for example when multiple retorts draw steam from a common manifold) etc.

Those deviations that occur prior to the start of the thermal process may be noted and appropriate adjustments made to the initial conditions of the model. The deviations that occur during the thermal process and are measured by the process control system may be used to modify the model boundary conditions of one computation cycle to the next with a corresponding influence on the temperatures calculated at each node point.

5.4.2 Real-time process optimisation

Whilst the heat applied to food during pasteurising or sterilising processes is necessary to achieve Commercial Sterility, it is also responsible for the degradation in certain quality attributes. Mansfield (1962) and Lund (1986) describe the concept of Cook Value, C_{100}, which relates quality loss during the commercial thermal process to the equivalent cooking process at 100°C (domestic cooking temperature). This concept links commercial process temperature T with the temperature dependent factor z_c (the temperature difference required for a ten-fold change in a given quality attribute) and provides a method to describe overall quality loss with time t.

$$C_{100} = \int_0^t 10^{-(100-T)/z_c} dt \tag{5.11}$$

The value of z_c generally ranges from 25°C to 47°C covering various texture and colour attributes. Lund (1986) suggests a value of 33°C may be used to describe the overall quality loss. This subject is also discussed at length in Chapter 1 by S. D. Holdsworth and the reader is encouraged to understand Cook Values further.

Equation 5.11 is very similar to the right side of equation 5.1 defining lethality, F_T in very similar terms. Given that the finite difference model for heat conduction in finite cylinders may be used to compute lethality at a point with time, it follows that a very similar model may also be used to compute Cook Value and hence the level of degradation in various quality attributes with time and temperature during a thermal process.

So here lies the possibility to optimise a thermal process to deliver an appropriate Achieved Lethality to render the food product commercially safe whilst adjusting the thermal process control variables to produce a product with maximised quality attributes. Naturally, food safety is paramount – but given that the Scheduled Process is the target and it is necessary to deliver sufficient Achieved Lethality to attain Commercial Sterility, there may be opportunities to avoid overprocessing (for instance avoiding extended time at sterilising temperature or cooling too slowly) by adjusting process control variables. Further, assuming the product of concern has several quality attributes that require maximising with different reaction kinetics (z_c), it may be possible to consciously 'trade' one attribute for another (for instance compromising on colour to retain flavour).

5.4.3 Real-time process simulation

The ability to compare actual with intended lethality at virtually any position within a can provides additional opportunities including:

- predicting the end point of a thermal process to assist multi-cooker control synchronisation – perhaps prompting faster cooling to avoid 'gaps' in an otherwise continuous outflow of cooked product from a synchronised row of batch cookers;
- adjusting the start or progress of a cooker in a multi-cooker installation to avoid peaks in steam or cooling water demand that would otherwise disturb other retorts (this may require forcing a process deviation which will later be corrected automatically by the control system);
- allowing for abnormally high initial microbial loads by increasing the target Achieved Lethality according to a prescribed system (to ensure safety) in the event that perhaps an alternate supply of raw ingredients, known to have a larger initial population, is used;
- replaying a given thermal process in the event that a post-process incident occurs, to either help identify the cause of the incident or, if the Achieved Lethality was sufficient, to eliminate the thermal process from the investigation – of course, assuming that the entire process is logged appropriately;

- forward computing in real-time, given assumptions about the future of the thermal process perhaps based on historical performance, to assess the benefit of continuing a very deviant process – for instance, to discover if sufficient quality attributes are retained in the event that the deviant but corrected process is allowed to continue.

5.4.4 Off-line modelling

Modelling thermal processes off-line offers a range of advantages to the processor including:

- defining the sensitivity of a product/container/cooker system to various product or process variables to identify which may require closer control or greater care during production: this may be achieved by running the model using ranges of variables as shown by Bown (1989);
- pre-process design; allowing the process engineer to consider alternate process strategies to perhaps organise a series of batch cookers to produce a continuous processing effect or to review alternate process cycles to compromise production output for the sake of product quality (or perhaps vice-versa!);
- investigating the effects or advantages of changing product recipes, container styles or sizes and alternate retort engineering (e.g. steam capacity);
- investigating a range of process variations to provide data for an HACCP analysis or to provide instructions for an operator's handbook in the event manual control is necessary;
- to define the contribution a thermal process may make in a situation where a number of microbial growth inhibiting processes are combined to form a minimised process regime (combining partial cooking prior to filling and sealing and post-retort chilling and storage with a minimal thermal process).

5.4.5 Precautions

DHSS (1994) and NFPA (1982) require an appropriate recording system is used to provide a paper record of the thermal process and offer advice to ensure thermal process accuracy and reproducibility. Plainly, a self-correcting or real-time optimising process control system would be required to provide additional documentation to detail any deviations, corrections or optimisations made during the thermal process cycle. The system would also be required to operate within predefined limits to avoid an under-process situation where sub-lethal product may be produced.

5.5 Achieving lethality using the pre-heating and cooling phases of the retort cycle

The traditional view of Achieved Lethality is that it is accumulated during the 'cook' or 'hold' phase of the retort cycle, beginning when the process control system first achieves and maintains retort temperature to better than −0.5°C and +1°C and ending when this degree of temperature control can no longer be maintained, usually when cooling begins. During this time, the retort is expected to provide an isothermal environment wherein no two points are more than 1°C apart (interpreted from DHSS, 1994).

However, recently designed and commercially available retorts equipped with modern process control systems and high quality steam, water and compressed air supplies, now routinely achieve accurate, predictable and reliable temperature control within these limits, during portions of the pre-heating and cooling phases of the retort cycle. This raises the question; *should those portions of the retort cycle also contribute to Achieved Lethality?* If this view were to prevail and contributions to Achieved Lethality during pre-heating and cooling were permitted, then the amount of thermal damage done to the product could be reduced significantly and the consequently shorter retort cycle would effectively increase retort capacity and reduce costs.

However, a number of practical issues need to be addressed and understood to avoid situations that may lead to sub-lethal processes, such as:

- *Pre-heating retorts containing packs with different initial temperatures*. It is possible to achieve the required retort temperature control limits during pre-heating even if the product load contains packs with variations in initial temperature – for instance those filled earlier in the 'batch'. The variations in initial product temperature will induce variations in the lethality achieved at the point of lowest lethal heat, within the container. Normally this situation exists – but in a system where contributions to Achieved Lethality do not start until 'cook' temperature falls into the accepted tolerance, the variations in the lethality achieved at the point of lowest lethality are likely to be smaller than those that would exist if the start point for the accumulation of Achieved Lethality is earlier in the cycle. This is likely to be reflected in a wider range of Achieved Lethality within the population of containers in a given retort load.
- *Cooling water distribution*. The permitted tolerance on temperature distribution within the retort must not be exceeded during that part of the cooling phase expected to contribute to Achieved Lethality. This may be especially difficult at the start of cooling, if cooling water is introduced directly into the process vessel since its cooling effects will be first felt close to its point if entry. Retorts that use high velocity recycling water, heated and cooled indirectly via an external heat exchanger are more likely to achieve the required tolerance than those that flood the retort with cooling water or those that displace water used during cooking.
- *Package format*. The geometry of semi-rigid or flexible containers during retorting is a function of the differential pressure established across the

container walls (difference between in-pack pressure and retort pressure – usually retort pressure is greater to ensure pack geometry is minimised and to assist seal integrity). The response of the container headspace to the start of cooling usually causes some change in internal pressure and hence pack-geometry. The differential pressure must be maintained to retain pack geometry to avoid variations in heat transfer that may induce variations in Achieved Lethality. This may complicate retort pressure control during the first few moments of the cooling phase and in any case would not necessarily address the behaviour of *all* containers – so a risk remains that conditions within some containers (perhaps those most susceptible to pressure-induced geometry change) may yield atypical Achieved Lethality.

- *Where to place the thermocouple.* The traditional concept that the point in the container where the probability of microbial survival was highest was also point of slowest heating no longer applies. If some part of the cooling cycle is permitted to contribute to Achieved Lethality, then the point of lowest lethal heat is neither the point of slowest heating or the point of fastest cooling. New thermal process evaluation works are required to establish the position of the point 'of most concern', prior to establishing the effect of the thermal process upon that point. This may provide an opportunity to use a numerical model off-line, to identify the point of lowest lethal heat and to describe the distribution of lethal heat with time and position within the container.

As the leading UK authority in these matters, this question has been addressed to CCFRA and is receiving attention from CCFRA experts who in turn are seeking input and opinion from industry experts including those practitioners who manage commercial retort systems on a daily basis and those who build this calibre of retort. To date (November 2003), there is no conclusion to the debate. The guidelines given in DHSS (1994) are considered paramount, i.e. that:

- retort temperature control within $-0.5°C$ and $+1°C$ be achieved and maintained for the duration of the period during which contributions to Achieved Lethality are made, *and*
- the isothermal environment within the retort is such that the temperature difference between any two points is less than 1°C.

A 'working interpretation' for this situation has been drafted:

> The use of cooling lethality in F_0 calculations is recommended only when the cooling process is controlled within specified limits (temperature and pressure), and these are fully taken into account when establishing the process. Specified limits should include definition of the intended cooling profile (including tolerances), the relationship between the retort instrumentation during cooling and the lowest temperatures in the retort. Once these relationships have been established processes may be established taking into account the worst case cooling profile either by placement of heat penetration samples in the cold zone (which may be different from the heat phase) and/or process heat transfer modelling.

If cooling lethality is used food manufacturers should be aware that process lethality may be affected by:

1) Loss of pressure control in the retort (resulting in boiling inside the containers).
2) Abnormally low water supply temperatures.
3) Any modification to the water flow rate.

It will be interesting to see how the debate concludes. The pressure to reduce thermal damage and to include contributions during pre-heating and cooling is strong but the Duty of Care to the consumer is probably stronger so that only in those cases where systems are shown to be very reliable are food processors likely to re-think the conventional view particularly given the magnitude of the associated risks.

5.6 Future trends

The current debate is likely to be resolved by way of a more formal statement from CCFRA, reflecting expert opinion and including a rigorous set of pre-conditions and recommendations for process evaluation. This may then encourage the wider use of 'high performance' retorts and may open the doors to retort process optimisation. As the cost of computing power falls and the use of dedicated microprocessor based control systems increases, it is likely that more and more food processors will have the potential to use numerical model systems to simulate food preservation in real-time

CCFRA and others have proposed a number of modelling systems similar to that outlined here and so it would appear that the necessary confidence in modelling systems is steadily growing. A real-time process control and deviation correction system is available from FMC Food Tech, known as Log-Tech, but it would appear that recent FDA regulations governing the use of electronic records (FDA 21CFR Part 11 Electronic Records: Electronic signatures) may limit the use of the latest versions of Log-Tech. The debate is ongoing but it does indicate a move towards intelligent automation and perhaps 'derived value control'.

In addition, the growing interest in 'hurdle' technology, where several gentle techniques and processes are combined to achieve Commercial Sterility, will not only reduce the heat damage otherwise suffered by foods processed by heat alone but will increase the need for tight and reproducible process control in order to avoid minor process deviations causing sub-lethal situations. 'Derived value control' appears to offer part of the control solution. These trends suggest a greater need to understand more fully the variables that influence Achieved Lethality to permit a closer control and hence a more optimised process. Given that retail competition is largely about differentiation via quality, it seems likely that numerical modelling and perhaps even 'derived value control' will play a part in future process strategies to help food companies deliver highest quality, Commercially Sterile foods.

5.7 Sources of further information and advice

For further reading, the bibliographical notes by Holdsworth and Overington (1975); Holdsworth (1979, 1982, 1985, 1988, 1990 and 1997) provide one of the most complete reviews of works on thermal process calculations.

For expert advice, the Campden and Chorleywood Food Research Association is a centre of technical excellence for the food industry and has a wealth of expertise available to support food processors in all aspects of food technology and related sciences, in particular in connection with thermal processing and retort use.

CCFRA
Chipping Campden,
Gloucestershire, GL55 6LD
UK.
Tel: +44 (0)1386 842 000 Fax: +44 (0)1386 842 100

5.8 Glossary of terms

Achieved Lethality. The amount of lethal heat delivered to the point of lowest lethal heat within a container, during the period of time when cooker conditions are within prescribed limits for such lethal heat to contribute to Achieved Lethality: normally that quoted to describe the lethality of a given thermal process.

Commercial Sterility (Appertisation) of food. The condition achieved by the application of heat that renders food free from viable microorganisms, including those of known public health significance, capable of growing in the food at the temperature at which the food is likely to be held during distribution and storage (DHSS, 1994).

Cook temperature. The temperature of a retort specified for the sterilising operation (DHSS, 1994).

Cook time. The length of time during which the sealed containers are totally exposed to the specified cook temperature (DHSS, 1994).

Delivered Lethality. The amount of lethal heat delivered to the point of lowest lethal heat within a container during the entire thermal process. Delivered Lethality is usually greater than Achieved Lethality.

Lethal heat. The effect of exposure to temperature, under specified conditions, transformed mathematically in order to give a measure of sterilisation achieved (DHSS, 1994).

Point of lowest lethal heat. The point in a container that receives the least lethal heat during the Scheduled Process. Note this point will be that where heating is slowest during the heating phase of the process but may not be the same point, if part of the cooling process is permitted to contribute to Achieved Lethality.

Post-process contamination. The contamination of a food product in a hermetically sealed container by the ingress of microorganisms after completion of the thermal process (DHSS, 1994).

Seal Integrity. The adequacy of a seal that ensures hermeticity (DHSS, 1994).

Scheduled Process. The heat process chosen by the processor, validated for a given product and container size, to achieve Commercial Sterility (DHSS, 1994).

Sub-lethal process. A thermal process in which Achieved Lethality is insufficient to achieve Commercial Sterility.

Thermal process. The heat treatment given during the sterilisation operation, expressed minimally as a combination of time and temperature (DHSS, 1994).

5.9 References

ARPACI, V.S. (1966). *Conduction Heat Transfer,* Addison-Wesley Publishing Co,.

BOWN, G. (1985). *Retort Control. The application of a microcomputer based control system.* Technical Memorandum 391. CFPRA, Chipping Campden, Glos. GL55 6LD, UK.

BOWN, G. (1989). *The Influence of Product and Process Variability on Process Control in the Thermal Sterilisation of Food.* Procter Department of Food Science, University of Leeds, Leeds.

BOWN G. (2003). Developments in conventional heat treatment, in *Food Preservation Techniques*, edited by Peter Zeuthen and Leif Bøgh-Sørensen, Cambridge, Woodhead.

CARNAHAM, B., LUTHER, H.A. and WILKES, J.O. (1969). *Applied Numerical Methods*. John Wiley and Sons, New York.

CARSLAW, H.S. and JAEGER, J.C. (1959). *Conduction of Heat in Solids.* 2nd Edition, Oxford University Press, Oxford.

CCFRA (1997a). *Guidelines for batch retort systems – full water immersion – raining/spray water – steam/air.* Guideline 13. CCFRA, Chipping Campden. Glos. UK.

CCFRA (1997b). *Guidelines for establishing heat distribution in batch overpressure retort systems*. Guideline 17. CCFRA, Chipping Campden. Glos. UK.

CCFRA (1997c). *Guidelines for performing heat penetration trials for establishing thermal processes in batch retort systems.* Guideline 16. CCFRA, Chipping Campden, Glos. UK.

CROFT, D.R. and LILLEY, D.G. (1977). *Heat Transfer Calculations Using Finite Difference Equations.* Applied Science Publishers, London.

DEINDOERFER, F.H. and HUMPHREY, A.E. (1950). Analytical methods for calculationg heat sterilising times. *Applied Microbiology*, 7, 256.

DHSS (1994). *Guidelines for the Safe Production of Heat Preserved Foods*. Department of

Health, HMSO, London.

HOLDSWORTH, S.D. and OVERINGTON, W.J.G. (1975). *Calculation of the Sterilising Value of Food Canning Processes.* Technical Bulletin 28, CFPRA, Chipping Campden, Glos. UK.

HOLDSWORTH, S.D. (1979). *A Bibliographical Guide to Thermal Process Calculations. Part 2.* Technical Memorandum 130. CFPRA, Chipping Campden, Glos. UK.

HOLDSWORTH, S.D. (1982). *A Bibliographical Guide to Thermal Process Calculations. Part 3.* Technical Memorandum 291. CFPRA, Chipping Campden, Glos. UK.

HOLDSWORTH, S.D. (1985). *A Bibliographical Guide to Thermal Process Calculations. Part 4.* Technical Memorandum 410. CFPRA, Chipping Campden, Glos. UK.

HOLDSWORTH, S.D. (1988). *A Bibliographical Guide to Thermal Process Calculations. Part 5.* Technical Memorandum 498. CFPRA, Chipping Campden, Glos. UK.

HOLDSWORTH, S.D. (1990). *A Bibliographical Guide to Thermal Process Calculations. Part 6.* Technical Bulletin 77. CFPRA, Chipping Campden, Glos. UK.

HOLDSWORTH, S.D. (1997). *Thermal Processing of Packaged Foods*, Glasgow, Blackie Academic and Professional.

LUND, D.B. (1986). 'Kinetics of physical changes in foods' in Okos, M.R. editor, *Physical and Chemical Properties of Food.* St Joseph, Mitch: American Society of Agricultural Engineers, 367–81.

MANSFIELD, T. (1962). 'High temperature–short time sterilisation', in Leitch, J.M. editor, *Food Science and Technology*, Proc. 1st International Congress of Food Science and Technology, 18–21 September 1962, Vol. 4, pp. 311–16, London. Gordon & Breach Science Publisher.

MAY, N.S. (2001). 'Retort Technology', in Richardson P., editor, *Thermal Technologies in Food Processing*, Cambridge, Woodhead, 7–28.

NFPA (1982). *Thermal Processes for Low-acid Foods in Metal Containers.* Bulletin 26-L. 12th edition. National Food Processors Association. Washington, USA.

TEIXEIRA, A.A., DIXON, J.R., ZAHRADNIK, J.W. and ZINSMEISTER, G.E. (1969a). Computer optimisation of nutrient retention in thermal processing of conduction heating foods. *Food Technology*, 23 (6), 137.

TEIXEIRA, A.A., DIXON, J.R., ZAHRADNIK, J.W. and ZINSMEISTER, G.E. (1969b). Computer determination of spore survival distributions in thermally processed conductive heating foods. *Food Technology*, 23 (3), 352.

6

Improving rotary thermal processing

G. S. Tucker, Campden and Chorleywood Food Research Association, UK

6.1 Introduction: the use of rotation for batch thermal processing

Rotation for batch thermal processing falls into two categories: firstly, end-over-end (EoE) processing, where a retort crate rotates around a central horizontal axis and the containers are loaded vertically; and secondly axial rotation, where cans are rotated in the horizontal plane. EOE rotation is most commonly found in batch retorts. Axial rotation is the mode of rotation encountered with continuous systems such as the reel and spiral cooker-cooler. Many thermally processed foods heat by convection to some degree, which can be used to the processor's advantage in reducing process times, increasing production efficiency and in some instances minimising the deleterious effects of heat. Examples are soups, sauces, vegetables in brine, meat in gravy and some petfoods. This is achieved by agitating the container of food during the process by rotation, and in doing so, inducing forced convection currents that mix and heat the food more effectively. A key factor in mixing the container contents is the headspace bubble that sits above the food until the container is rotated.

The present food industry approach to the selection of rotary processing conditions for products is based on experience and measurements from limited numbers of samples. Most rotation rates are based on existing steriliser operating values, although these can correspond to different products, processes and containers. These are likely to be far removed from the optimal conditions. Variations in factors such as container shape, product viscosity and rotation speed can have a great effect on the internal mixing within a container and therefore on the rate of heat penetration into the product.

This chapter reviews some of the published work on container rotation effects that have influenced commercial applications for rotary retorts. It concludes by focusing on recent efforts to combine measurements of in-container mixing efficiency with computational modelling to find optimum rotational conditions.

6.2 The effectiveness of rotation in improving heat transfer

Advantages of rotary over static processing have been known for many years. In one of the first studies on rotation, Clifcorn *et al.* (1950) studied the advantages of agitating cans during EoE and axial rotation methods. High rotation rates and high temperatures gave an overall improvement in product quality, particularly for viscous and heat sensitive products. EoE methods gave better results at lower rotation speeds (less than 120 rpm) than axial; above this the results were equal.

Several studies have identified that mixing via the headspace bubble can have a significant effect on the heat transfer within the product. Conley *et al.* (1951) showed that, as the speed of rotation increased beyond the optimal centrifugal force, the movement in the product reduced; this effect was dependent upon the radius of rotation. Parchomchuk (1977) and Naveh and Kopelmann (1980) studied this effect for both EoE and circular agitating product. Rotation speed, viscosity and headspace were investigated and it was found that maximum heat transfer was achieved during EoE rotation with large headspaces at speeds from 40 to 80 rpm. Berry *et al.* (1979) studied the effect of an increasing headspace during the processing of cream style corn, where gases are released from the product, thus increasing the headspace. This increase in headspace was found to increase dramatically the sterilisation value achieved. Decreasing the headspace decreased the rate of heat penetration and therefore the sterilisation value F_0. For example, at 10 rpm the heating factor, f_h, was increased from 10 to 50 minutes with a decrease in headspace from 10/32 to 4/32. In all cases, the headspace bubble flow pattern was given as the reason for the difference in heat transfer between the cans. Theoretically, as the size of a bubble increases so its relative velocity increases, thus improving the mixing. However, beyond a certain point, due to an increase in bubble deformation, the increased bubble size will probably increase drag on the walls and therefore decrease the mixing effects of the bubble. Secondly, an increase in heat penetration rates may be offset by viscous forces within the can which oppose product movement at higher oscillating speeds.

Studies by Javier *et al.* (1985) and Ramaswamy *et al.* (1993) looked further into the effect of viscosity, mode of rotation, can size and rotation speed on the f_h of products. In both studies, the results showed an increase in f_h with an increase in viscosity of the product and a decrease in f_h with increasing rotation speed. In all cases, higher heating rates were achieved in EoE than axially rotating or reel and spiral simulation cans. A further finding was that in EoE and reel and spiral simulation, the f_h value decreased with increasing rotation speed; however, in axial rotation there was a zone where an increase in rotation speed led to an increase in f_h value. After this zone, the heating rates increased with increasing rotation speed. In axially rotating cans, bubble movement had two effects: increasing the rotation disrupted the streamlines, therefore promoting convective transfer to the centre, but this also meant that the bubble moved to the centre of the pack, thus slowing down the heating rate. The movement of the bubble to the centre of the can did not occur in the EoE cans; this was thought to be due to the strong end effect of the packaging as it rotates.

Britt *et al.* (1994) studied the influence of the radial position during rotation and varying processing conditions on the uniformity of product heating. A 1% (w/v) bentonite solution was filled into 76.2 × 115.9 mm cans with a 5% headspace and closed without vacuum. Different positions were studied within a retort basket, plus rotational speeds of 0, 10 and 20 rpm. It was found that whilst rotation resulted in higher F_0 values, this was not consistent through the product load. The lowest F_0 value varied throughout the load, with no one position being identified as the worst place within the basket; this agreed with the findings of Anantheswaran and Rao (1985). In contrast, Knap and Durance (1998) investigated the effect of radial position on f_h in a potato in water product. They concluded that radial position did influence heating rates in both the liquid and particle phase, with f_h decreasing as the distance from the axis of rotation increased. A possible explanation could be that moving the can away from the central axis improved the heat transfer coefficient, because the centripetal acceleration increased with increasing radial position, giving rise to greater relative bubble velocity.

Stoforos and Merson (1991, 1992) studied the use of aluminium and Teflon particles to determine the uniformity of heating under axial can rotation with different viscosity liquids. Liquid crystals were used as temperature sensors for measuring the particle surface temperatures. In general, it was found that liquid-particle film heat transfer coefficients increased with increasing rotational speed and increasing fluid viscosity; however, the overall heat transfer coefficient (heating medium/container wall/internal liquid) decreased with increasing fluid viscosity but increased with increasing rotational speed. Knap and Durance (1998) investigated the difference in sterilisation values between particles moving freely in a low viscosity liquid solution and a particle suspended on a mock thermocouple. In all cases, a higher sterilisation value was achieved on the suspended particle; this was thought to be due to the relative velocity between the fluid and the particle. Where a particle is free to move and the solid and liquid phase have similar densities, there will be less drag force on these particles than on the suspended particle, so resulting in a lower sterilisation value in the freely moving particles.

Sablani and Ramaswamy (1995) studied the fluid-to-particle heat transfer coefficients in cans during EoE processing. Using water and oil as the liquid phases and polypropylene spheres as the particles, heat penetration trials were carried out at three retort temperatures and four rotation speeds. The results showed that rotation speeds had a greater influence than temperature on the fluid to particle heat transfer coefficient. On average, the fluid-to-particle heat transfer coefficient increased fourfold as the rotational speed increased from 0 to 20 rpm, whilst an increase in retort temperature from 110 to 130°C resulted in only a 10% increase. It was also found that an increase in rotation speed resulted in an increase in the overall heat transfer coefficient, as did an increase in retort temperature. For all of these effects it was concluded that increases in heat transfer coefficient due to rotation speed could be attributed to the increased turbulence within the container. Increasing the retort temperature probably led to a reduction of the viscosity of the product, so increasing the heat transfer coefficient.

6.3 Optimising mixing during rotation to improve heating rates

A wide range of work has been carried out in this area; however, most research considered materials of single viscosity only and did not take into account viscosity changes that occurred in the product as it was sheared and heated. Viscosity changes within the container are a critical component in determining optimum rotation conditions, because this dictates how successfully the headspace bubble can move through and mix the food. Adjusting the rotation speeds to maintain optimum mixing, in response to viscosity changes, was a key part of the experimental and simulation programmes in the research findings reported here.

To quantify these effects, processing trials were carried out using commercially available sauces at a range of temperatures (Emond and Tucker, 2000). Rheological characteristics of each product were measured in order that a range of transparent and semi-transparent solutions could be identified with similar viscosities to those of the products. Of key importance was the need to operate visual trials at ambient temperature, so viscosities of the 'simulant' material at ambient were needed to match those of the product at different temperatures.

The objectives of this study were to:

- calculate the shear rates within a rotating container using computational fluid dynamics models, in order to
- match the rheological characteristics of selected food products at different temperatures with ambient temperature simulants;
- visualise the mixing patterns that occur during EOE rotation of a glass jar, using varying rotation speeds and simulant viscosities;
- optimise the EOE rotation speeds for the commercial products using the visualisation results;
- verify the thermal process by carrying out heat penetration trials on the product using the optimised process, and compare with the standard single rotation speed process.

6.3.1 Numerical simulation of the flow

Numerical simulation of the flow inside a rotating container was carried out in order to establish the shear rates in the containers. This was necessary to ensure that the rheological models were applied over the correct shear rate range. Magnitude of shear rates during processing was an unknown variable but critical in determining the solution viscosity. With most in-pack processed foods exhibiting non-Newtonian rheology, this step was essential in order to define the shear rate range for measuring rheological properties. To achieve this, the flow of liquid inside a can, with a 10% gas headspace, rotating about its axis was numerically simulated using the CFX 4.1 Computational Fluid Dynamics software (AEA Technology, Harwell, UK). Newtonian and generalised non-Newtonian fluids were considered and the results validated in two ways.

Firstly, experimental observations of the headspace bubble location and shape were compared with those predicted numerically. The experiments were carried out for a range of rotational speeds (10–50 rpm) and for several fluids. Secondly, measurements of wall shear stress were made, using hot-film probes and constant temperature anemometry, on the curved and flat end-walls of the can as it rotated. The numerically predicted wall shear stresses were then compared with the measurements, again for different fluids, a range of rotational speeds and for on- and off-axis rotation.

The numerical solutions, once validated, allowed details of the velocity and shear rate fields to be examined. The working range for the shear rates was found to be from 0 to 6 s^{-1}, information that was essential for positioning the rheological characteristics of the chosen transparent simulants.

Rheological tests, using a controlled shear stress rheometer (Carrimed, UK), were undertaken with a double concentric cylinder cell with wide gap to accommodate any small particles in a product. The cell was surrounded by a heating coil on the inner and outer cylindrical surfaces in order strictly to control the temperatures in the test cell. Rheological tests for simulants (products containing no particulate matter) were tested using a Contraves Rheomat-30 (Contraves AG, Zurich).

A tomato-based sauce containing small vegetable particles and modified maize starch thickener was tested at a range of temperatures in the double concentric cylinder cell. Over the shear rates of interest, 0 to 6 s^{-1}, the Careau model was applied. However, the shear rate region between 0.06 and 6 s^{-1} included most of the flow volume inside the rotating can (predicted from the numerical solutions) and so this power law region was used to compare the product rheology to that of the simulants. Figure 6.1 shows the change in apparent viscosity with shear rate for temperatures between 40 and 80°C. Effects that may have occurred from zero shear viscosity properties were not included.

Several transparent and opaque solutions were tested at ambient conditions (25°C) using either a Carrimed cone and plate or a Contraves concentric cylinder rheometer. Different concentrations of simulant solution were tested to find solutions matching the rheological characteristics of the tomato sauce at elevated temperatures. Table 6.1 summarises the rheological results for solutions

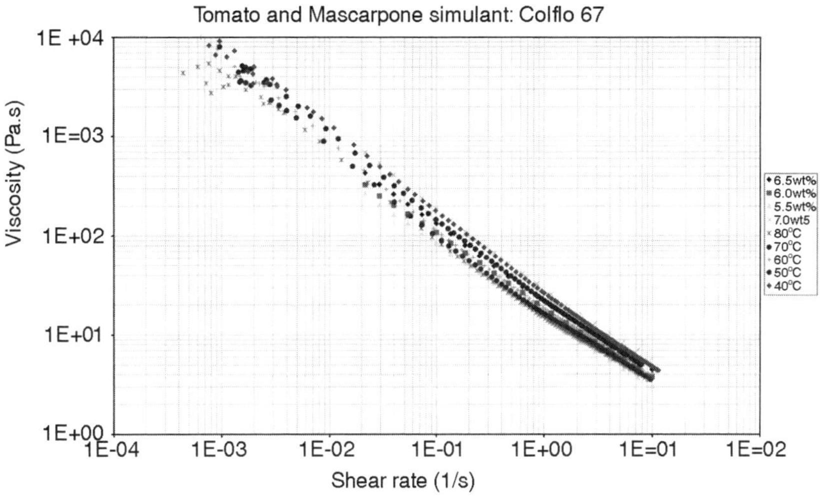

Fig. 6.1 Rheological data for Colflo 67 solutions at 25°C from which the concentrations to match the tomato sauce at temperatures of 40, 50, 60, 70 and 80°C were estimated.

of Colflo 67 starch at 25°C, which was found to be the closest match between sauce and simulant rheology. Colflo 67 is a waxy maize starch that is widely used in heat preserved foods (National Starch & Chemical Ltd., Manchester) because of its heat stability. Figure 6.1 also shows the power law model applied to Colflo 67 solutions at 25°C from which the concentrations to match the tomato sauce at temperatures of 40, 50, 60, 70 and 80°C were estimated. Table 6.2 gives a summary of the consistency coefficients (k), flow behaviour indices (n) and apparent viscosities at a shear rate of 10 s^{-1}. Table 6.3 shows the concentrations of Colflo 67 solutions that matched the rheological behaviour of tomato sauce at the chosen elevated temperatures.

6.3.2 Flow visualisation of mixing in containers

The tests reported used a glass jar of dimensions 152 mm height, 64 mm neck width, and 76 mm body width. This was first filled to 490 g with a 6.29% w/w

Table 6.1 Product characteristics for tomato sauce at elevated temperatures, using power law constants k (consistency coefficient) and n (flow behaviour index)

Product temperature (C)	k (Pa.s^n)	n	Apparent viscosity at shear rate 10s^{-1} (Pa.s)
40	25.56	0.27	4.76
50	21.94	0.28	4.21
60	18.46	0.32	3.84
70	16.47	0.34	3.57
80	14.62	0.37	3.44

Table 6.2 Rheological results for Colflo 67 solutions at 25°C, using power law constants *k* (consistency coefficient) and *n* (flow behaviour index).

Colflo 67 solution % (w/w)	k (Pa.s^n)	n	Apparent viscosity at shear rate 10 s^{-1} (Pa.s)
5.5	19.74	0.24	3.44
6.0	25.27	0.23	4.34
6.5	28.44	0.27	5.24
7.0	42.72	0.24	7.51

Table 6.3 Concentration of Colflo 67 solutions to simulate the power law values for tomato sauce for a range of temperatures

Product temperature (°C)	Colflo 67 solution (% w/w)
40	6.29
50	5.97
60	5.73
70	5.58
80	5.48

Colflo 67 starch solution plus 10 g of 6.29% w/w Colfo 67 starch solution mixed with blue food dye giving a headspace of 10 mm at ambient temperature (25°C). Jars were placed in a pilot scale Stock Rotomat PRU 900 (Hermann Stock Maschinenfabrik GmbH, Germany) on the central axis of rotation, and rotated under end-over-end conditions at 15 rpm. Movement of the headspace bubble and blue dye was recorded using a digital video camera over a three-minute period to assess the homogeneity of dye incorporation into the starch solution. The images were captured at one-minute intervals in order to assess the mixing efficacy. This was repeated at 20, 25 and 30 rpm and at intermediate speeds to find the optimum mixing speed for 6.29% w/w starch.

The trial was then repeated with starch solutions of 5.97, 5.73, 5.58 and 5.48% w/w under the same conditions. These percentage Colflo 67 concentrations represented the changing viscosity of the tomato-based sauce at different

Table 6.4 Optimised rotation speeds to achieve mixing in the tomato-based sauce product at different temperatures during pasteurisation

Product temperature (°C)	Rotation speed (rpm)
20	15
40	22
50	22
60	27
70	27
80	30

temperatures relevant to pasteurisation. The images for each of these solutions were assessed, giving optimum process rotation conditions for the product as its viscosity changed with temperature throughout a process.

From the results of the visualisation trials (assessed by the degree of incorporation of blue dye into the solution) an optimised rotation profile was established that corresponded to increasing sauce temperature. Table 6.4 shows the optimised rotation rates determined for the sauce at different temperatures.

6.4 Testing changes in rotation rate to improve heat transfer

Having determined the optimum end-over-end rotation speed that corresponded to the decreasing sauce viscosity during the process, thermal tests were conducted to prove that this approach did have a beneficial effect on the process. Tests were set up first using pilot-scale processing equipment and then taken into a food production factory where full-scale equipment and commercial product were used. For this tomato product, the target microorganisms were butyric anaerobes, for which the target P-value (reference temperature 93.3°C, z-value 8.3°C) was 10 minutes at the end of heating.

6.4.1 Pilot-scale tests

Tomato-based sauce, taken from jars of sauce that had been processed, was filled into jars to a fill weight of 500 g giving 8 mm headspace. The jars were placed on the central axis of rotation in a pilot scale Stock Rotomat PRU 900 and processed using the standard processing conditions for the product (full water immersion, 100°C at 15 rpm for 30 minutes). This was repeated using the modified (optimised) rotational process where the rotation rate was increased in steps during the process.

Minor changes in retort processing conditions between runs (e.g. initial temperature, sterilisation temperature and retort come-up time) were removed from the analysis by standardising the sterilisation times and conditions for each of the trials using the CCFRA CTemp model (Tucker *et al.*, 1996). Heating factors (f_h values) were calculated from the time and temperature data and used as measures of the product heating rates in the CTemp model. They represent a relative rate of heating, which was used to compare the effectiveness of each thermal treatment.

Results from the pilot-scale tests for the standard and for the optimised rotation conditions can be seen in Table 6.5. Heating factors were calculated together with end of heating pasteurisation values (P values) and the holding time to exceed P 10. Time-temperature data from the optimised conditions showed broken heating characteristics due to the changes in the heating rate. The reduction in process times achieved with the optimised process was between 12 and 24%. Industrial processes that do not contain such a lengthy come-up contribution should allow even greater savings in process time.

Table 6.5 Heating factors and P values calculated for jars of tomato sauce processed under optimised conditions in a pilot-scale retort. Butyric anaerobes were the target microorganism

Process replicate	f_{h1} (mins)	f_{h2} (mins)	End of heating P value for a 25 minute hold (minutes)	Time to P 10 (minutes)
Standard 1	9.5		52.6	16.5
Standard 2	11.7		59.6	16.0
Standard 3	13.5		43.7	18.5
Optimised 1	11.2	6.1	76.2	14.0
Optimised 2	12.9	6.5	76.6	13.5
Optimised 3	14.8	5.1	75.7	14.0

6.4.2 Full-scale tests

The tomato-based sauce was prepared as a standard batch and filled into the jars to a maximum fill weight of 330 g. Standard product jars of the same diameter as in the pilot-scale tests were used of dimensions 132 mm height, 64 mm neck width and 76 mm body width (headspace 8 mm). Jars were processed in a four-basket Stock Rotomat using a full water immersion process at a process temperature of 100°C, with a rotation rate of 15 rpm. The product was held at process temperature for 30 minutes according to the production scheduled process. This gave a standard set of data against which the processes with modified rpm were compared.

A second trial was carried out using the same preparation procedures; however, the rotation rate was manually changed in response to measured sauce temperatures. Each change took about one-minute to implement and required the rotation to be first stopped before being increased to the target. A third trial was carried out in which the retort was programmed to automatically increase rotation rate based upon the times documented from the second trial. This did not require the rotation rate to be stopped at each change.

The results for each of the three trials were analysed using the CTemp programme to assure consistency between conditions and calculate the heating factors (f_h) for each replicate. Results were compared by using a consistent retort profile and initial sauce temperature to establish if there had been any reduction in the required process times. Observation of differences in product quality were carried out immediately after the trials and after 8 weeks' storage at ambient temperature. The key aim was to avoid separation. In addition, a taste panel was set up to blind-taste the three products.

Table 6.6 illustrates the improvements in heating rates that were achieved by optimising the rotation rates. Heating factors, end of heating P values and time to P 10 were again used as measure of the effectiveness of each process. High f_h values were estimated at the start of each process because of the period to fill the

Table 6.6 Product heating rates and P values for the tomato-based sauce processed under standard and both manually and automatically optimised conditions. Butyric anaerobes were the target microorganism

Process	f_{h1} (mins)	f_{h2} (mins)	End of heating P value (mins)	Time to P 10 (min:sec)
Standard 1	5.6	6.7	33.4	13:30
Standard 2	6.7	7.4	24.5	15:00
Standard 3	5.8	8.6	24.9	14:30
Manual adjustment 1	4.2	6.8	47.6	11:00
Manual adjustment 2	4.0	6.1	54.5	10:00
Manual adjustment 3	4.2	6.1	49.4	11:00
Automatic adjustment 1	2.6	8.2	39.2	11:30
Automatic adjustment 2	2.7	7.5	45.1	11:30
Automatic adjustment 3	3.7	8.0	42.9	11:30

Table 6.7 Taste panel results for the tomato-based sauce immediately after processing and after eight weeks storage

Process	After processing	After storage
Standard	As expected	As expected
Manually timed	Similar colour to standard, possibly a little paler No visible separation	Similar colour to standard No visible separation No perceptible difference in flavour attributes
Automatically timed	Similar colour to standard No visible separation	Similar colour to standard No visible separation No perceptible difference in flavour attributes

retort when the retort rpm was zero. These were not given in the table. Both types of optimised processes delivered a faster rate of heat penetration than with the standard processes.

Sensory taste panel assessments immediately after processing and after eight weeks' storage did not identify any quality benefits but more significantly did not show that the sauce had separated. Concerns were expressed that the faster rotation speeds may have caused the oils to separate, giving the product an undesirable appearance. Table 6.7 summarises the results of the informal sensory analysis after processing and after storage.

6.5 Optimising rotation speeds in thermal processing

The research work reported above highlighted the importance of matching rotation conditions to the flow characteristics of the food products. The ideal scenario is for the headspace bubble to move through the centre of the food so that it imparts the maximum amount of disturbance to the food. This is rarely achieved and would be too difficult to control on a routine basis where variations in fill temperature and rheological behaviour occur. However, the principle of increasing rotation speed as viscosity decreases is important.

The flow visualisation trials with gelled starches never attained the perfect mixing pattern of the bubble moving through the centre of the jar as the jar turned end-over-end. As CFD simulations become more advanced it should be possible to predict the combination of product, package and process condition that can result in the bubble moving through the centre. This would avoid the need for lengthy experimental tests to find the optimum rotation speeds. At the time of writing this chapter, CFD simulations with a free surface are at the leading edge, and so the use of CFD for optimising retort technology is in its infancy.

From the reported data it was seen that as the rotation speeds increased beyond 15 rpm, the rate of incorporation of dye into the solution improved. Since most industrial processes operate at or around 15 rpm, this indicated that most were not operating close to optimal values. However, the centrifugal action that causes the bubble to move through the solution can also work in the opposite direction. If rotation speeds are too fast then the bubble will not have time to move from top to bottom of the jar and the mixing will be less than optimal. This can occur at the start of a thermal process when the food is at a low temperature and in consequence its viscosity is at its highest. There is a balance to be reached in which significant improvements can be achieved but the operating window for rpm needs to be wide enough to allow for production variations in the product and processing factors that effect heat transfer.

One further point of caution is that the high rotation speeds used in the reported work are best suited to water immersion retorts. These use the buoyancy of packs in water to soften the potential damaging effects caused by rotating crates at high speed. Other processing systems, such as steam, steam/air, sprayed water and raining water, do not possess these buoyancy effects and so could give rise to greater pack damage and wear on shaft bearings. In addition, it is necessary to prove that the increased centrifugal forces do not act to reduce the heat transfer efficiency at the centre of a retort crate by entrapping air. This has implications for all types of heating media but it is the raining water and steam/air mixtures that may be the most affected. Little published work is available to substantiate this theory, but the common view in the industry is that 15 rpm is towards the limit for all systems other than water immersion.

One further factor that has an impact on the efficiency of heat transfer is the shape of the container. Observations of headspace bubble movement showed that in most cases the bubble deformed and moved along the jar surfaces rather than going through the jar centre. A jar shaped like a globe would be a poor shape for heat transfer because the bubble would not have much effect on the food at its centre, never mind the fact that the surface area to volume ratio of a globe is low. Jars shaped like hour-glasses or of narrow diameter are likely to provide effective heat transfer from bubble-induced mixing.

6.6 Future trends

Rotary thermal processing in batch retorts is becoming increasingly popular with manufacturers of ambient and chilled foods. The main reasons for this are that it offers:

- flexibility in processing;
- lower cost of purchasing new equipment compared with continuous retorts;
- it is suitable for processing most pack types; and
- meals can be processed in their packs, which reduces the chances of contamination.

Many of the heat processed foods on retailers' shelves have short product lifetimes because of the drive for product innovation and the need to present new products to consumers. This lends itself to small production runs that can be produced more economically in batch retorts than with continuous systems that must be set up for long production runs under the same conditions. Thus, the trend is for companies to replace worn out and corroded continuous retorts such as reel and spiral and hydrostatic cooker-coolers with rotary batch retorts. This has been recognised by the companies that manufacture continuous cooker-coolers, and they all now offer a batch retort alternative within their portfolio of products.

Ready-meal production of high quality foods is a growing sector that is moving towards in-pack pasteurisation to help increase chilled shelf-life beyond a few days. Since most of these products contain meat or vegetable particulates in a sauce, there is scope to increase production efficiency by applying rotary thermal processes. Both particulates and headspace bubble will help mix the product in the pack and so reduce the thermal damage. Thus, hot-filling of cooked ready meals into plastic trays, pots or pouches has been replaced in some factories by retorting, which requires less preparation because the product is cooked within the pack during the retort process. By processing a sealed pack, this ensures that both the product and the package receive greater guarantees of achieving a minimum level of pasteurisation. Shelf life will be extended upwards of a few days under chilled storage.

One innovation that can be used to speed up production time is to formulate a sauce-based product so that the sauce is water-like at the start of processing. This is achieved by adding the thickeners as powders and allowing them to thicken during the in-pack process. At a specific temperature in the pack, the thickeners will gelatinise and convert the product viscosity from low to high. This is known as broken heating and is a method of getting the heat into the product faster. When combined with rotary processing it gives rise to a very efficient way of manufacturing formulated food products. This has been used by canned soup manufacturers, in which the residence time in the sterilising chamber of a reel and spiral cooker-cooler can be as low as 10–15 minutes at 125–130°C. This processing advantage can be realised with batch retorts for any pack type using end-over-end rotation.

6.7 Sources of further information and advice

HOLDSWORTH, S.D. (1997). *Thermal Processing of Packaged Foods*. London, Blackie Academic and Professional.

DEPARTMENT OF HEALTH (2000). *Guidelines for the Production of Heat Preserved Foods*. London: The Stationery Office. Second edition.

RICHARDSON, P.S. (2001). *Thermal Technologies in Food Processing*. Cambridge, Woodhead Publishing.

CCFRA, (1992). *Pasteurisation Heat Treatments*. CCFRA Technical Manual No. 27.

6.8 References

ANANTHESWARAN, R.C. and RAO, M.A. (1985). Heat transfer to model Newtonian liquid foods in cans during end over end rotation. *Journal of Food Engineering* 4 (1), 1–19.

BERRY, M.R., SAVAGE, R.A. and PFLUG I.J. (1979). Heating characteristics of cream style corn processed in a steritort: effects of headspace, reel speed and consistency. *Journal of Food Science* 44, 831–835.

BRITT, I.J., PAULSON, A.T., STARK, R. and TUNG, M.A. (1994). Heat transfer in liquid filled containers during end over end rotation. In: *Developments in Food Engineering*. Ed. Yano, T., Matsuno, R. and Nakamura, K. 716–718.

CLIFCORN, L.E., PETERSON, G.T., BOYD, J.M. and O'NEIL, J.H. (1950). A new principle for agitating in processing of canned foods. *Food Technology* 4, 450–460.

CONLEY, W., KAAP, L. and SCHUHMANN, L. (1951). The application of 'end-over-end' agitation to the heating and cooling of canned food products. *Food Technology* 5, 457–460.

EMOND, S.P. and TUCKER, G.S. (2000). *Optimising In-container Mixing Conditions to Enhance Heat Transfer to Thermally Processed Foods*. CCFRA R&D Report No. 116. Chipping Campden, Glos.

HOLDSWORTH, S.D. (1997). *Thermal Processing of Packaged Foods*. Blackie Academic and Professional, London, UK.

JAVIER, R.A., NAVEH, D., PERLSTEIN, E. and KOPELMAN, I.J. (1985) Convective heating rate parameters of model solutions in an agitating retort simulator. *Lebensmittel-Wissenschaft und Technologie* 18, 311–315.

KNAP, R.P. and DURANCE, T.D. (1998). Thermal processing of suspended food particles in cans with end-over-end agitation. *Food Research International* 31 (9), 635–643.

NAVEH, D. and KOPELMAN, I.J. (1980). Effect of some processing parameters on the heating transfer coefficients in a rotating autoclave. *Journal of Food Processing and Preservation* 4, 67–77.

PARCHOMCHUK, P. (1977). A simplified method for agitation processing of canned foods. *Journal of Food Science* 42, 265–268.

RAMASWAMY, H.S., ABBATEMARCO, C. and SABLANI, S.S. (1993). Heat transfer rates in a canned food model as influenced by processing in an end over end rotary steam/air retort. *Journal of Food Processing and Preservation* 17, 269–286.

RAO, M.A. and ANANTHESWARAN, R.C. (1988). Convective heat transfer to fluid foods in cans. *Advances in Food Research* 32, 38–84. Academic Press Inc. London, UK.

SABLANI, S.S. and RAMASWAMY, H.S. (1995). Fluid to particle heat transfer coefficients in cans during end over end processing. *Lebensmittel-Wissenschaft und Technologie* 28, 56–61.

STOFOROS, N.G., AND MERSON, R.L. (1991). Measurement of heat transfer coefficients in rotating liquid/particulate systems. *Biotechnology Progress* 7, 267–271.

STOFOROS, N.G. and MERSON, R.L. (1992). Physical property and rotational speed effects on heat transfer in axially rotating liquid/particulate canned foods. *Journal of Food Science* 57, 749–754.

TUCKER, G.S., NORONNA, J.F. and HEYDON, C.J. (1996). Experimental validation of mathematical procedures for the evaluation of thermal processes and process deviations during the sterilisation of canned foods. *Transactions of the Institute of Chemical Engineers* 74, Part C, 140–147.

7

Developments in packaging formats for retort processing

N. May, Campden and Chorleywood Food Research Association, UK

7.1 Introduction: requirements for low- and high-acid foods

Before going on to discuss the developments taking place for packaging of heat preserved foods it is worthwhile discussing the basic requirements for all such packaging. Firstly a pack to be processed in a retort system needs to be sufficiently robust to withstand the mechanical and thermal conditions created by the retort. Retorted food may be processed at external temperatures of up to 135°C, though the majority will be in the range of 115–125°C. For pasteurised foods lower external temperatures may be applied, typically in the region of 80–100°C. Mechanical stresses are more difficult to define, but both the ability to withstand pressure differentials and puncture resistance during mechanical handling should also be considered. The extent of pressure differential that a pack might be exposed to will depend upon the type of retort system being used. For example, a glass jar containing a product filled at 60°C, initial vacuum of 500 mmHg and 5% headspace processed in a retort at 121°C (1 Bar) may develop an internal pressure of 1.5 Bar. Less inherently strong pack formats, e.g. glass, pouches or semi-rigid pots, must be processed in an overpressure retort, i.e. one in which an external pressure above that of steam at the sterilising temperature can be applied. This external pressure is used to balance the internal pressure which develops in the container during sterilisation due to:

- expansion of the pack contents;
- dissolved gases being released into the headspace;
- expansion of headspace gases.

For plastic packs failure to apply an (adequate) overpressure will typically result in pack 'ballooning'. When severe this ballooning may result in the pack bursting. The effect of the pressure differential being exaggerated by the weakening/softening of the plastic films/laminates by the heat. Even if bursting does not occur ballooning can have a damaging effect upon the finished pack appearance as distortion of the films/laminates in the deflated balloon gives a shrivelled appearance, or barrier layers can be fractured.

For heat preserved foods safety depends heavily upon the reliable creation of a hermetic seal and ensuring that this can be maintained at all handling stages. In addition it should not be forgotten that, as with any packaging in contact with food, migration of packaging component chemicals into the preserved food must be within legal limits.

Once the basic technological requirements have been achieved then economic factors become important in selection of packaging. The economic driving forces for selection of a packing format are never static and therefore this text cannot evaluate them. It can be stated that the following need to be considered:

1. The relative changes in the costs of the packaging raw materials, e.g. steel, aluminium, tin, glass or the various plastics.
2. The cost of converting these raw materials into a completed pack.
3. The cost for the food processor in filling/sealing and sterilising the pack.

Technological developments at stages 2 and 3 can significantly affect the relative balance of cost of raw packing materials, which are also subject to their own volatility. For high volume-low cost food items these balances will be significant, though perhaps if the market for heat preserved food should shift towards premium 'high added value' items these factors will diminish in importance.

Possibly the most fundamental choice in selection of a heat preserved food container is what sealing technology is considered appropriate, the choice being between heat seals and double seams. Heat sealing is currently relatively slow and more likely to be affected by food debris than a double seaming. A second major decision in packaging selection will depend upon the type of retort systems available: saturated steam retorts can be used for containers with a degree of inherent strength, e.g. metal cans which withstand the internal pressure built up during sterilisation. Other weaker packing formats, e.g. pouches, plastic pots or glass jars, must be protected by the use of an overpressure retort system (Fig. 7.1). At ambient pressure there is no relative pressure in the containers, but when heated in steam the contents expand to make the pressure inside the container greater than the pressure outside. The metal can withstand the pressure but the pouch swells and may burst. In the overpressure retort a greater external pressure is applied which prevents either the can or pouch from expanding. These technological requirements hide one of the biggest current dilemmas for the canning industry because the reel and spiral retorts used world-wide for production of high volumes of heat sterilised foods are currently operated as non-overpressure systems, so are restricted to strong metal packing formats.

Within the current framework of pack formats available for heat preserved foods the following are the most significant driving forces for change:

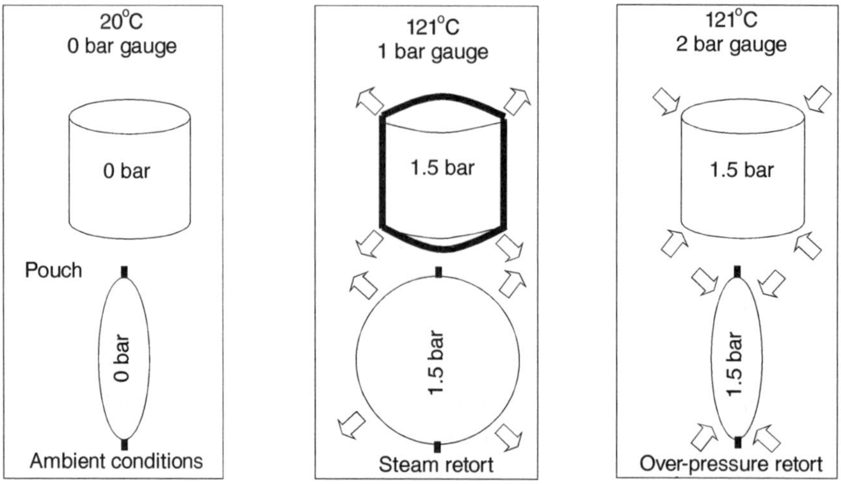

Fig. 7.1 When is overpressure required?

1. Reduction of packaging/production costs. For widely accepted packing formats like metal cans there is a constant driving force to reduce packaging costs, which largely means design modifications to reduce the amount of material in each container. For example, with cans we have seen a gradual thinning (down gauging) in the can walls, or 'necked in' designs for ends.
2. Improvement in openability of containers without compromising security. There is a balance to be achieved in the packaging of retort processed foods between protecting the food from recontamination and ease of opening of the finished container. The requirement for ease of opening is particularly applicable to retail containers, though cannot be ignored for catering containers. It is believed that thousands of consumers per year are injured by 'easy open' cans in the UK.
3. Creation of a new/improved image for heat preserved foods. In particular the image of the metal can is seen as tarnished and old fashioned, consequently it is believed that sales can be boosted by novel packaging. Interestingly this may be done at an increased final product cost. This need has led to the use of relatively complex printing technology and container shaping to make packs more attractive.
4. Microwaveability is an important consideration in the selection of packing formats for convenience food products.
5. Compliance with changing legislation. For example, the lacquer systems used on metal cans have been evolving quickly in the last decade to comply with legislation brought in to minimise migration risks and make production more environmentally friendly.

7.2 Developments in packaging formats: the metal can

It is estimated that the number of metal food cans consumed world-wide per year is of the order of 160 billion. Although the metal can might be seen as a standard item on the consumer's shelf there are constant developments in the technology and shifting tides in the economics of the various means of fabrication. There is always potential for ebb and flow between the relative cost effectiveness of tin plate, tin free steel and aluminium container constructions. The following are developments in metal cans for heat preserved foods in recent years:

1. Shifts towards two-piece cans, with the main body being manufactured by a drawing process from a single piece of metal, eliminating the need for a second end to be added. There is an ongoing process of updating the forming technologies making a wider range of formats possible and based upon a wider range of sheet raw material.
2. The virtual elimination of soldered side seams by welded seams, except in a few specialist markets.
3. 'Necked-in' designs for cans which are beneficial in terms of cost savings on metal and can also be technologically advantageous in preventing seam-to-seam impacts during container conveying (Fig. 7.2).
4. Reduced seam dimensions. The traditional can seam has a record for safety which depends upon the physical barrier provided by the five overlapping layers of metal in the seam between the can body and end. There have been moves to reduce the amount of metal in the seam by reducing seam lengths, such as the Euroseam or more extreme Kramer seam (Anon, 1994).
5. The replacement of solvent based coatings with water based coatings to reduce volatile emissions during the can making process.

One of the more negative points for retail cans has been the perceived difficulty of opening. This has led to the development of the 'easy open' can end in which one end of the can is fitted with a special end in which a (near) circular score has been made in the metal reducing it to a fraction of the thickness of the original sheet, e.g. 40% for aluminium and 30% for tinplate (Montanari *et al.*, 1995). In addition, a tab is added to a rivet formed in the end. The ultimate objective being that the consumer pulls up the tab which breaks the adjacent score and allows the entire circular centre of the scored region to be removed. The created aperture may be virtually the full diameter of the can or a smaller proportion (a factor which can have a significant impact upon the ease with which the consumer removes the product from the can).

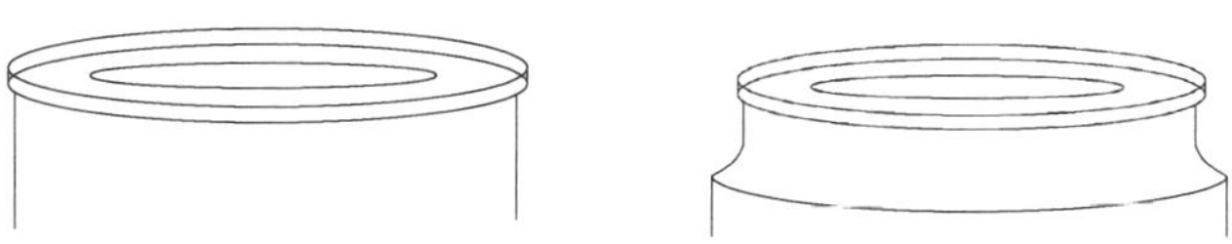

Fig. 7.2 Standard and necked-in can.

The disadvantage of the scored easy open feature is that opening creates very sharp edges that are a danger to the consumer. Secondly, depending upon the exact construction of the easy open feature the opening force required may still make it difficult for some consumer groups. It is not unreasonable to say that the greater the force required for opening an end of this type the greater the risk to the consumer from the sharp edges.

The easy open end has additional drawbacks in terms of food safety risks from recontamination of the pack as the scoring of the end is a deliberate weakening that could affect integrity. It is well known that canners switching from standard to easy open ends tend to see an increase in the level of mould complaints from consumers. Mould growth will only take place in aerobic conditions, which indicates that these complaints result from partly opened cans. The fact that the predominant microbial growth is mould rather than bacterial suggests that contamination is taking place during dry conditions in the distribution chain rather than in wet conditions of manufacturing where a relatively high risk of bacterial contamination would occur. These patterns indicate that a proportion of easy open ends can be expected to open during distribution.

An advantage of scored metal easy open can ends is that they can be retorted in traditional saturated steam retort systems, though under some circumstances more protection can be offered by an overpressure retort. Easy open ends can be handled through the high volume retort types like 'reel and spiral' retorts where the cans are rolled on their side through the retort. However, a reel and spiral retort must be configured in such a manner that the easy open tabs do not impact upon either the pushing mechanisms or each other, thus converting the retort into a large-scale can opener.

A possible step forward from the easy open end is the concept of applying a double seamed end onto a can that has at its centre a heat sealed foil end. For the canner this type of end has the advantage of a relatively high speed filling-seaming line, whilst the consumer has the benefit of a low force easy open end. In the best implementations the idea is further enhanced by elimination of all sharp edges from the open end, making the format more suitable for a wider range of consumers, e.g. children and the elderly. This concept has been embodied in the Impress Easy Peel® system (Fig. 7.3).

The current limitation for this type of closure is that the foil seal cannot be processed in the non-overpressure high volume reel and spiral retort systems used for scored easy open cans; so, for the moment, we are not likely to see beans in tomato sauce in this type of can. A second possible drawback is that the foil end is tightly stretched over the metal ring of the can end, meaning that the foil will be susceptible to impact damage. A secondary protective cap might be considered during distribution.

An interesting hybrid is a can with the opening features of a jar produced by Rojek of Brazil (Fig. 7.3). The 'Abre-Facil' (easy open) end is held in place not by a double seam but by a vacuum like a jar lid. The can-jar is opened by removal of a section of rubber compound from the centre of the lid which reveals a hole and releases the vacuum, and hence frees the end.

Fig. 7.3 Left to right: Rojek 'Abre-Facil' can; Winalot Tetra Recart retortable carton; an unseamed Easy Peel® can end (partially opened); and a fish can fitted with an Impress Easy Peel® end.

Marketing forces are pushing can makers toward more unusual can shapes. The steel makers Corus have been experimenting with cans which are square in vertical cross section. They claim benefits in terms of retail display (presenting a flat surface to the buying customer, and reduced shelf space) and potentially in terms of heat transfer as the cold points in the can are likely to be closer to the pack surface than for a cylindrical can of equivalent volume. Other more complex blown can shapes are also on the market but it is worth remembering the difficulty of manufacturing and distributing, without impact damage, such complex shapes.

There is a niche market for self-heating cans which can be used in environments where reheating apparatus is not available, e.g. outdoor sports. These cans are constructed to include a unit composed of two divided chambers which the consumer breaks the seal between to allow an exothermic reaction to take place and heat the can contents.

As a final point on metal cans it is worthwhile being aware of changes in metal packing formats for non-heat preserved food and beverage containers, as these developments, if made suitably rugged, may cross over into the heat preserved sector. Examples are recloseable outer lids over other forms of inner closure. These may be either twist on/off or push on types. However, when considering this approach it should be noted that long-term consumer resealing and storage of food in metal cans could create metal contamination or microbiological risks, particularly if food is not refrigerated. Non-heat preserved food cans have included plastic windows which could be a novel feature. Perhaps more realistically there is potential for a wider use of embossing and designer metal finishes, where the product can support the cost. A technology that may cross over from beverage cans to food cans is the use of internal pressurisation with inert gases to add strength to the pack and allow reduced metal contents.

7.2.1 Other metal containers

A specialist market has developed for heat sealed aluminium or steel trays as single serve pet food containers. These are formed trays with a heat sealed lid.

7.3 Developments in packaging formats: the plastic can, pot and bottle

Several attempts have been made to commercialise a plastic or partly plastic can. For example, Letpak, developed by Akerlund and Rausing, featured a polypropylene/aluminium laminate for the cylindrical can body with moulded ends, the three components being joined by a welding process. Another was the Omni can from Nacanco which was based upon an ethylene vinyl alcohol (EVOH) barrier laminate and double seamed ends. While packs of this type appear regularly, they have yet to gain a strong foothold in the marketplace.

Some packaging manufacturers preferred to adhere to double seaming technology for closure of their plastic cans, and seamed a metal end onto a plastic body, the double seaming technology having the potential benefits over heat seals of relatively high speed through the seamer and reduced impact on the seal from food debris. Examples are the Lunchbowl produced for Heinz (Fig. 7.4) and the pasteurisable Stepcan produced by the then Metalbox, the latter having the benefit that the can body was a clear laminate so the container could be used to show off premium quality products.

Blow moulding has been used for many years for the production of bottles for in-container sterilised milks, and increasingly this type of technology is attracting interest from other producers of liquid foods such as soups and sauces. The containers are typically laminated PP (moisture barrier) and EVOH (oxygen barrier) and are commonly heat sealed with a foil laminate cap. Such containers offer potential for convenient fridge door storage and if the cap can be removed cleanly, microwaveability.

A significant amount of R&D money in the heat preserved food industry has gone into retortable plastic bowls/trays with heat sealed plastic or foil lids, particularly where these can be marketed as convenient ready meals which are microwaveable. Such bowls/trays are generally laminates, e.g. polypropylene/EVOH or PVDC/polypropropylene, to give a shelf life of approximately 12 months. These laminates are coextruded with tie layers to bind them together. The exact number of layers in the laminate varies. The finished packs are commonly distributed in an outer cardboard display sleeve which also provides protection.

Fig. 7.4 Back row left to right: Heinz Lunchbowl – a plastic laminate bowl with metal double seamed easy open closure; Marks & Spencer gusset style pouch with resealable pouring nozzle; form fill seal vacuum pack used for heat sterilised beetroot. Front row left to right: John West pillow style pouch; and Friskies Gourmet foil tray.

For pasteurised products PET bottles are available that can withstand, without deformation, heating to temperatures approaching 92ºC for hot filling or a pasteurisation environment of up to 75ºC.

7.4 Retort pouches: construction, sealing, processing and packaging

The retortable pouch made of laminated plastic films has existed as a concept since the 1950s, largely being championed by the military for field rations. After significant development effort in the 1960–70s the technology fell out of favour for economical and technological reasons, an exception to this being Japan where the market for retorted pouches has developed differently from other markets. The 1990s has seen the retortable pouch return to favour as a retail pack. This change has been largely marketing led with the pouch revitalising sales though modified image, especially where high quality printing is used. A second area of increased interest has been the use of pouches as an alternative to large volume cans for catering applications, the pouch having the benefit of minimised food contamination risk from the opening operation (can openers have the potential to introduce metal shards) and also relatively low volume of packaging waste disposal, although it should be pointed out that a laminated pouch is not as recyclable as a can.

7.4.1 Pouch construction

Interestingly, except for vacuum packs, heat processed pouch manufacture has been dominated by prepared pouches, i.e. those supplied with one end open, that is sealed by the canner. As a general point, in construction it is best to avoid open flaps or any features that can entrap water/dirt and therefore encourage post-process recontamination. For example, having a heat seal right up to the edge of the pouch is advisable.

The laminates within the films used for pouch construction depend on the application. The first generation of pouches used for heat sterilisation applications were 3 ply. However, the current generation of pouches are more robust 4 ply constructions with an additional layer of nylon. For details on constructions suitable for an application it is best to approach a pouch supplier. Typical elements within a long life pouch laminate are shown in Fig. 7.5. Between the main film layers there will be adhesives to hold the laminate together. When purchasing any film/pouch it should be ensured that the materials comply with EU regulations (90/128/EEC) for migration of packaging components into foods.

For long shelf life products laminates may have an aluminium layer as the ultimate gas and light barrier. However, where a clear pouch is desired for product display or microwaveability, the aluminium layer can be replaced by a silicate layer. This silicate layer is effectively a very fine flexible glass layer that acts as a gas/moisture barrier (though not a light barrier). The oldest format of pouch is the pillow style (see John West pouch in Fig. 7.4), which have been used for military/backpackers rations. For retail applications this type of pouch has often been distributed and sold in a cardboard box which gives additional protection. Increasingly, gusset style pouches are coming to the fore, this being a construction which includes an end on which the pouch can be stood (see Marks and Spencer pouch in Fig. 7.4). This style pouch has display advantages over the pillow style as the pouch surface can be used to advertise the product in a free standing retail display. This feature requires the use of robust laminates that allow pouches to be retailed without an outer carton.

There is a niche market for vacuum packed heat sterilised vegetables, the most significant being beetroot (Fig. 7.4), but potatoes and corn on the cob are also sold in this style of pack. In these cases the prepared vegetables are filled into a bottom web (film) in which a cup is formed just prior to filling. The top web is fed in over the product and heat sealed inside a vacuum chamber. A

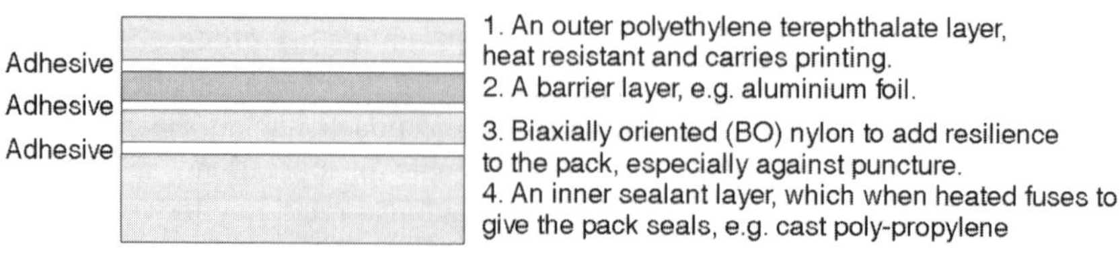

Fig. 7.5 Laminate construction for a long shelf life retortable pouch.

common misconception is that some juice is added at filling, but this is the juice that comes out of the vegetables during sterilisation. For non-heat sterilised products closures which allow a higher degree of user convenience are being introduced, such as re-sealable or easy pour features (see the Marks and Spencer pouch in Fig. 7.4).

7.4.2 Pouch sealing

Pouch seals are formed by the application of heat to the outer pouch surfaces which causes fusion of the inner sealing layer. This heat can be applied by hot jaw or impulse sealing methods.

One possible explanation of the wide use of pre-formed pouches is that pouch seals are much easier to form in the absence of product. Without product, seals can be formed without the contamination resulting from the product itself or condensation resulting from hot filling. In addition, once filled, pouches become distorted making presentation of a flat surface to the sealer more difficult. The presence of product in the seal area is the greatest obstacle to seal quality, particularly particles. Where product is present on the inside of the seal this can cause a seal failure; where product has contaminated the sealer jaw this will cause largely cosmetic embossing defects. Manual wiping of pouch sealing surfaces is sometimes used, but this is unlikely to be as effective as clean filling, especially as contamination tends to be wiped to the seal corners. Mis-formed, e.g. wrinkled seals, are to be avoided as they may result in an increased risk of post-process recontamination (as indicated by bio-testing) and reduced burst test strength. However, a wrinkle does not automatically mean that there is a channel for post-process recontamination as it may fill with molten sealant.

The effectiveness of the sealing operation can be assessed by the following methods:

1. *Burst testing* – probably the primary indicator of sealing faults, especially where product contamination is likely (Fig. 7.6).
2. *Seal thickness measurements* – there is some correlation between seal thickness and the effectiveness of the seal. This is a good measure of consistency of the sealer where product contamination is not the issue. A typical percentage of the seal thickness compared with the thickness of the original films is 90–95% (though not if the sealing jaws raise a pattern).
3. *Seal strength tests* – possibly the most useful information is that obtained at elevated temperatures corresponding to those that pouches will be exposed to during retorting.

7.4.3 Heat processing

Because of their lack of inherent strength, pouches, unlike cans, must be sterilised in an overpressure retort system. For relatively low temperature pasteurisation processes, e.g. <100°C, processes may be carried out without an

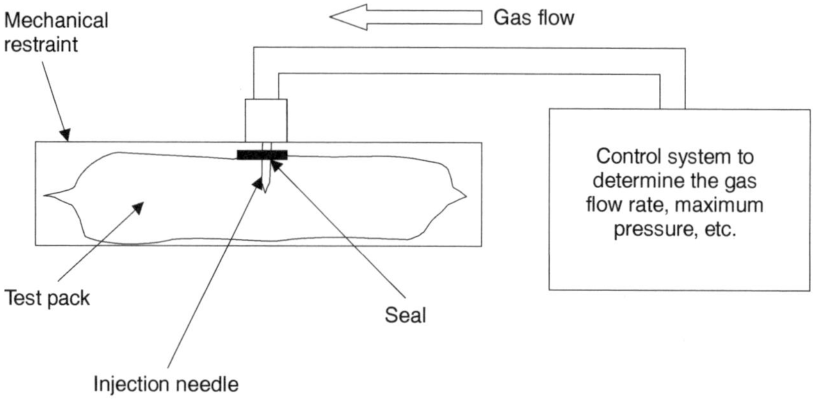

Fig. 7.6 Burst testing.

overpressure; however, this is highly dependent upon the residual gas content in the pack. The degree of evacuation of packs at filling impacts upon the product heating rate, because air pockets reduce the heat transfer rates through adjacent surfaces. The sequence of filling components also has some effect upon the level of entrapped air, and so during testing fill sequence should match that planned for production.

7.4.4 Post-process handling

Pouches, like other container formats, will be susceptible to handling damage. For pouches there is a greater risk of pin holing as result of snagging on machinery. Hence all machinery used to handle pouches should be free from sharp edges. Another area requiring attention is that pouches exposed to excessive flexing may undergo weakening.

The mechanics for post-process recontamination in pouches are very different from the well-known risks from cans. The pathway through heat seals is normally longer than through a double seam and the wetting properties of the packaging materials are different, which may also have some effect on bacterial passage. In addition, pouches have no internal vacuum, so some of the driving force sucking bacteria into the pack is absent. Balanced against the absence of vacuum is the fact that as pouches are moved during processing or distribution negative and positive pressures may be created by a bellows like effect.

The absence, for pouches, of historical evidence of post-process spoilage/food poisoning is not sufficient reason for not taking a precautionary approach and applying the same controls expected for canned goods, e.g. cooling water sanitation and avoiding wet handling. This precautionary approach appears particularly valid when the relatively high level of pinhole defects in pouches compared with cans is considered.

7.4.5 Outer packaging

Pouches may be distributed in individual outer cartons or in multiple presentation boxes, to present the pouches. Ideally pouches should be distributed in a horizontal orientation as this minimises the effect of the product upon pack distortion. Concerns have been raised about hydraulic shock taking place and causing damage to pouches, particularly with watery products. The laminates within pouches are particularly prone to damage due to flexing movements, so they should be avoided wherever possible in manufacturing/distribution operations. The stacking weights that will be withstood by boxes of pouches will be less than those for cans, so this should be considered when planning any storage facilities.

7.4.6 Tetra Recart

Tetrapak have moved into the area of retorted packaging because of the huge size of the market. Their retortable carton (Recart) is an aluminium laminate based board pack (Fig. 7.3), rooted in their experience with aseptic packs. The Recart project has been the biggest ever undertaken by Tetra; it has been underway for approximately ten years with the first field tests in 2001 and a wider launch in 2002 (CCFRA, 2003). They can demonstrate the following benefits over cylindrical cans and jars:

- The flattened packs (blanks) can be distributed with 35,000 packs/pallet, reducing the number of vehicles required for transportation compared with cans/jars by a ratio of 15:1.
- In shelf display 60% more product can be held in the same shelf area compared with cylindrical containers (this is also a benefit for home storage).
- The pack surfaces are easily readable in display.
- The pack is easy and safe to open.

The basics of the process are like those of canning as a non-sterile product is filled and the sealed pack is retorted. The retort systems used are fully automated steam overpressure systems operating at 120–130°C for up to 2 hours. The heat process is currently applied without agitation but this should not be a huge technical leap.

The use of steam sterilisation meant that a new stable paper board had to be developed. The polymer content of the board had to be modified from that used for aseptic application as polyethylene melts at the retort temperatures, but the replacements had to be both sealable and openable. The pack format is a rectangular carton with a laser perforated easy open feature. Initial development has been on a 400 ml pack (product fill approximately 375 g) but larger pack formats are planned. Flexographic printing is used, giving potentially very high print qualities.

The standard production line will comprise a forming machine, fillers (several fairly conventional fillers may be used), a rack handling system which handles product into and out of the retorts, and a secondary packaging machine.

The target production speeds are 24,000 packs/hour. Headspace is controlled at filling and packs must always have a headspace. Cold filling temperatures are used in the region of 40°C, as hot filling is not required for vacuum. Sealing is achieved with the same high frequency sealers as other Tetra packs though machines are adapted to allow for the different polymer technology. The pack is easy open with high precision laser technology used for cutting through the board whilst the foil layer is not perforated.

Typical shelf lives are shorter than for metal cans but the aim is for 24 months. The packs are not as gas impervious as a can, though they are so low as to require the development of new evaluation methods. Field tests have been conducted with EU countries, with no major technical problems, good market performance and retailer interest.

7.5 Methods of improving glass packaging

Glass could claim to be the oldest packing format used for heat preserved foods, as Nicolas Appert, the father of the technology, did his experiments in glass containers. It is difficult to see radical developments in the format of glass containers but we might look for improvements in openability and should expect weight reduction. The difficulties in opening glass jars of heat preserved foods apparently lie not only in the vacuum used to secure the lid in place but also in the adhesion between the glass neck finish and the compound in the cap, so changes might be made in this chemistry.

Changes in glass pack formats are not likely to be dramatic and will probably be limited to improved presentation such as decorative sleeves, highly decorated cap designs or attractive jar shapes. Although radical developments might not be expected for glass packaging, glass should not be discounted as a possible format for heat preserved foods because it is a format understood by consumers and permits good visual presentation for high quality products.

7.6 Future trends

It is anticipated that all the trends toward increased consumer convenience will continue, e.g. microwaveability, easier opening, etc. Likewise there will always be driving forces for reducing packaging raw material and conversion costs. Perhaps the most significant changes facing the industry in future will come from legislative requirements, specifically the ease of recycling packaging waste and the impact of the manufacturing processes upon the environment. In addition, as shown by the can making industry in minimising BADGE and NOGE contamination from lacquers, packing manufacturers must be prepared to move quickly when potential safety issues are identified.

Maybe one of the greatest challenges facing the whole heat preserved foods industry is consumer perception that such foods are overprocessed and lack fresh qualities. In the future packaging formats may play a part in overcoming this problem where packaging allows for minimal heat processing, e.g. low resistance to heat transfer and small pack dimensions.

7.7 Sources of further information and advice

The first source for information on packaging should always be the supplier, but for more general background research organisations such as the following might be contacted:

Campden & Chorleywood Food Research Association
Chipping Campden
Gloucestershire
England
GL55 6LD
Tel: +44 (0) 1386 842000
Fax: +44 (0) 1386 842100
E-mail: info@campden.co.uk

National Food Processors Association
1350 I Street, NW Suite 300
Washington, DC 20005
USA
Tel: + 202/639–5900
Fax: + 202/639-5932
E-mail: nfpa@nfpa-food.org

Pira International
Randalls Road
Leatherhead
Surrey, KT22 7RU
UK
Tel: +44 (0) 1372 802000
Fax: +44 (0) 1372 802238

7.8 References

ANON (1994) Microseam goes worldwide. *The Canmaker*, 7, November, 21.

CCFRA (2003) Tetra Recart – New angles on canned food. Presentation given by Mr Olle Bergman of Tetra Recart. Minutes of CCFRA Heat Preserved Foods Panel 28th January 2003.

MONTANARI A., MARMIROLI G., PEZZANI A, CASSARA A and LUPI R (1995) Easy open ends for food and beverage cans: Definition, manufacture, coating and related problems. *Industria Conserve*, 70, 410.

8

Developments in cook-chill and sous vide processing

S. Ghazala, Memorial University of Newfoundland, Canada

8.1 Introduction: sous vide, cook-chill and home-meal-replacement technologies

This chapter addresses the use of minimal thermal processing combined with the use of modified packaging such as sous vide (under vacuum) and cook-chill techniques for preservation of foods. Consumers demand convenient, innovative, fresh-like foods, including new 'minimally processed' products. Meanwhile, there have been remarkable advances in packaging technologies, especially as it is related to improving thermal processing and thus the quality and shelf-stability of highly perishable food products. However, these foods have received minimal processing or pre-cooking and therefore they present a challenge to ensure microbiological quality and safety. The food industry has recently been utilizing novel food processing technologies with the purpose of providing the desired quality demanded by consumers in the twenty-first century while ensuring that all food safety standards are met.

Sous vide is a French term meaning 'under vacuum'. It was developed in France in the 1970s, and it means processing foods under reduced air pressure (vacuum) packaging. This technique provides an environment that contains little or no oxygen. Sous vide is a special process of reduced oxygen packaging for partially cooked ingredients alone or combined with raw foods that require refrigeration or frozen storage until the package is thoroughly heated immediately before service. Generally, sous vide processing offers unique advantages and opportunities for the food industry but it also raises several microbiological concerns. Products processed using the sous vide technique can be produced safely if proper controls are in effect. The main thermal processing in the sous vide is a pasteurization step that reduces the microbial load but is not sufficient to make the food shelf-stable.

Cook-chill is a process that uses a plastic bag from which air is removed and immediately filled with hot cooked food and then the bag is closed with a plastic or metal crimp. The Nacka system, introduced in a Swedish hospital in the 1960s, was an early attempt at cook-chill. The procedure consisted of heating food to 80°C, filling it into plastic bags which were then sealed using a vacuum, placed in boiling water for 3 minutes to pasteurize the food and then cooled to around 4°C. Generally in the cook-chill system, foods are cooked conventionally and then portioned and cooled before chilled storage and distribution for re-heating in the satellite kitchen.

Home-meal-replacement (HMR): HMR represents meals as 'ready-to-eat, ready-to-heat or ready-to-serve hot or cold entrées prepared or packaged outside the home which are brought or delivered to the home', or 'HMR is the provision of a meal solution for time pressed consumers'. HMR has grown steadily since the early 1990s. The exact definition of HMR continues to change, but one thing that has not changed is consumer interest. It is expected that HMR will account for as much as 80% of the food industry by 2005 (http://www.foodfactory.sik.se/Proceedings/SKIPNES.pdf). The main reason for this growth is health conscious consumers who are demanding convenient (ready-to-heat-and-eat), home-style, fresh, innovative, low-fat, low sodium, low sugar and so on foods. New methods of processing and packaging that can extend the shelf-life and freshness of perishable foods have been developed over the past decade. Significant steps are being taken by the food industry towards understanding and managing risks that exist, or that are anticipated, and the development of indicators and methods for identifying health hazards and predicting food safety is a high priority.

8.2 The pasteurization process

The pasteurization process was named after Louis Pasteur who discovered that spoilage organisms could be inactivated in food by applying heat at temperatures below the boiling point of water. Therefore, pasteurization is a mild heat treatment, usually below 100°C, to destroy or reduce vegetative microorganisms and reduce enzymatic activities. It is usually combined with other preservation techniques such as refrigeration, MAP, low pH, low water activity, vacuum packaging, etc. Thermal pasteurization for sous vide and cook-chill systems should be designed so that, at a minimum, all vegetative pathogens are destroyed by this process.

Generally, pasteurization is usually performed on food products after the product is placed in the hermetically sealed finished product container. The most crucial factors in pasteurization are temperature and time. When heating times are chosen, the focus is usually only on the holding time at the stated process temperature for convectively heated food products. However, equally important is the time needed to heat and cool the product to and from the desired process temperature.

The main purpose of pasteurizing food is essentially to attain the following:

- reduction of the microbe charge (bacteria, yeasts, molds); and
- inactivation of enzymes.

A series of final stages are met upon achieving these two goals, the most important of which are:

1. product sanitation (killing germs which are directly or indirectly pathogenic for mankind);
2. increase of the product shelf-life (reduction of the microorganisms causing the product degradation, from edible to inedible); and
3. stability of the chemical and organoleptic characteristics (block off the enzymatic activities that cause alterations in the product color, taste and texture).

The pasteurization has therefore to take into account not only the necessary sanitation level or the shelf-life that has to be ensured to the product; it has also to consider the product nature (kind of filling, shape, weight of the single pieces, etc.). If this doesn't happen, two different mistakes can occur:

1. making the product undergo an extreme thermal treatment, so to sacrifice its original organoleptic characteristics more than is necessary;
2. exposing the product to a thermal treatment which is inadequate to give the demanded sanitation level or shelf-life.

This means that an optimum pasteurization process:

1. exposes the product to a high temperature for the shortest possible time to result in that target bacteria are quickly killed;
2. utilizes the shortest possible time to heat and cool the product in order to minimize any chemical changes that result in altering the original organoleptic characteristics (color, taste, smell, appearance, etc.).

8.2.1 Types of thermal pasteurization

1. heating liquid and semi-liquid foods in large scale (50–70 liters) kettles or cook-tanks (with some form of continuous, controllable agitation device for mixing) and then pumping them into heat-stable packs at high temperature (85–88°C);
2. the pre-packaged solid foods are pasteurized in a large-scale jacketed tank at specific time-temperatures and then chilled in a chill tank (to about 3–4°C).

8.3 Cook-chill systems: process stages

1. *Conventional (traditional) catering system.* This system can be adapted to various forms of application from the gourmet restaurant, hotel dining room

and coffee shop through to the fast food outlet and the snack bar. Briefly, raw or pre-prepared food (e.g. frozen or dried) is received and stored prior to service. The next step involves the cooking of the food, and finally the cooked food is served. Information relating to the system's control and its quality flows back as the food flows through the system, to help maintain processing conditions (e.g. time, temperature, quality, and so on).

2. *Simple catering system*. This is similar to a conventional catering system, except the addition of the time buffer (distribution) step. This step involves a chilling method of storing the cooked food in conditions designed to extend the product's shelf-life. In this system, the place of service is not assumed to be the same site at which the food was produced. This evidently improves the caterer in terms of planning and optimization of production; however, the caterer must exercise a far greater degree of control and monitoring of the cook-chill system than in conventional system.

8.3.1 Cook-chill process

A standard (traditional, i.e. cook/warm-hold/serve) cook-chill catering process involves the following steps (Fig. 8.1):

1. *Choice and preparation of raw food materials*. A reasonable quantity of raw materials may be purchased as pre-prepared goods since the emphasis is on the manufacture of a set of menu items in bulk and to a reproducible standard.
2. *Storage of the raw materials*. Raw materials may be stored in one of four ways: dry store, frozen storage, chilled storage (0–3°C) or refrigeration (3–7°C). Pre-preparation before cooking is common that may include cutting, boning, washing, trimming, chopping, beating, mixing and so on.
3. *Cooking*. The first step in cook-chill process is to cook the prepared food materials. Pasteurization employing heat treatment is the thermal process used in this case in which prepared foods are heated to kill only the vegetative (growing) forms of the bacteria which contaminate them. Temperatures of less than 100°C (i.e. 63–95°C) are generally used. The time that it is held at the pasteurization temperature is as important as the temperature that also determines the extent of bacterial death rate. Control of critical parameters of pasteurization time and temperature is a requirement for the optimal retention of quality throughout the chilled storage. Thus, there is a trend to utilize programmable and self-monitoring with time and temperature settings and recording, automatic cut-off and alarm devices. There are four main types of machinery used in food manufacturing of cook-chill industry: cook-tanks for pumpable foods; combined cooker/chillers; continuous conveyer cooker (where packs of pre-cooked food can be vacuum sealed on-line); and batch steaming/water heating and cook-chill tanks (in container pasteurization) units in which pre-packaged food products are cooked and chilled. Most of the pre-packaged containers are generally vacuum packed and then pasteurized.

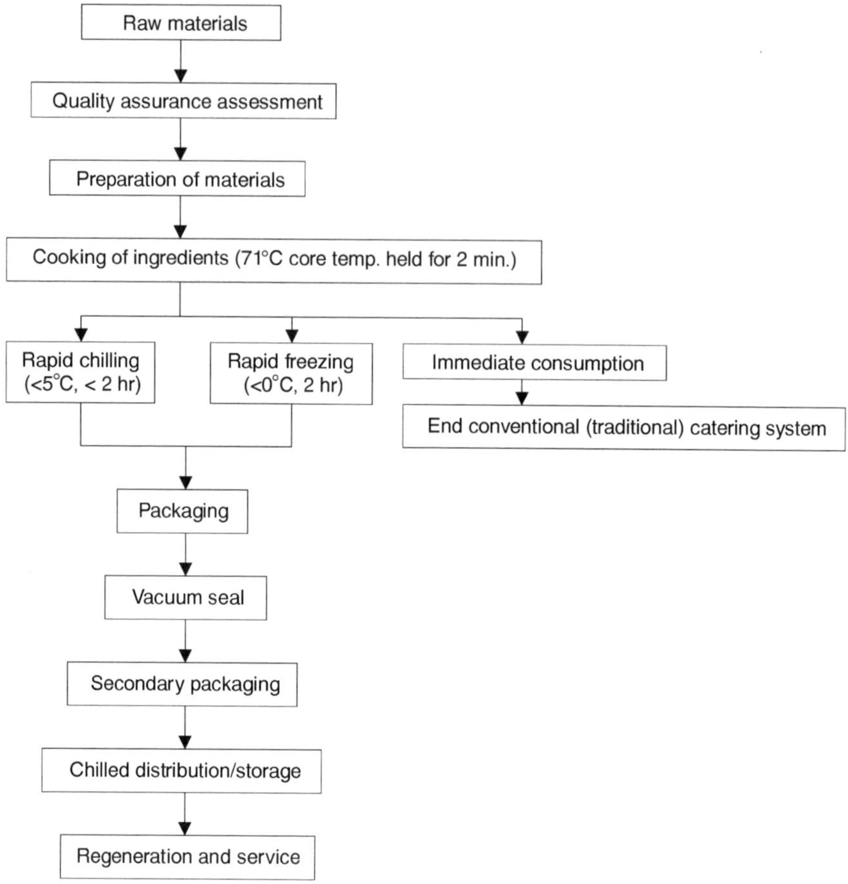

Fig. 8.1 The cook-chill systems.

4. *Portioning*. Bulk cooked items are split down into single or multiple portion sizes using automatic or semi-automatic machinery to ensure hygienic conditions and thus reduce the risk of food contamination and also to increase the degree of control on portion size/weight. Automatic filling and container-forming, special pumps, vacuum pressure pipeline (in-line chilling) and vacuum/modified atmosphere packaging are the types of equipment used at this stage.
5. *Rapid chilling*. Cooked and portioned food products are rapidly chilled below the effective minimum growth temperature needed for the food spoilage and food poisoning bacteria. It is paramount that the chilling system should be able to chill different types of food products from pasteurization temperature to below 3°C within a set time (usually 90 min). Three main methods are used for rapid chilling; these are air-blast chilling, cryogenic chilling and iced water chilling (Hutton *et al.*, 1991). Other

methods, less widely used, are cold water spraying within the cooking chamber and vacuum chilling (mostly suitable for unpackaged products).

6. *Chilled storage.* Chilled food products are stored at a constant low temperature (0–3°C) in order to inhibit or minimize microbial growth during chilled storage. Chilled storage includes cold rooms and chill cabinets that are designed to retain cold air at a temperature between 0 and 3°C, and allow the 'roll-in-roll-out' types to take food on trolleys either free standing or on wheels. The chilled storage must be fitted with continuous recorder for the temperatures profile within the room or cabinet through each 24 hr period; an alarm system if the actual storage temperature falls below or rises above the desired set levels or if power fails to the storage; and a digital logging real-time read-out of storage temperature for in-line monitoring.
7. *Distribution.* Cook-chilled food products may be immediately distributed to satellite kitchens or stored for long or short period in a central production unit before distribution prior to shipment to end kitchens. In each case, the temperature of chilled food must be maintained between 0 and 3°C during transport to the end kitchen (Harris, 1989). Insulated mobile cabinets for on-site distribution and insulated containers for off-site deliveries insulated refrigerated trailers or vehicles are required. It should be emphasized that proper control and monitoring systems are incorporated so that food temperature can be adequately followed during the distribution (Bryan *et al.*, 1978; Kalish, 1991).
8. *Regeneration (or re-heating).* The chilled food must be adequately re-heated before service, to kill properly any vegetative forms of bacteria that may have grown in or on the food, while at the same time maintaining the organoleptic quality of the food. Bulk regeneration oven, combined oven (low pressure steamer and dry heat), mobile regeneration trolleys and microwave oven are the most common equipment used for the regeneration of foods in cook-chill systems.

8.4 The sous vide system: process stages

Sous vide is a variant of the classic cook-chill catering system (http://www.kolva.dk/eng/sousvide00.htm). This system involves cooking foods sealed under vacuum in multi-laminate plastics. The sous vide system involves: preparation of raw ingredients and the food for cooking as in standard cook-chill system; pre-cooking or browning; vacuum packaging; pasteurization; rapid chilling and chilled storage; and re-heating and service (regeneration).

8.4.1 Sous vide process

Sous vide is the 'under-vacuum' special processing of foods completed by placing the food in evacuated 'vacuumed' containers. The key phase in this process is sous vide cooking, which involves the pre-packaging of foods in

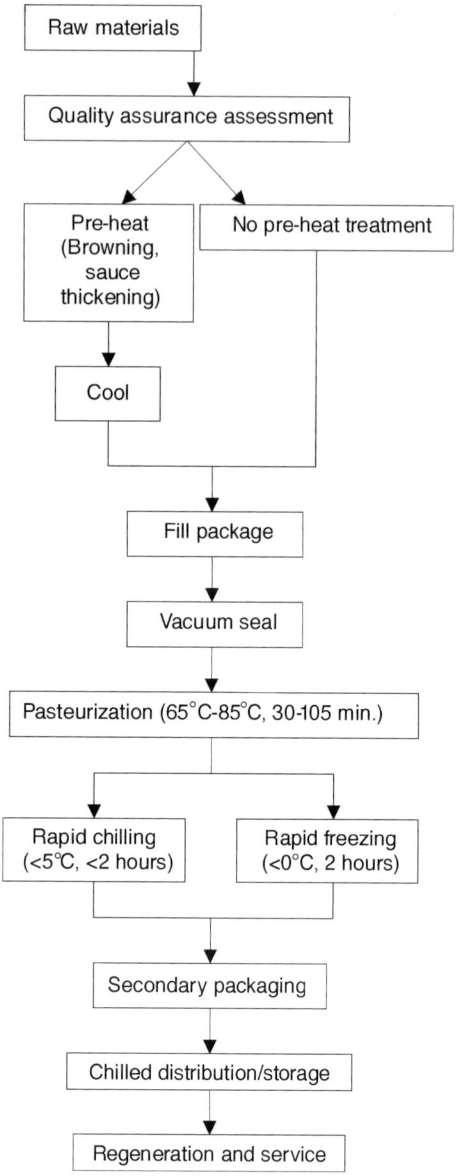

Fig. 8.2 The sous vide system.

plastic bags or pouches, forming a vacuum, and cooking the food within the bags at controlled pasteurization temperatures. The basic sous vide system involves the following stages (Fig. 8.2):

1. *Preparation*. The raw ingredients are prepared by such steps as washing, peeling, trimming, seasoning and so on, depending on their sources prior to

cooking. A traditional approach as in standard cook-chill methods can be followed at this stage.

2. *Pre-cooking or browning.* As the food is not directly heated in sous vide processing since it is heated inside a plastic container, it may be necessary for some products to be pre-browned by traditional cooking; some vegetables have to be blanched, and some strong aroma and flavors to be mimicked before packaging.
3. *Vacuum packaging.* The prepared food is placed into special heat resistant, air impermeable plastic bags or pouches. These bags and pouches are produced from multi-laminate plastic materials that give high gas impermeability and heat-resistant capabilities. After the food is packed into the container, a vacuum is produced using a specially designed vacuum chamber which is equipped with a vacuum pump to remove air from the container. The pack is then heat sealed immediately before any loss of vacuum occurs. Vacuum packaging is the most important stage in sous vide processing and thus the amount of residual air removed from the bag must be adequately achieved.
4. *Pasteurization.* The packaged food is pasteurized for a set time and at a set temperature appropriate to the individual characteristics of the food to be cooked. Water bath or stem combination oven can be used for this stage. The latter unit allows more accurate temperature control needed in sous vide processing, since pasteurization is one of the most important steps in this process to produce highly flavored, tender products.
5. *Service.* Sous vide cooked products may be served immediately or the chilled (0–3°C) pre-packaged, unpasteurized food (single or multi-portion) packs are taken from the chill store and immediately pasteurized prior to service.

Standard sous vide system can also be adapted for use as a cook-chill system with very small differences from the standard cook-chill process except that the raw materials may have to be pre-processed prior to pasteurization, and they are pasteurized in vacuumized bags and pouches. In this case steps 3–5 would be as follows:

6. *Rapid chilling.* The packaged pasteurized foods are chilled to between 1 and 3°C within 90 min. Equipment used for this stage is iced water bath (direct chilling) which is both cheap and efficient as compared to blast chilling (indirect chilling) which is usually used in standard cook-chill process.
7. *Chilled storage.* Chilled storage at 0–3°C for a period of time prior to service. The chilled storage must be fitted with continuous recorder for the temperatures; an alarm system if the actual storage temperature falls below or rises above the desired set levels or if power fails to the storage; and a digital logging real-time read-out of storage temperature for monitoring. The fact that the food in this case is stored under vacuum sealed containers helps to inhibit aerobic bacterial growth and also slows down the rate of oxidation and other chemical changes that may occur during this stage.

8. *Regeneration*. This involves re-heating and servicing of the food product. The food must be re-heated to at least 70ºC at its center based on standard cook-chill process. Different re-heating methods can be used for this purpose such as using heated water bath or a combination steam oven to heat the food in the bag that is still under vacuum, or using traditional top of the cooker or inside the oven to heat the food after the contents of the bag have been emptied in a heat-resistant dish. More common re-heating process is the use of a microwave oven where the bag is punctured and heated; however standard microwave ovens are not recommended as they may not re-heat food evenly unless the food can be re-heated within 30 min of being removed from the chill store. A maximum of 5 min between re-heating and service is allowable.

8.5 Advantages and disadvantages of cook-chill and sous vide systems

Table 8.1 lists the advantages and disadvantages of cook-chill and sous vide (Armstrong, 2003).

Generally, cook-chill systems are those for caterers who serve many meals per day (i.e. more than 600–800 meals/day); who have problems with temperature control due to long distance and time needed for transportation of meals; who wish to control the consistency and quality of their meals; who have very limited production space yet serve a large number of meals; where their needs expand rapidly and they have a great deal of skilled labor competition. The main advantages are:

- labor saving since it deals with one large batch; one large kitchen; mechanical mixing and packaging; and increased productivity;
- reduced food cost since there is reduced food loss, less food shrinkage, and economics of one order purchase, ingredient control, reduced food wasting;
- reduced management cost;
- reduced purchasing cost;
- reduced energy cost; and
- improved food quality.

However, the main disadvantages of cook-chill systems are:

- adaptability of poor recipes to this process;
- improper re-heating of food is a safety risk;
- inadequate personnel training; and
- poor inventory control of stored food.

The advantages of sous vide system generally are:

- yields consistent and high quality convenience meals with intense flavors and aromas, desirable taste, reduced shrinkage, prevented dehydration, increased tenderness and retained nutritional value;

Table 8.1 Sous vide vs. cook-chill systems: a comparison

	The cook-chill system	The sous vide system
Definition	Cook-chill technology is based on pasteurizing, rapidly cooling and holding food under chilled conditions for an extended period before reheating and consumption.	'Sous Vide is the term used to describe the process of vacuum packaging food before the application of low-temperature thermal processing under chill conditions.'
Advantages	Minimal processing Production separated from consumption Value/convenience Extended shelf life in the distribution chain Freshness Minimizes processing impact on sensory and nutritional qualities Anaerobic environment prevents the growth of aerobic spoilage organisms (Gram negative bacteria such as aerobic yeasts and molds) Reduced Oxygen Packaging retards the oxidative rancidity of fats and oils Allows for a foodservice director to have coveted control in all arenas-cost, convenience, labor and quality	Minimal processing Enhanced nutritional quality Extended shelf life in the distribution chain Less additives/preservatives needed More convenient, rapid service Centralized production Better portion control Reduced risk of post-process contamination Wider variety of produced goods, and therefore better presentation of food Minimizes processing impact on sensory and nutritional qualities (*ex*: decreases loss of water soluble vitamins through leaching and oxidation) Anaerobic environment prevents the growth of aerobic spoilage organisms (Gram negative bacteria such as aerobic yeasts and molds) Reduced Oxygen Packaging retards the oxidative rancidity of fats and oils

Table 8.1 (continued)

	The cook-chill system	The sous vide system
Disadvantages	Microbiological spoilage due to: temperature abuse, insufficient pasteurization, or insufficient vacuum (aerobic spoilage) Product shelf lives that are shorter than those demanded by the retailer, consumer and commercial caterer, due to exposure to oxygen during various manufacturing stages. Loss of sensory quality due to non-rapid chilling (>2hrs. to reach <5°C) Chilled storage is product dependant (i.e. vegetables may develop acidy/pungent flavors within 2 days Adequate refrigeration must be maintained at all times	Spoilage often not visible Microbiological spoilage due to temperature abuse, insufficient pasteurization, or insufficient vacuum (aerobic spoilage) Product safety is dependant on proper handling and processing Adequate refrigeration must be maintained at all times Extra equipment costs Extra cost to consumer
Food type	Meat Vegetables Poultry Potatoes Rice Pasta Fish Cheese, etc.	Meat Vegetables Poultry Potatoes Rice Pasta Fish Cheese, etc.
Main process	See Fig. 8.2	See Fig. 8.1

Cost	Considerable savings for the food service industry over conventional methods	12–26% less then that of a normal kitchen operation (in relation to total operating costs)
Shelf-life	14 days under refrigerated conditions (<3.0°C) Generally up to 8 weeks if frozen	42 days under refrigerated conditions (<3.0°C)
Safety	Special labeling to ensure adequate awareness of the necessity of refrigeration for consumers (5°C), and fully visible 'best before' dates	Special labeling to ensure adequate awareness of the necessity of refrigeration for consumers (5°C), and fully visible 'best before' dates The NACMCF*, chartered by the USDA** and the HHS*** has recommended a proposed requirement of a heat treatment sufficient to achieve a 4 decimal log reduction of *L. monocytogenes*.
Future	MAP cook-chill Widespread use of this technology, from small to large scale food production Potential for food costs to drop, as there will be less spoilage (from producers to consumers)	Use of Time-Temperature Indicators as indicators of temperature abuse in modified atmosphere packaging 'Active packaging' that indicates microbial spoilage to consumer Potential use of additional hurdles during production stages

*NACMCF: National Advisory Committee on Microbiological Criteria for Foods.
**USDA: US Department of Agriculture.
***HHS: Department of Health and Human Services.

- offers a varied and interesting menu;
- precise cost and portion control:
- reduced preparation costs and manpower savings;
- minimal wastage;
- quicker serving times;
- extends product shelf-life and allows for convenient post-preparation storage;
- no need for flavor enhancers;
- can be readily adapted for use as a cook-chill system; and
- ancillary items such as soups and vegetables may be used apart from main packaged meals.

However, sous vide has a number of disadvantages such as:

- has limited applications on its own since it cannot provide an entire menu range;
- the need to use very high quality raw materials, extra special equipment, and the need to package each item;
- labor-intensive since it involves extra steps compared to standard cook-chill process; and
- may pose a considerable threat to public safety if not properly used since the sous vide container provides the ideal environment for the growth of anaerobic bacteria capable of growth at chill temperatures.

8.6 Requirements for cook-chill and sous vide processes

8.6.1 Selection and storage of raw materials

Raw materials must be quality, good grade products. Materials that are below standard before using are not going to improve throughout processing. It is vital, therefore, that you check your supplies and, if necessary, check on your suppliers' handling and distribution methods. After purchasing quality ingredients, proper care must be taken to ensure that the value of the materials does not degrade during storage. Upon preparation of materials, it is extremely important to follow basic food safety principles, and to ensure that the appropriate temperature and humidity levels are met for each particular food.

8.6.2 Cooking

Types of equipment include:

- Microwaves
    1. Light wave: visible and non-visible.
    2. Light wave and microwave combination.
    3. Convection microwave: cooks using convectional heating, microwaves or both.
    4. Microwave.

- Ovens
  5. Moving air oven.
  6. Combination oven has three processes: convection/steam/roast or a combination. These ovens have humidity controls, roast and hold options for off-peak cooking, and come in a variety of different models, sizes and costs, depending on operation size.
- Steamers
  7. Boilerless steamer: saves energy and water.
  8. New generation steamers have no drain line, are lime free, have a longer element life and save on water consumption.
  9. Vacuum steamer: uses lower temperatures (60–100°C).
- Induction technology
  10. Magnetic wave technology uses magnetic coils to create a magnetic field to conduct heat.
- Steam griddles
  11. Steam griddles have no hot spots, which allows for consistent heating, therefore improving quality, safety and productivity.

8.6.3 Chilling methods for solid food (James and Bailey, 1990)

Batch air chillers

This method involves placing warm food items into a large refrigerated room, and is considered the most common method of chilling. The placement of ready meals in a cooler is complicated by the different thermal properties of the meal components. The risk of surface freezing limits the lowest air temperature of the batch chilling system.

Moving air

A widely used method that is cost effective, incurs little damage to equipment and is hygienic. Systems range from a sophisticated conveyerized blast chilling tunnel or spiral, to a simple fan that pushes air through a refrigerated coil that blows the cool air into an insulated room. One major disadvantage within this type of cooling system is surface dehydration of the food.

Continuous air chillers

In a simple continuous air chilling system the product is usually hanging from an overhead conveyer and moved through a refrigerated room. In a more technical system, the food may be passed through a chilling tunnel. This allows for even air distribution, as well as having the ability to vary the refrigeration capacity and air conditions within the tunnel.

Ice/ice water chilling

Involves packing the warm product in boxes and placing them between layers of crushed/cubed ice. As heat is drawn from the food, the temperature of the ice remains at 0°C while melting. This method involves considerable labor, and may not be as time efficient as other methods.

Vacuum cooling
The solid products, which have a large surface area and are readily able to dispel internal water, are placed in a vacuum chamber, which removes heat from the food through evaporative cooling. This form of cooling is very quick and cost effective to operate, but there are considerable capital costs due to expensive vacuum equipment.

Cryogenic cooling
This method involves the use of liquid nitrogen to freeze the product. This system tends to be inefficient, as the product is subjected to high thermal shock and only the latent heat of vaporization is used.

Immersion cooling/hydrocooling
A cost effective cooling method suitable for small products that involves immersing or spraying the product in cool water at or near 0°C. These systems can vary from straightforward stirred and unstirred tanks to a more complicated system whereby a product is conveyed on a belt through sprays or agitating tanks. Some small water weight gains are often observed during hydrocooling.

Plate cooling
Can involve continuous horizontal and rotating plate systems, or by a belt cooling system in which the product is cooled on its reverse side by liquid sprays or refrigerated air.

8.6.4 Chilling methods for liquid food

Batch cooling of liquids
A jacketed stainless steel vessel is usually used for this type of chilling. These vessels range in capacity from 100 to 10,000 L, and may contain an agitator to improve heat transfer. The coolant may circulate through the jacket of the vessel or through a coil placed in the liquid foodstuff, or both. A typical coolant using this chilling method is ambient or iced water. A common technique that is utilized to decrease cooling times of liquid foodstuff is to induce evaporative cooling by applying a vacuum to a closed vessel.

Continuous cooling of liquids
The continuous cooling of liquids can involve multi-plates and tubes, aeration and double pipe coolers. The most widespread piece of equipment is the multi-plate cooler, which has the best efficiency, most surface area for heat exchange, easy to clean, and require less materials than the others.

8.6.5 Chilling methods for solid/liquid mixtures

Batch cooling

A combination of liquid and solid food materials, it results in a combination of chilling methods. In larger-scale operations vacuum and water-cooling systems similar to those used for liquids are often employed. However, it is not as efficient time-wise as liquid cooling, as the solid particles are heated by conduction. Temperature stratification is a problem in non-agitating vessels, while stirring the solid and liquid mixtures may break down the delicate solid particles within the mixture.

Continuous cooling

The simple method in the continuous cooling of solid/liquid mixtures is to pass a liquid coolant over the pipe in which the mixture was previously heated. However, due to the solid particles being conduction controlled, and the increased viscosity of the mixture during cooling, there is a limit to the rate of cooling that can be reached. Some new methods are being developed and show promise in the continuous cooling of these mixtures. Some developments include the addition of liquid carbon dioxide into the piping, vacuum cooling, or separation of the two phases and cooling them individually and recombining afterwards.

8.6.6 Chilled storage

Bulk storage rooms

Used for most unwrapped meat and poultry and all wrapped foods. The bulk storage room is a large refrigerated air circulated room in which the unwrapped product is the smallest amount in keeping a constant temperature, as to minimize weight loss and appearance changes associated with air movement.

Controlled atmosphere rooms

Developed for storage of fruits, but interest has expanded to utilization of this type of storage for other food products. The controlled atmosphere is a result of gas tight seals, which maintain the desired atmosphere, usually lower in oxygen and higher in carbon dioxide and nitrogen than the air. The controlled environment works to prevent oxidation.

Jacketed cold stores

The refrigerated jacket is produced by inserting pipe coils in the floor, or using a double skin construction. This type of storage allows for little air movement, meaning that little heat can escape from the product. Thus, the product must be at its desired temperature prior to storing.

8.6.7 Distribution

Overland transport

These systems can range from a small un-insulated van that supplies food locally, to large refrigerated containers that are for road or rail movement.

Currently, most large refrigeration containers use either mechanical, eutectic plates or liquid nitrogen to keep cool.

Mechanical units
A popular mechanical cooler is a ‘plug’ unit that plugs into an opening in the wall of the vehicle.

Liquid nitrogen
This type of transport system involves an insulated liquid nitrogen storage tank connected to a spray bar, which runs along the ceiling of the vehicle. The thermostatically controlled valve initiates the spraying. Once the proper air temperature is met the valve turns off the nitrogen flow. This is a repetitive cycle during transport.

Eutectic plates
Developed for local distribution, the eutectic plate cooling system consists of a coil in which a primary refrigerant is passed, which is mounted on thin tank filled with eutectic solution. The plates are charged before transport, and air circulation is used to provide the optimum cooling capacity. Eutectic plates are considered to be low maintenance, simplistic and quiet; there is however a risk invoked as they can suffer from poor temperature control.

Air transport
The following recommendations have been established for the transport of chilled foods in the air:

1. Insulated containers should always be used to reduce heat gain.
2. Products should always be pre-cooled before loading onto the plane.
3. Dry ice should not be used on products that deteriorate after any surface freezing.
4. Containers should be full.
5. A thermo-graph should be with each load.

Sea transport
Sea transport involves insulated containers as well as refrigeration units that operate electrically, either through a generator or external power source on board the ship.

8.7 Microbial safety and barrier technology for cook-chill and sous vide processing

Cook-chill and sous vide processes relay on minimal heat treatment that is usually at 70 to 95ºC and then storage at a chilled temperature (≤3ºC) for safety and preservation. Therefore, these foods are not sterile and intended to have an extended shelf-life, often up to 42 days.

Use of vacuum packaging in sous vide foods can markedly increase safety concerns. Unless potentially hazardous foods are protected inherently, simply placing them in vacuum without regard to microbial growth will increase the risk of foodborne illnesses. Therefore, at least one barrier or multiple hurdles resulting in a barrier need to be incorporated into the production process for cook-chill and sous vide products using vacuum packaging. The incorporation of several sub-inhibitory barriers, none of which could individually inhibit microbial growth but which in combination provide a full barrier to growth, is necessary to ensure food safety (Association of Food and Drug Officials, 1990; Nolan *et al.*, 1992).

An anaerobic environment, usually created by vacuum packaging, provides the potential for growth of several important pathogens (Knabel *et al.,* 1990). Some of these are psychrotrophic and grow slowly at chilled temperatures. Additionally, the inhibition of the spoilage bacteria is significant because without these competing organisms, color and flavor changes signaling that the product is no longer fit for consumption will not occur. Non-spore bacteria are mainly eliminated by pasteurization. However, pathogens may survive in the final product if pasteurization is inadequate, poor quality raw materials or poor handling practices are used, or post-processing contamination occurs. Even if vacuum packaged foods receive adequate thermal processing, a potential for post-processing contamination by pathogens can occur (Doyle, 1991).

Since sous vide products are subjected to mild temperature abuse, i.e., 5–12°C, at any stage during storage or distribution, foodborne pathogens, including *Bacillus cereus*, Salmonella *spp.*, *Staphylococcus aurous*, and *Vibrio parahaemolyticus* can grow slowly. Marginal refrigeration that does not facilitate growth may still allow Salmonella *spp.*, Campylobacter *spp.*, and Brucella *spp.* to survive for long periods of time.

Temperature abuse is common throughout distribution and retail markets. Strict adherence to temperature control and shelf-life must be observed and documented by the establishment using vacuum packaging. Information on temperature control should also be provided to the consumer (Rhodehamel, 1992). Currently these controls are not extensively used. Additionally, some commercial equipment is incapable of maintaining foods below 7.2°C because of refrigeration capacity, insufficient refrigerating medium, or poor maintenance. Most warehouses and transport vehicles in distribution chains maintain temperatures in the 0–3.3°C range. It must be assumed, however, for purposes of assessing risk, that occasionally temperatures of 10°C or higher may occur for extended periods (Scott, 1989). At retail, further temperature abuse must also be assumed. For instance, retail display cases can be as high as 13.3°C for short periods and some refrigerated foods are provided no refrigeration for short periods of time. These realities point to the need for establishments to implement controls, such as buyer specifications, over refrigerated distribution systems so that better temperature control can be ensured.

Clostridium botulinum is the contributing agent of botulism, a severe food poisoning characterized by double vision, paralysis, and occasionally death. The organism is an anaerobic spore-forming bacteria that produces a potent neurotoxin. The spores are ubiquitous in nature, relatively heat-resistant, and can survive most minimal heat treatments that destroy vegetative cells. Certain strains of *C. botulinum* (type E and non-proteolytic types B and F), which have been primarily associated with fish, are psychrotrophic and can grow and produce toxin at temperatures as low as 3.3°C (Eklund *et al.*, 1967). Other strains of *C. botulinum* (type A and proteolytic types B and F) can grow and produce toxin at temperatures slightly above 10°C. If present, *C. botulinum* could potentially grow and render toxigenic a food packaged and held under vacuum because most other competing organisms are inhibited by vacuum packaging. Therefore, the food could be toxic yet appear organoleptically acceptable (Conner *et al.*, 1989). This is particularly true of psychrotrophic strains of *C. botulinum* that do not produce tell-tale proteolytic enzymes. Because botulism is potentially deadly, foods held in anaerobic conditions merit regulatory concern and awareness (Ghazala and Trenholm, 1998; Nolan *et al.*, 1992).

The potential for botulism toxin to develop also exist after pasteurization of sous vide foods since pasteurization (which is a mild heat treatment) will not destroy the spores of *C. botulinum*. Pasteurization in combination with vacuum packaging (as in sous vide) may actually opt for *C. botulinum* by killing off its competitors. If the applied heat treatment does not produce commercial sterility, the food requires refrigeration to prevent spoilage and ensure product safety. For this reason, sous vide products are frequently flash frozen in liquid nitrogen and held in frozen storage until use.

There is a further microbial concern with sous vide products at retail. Processed products such as meats and cheeses which have undergone an adequate cooking step to kill *Listeria monocytogenes* can be contaminated when opened, sliced, and repackaged at retail. Thus, a simple packaging or repackaging operation can present an opportunity for recontamination with pathogens if strict sanitary safeguards are not in place. If extended shelf-life is sought, a temperature of 3.3°C or lower must be maintained at all times to prevent outgrowth of *C. botulinum* and the subsequent production of toxin. *Listeria monocytogenes* can grow at even lower temperatures; consequently, appropriate use-by dates must be established and readily apparent to the consumer (Berang *et al.*, 1989). Since refrigeration alone does not guarantee safety from pathogenic microorganisms, additional growth barriers must be provided. Growth barriers are provided by hurdles such as low pH, a_w, or short shelf-life, and constant monitoring of the temperature. Any one hurdle, or a combination of several, may be used with refrigeration to control pathogenic outgrowth (Gorris and Peck, 1998).

Heat processes for cook-chill or sous vide productions should be designed so that, at a minimum, all vegetative pathogens are destroyed by a pasteurization process. Special labeling of these products is necessary to ensure adequate warning to consumers that these foods must be refrigerated at 5°C and consumed by the date required by the Code for that particular product (Moberg, 1989).

The National Advisory Committee on Microbiological Criteria for Foods

(NACMCF) chartered by the US Department of Agriculture (USDA) and the USA Department of Health and Human Services (HHS) recommended guidelines for evaluating the ability of thermal processes to inactivate *L. monocytogenes* in extended shelf-life refrigerated foods (NACMCF, 1991a). Specifically, it recommended a proposed requirement for demonstrating that sous vide process provides a heat treatment sufficient to achieve a 4 decimal log reduction (4D) of *L. monocytogenes* (Brown, 1991). Other scientific reports recommend more extensive thermal processing. Thermal processes for sous vide practiced in Europe are designed to achieve a 12–13 log reduction (12–13D) of the target organism *Streptococcus faecalis*. It is reasoned that thermal inactivation of this organism would ensure destruction of all other vegetative pathogens (FDA, 1997; NY DAM, 1993).

8.8 Good manufacturing practices and HACCP planning for safe cook-chill and sous vide processing

1. *Employee training*. Employees must have documented proof that demonstrates familiarity with vacuum packaging guidelines and the potential hazards associated with these foods.
2. *Refrigeration requirements*. Cook-chill and sous vide foods have only one barrier (i.e., refrigeration) to *C. botulinum*, must be refrigerated to 5°C or below and marked with a use-by date within either the manufacturer's labeled use-by date or 14 days after preparation at retail, whichever comes first. Foods that are intended for refrigerated storage beyond 14 days must be maintained at or below 3°C.
3. *Labeling – refrigeration statements*. Sous vide foods which rely on refrigeration as a barrier to microbial growth must bear the statement 'Important – Must be kept refrigerated at 5°C'. The statement must appear on the principal display panel in bold type on a contrasting background.
4. *Labeling – 'Use-by date'*. Each container of food in vacuum packaging must bear a 'use-by' date. The date assigned by a repacker cannot extend beyond the manufacturer's recommended 'pull date' for the food. The 'use-by' date must be listed on the principal display panel in bold type on a contrasting background. Any label must contain a combination of a 'sell-by' date and use-by instructions which makes it clear that the product must be consumed within 14 days of retail packaging or repackaging, as an acceptable alternative to a 14 day 'use-by' date.

8.8.1 Hazard analysis and critical control point (HACCP) operation

All cook-chill and sous vide food establishments packaging food in a reduced oxygen atmosphere must develop a HACCP plan and maintain the plan at the processing site for review by the regulatory authority (Daniels, 1991; NACMCF, 1991b). The plan must include:

1. a complete description of the processing, packaging, and storage procedures designated as critical control points (CCPs), with attendant critical limits, corrective action plans, monitoring and verification schemes, and records required;
2. a list of equipment and food-contact packaging supplies used;
3. a description of the lot identification system;
4. a description of the employee training program;
5. a listing and proportion of food-grade gases used; and
6. a standard operating procedure for method and frequency of cleaning and sanitizing food-contact surfaces in the designated processing area.

8.9 Conclusions

There is enormous potential to improve the microbiological safety of minimally thermally processed foods, such as cook-chill and sous vide products, using combinations of traditional and novel preservation approaches. Cook-chill and sous vide systems provide great prospects as well as challenges to food scientists, processors, distributors, retailers, consumers and lawmakers. Nonetheless, demands for minimally processed, ready-to-eat, high quality and convenient foods will continue to expand. The application of hurdle technology in conjunction with minimum heat treatment is necessary to ensure sufficient shelf-life, stability at low prices while keeping sensory quality at a maximum. Hurdles such as adjusting pH and water activity, and adding biological and chemical antimicrobial agents, can be incorporated into the products before packaging, using modified atmosphere or vacuum techniques, and then maintain the products during the shelf-life at refrigerated temperatures.

The US Food and Drug Administration (FDA) examined the sous vide system to produce ready-to-eat, high quality and convenient products by the food industry. They concluded that sous vide used alone was inadequate to destroy many bacterial species and thus would not be accepted as a method of food preservation. However, using traditional preservation techniques, such as refrigeration and incorporation of hurdle effects with sous vide system, is considered adequate and practical food preservation techniques. They established guidelines for evaluating the ability of thermal pasteurization process to inactivate *L. monocytogenes* in cook-chill and sous vide refrigerated products. These processes should provide a heat treatment sufficient to achieve a 4 decimal log reduction (4D) of *L. monocytogenes.* Other scientific reports recommend more extensive thermal processing. Thermal processes for sous vide practiced in Europe are designed to achieve a 12–13 log reduction (12–13D) of the target organism *Streptococcus faecalis*.

The food safety challenge that associates minimal thermal processing and modern packaging, designed to retain quality and microbiological safety, demands continuous research and development by scientists as well as regulators who are mainly responsible for setting rational microbiological specifications. It is greatly important that food processors become familiar with the different techniques and tools that allow them to successfully process, package and distribute their minimally processed foods. An integrated design of minimal heat treatment and hurdle technology has been suggested by several scientists and processors, and therefore a number of manufactures have already applied its principles, which comprise hurdle technology, predictive microbiology, and HACCP guidelines. Although, currently the principles of combined preservative factors are considerably examined, its practical application and control is still progressing.

8.10 References

ARMSTRONG, G. A. (2003). Minimal thermal processing: cook-chill and *sous vide* technology. In: *Handbook of Vegetable Preservation and Processing*. Chapter 24, pp. 409–423. Marcel Dekker, Inc. New York, NY. September 2003.

ASSOCIATION OF FOOD AND DRUG OFFICIALS. (1990). Retail guidelines – refrigerated foods in reduced oxygen packages. *J. Assoc. Food Drug Offic.* 54(5): 80–84.

BERANG, M.E., BRACKETT, R.E. and BEUCHAT, L.R. (1989). Growth of *Listeria monocytogenes* on fresh vegetables stored under controlled atmosphere. *J. Food Prot.* 52: 702–705.

BROWN, W.L. (1991). Designing *Listeria monocytogenes* thermal inactivation studies for extended-shelf-life refrigerated foods. *Food Technol.* 45(4): 152–153.

BRYAN, F.L., SEABOLT, L.A. PETERSON, R.W. and ROBERTS, L.M. (1978). Time-temperature observations of food and equipment in airline catering operations. *J. Food Prot.* 41: 80–92.

CONNER, D.E., SCOTT, V.N. BERNARD, D.T. and KAUTTER, D.A. (1989). Potential *Clostridium botulinum* hazards associated with extended shelf-life refrigerated foods: a review. *J. Food Safety* 10: 131–153.

DANIELS, R.W. (1991). Applying HACCP to new-generation refrigerated foods at retail and beyond. *Food Technol.* 45(4): 122–124.

DOYLE, M.P. (1991). Evaluating the potential risk from extended-shelf-life refrigerated foods by *Clostridium botulinum* inoculation studies. *Food Technol.* 44(4): 154–156.

EKLUND, M.W., WIELER, D.I. and POLSKY, F. (1967). Growth and toxin production of nonproteolytic type B *Clostridium botulinum* at 3.3 to 5.6°C. *J. Bacteriol.* 93: 1461–1462.

FOOD PROCESSING (1997). *1997 Food Code*. US Department of Health and Human Services. Public Health Service Food and Drug Administration. Annex 6, pp. 457–476.

GHAZALA, S. and TRENHOLM, R. (1998). Hurdle and HACCP concepts in *sous vide* and cook-chill products. In: *Sous Vide and Cook-Chill Processing for the Food Industry*, ed. S. Ghazala, pp. 294–310. Aspen, Maryland.

GORRIS, L.G.M. and PECK, M. W. (1998). Microbiological safety considerations when using hurdle technology with refrigerated processed foods of extended durability. In: *Sous Vide and Cook-Chill Processing for the Food Industry*, ed. S. Ghazala., pp. 207–233. Aspen, Maryland.

JAMES, S.J. and BAILEY, C. (1990). Chilling systems for foods. In: *Chilled Foods: The State of The Art*, ed. T.R. Gormley, pp. 1–35. Elsevier Science Publishers, England.

HARRIS, R.D. (1989). Kraft builds safety into next generation refrigerated foods. *Food Proc.* 50(13): 111–112, 114.

HUTTON, M.T., DHEHAK, P.A. and HANLIN, J.H. (1991). Inhibition of botulinum toxin production by *Pedicoccus acidilacti* in temperature abused refrigerated foods. *J. Food Safety* 11: 255–267.

KALISH, F. (1991). Extending the HACCP concept to product distribution. *Food Technol.* 45(4): 119–120.

KNABEL, S.J., WALKER, H.W. HARTMAN, P.A. and MENDONCA A.F. (1990). Effects of growth temperature and strictly anaerobic recovery on the survival of *Listeria monocytogenes* during pasteurization. *Appl. Environ. Microbiol.* 56: 370–376.

MOBERG, L. (1989). Good manufacturing practices for refrigerated foods. *J. Food Prot.* 52: 363–367.

NACMCF. (1991a). National Advisory Committee on Microbiological Criteria for Foods. *Listeria monocytogenes. Int. J. Food Microbiol.* 14: 185–246.

NACMCF. (1991b). National Advisory Committee on Microbiological Criteria for Foods. I HACCP Principles, II Meat and Poultry, III Seafood. *Food Control* 2(4): 202–211.

NY DAM. (1993). New York Department of Agriculture and Markets. Proposed Reduced Oxygen Packaging Regulations. Division of Food Safety and Inspection, 1 Winners Circle, Albany, NY, 12235, 6 pp.

NOLAN, D.A., CHAMBERLIN, D.C. and TROLLER J.A. (1992). Minimal water activity of *Listeria monocytogenes* and *Listeria innocua*. *Int. J. Food Microbiol.* 16: 323–335.

RHODEHAMEL, E.J. (1992). FDA concerns with sous vide processing. *Food Technol.* 46(12): 73–76.

SCOTT, V.N. (1989). Interaction of factors to control microbial spoilage of refrigerated foods. *J. Food Prot.* 52: 431–435.

8.10.1 Websites references

http: //www.foodfactory.sik.se/Proceedings/SKIPNES.pdf. Improving efficiency of meal production in public food service. Food Factory of the Future. Accessed 26 August 2003.

http: //www.kolva.dk/eng/sousvide/sousvide00.htm. Traditional Kitchen vs. Sous Vide Kitchen. Kolva A/S. Published: 1995. Accessed 26 August 2003.

http: //www.cfsan.fda.gov/~dms/fc01-a6.html. Food Processing. (2001). *2001 Food Code*. U.S. Department of Health and Human Services. Public Health Service Food and Drug Administration. Annex 6 Last Updated: August 19, 2003. Date Accessed: August 27, 2003.

Part III

Developments in continuous heat processing

9

Developments in aseptic processing

K. P. Sandeep, J. Simunovic and K. R. Swartzel, North Carolina State University, USA

9.1 Introduction: key issues in aseptic processing

Aseptic processing (or UHT, as it is often referred to as) involves the processing of a food (by heat or other means) in a manner that leaves the food free of microorganisms of public health significance and packaging the product in a sterile environment in a package whose food-contact surface has been sterilized separately. Typically, the food product is rendered commercially sterile by subjecting it to high temperatures (130 to 150°C) for a few seconds. The combination of the heat process and packaging renders the product shelf stable. During an aseptic process, high quality raw product (usually pre-heated) is pumped using a positive displacement pump through an air removal system to the heating unit. After heating, the product then passes through the holding tube (where it receives the required time-temperature effect), cooling unit (where the product is cooled to prevent loss of nutrients and flavor components), and the packaging unit.

Some of the advantages of an aseptic process are better product quality (in terms of nutritional value and flavor retention), easy automation, possibility of using cheaper packaging material, use of thermally sensitive microwave-friendly packaging for consumers, energy savings, and unlimited package size. Some of the disadvantages of aseptic processing include the need for better trained personnel, higher capital costs, slower filler speeds, possible detrimental reactions that could take place at the high temperatures attained during aseptic processing, and the more stringent requirements from regulatory agencies. The chemistry of aseptically processed foods is significantly different from those that take place during pasteurization. A detailed description of the chemical changes and kinetics of aseptically processed food products has been presented by Nielsen *et al.* (1993) and van Eijk (1993).

Other than the processing of the product, the two important facets of aseptic processing are sterilization of the equipment and process validation. Sterilization of the equipment (processing and packaging) is achieved by heat, chemicals, radiation, or a combination of them. Process validation involves the assurance of commercial sterility of the product. This entails proper documentation of the temperature of the product at the end of the holding tube and determination of the fastest moving fluid element or particle in the system. The overall goal of the validation process is to ensure that the process adheres to the previously determined requirements (scheduled process) set forth by a process authority.

A thorough understanding of aseptic processing of particulate foods involves the understanding of the fluid flow (to determine the RTD of fluid elements and particles), kinetics (to determine the extent of microbial destruction, enzymatic inactivation, and nutrient retention), heat transfer (convection at the surface of particles and conduction within particles), chemistry (browning, oxidation, and flavor changes), and microbiological (kinetics of destruction of the target organism) aspects of the process. The technical details of these aspects have been presented by Sandeep and Puri (2001). Many of these aspects can be analyzed by mathematical modeling of the entire process. It not only facilitates understanding of the process, but also aids in quantifying the effects of different parameters on process lethality and product quality. It can also be used to minimize the number of experimental runs (which are expensive) required for process validation.

9.2 Components of an aseptic processing system

The major components of an aseptic processing unit are the pump, deaerator, heating unit, holding tube, cooling unit, back pressure device, an aseptic surge tank, and a packaging unit. Each of these components serves a specific need.

The pump is the component that facilitates the flow of the product throughout the system. If there is only one pump in a system, it serves as the metering pump. If there is more than one pump in the system, one serves as the stuffing pump and the other as the metering pump. The metering pump determines the flow rate of the product. For operating pressures up to 68.94 kPa, rotary pumps are used. For higher pressures, a positive displacement pump (plunger or piston type) is used. If the pump has a speed control mechanism, care is taken (usually a lock on the speed control device) to ensure that the flow rate of the product does not exceed the set-point value, since a higher flow rate translates to a lower heat treatment and hold time. Depending on the product and process, a homogenizer may be used.

The deaerator serves the purpose of removing air from the product. This not only saves energy in the heat exchanger and cooling unit, but also ensures a constant specific volume of the product in the holding tube (thereby preventing a reduction in holding time due to product expansion), a constant fill rate during packaging, and a longer shelf life of the product under ambient storage conditions. The deaerator is generally located before the pre-heating unit in situations where loss of volatiles (at elevated temperatures) is of concern and after pre-heating in other instances since it is easier to deaerate the product at higher temperatures due to its expansion.

The heat exchanger used to heat the product may be a direct (steam injection or steam infusion) or indirect contact (plate, SSHE, tubular, shell and tube, ohmic, or dielectric) heat exchanger. Steam (non-regenerative heating medium) or hot water (regenerative heating medium) is the heating medium usually used. In a direct contact heating unit, heat is transferred rapidly from steam to the product and most of the required time-temperature effect for the product to be rendered commercially sterile is achieved in the holding tube. In an indirect contact heating unit, a significant amount of time-temperature effect for the product to be rendered commercially sterile may be achieved in the heating and cooling units. However, the time-temperature effect in the holding tube is only considered from a public health safety standpoint. The type of heating and cooling equipment used depends on factors such as the type of product (acid or low-acid, viscous or non-viscous, fluid or particulate, heat sensitive or heat stable), potential for fouling of the product on the heat exchanger surface, ease of cleaning, and the cost of the heat exchanger.

The holding tube is where the product receives the required time-temperature effect. Its design is thus critical from a public health safety standpoint. The holding tube is inclined upwards such that there is at least a ¼″ vertical rise in the tube per foot length of the tube (required by law). This ensures that the flow in the tube is full (no air pockets) and that the product drains back to the pump when the system is shut-off, thereby minimizing the potential for microbial contamination problems.

The cooling unit cools the product by evaporative cooling using a vacuum chamber (in the case when a direct contact heater is used to heat the product) or an indirect contact cooler. Indirect contact heat exchangers used for cooling are similar to those used for heating, with the difference being that the medium used for transfer of heat is chilled water or glycol instead of steam or hot water.

The back pressure device (piston-type air actuated valve, diaphragm valve or pressurized tank) provides the means of maintaining a high enough pressure in the system to prevent boiling (flashing) of the product at the high temperatures encountered in aseptic processing. The product attains very high temperatures (130 to 150°C) in the heat exchanger and holding tube and continues to be at these high temperatures during the initial stage of cooling. Thus, the back pressure device is placed after the cooling unit to prevent product flashing in the cooler.

The various types of packaging systems (can systems, bottle systems, sachet and pouch systems, cups systems, carton systems, and bulk packaging systems) in the food industry have been described in detail by Reuter, 1989. Sterilization of the food contact surface of the packaging material (usually by chemical means such as the use of hydrogen peroxide) and maintenance of an aseptic environment at the filling station (usually achieved by positive pressure of sterile air) are two of the most important factors that define aseptic packaging.

An aseptic surge tank serves as a buffer reservoir for the product between processing and packaging. After sterilization of the surge tank, it is kept pressurized with sterile air or an inert gas such as nitrogen. The pressure of the gas provides the means of transferring the product from the surge tank to the packaging unit without using a pump.

9.3 Equipment sterilization and process validation

9.3.1 Sterilization of equipment

Sterilization of the processing, packaging, and the air flow system prior to processing is of utmost importance. Sterilization of the air system is done by high efficiency particulate arresting (HEPA) filtering or incinerated air. For equipment, it is accomplished by steam, hydrogen peroxide, radiation, or combinations of them. For filling lines, sterilization is done with steam or water at high pressure. The recommended heating effect for sterilization (using hot water) of the processing equipment for low-acid foods is the equivalent of 250ºF for 30 minutes. The corresponding combination for acid or acidified products is 220ºF for 30 minutes. This often involves acidification of the water (to below a pH of 3.5 for acid products) used for sterilization. Sterilization of an aseptic surge tank is usually done by saturated steam and not hot water due to the large volume associated with the surge tank.

Once the product is processed, the system has to undergo a clean in place (CIP) operation. The CIP cycle for low-acid foods involves the use of hot water, alkali, hot water, acid, and hot water sequentially. The CIP cycle for high-acid foods is hot water, alkali, and hot water sequentially. The details of product fouling on the heat exchanger surface, cleaning, and disinfection of the entire equipment have been presented by Lewis and Heppell (2000).

Sterilization of the food contact surface of packaging material is the next point of consideration. For non-sterile acidic products ($pH < 4.5$), a 4D process is required. For sterile, neutral, low acid products ($pH > 4.5$), a 6D process is required. However, if there is a possibility that *C. botulinum* is able to grow in the product, then a full 12D process is required. It has been suggested that only 3% of the total number of microorganisms on the package surface are spores. An upper value of 1,000 microorganisms per m^2 (30 spores per m^2) has been assumed for plastic films and paperboard laminates on reels, and 3,000 microorganisms per m^2 (90 spores per m^2) for prefabricated cups.

Some of the techniques used for sterilization of the packaging equipment and packaging materials are the use of radiation (UV, infrared, or ionizing), heat (saturated steam, superheated steam, hot air, hot air and steam, or extrusion), or chemicals (hydrogen peroxide, peracetic acid, or ethylene oxide). Verification of sterilization is done by inoculating the surface of the web, cup, or lid stock with a pre-determined concentration of the test organism and allowing it to dry. The entire process is then run as in a commercial run and the finished containers are filled with a pre-determined growth medium and observed for growth. Two of the most important factors affecting the success of the tests are the choice of the indicator organism and the physical state of the microorganisms used. The indicator organisms used are: *B. stearothermophillus* – strain 1518 (superheated steam, peroxide + steam, extrusion), *B. polymyxa* – PSO (dry heat), *B. subtilis* strain A (peroxide + UV), *C. sporogenes* – PA 3679 (ethylene oxide), or *B. pumilus* (gamma radiation) depending on the process used.

9.3.2 Validation of an aseptic process

Validation of an aseptic process relates to the development of an adequate process, assurance of adequate heat treatment to the entire product, assurance of adequate packaging standards, and appropriate record-keeping mechanisms. In developing an adequate process, several factors are considered. Some of those factors are the product characteristics (ingredients, properties, pH, and homogeneity of the product), equipment issues (type of heat exchanger and holding tube), and location of the critical point (the slowest heating point in the system).

Identifying the critical point in a system can be a daunting task by itself. For liquid or homogeneous products flowing in a conventional (straight) holding tube, the location of the critical point is the center of the tube. However, for other situations, that is not the case. To illustrate this point two examples are offered here. In the case of a helical holding tube, the location of the critical point would not be the center of the holding tube, but at a radial location away from the center of the tube (towards the wall of the tube). In the case of a liquid product containing discrete particles of different thermal diffusivities, the critical point would be the center of the slowest heating particle. The slowest heating particle is not necessarily the fastest moving particle since it is possible that faster moving particles may have a higher thermal diffusivity than slower moving particles and hence heat up faster. Thus, a careful analysis of the residence time distribution of the particles and the heat transfer characteristics of the particles must be conducted to arrive at the critical point. This is usually done by means of mathematical modeling coupled with experimental verification.

Biological validation tests are performed at various stages of the process – just after start-up, during the middle of the process, and just before shut-down. These tests account for variations during the process as a function of time and also for factors such as fouling which affect the rate of heat transfer and residence time of the product.

In European countries, regulations (Rose, 1986, 1987) are based on spoilage tests. However, in the US, the FDA requires microbiological tests to prove the safety of a process with sufficient latitude for variability in process conditions. In fact, different regulatory agencies and rules apply to different products. For example, UHT milk processing is covered under title 21 (parts 108, 113, 114) of the code of federal regulations (CFRs). The process should also adhere to the pasteurized milk ordinance (PMO). When meat is involved, the regulations are imposed by the USDA. In addition to these regulations, certain states have state regulations imposed on certain processes. During the past few years, HACCP has gained tremendous importance and its implementation has been extended by the FDA to various products after its initial application to certain acidified and low-acid canned foods. Further details of requirements have been presented by David *et al.* (1996).

Shelf-stable low-acid food products are of special concern from a regulatory standpoint since the conditions of storage and the chemical composition of the food product are conducive to the growth and toxin formation by various strains of *Clostridium botulinum*, microorganisms capable of producing one of the most potent toxic substances known (resulting in botulism poisoning). Processes for treating these types of products need to be designed and validated to consistently deliver a 12D reduction of spores of the most resistant proteolytic strains of *C. botulinum*.

To aid processors in developing a validated process schedule for low-acid particulate foods, a series of industry-university-government workshops on aseptic processing of multiphase foods were conducted by the National Center for Food Safety and Technology in Chicago, University of California at Davis, and the Center for Aseptic Processing and Packaging Studies at North Carolina State University in Raleigh. This resulted in the publication of the *Case Study for Condensed Cream of Potato Soup from the Aseptic Processing of Multiphase Foods Workshop* (Anonymous, 1996). Other results of the workshop were summarized in a series of articles that appeared in the *Food Technology* magazine (Damiano *et al.*, 1997). Based on the recommendations of the workshops, Tetra Pak Inc. (with the assistance of the NFPA) developed the necessary data required for filing a scheduled process for aseptic processing of a low-acid product (cream of potato soup). The filing resulted in a 'no-objection' letter from the FDA in May of 1997. It was thus demonstrated that it was indeed possible for processors to adopt the recommendations of the workshop to gain the 'approval' of the FDA for aseptically processing low-acid foods containing large particulates. A detailed description of process filing using the FDA form 2541c (for aseptic processing of low-acid foods) has been described by Sastry and Cornelius (2002).

General FDA Requirements for Establishment Registration, Thermal Process Filing, and Good Manufacturing Practices for Low-Acid Canned Foods and Acidified Foods are covered in 21 CFR 108, 21 CFR 110, 21 CFR 113, and 21 CFR 114. These and other listed regulations and forms are also accessible through contact with FDA directly or from their website (http://www.cfsan.fda.gov).

9.4 Recent developments in aseptic processing

Recent developments in aseptic processing include the development of pumps, back pressure devices, heat exchangers and sensors to efficiently process, record, and modify various aspects of the process. Some of the challenges have been related to the processing of particulate foods – specifically, handling particulates in a continuous flow system without physical damage to the particles, non-invasive determination of the internal temperature of particles, and efficient heat exchangers to rapidly heat and cool the products.

As far as the handling of particulates is concerned, in the past, systems were designed so that they handle the liquid and solid portions of a particulate food separately. Some of those systems were the Jupiter system, rotaholder, and the fluidized bed system (Willhoft, 1993). However, there were practical limitations to those techniques. Thus, new back pressure devices and pressurized tank systems are being developed to overcome this problem.

As far as non-invasive temperature measurement is concerned, the main issue is determination of the critical point in the system and the overall thermal treatment received at this point. The use of a time-temperature integrator (TTI) is one of the means of determining the thermal treatment received at any point in the system. However, the concerns of using this technique are the leaching of the TTI out of the critical point and the accurate determination of the kinetics of the TTI under the given process conditions. A detailed discussion on the use of enzymatic TTIs has been presented by van Loey *et al.* (1999). A better technique to determine the thermal treatment received by different particles is by actually determining the internal temperature of particles. One such technique involves the use of thermo-magnetic switches and has been developed by Palazoglu *et al.* (2003).

As far as improved heat exchange designs are concerned, helical, corrugated, and volumetric heating mechanisms (ohmic, microwave, and radio frequency) have been developed. The use of helical and corrugated heat exchangers results in enhanced rate of heat transfer by either inducing secondary flow or turbulence. The use of ohmic, microwave, or radio frequency heating units results in the entire product heating up at the same time rather than having to rely on heat conduction within the particles. These and other designs of heat exchangers have been discussed by Afgan *et al.* (1996).

Some of the companies that manufacture plate heat exchangers are Tetra Pak Inc (Vernon Hills, IL, USA), Alfa Laval, Inc. (Glen Allen, VA, USA), FMC FoodTech (FranRica, Madera, CA, USA), Waukesha Cherry-Burrell (Delawan, WI, USA), Invensys/APV (New York, NY, USA), and Stork Food and Dairy Systems (Gainesville, GA, USA). Some of the companies (other than those listed previously) manufacturing tubular systems include Feldmeier Equipment Inc. (Syracuse, NY, USA) and Rossi & Catelli (Parma, Italy). Helical heat exchangers have been manufactured by VRC Co. Inc. (Cedar Rapids, IA) and GEA-AG (Bochum, Deutschland). Some of the companies dealing with volumetric heating techniques include Radio Frequency Co., Inc. (Millis, MA), Invensys/APV (New York, NY, USA), Raztek (Sunnyvale, CA, USA), Industrial Microwave Systems (Morrisville, NC, USA), Keam Holdem Associates (Auckland, NZ), and Armfield Limited (Ringwood, England).

Among the above listed types of equipment and companies, the continuous flow microwave unit manufactured by Industrial Microwave Systems has shown considerable promise in uniformly processing liquid and possibly particulate foods. Further studies on this equipment are being conducted to possibly use it for commercial application to aseptic processing. Despite these developments, there is still plenty of room for further improvement in various aspects of aseptic processing. Some of these potential applications and the future of aseptic processing are discussed in the following section.

9.5 Future trends

Optimistic predictions from the early days of aseptic processing and packaging about this technology replacing most of the conventional canning products have not yet materialized. However, aseptically processed products have gradually grown to occupy a significant segment of the commercially processed food product range, and have indeed, in some areas, such as shelf stable milk and milk beverages, almost completely replaced the previously dominant processing and packaging forms.

Currently, aseptic processing industry is going through another period of dynamic growth and new product development and distribution, brought on by the introduction of aseptically packaged polymer (primarily polyethylene and polyethyleneterepthalate) containers to the world market. These packages have been in commercial use for several years for packaging of high acid products such as fruit juices and spaghetti sauces and refrigerated low acid products like fruit smoothies and white and flavored milk beverages, but are being used with increasing frequency for low acid shelf stable fluid products like coffee and protein-fortified beverages.

Developments on the filling and packaging side of the aseptic technology will continue to be the major driving force in the further expansion of this technology. Emerging package and packaging material sterilization processes like plasma generation and glass lining will expand the number and variety of polymers in use for aseptic packaging as well as available package shapes and sizes. Larger (institutional and industrial ingredient) sizes of aseptically packaged products currently have the most favorable (low) ratio of packaging material used per unit of product weight and volume and this advantage will continue to grow, especially for products yet to make a significant impact in the aseptic processing area, such as low acid particulate products. Uniform quality, reduced need for frozen and refrigerated distribution and storage as well as progressively more favorable ratios of packaging material used per unit product will positively impact the environmental and economic advantage of this technology over currently dominant conventional technologies.

Developments in conventional tube in tube heat exchanger design will also continue. Helical, dimpled, corrugated, and agitated heat exchangers will be introduced in commercial production with increasing frequency. Emerging thermal processing technologies, specifically volumetric heating methods like continuous flow microwave heating, radio frequency and ohmic/electric resistance heating will have a major impact on quality and variety of available aseptic food products in the near future. Applications of emerging non-thermal and thermally-assisted technologies like ultra-high pressure processing, pulsed electric field treatments, irradiation, sonication, thermo-sonication, and mano-thermo-sonication will expand and integrate with other aseptic processing operations in single and multiple concurrent and sequential bactericidal treatments to achieve extended refrigerated shelf life and shelf stability.

At the temperature reduction end of the processing chain, emerging cooling methods will be introduced and integrated into aseptic processes over an extended period of time. Evaporative, cryogenic, electro/peltier cooling and newer volumetric cooling technologies such as magnetic field cooling and ultrasonic cooling will have an impact on quality and economy of aseptic product processing.

Incremental adjustments will be made to all processing stages and equipment in order to accommodate multiphase products, especially low acid products containing large particulate components. Particle-compatible pumps, heat exchangers, tanks, back pressure valves, filler heads and package fitments will continue to be improved and integrated into aseptic production lines.

The most significant improvements and changes in aseptic processing will take place in the area of process delivery design, monitoring, and validation. New techniques of real-time and post-process measurement and monitoring will be implemented to accurately and reliably monitor and quantify all lethality-delivering segments of aseptic processing systems. These emerging techniques and tools will take advantage of miniaturization of sensing elements and nano-technology level developments currently taking place in other areas of research and development. Aseptic processing technology will be one of the first areas of commercial food processing to implement new methods of non-contact, real-time temperature monitoring, subsurface temperature monitoring in natural and simulated food solids and miniaturized, implantable time-temperature recording devices. Similar alternative tools and techniques will be developed and implemented for monitoring, recording and validation of bactericidal lethal treatments delivered by non-thermal and thermally-assisted technologies. These tools will concurrently measure, quantify, communicate and record all effective components of bactericidal treatments such as temperature, time, pressure, voltage, frequency, wavelength and pulse number as appropriate. Unavailability or inconsistent application of process establishment, monitoring and validation tools have been one of the most significant hurdles in expanding the range of aseptic processing to more difficult, particularly multiphase food products. This will gradually change as the industry integrates new tools of process monitoring to supplement and improve over the traditional fixed and wired sensor based monitoring and control. In many cases, this could potentially prove to be the enabling technical elements, leading to creation and commercial introduction of new products and packages and expansion into new markets and applications.

Aseptic processing techniques and methods will also continue expansion into other, non-food processing areas. Biomaterials such as pharmaceutical and fermentation materials and products are preserved, packaged and distributed using the aseptic processing methods and techniques originally developed for foods. These trends will continue to expand in volume, variety, and frequency of implementation.

Disclaimer

The use of trade names in this chapter does not imply endorsement by the North Carolina State University of the products named nor criticism of similar ones not mentioned.

9.6 Abbreviations

CFR	Code of federal regulations
FDA	Food and drug administration
HACCP	Hazard analysis of critical control points
HEPA	High efficiency particulate arresting
PET	Polyethyleneterepthalate
PMO	Pasteurized milk ordinance
RTD	Residence time distribution
SSHE	Scraped surface heat exchanger
TTI	Time temperature indicator
UHT	Ultra high temperature
USDA	United States Department of Agriculture
UV	Ultraviolet

9.7 References

AFGAN, N., CARVALHO, M.D.G., BAR-COHEN, A., BUTTERWORTH, D. and ROETZEL, W. 1996. *New Developments in Heat Exchangers*. Gordon and Breach Publishers, Amsterdam, Netherlands.

ANONYMOUS. 1996. *Case Study for Condensed Cream of Potato Soup from the Aseptic Processing of Multi-phase Foods Workshop*. Published by the National Center for Food Safety and Technology, Chicago, IL and the Center for Advanced Processing and Packaging Studies, Raleigh, NC.

DAMIANO, D., DIGERONIMO, M., GARTHRIGHT, W., MARCY, J., LARKIN, J. and SASTRY, S.K. 1997. Workshop targets continuous multiphase aseptic processing of foods. *Food Technology*. 51(10): 43–62.

DAVID, J.R.D., GRAVES, R.H. and CARLSON, V.R. 1996. *Aseptic Processing and Packaging of Food: A food industry perspective*. CRC Press, Boca Raton, FL. pp. 41–48.

LEWIS, M. and HEPPELL, N. 2000. *Continuous Thermal Processing of Foods: Pasteurization and UHT sterilization*. Aspen Publishers, Gaithersburg, MD. pp. 331–368.

NIELSEN, S.S., MARCY, J.E. and SADLER, G.D. 1993. Chemistry of aseptically processed foods. In *Principles of Aseptic Processing and Packaging*, 2nd edition. Edited by

Chambers, J.V., Nelson, P.E. The Food Processors Institute, Washington, DC. pp. 87–114.

PALAZOGLU, T.K., SIMUNOVIC, J., SANDEEP, K.P. and SWARTZEL, K.R. 2003. Conservative process evaluation for multiphase aseptic processing: Magnetic sensor arrays and thermo-magnetic switch implants. IFT Annual Meeting, Chicago, IL 2003.

REUTER, H. (ed.) 1989. *Aseptic Packaging of Foods*. Technomic Publishing Company, Lancaster, PA.

ROSE, D. 1986. Guidelines for the processing and aseptic packaging of low acid foods. Part I: Principles of design, installation and commissioning. *CCFRA Technical Manual 11*.

ROSE, D. 1987. Guidelines for the processing and aseptic packaging of low acid foods. Part II. *CCFRA Technical Manual 11*.

SANDEEP, K.P. and PURI, V.M. 2001. Aseptic processing of foods. In *Food Processing Operations Modeling*. Edited by Irudayaraj, J. Marcel Dekker, New York, NY. pp. 37–81.

SASTRY, S.K. and CORNELIUS, B.D. 2002. *Aseptic Processing of Foods Containing Solid Particulates*. Wiley-Interscience, New York, NY.

VAN EIJK, A. 1993. Flavorings for UHT-treated and aseptically packed soups and sauces. In *Aseptic Processing of Foods* edited by Reuter, H. Technomic Publishing Company, Lancaster, PA. pp. 145–152.

VAN LOEY, A., HAENTJENS, T., SMOUT, C. and HENDRICKX, M.E. 1999. Enzymatic time-temperature integrators for the quantification of thermal processes in terms of food safety. In *Processing Foods: Quality optimization and process assessment*. Edited by Oliveira, A.R. and Oliveira, J.C. CRC Press, Boca Raton, FL. pp. 13–40.

WILLHOFT, E.M.A. 1993. *Aseptic Processing and Packaging of Particulate Foods*. Blackie Academic and Professional, London, England.

10

Developments in tubular heat exchangers

G. S. Tucker, Campden and Chorleywood Food Research Association, UK, and U. Bolmstedt, Tetra Pak Processing Components AB, Sweden

10.1 Introduction: applications of traditional tubular heat exchangers

This chapter provides a brief discussion of the parameters that determine the proper choice of heat exchanger for each specific duty, recent developments in tubular heat exchangers, and how to model heat exchanger performance. Section 10.2 considers the design principles that are required, which includes flow behaviour of the product and how to measure this for foods that are usually shear-thinning. Section 10.3 introduces plate, tubular and scraped surface heat exchangers and describes which food products they are best suited to processing. Most of the focus for this section is on the different types of commercial tubular heat exchangers. Section 10.4 describes the methods used to quantify the relative heat transfer efficiency of heat exchangers and introduces some of the complexities involved with sizing the correct heat transfer area. Section 10.5 presents information on emerging tubular heat exchanger designs from a variety of companies world-wide and goes on to discuss some future trends and work that is ongoing to help design innovative exchangers that can recover heat more effectively.

10.2 Improving exchanger design: product flow behaviour

Continuous processing of liquid food products as compared to in-container processing offers many economic advantages to food processing companies. By minimising the product's exposure to the adverse effects of high temperatures, long processing times and high shear preparation methods, commercial benefits can be realised in improved product quality, reduced processing costs, increased safety and increased plant throughput. A classic example of this is with UHT milk where an in-pack sterilisation process usually results in excessive browning reactions taking place and undesirable product. By operating a HTST process (high temperature short time), it is possible to reduce these browning effects and produce milk that has a more acceptable appearance and taste to the consumer. A heat exchanger such as a plate pack or narrow bore tubular will be the core of this process and will deliver a rapid rise and fall in milk temperature. The heat exchanger part of the HTST process for milk is not now a difficult technical challenge, with the exception of minimising the fouling that builds up over the run-time and reduces heat transfer efficiency.

A challenge in the design of heat transfer equipment for the liquid food industry is with the so-called prepared food products, such as tomato products, soups and sauces and dessert products (Tucker and Bolmstedt, 1999). These products are normally of high viscosity as well as of complex composition. In each case the optimal heat processing equipment has to be chosen in order to retain particulate integrity, flavour and colour of the end product. There is not one heat exchanger that will process all of these types of products from fibrous fruit juices to soups with particulates. These are the product types for which it is necessary to understand how their flow behaviour interacts with the heat exchanger choices.

A typical continuous processing line consists of a preparation sub-module, the actual heat exchanger and a filling machine. The preparation sub-module is used mainly for formulated food products, e.g. vanilla puddings, salsa sauces, where the product is prepared and exits this section ready to be pasteurised or sterilised. The processing module comprises a heat exchanger that can be a series of tubes of varied design (see section 10.3), where the product flow is counter-current to the heating or cooling media. Thermal and shear damage are minimised using tubular systems because of the simple once-through flow path taken by the product. Filling systems will operate either hot or cold, and require a balance tank or intermediate aseptic storage tank that acts as the buffer between the processing and filler modules.

With respect to flow behaviour or rheology, most of the formulated products are typically non-Newtonian, showing in many cases quite extraordinary behaviour. Few foods flow as Newtonian liquids, apart from milk and thin fruit juices. Most food products are shear-thinning and some also display time-dependent properties caused by the complex macromolecular structures that are used in their formulation. These properties must be considered when designing a processing system in order that excessive damage is avoided to the delicate structures. Depending on the rheological properties of the product and the possible presence of particulates, the design and choice of equipment can vary significantly from case to case. Various additives, e.g. thickeners and stabilisers, often also change the physical and rheological properties of the product.

The microbiological demands are basically to reach commercial sterility, i.e. the end product must be free from pathogens, free from toxins and free from micro-organisms capable of multiplication under normal storage and distribution conditions.

10.2.1 Flow behaviour

In the design of heat exchangers and choice of heat exchanger configurations, the flow behaviour of the product to be processed has to be taken into consideration. For instance, the flow behaviour will affect the residence time distribution and hence the design of heat exchangers and holding cells to obtain sufficient thermal treatment. The basic difference between laminar (streamline) and turbulent flow is well known, as is the effect on the velocity profile from heating or cooling of the product. For example, the maximum velocity in laminar flow, originating from the parabolic velocity profile, is theoretically twice the mean velocity, and in turbulent flow it is around 20% higher than the mean velocity (see Fig. 10.1). Laminar flow is assumed to occur up to Reynolds numbers of around 2,100 whereas turbulent flow occurs at greater than 10,000 (see equation 10.1). The region in between 2,100 and 10,000 is referred to as the transitional region, because the flow regime is changing from laminar to turbulent. This is a region that equipment designers will try to avoid because of the uncertainties in flow behaviour and the key relationship between fastest and mean velocity. Should a heat exchanger be operated under transitional flow then it would be safe to assume that the fastest liquid along the pipe centre could be travelling twice as fast as the mean velocity. For viscous products, however, the flow conditions are nearly always laminar and so the above assumption should always be applied. A tomato paste steriliser, for instance, operates at Reynolds numbers (Re) around 1. For milk and juice products, flow conditions are almost always turbulent.

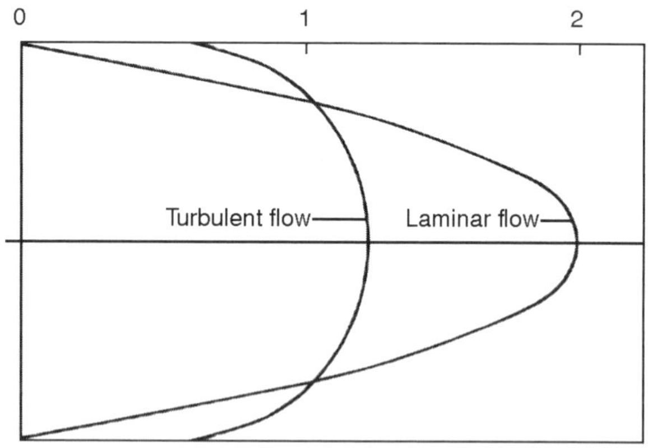

Fig. 10.1 Velocity profiles for turbulent and laminar (streamline) flow showing the ratio of maximum velocity to average velocity.

$$\mathrm{Re} = \frac{d_h.\rho.v}{\mu} \tag{10.1}$$

where, d_h = hydraulic diameter, m; μ = dynamic viscosity of liquid, Pa.s; ρ = density of liquid, $kg.m^{-3}$; v = velocity, $m.s^{-1}$.

For non-Newtonian conditions prevailing in liquid food processing, the velocity profiles are still more complex. With decreasing flow behaviour index (n), i.e. increasing degree of non-Newtonian behaviour, the velocity profile increases in flatness. This means in practice that the maximum velocity decreases from its Newtonian value of twice the mean velocity, which affects the design of especially holding cells. Maintaining the factor 2 for the calculation of the necessary holding cell length thus creates overcooking of the product. However, there are few commercial operations that do not apply the factor 2 when designing the holding tube length, irrespective of the measured flow behaviour index.

Additives giving viscoelastic properties, e.g. xanthan or gellan gum, are sometimes used to enhance the particulate carrying properties for a carrying fluid in continuous processing. The so-called yield value, which normally is a measure of the product's willingness to flow by itself, e.g. from a storage tank, is also a measure of the particulate carrying abilities. A significant yield value, typical of paste-like products, also adds to the flatness of the velocity profile and hence further increases the deviation from the parabolic shape. Again, the factor 2 would be used in the holding tube calculations even though the velocity profile may be almost flat.

10.2.2 Viscosity measurements and flow behaviour modelling

Non-Newtonian properties are normally examined and described by a viscometric or rheometric analysis. With a viscometric analysis, a shear rate sweep is performed on the product. When the data are shown in a log shear stress versus log shear rate graph, or a log apparent viscosity versus log shear rate graph, the basic shearing rheological behaviour can be determined.

The results from a viscometric analysis can be fitted with the aid of a suitable model, depending on product characteristics and intended use. Normally, the so-called power law or Ostwald de Waele model is employed for fitting of viscometric measurement data. The main benefits of the power law model are its simplicity and its applicability to most liquid food products within the shear rate range of interest. The simple application of the power law in a double log graph provides a quick but still widely applicable way of retrieving the necessary viscosity data parameters for use in equipment design. In the power law case the parameters are the consistency value K ($Pa.s^n$) and the power law index n (dimensionless), respectively.

The viscometric analysis provides the necessary information for design of process equipment. Normally the power law is used for description of the behaviour of the product under shearing conditions and therefore the power law index n and the consistency K are determined. Regarding liquid foods, practice has shown that the power law can be used to model most products within the limited range of shear rates prevailing in most processing equipment.

Issues can sometimes arise when attempting to fit power law models over a wide shear rate range because of the tendency of some high viscosity foods to possess a yield stress. More complex models are required to extend the flow behaviour into the range of shear rates encountered in flowing systems, which can drop down close to zero and reach several hundred reciprocal seconds. A model such as Cross or Carreau that contains a zero shear viscosity and/or infinite shear viscosity may be useful.

10.3 Selecting the right type of tubular heat exchanger

The choice of optimal heat exchanger depends a great deal on the flow conditions. Fluids with low viscosities and no particulates are preferably treated in a plate heat exchanger (see Fig. 10.2). This should be the first choice of exchanger because of its low cost and high heat transfer rate. For fruit juices with pulp and fibres of up to 5 mm length, special types of plate are available

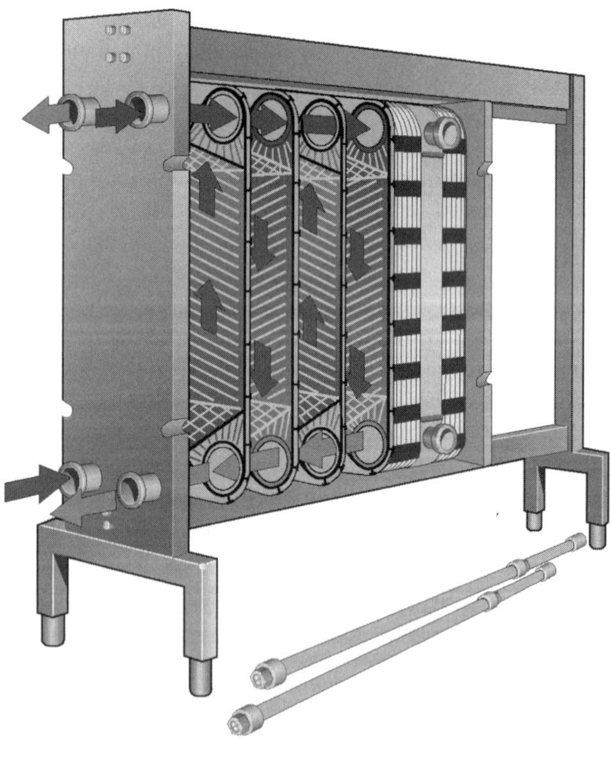

Fig. 10.2 Plate type heat exchanger. Courtesy of Tetra Pak Processing Components AB.

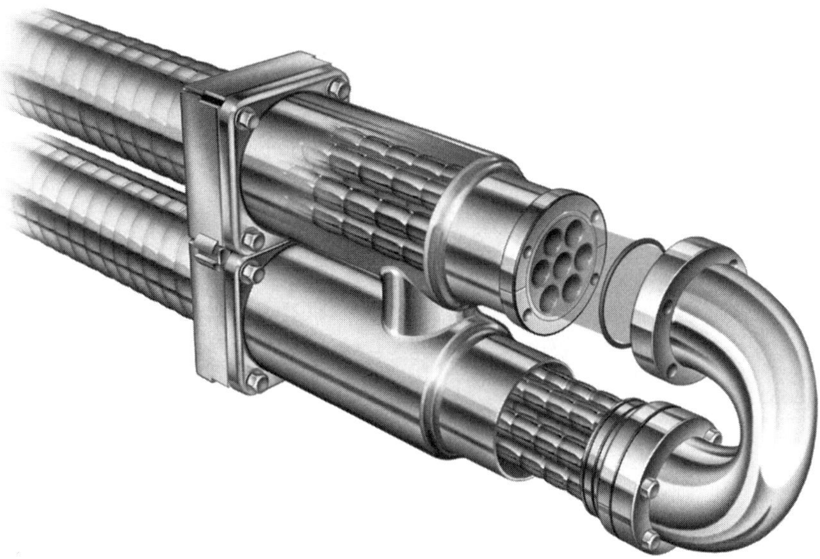

Fig. 10.3 Multitube type tubular heat exchanger. Courtesy of Tetra Pak Processing Components AB.

with more open channels that allow the fibres to pass through unhindered. In addition, even with high viscosity foods, the plate heat exchanger can be utilised as long as the pressures developed are not too high and the rheological behaviour does not indicate that a yield stress is present. Foods that contain yield stresses can experience maldistribution of flow between the plate gaps and in some instances this can lead to flow stagnation and blockages.

For fruit juices with fibres of up to 15 mm length and for relatively water-like foods, a multitube tubular heat exchanger is preferably used (see Fig. 10.3). Also, fluids of moderate to high viscosity with only small particulates (< 5 mm) will flow through a multitube heat exchanger without problems. Despite being less thermally efficient than a plate heat exchanger, multitubes can be configured in various options of numbers of tubes in parallel and of tube diameter. This gives them a high degree of flexibility. A common application for the multitube exchanger is milk sterilisation, despite the heat transfer disadvantagcs of tubcs comparcd with platcs. Tubc bundlcs containing largc numbers of small diameter tubes in parallel offer adequate thermal efficiency with the benefit of cleanability without dismantling. A plate heat exchanger would need to be taken apart to remove the fouling deposits that build up over several hours of processing. This increases the downtime and increases the chances of contamination being introduced with poorly fitted gaskets.

Most tubular heat exchangers now use corrugations on the shell and tubes to enhance heat transfer with the heating and cooling media, typically water. Design of the corrugations will be specific to the company supplying the heat exchanger but they will each create the same effect, which is to generate

Fig. 10.4 Fibretube tube plate showing the curved surfaces that allow fibres to enter the tubes without damage. Courtesy of Tetra Pak Processing Components AB.

turbulence in the water flows. This reduces the resistance to heat transfer caused by boundary layers that can be set up adjacent to the tube walls. In effect, it ensures that the media, whether it is heating water or cooling water, does not restrict the heat transfer performance of the exchanger. The limit to performance is therefore within the tubes.

One modification to the multitube is the fibretube, which has a specially designed tube plate where the tube inlets are shaped to avoid any risk of fibre blocking (see Fig. 10.4). For juices with long fibres, i.e. $>$ 15 mm in length, or fruit pulp with a very high concentration of pulp and fibres, the fibretube is the best choice. Fibres are gently directed into the tubes and once within will flow with ease along the straight tubes. The curved entry surfaces take up more space in the tube plate and so fibretubes typically have fewer tubes in parallel within the tube bundle. For example, a 7×16 multitube would contain seven parallel tubes of 16 mm diameter enclosed within an 85 mm diameter shell. The corresponding fibretube may contain only four parallel tubes. The rest of the exchanger is identical.

The length of most commercial tubular systems has been standardised at six metres, therefore, tubular heat exchangers are long and thin in terms of their geometry. This makes them suitable for placement next to factory walls or even above head height. The need for access to the tubes is rare because the tubes themselves were designed to be cleaned in place and access will only be required to the connecting pipes and equipment. These can be positioned at ground level where access is easy.

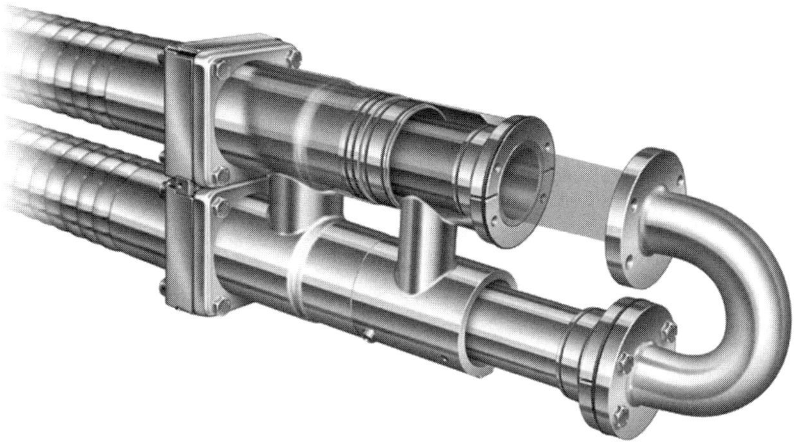

Fig. 10.5 Concentric channel type heat exchanger. Courtesy of Tetra Pak Processing Components AB.

One of the greatest challenges to heat exchanger design is when the food fluids are significantly viscoelastic, i.e. exhibit a large yield value, often in combination with a high viscosity. For these fluids, there is a risk of maldistribution across the inner tubes of a multitube heat exchanger. In the worst case, the product flow will stop in some of the tubes causing overcooking of parts of the product and also cleaning problems where the food will burn onto the inner tube surfaces. Examples of such products are hot break tomato pastes or stiff dessert puddings, where the multitube is not suitable and concentric tubes are a preferred choice (see Fig. 10.5).

Concentric tubes have only one product channel, which eliminates the risk of maldistribution. Here, the product flows in a gap between two concentric tubes with the media on both sides, therefore increasing the heat transfer efficiency. Particulate products can be processed in wide gap modules, with particulate sizes up to 5–6 mm. Concentric tubes are a common choice of heat exchanger for tomato ketchup manufacture because of the yield stress that is an essential feature of the ketchup.

If large particulates are present in the food product, the monotube is probably the optimal choice (see Fig. 10.6). The drawback with a monotube compared to a concentric tube or multitube is reduced thermal efficiency due to the thicker product layer. However, the particulates present in the product will to a great extent work as 'internal mixers' and will hence promote heat transfer. This makes design of monotube exchangers difficult unless prior knowledge on the heat transfer behaviour of that food is available. The limiting design criterion for monotubes is the heat treatment given to the particulates, because of the need for heat to conduct into the particulate. The need for sufficient contact time between the carrier liquid and the particulates can be an advantage of the tubular concept,

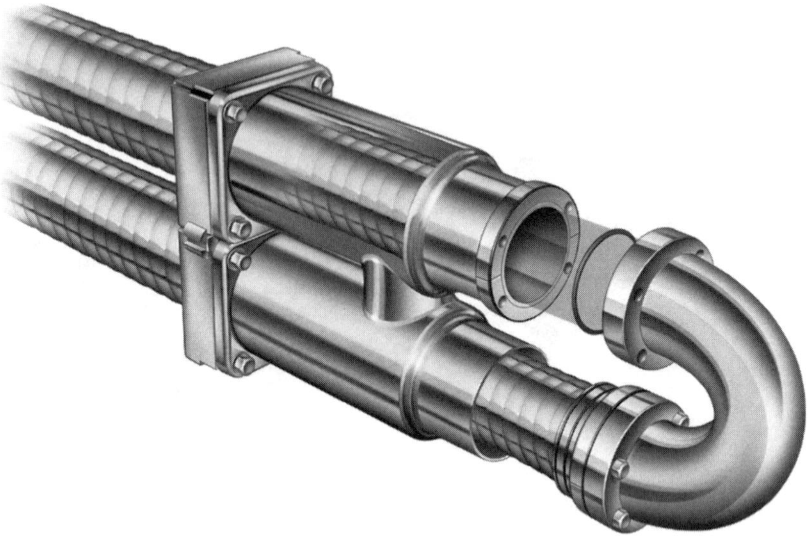

Fig. 10.6 Monotube type tubular heat exchanger. Courtesy of Tetra Pak Processing Components AB.

in which the food product has sufficient residence time in the exchanger to equilibrate towards a uniform temperature.

In case none of the tubular types are suitable, the scraped surface heat exchanger must be employed. In principle, a scraped surface heat exchanger is a monotube equipped with a rotating internal scraper. The scraper keeps the heating surface free from any deposits and also promotes turbulence. Hence this type of heat exchanger is ideal for products of very high viscosity, possibly also containing large particulates. Unlike tubes that usually operate with water as the medium, scraped surface heat exchangers tend to use steam for greater heat transfer efficiency. There are drawbacks with scraped surface heat exchangers in that the costs are high for purchasing a system and ongoing maintenance is higher than tubular heat exchangers because they have moving parts that wear. Therefore, they are seen as the last resort in heat exchanger choice.

10.4 Heat transfer efficiency in tubular heat exchangers

Design models for heat exchangers are normally based on empirical correlations of the dimensionless Nusselt, Prandtl and Reynolds numbers. As a generalisation, the Nusselt number is a heat transfer number (see equation 10.2), Reynolds is a flow number (see equation 10.1) and Prandtl a physical properties number (see equation 10.3). The derived equation is basically of the form $Nu = f(Re, Pr)$. By using dimensionless numbers only a limited number of experiments have to be performed in which the product and heating/cooling medium flow rates and the product physical properties are varied in order to cover a large range of Reynolds as well as of Prandtl numbers. The physical properties are changed preferably by changing the temperatures of the fluids involved. Decreased and increased temperatures can normally vary the viscosity of the product significantly. This is easier experimentally than changing the liquid.

The definition of the Nusselt (Nu) and Prandtl (Pr) numbers is as follows:

$$\mathrm{Nu} = \frac{\alpha \cdot d_h}{\lambda} \tag{10.2}$$

$$\mathrm{Pr} = \frac{c_p \cdot \mu}{\lambda} \tag{10.3}$$

where, α = individual heat transfer coefficient, W.m^{-2}.K^{-1}, λ = thermal conductivity of liquid, W.m^{-1}.K^{-1}, c_p = specific heat of liquid, J.kg^{-1}.K^{-1}.

A comparison of tubular heat exchangers with plate heat exchangers shows that the thermal performance as read off in a typical Nu-Pr-Re graph is better for the plate heat exchanger, mainly due to the complex corrugated pattern including a large number of contact points in the product channel. Figure 10.7 presents a comparison of the Nu-Pr-Re correlations for plate and tubular heat exchangers, which shows the lines for plates above those for tubes. This represents a higher Nusselt number for equivalence in Reynolds number, and hence a higher overall heat transfer coefficient. There is, however, a penalty for the good heat transfer in the form of higher friction factors at the corresponding conditions. This will result in a greater operating pressure drop. Also, due to the contact points, the plate heat exchanger is less suitable for particulate products.

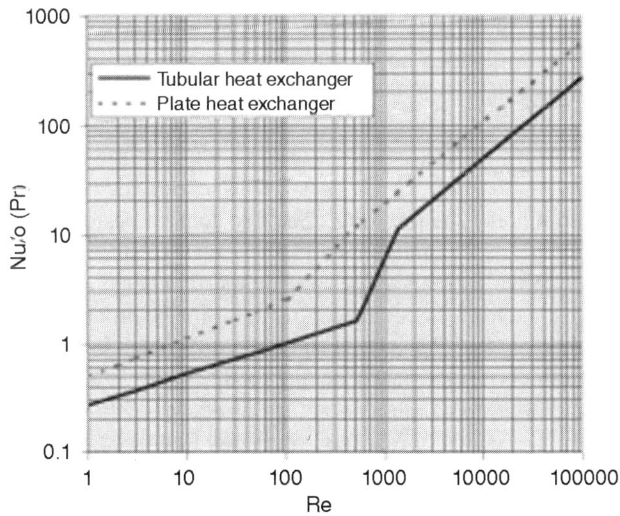

Fig. 10.7 Comparison of Nu-Pr-Re correlations for plate and tubular heat exchangers. Courtesy of Tetra Pak Processing Components AB.

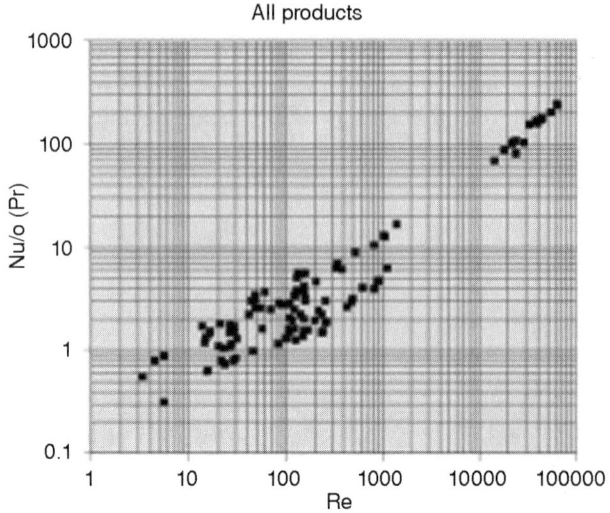

Fig. 10.8 The Nu-Pr-Re correlation as a result from tests on real food products. Courtesy of Tetra Pak Processing Components AB.

From systematic laboratory tests and from experience of commercial plants it has been determined, however, that to model completely the heat transfer of liquid food products some more parameters have to be included. Such parameters are for instance particulate content, particulate shape and size as well as type of particulates. In addition, the softness or hardness of the particulates has to be taken into consideration. This is a complexity observed with fruit and vegetable purée products in which the presence of fibres, cells and pulp can exert and influence on the predicted pressure drop and inversely the heat transfer performance. For example, tomato paste products can exhibit pressure drops some 50% greater than predicted from rheological measurements and pipe dimensions, whereas the heat transfer performance in monotube systems can be up to three times as large as expected. Some orange juice concentrates can display the opposite effects to tomato paste.

The major suppliers of innovations in the heat exchanger markets conduct extensive tests in order to analyse the specific thermal behaviour of a number of food products, e.g. fruit juice concentrates, tomato products, starch based puddings and fruit purées. The results show the difference in thermal behaviour between the products tested based on the traditionally used correlations and hence also show the need of more complex correlations taking into account the detailed physical properties of real food products. Figure 10.8 shows the spread in heat transfer correlation for a variety of foods. Little of this work gets published because it is specific to one company's equipment and represents commercial gains within a market where many competitors exist.

10.5 Emerging designs and future trends

In order to analyse the function of specific construction details before the actual manufacturing of expensive tools and prototypes, computational fluid dynamics (CFD) analysis is preferably performed. CFD is an analytical technique that is increasing in popularity as the CFD vendors develop their software codes to be more easily understood by the design engineers. It has long been used in the aeronautical and automotive industries to evaluate wind resistance and is commonplace in motor racing's Formula 1 as a prior step to wind tunnel testing.

The example shown here concerns a new header to a concentric tubular heat exchanger (see Fig. 10.9). The heat exchanger studied has a modular construction with the header in either a single or double design. With the double design the heating or cooling water at the outer service side is transferred from one module to another in series as is the product flow. The water flow at the inner service side is transferred via a normal bend connected to the tubes. The concentric tube design provides good heat transfer due to the narrow product gap giving excellent conditions for laminar heat transfer. In addition, the two service side flow rates are balanced to give an even thermal treatment of the product from the two sides. Design of the headers provides easy dismantling for full inspection of the product channel.

The main purpose of the CFD analysis was to pinpoint any possible stagnation areas in the headers. Stagnation in the product flow will lead to overcooking of part of the product and hence to a decrease in product quality and may also result in cleaning problems. In order to create the proper boundary conditions that were critical to the complete solution, modelling of the headers included also a significant part of the annular product channel. Simulations were made for rheological behaviour from water to tomato paste, the latter including models with as well as without yield value incorporated.

Fig. 10.9 Result from CFD analysis of a concentric channel heat exchanger header, with the flow pattern from tomato paste data. Courtesy of Tetra Pak Processing Components AB.

There are many configurations for tubular heat exchangers that make them highly adaptable to almost all types of flowing foods. Those described in section 10.3 are for straight tubes, which tend to be the food industry standard. Designs from different suppliers tend to differ only in the nature of the corrugations, inlet and outlet manifold design, tube bundle dimensions and ancillary equipment. However, there are tubular versions designed as coils that are claimed to possess a greater uniformity in residence time distribution, specifically with particulates. The tube is coiled into a helix and immersed in a heating and/or cooling bath to effect the heat transfer to the outside tube surface. The ratio of the diameter of the tube to the diameter of the spiral sets up a flow pattern within the tube that continually exposes new product to the walls of the tube for efficient heat transfer. There are applications to wine and juice processing where the presence of twigs and pips can create problems in multitubes as the liquid passes from one six-metre length to another via a header. In a continuous helical pathway this issue is overcome.

Recovery of heat in tubular heat exchangers is presently used in the food industry only with low viscosity food products such as milk, fruit juices and beverages, where it is referred to as regeneration. The shell of a tubular heat exchanger is designed to take water in turbulent flow and so little attention has been paid to flow distribution issues that are likely to occur with liquids of higher viscosity and flow regimes that are laminar. With the world-wide need to conserve energy supplies and reduce carbon dioxide emissions, this may be one of the key areas in which future exchanger developments are concentrated.

If a tubular heat exchanger was set up to recover heat, the hot 'processed' product would be cooled down in the tubes by the cold 'unprocessed' product, which in turn gets heated in the shell. This arrangement ensures that sterilised product flows only through the tubes where the flow paths are well defined and cleanability issues are minimal.

In order for food processing companies to adopt heat recovery technology for foods of viscosities greater than water-like liquids, a project was set up at CCFRA to demonstrate the viability of heat recovery (Tucker and Shaw, 2001). This was achieved using starch solutions of varying viscosity that represented the flow behaviour of food types processed in tubular heat exchangers, for multitube, concentric tube and monotube designs. Tetra Spiraflo (Tetra Pak Processing Components AB, Sweden) was the tubular heat exchanger used for the project work. Determination of the limitations on maximum viscosity were critical in that (a) below this viscosity heat recovery could be realised with existing commercial exchangers, and (b) above this viscosity re-design of the shell-side would be required to prevent stagnation and poor distribution of flows. This limit was found to be at equivalent viscosities of approximately 20 cP (estimated at 10 s^{-1} shear rate).

The data generated within this project on energy and cooling water savings highlighted the considerable potential for increasing the operational efficiency by adopting heat recovery. Minimal changes to the current shell designs for a Tetra Spiraflo were required if the viscosity of the food was below the

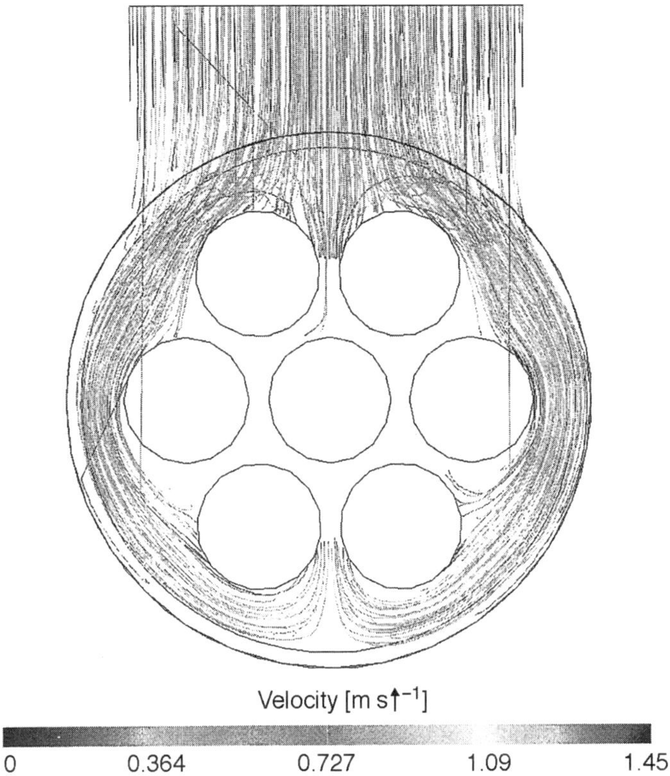

Fig. 10.10 Flow streamlines for a 3 wt% Colflo entering the shell of a 7 × 16 multitube exchanger.

equivalent viscosity to 20 cP. This was the critical viscosity above which it was thought necessary to improve the flow distribution in the shell. Figure 10.10 illustrates the flow streamlines for a 3 wt% Colflo starch solution that had an equivalent viscosity of 200 cP. At this level of viscosity the flow was not distributed evenly around all tubes and had difficulty in penetrating the centre tube of the bundle. Changes such as increasing the tube spacing and streamlining the tube supports could be made without recourse to complex computational design techniques. These would allow heat recovery to be utilised for foods in the equivalent viscosity range between 20 to 200 cP. However, to extend the use of heat recovery to foods with equivalent viscosities closer towards 1,000 cP, a more radical design approach would be required.

To address the issue of foods with viscosities in the range 200 to 1,000 cP, CFD techniques are being investigated as tools to design novel tubular heat exchangers. The objective is to arrive at a design that will recover heat from variable viscosity foods and provide thermal efficiencies similar to those for thin liquids. The types of tubular exchanger appropriate for this include multitubes and fibretubes. Some of the simpler design ideas mentioned above will be accommodated in addition to using CFD to evaluate more innovative designs such as twisted tubes and flow disrupters. This is an ongoing development illustrating how CFD techniques can be used to improve the design of commercial heat exchangers. By using the CFD codes, many novel design ideas can be tested and their improvements to heat transfer and cleanability determined computationally. This should result in the need for only limited prototypes to be constructed for experimental analysis.

10.6 Sources of further information and advice

Much of the information provided in this chapter uses Tetra Spiraflo as the example for different tubular heat exchanger types. Some website addresses of relevant companies that manufacture heat exchangers or are agents on behalf of other companies are given below:

www.alfalaval.com
www.gotoaps.com
www.apv.invensys.com
www.bertuzzi.it
www.fbr-elpo.it
www.tuchenhagen.com
www.hrs-spiratube.com
www.sigmanzini.com
www.pietribiasi.it
www.redaspa.com
www.sudmo.de
www.terlet.com

10.7 References

TUCKER, G. and BOLMSTEDT, U. (1999). Gently does it. *Liquid Foods International*, 3(3), 15–16.

TUCKER, G.S. and SHAW, G.H. (2001). Heat recovery in tubular heat exchangers for medium viscosity food products. *CCFRA R&D Report No.140*. Chipping Campden, Glos., GL55 6LD.

11

Optimising plate heat exchanger design and operation

L. Wang and B. Sundén, Lund Institute of Technology, Sweden

11.1 Introduction: plate heat exchangers (PHEs)

The food industry is perhaps the most important one in the world, in that it provides food for all people. Its major segments include dairy products, meat and poultry products, canning and preserving fruits and vegetables, grain mill products, bakery products, sugar and confections, fats and oils, beverages, and other miscellaneous and kindred products (Valentas *et al.*, 1990). Thermal processing is of great interest to food industry, because food processing involves heating, cooling, sterilisation or pasteurisation, etc. Therefore, heat exchangers are very important to the food industry, and are widely used.

Although many different types of heat exchanger are available, PHEs have proved to be the most successful in food industry, thanks to their characteristics of being very compact, relatively low in cost, highly efficient, versatile, easily cleaned, etc. (Wang *et al.*, 2003). In fact, the earliest development and usage of PHEs was in response to increasingly stringent statutory requirements from foodstuffs, particularly dairy products, in the late nineteenth century (Magnusson, 1985). At the beginning of the 1880s, there was growing public awareness that diseases, including tuberculosis, were spread by 'raw' or untreated milk. This initiated the early experiments with milk pasteurisation, which involved heating the milk to a certain temperature that does not influence the taste, holding it at this temperature for a short time, and then immediately cooling it. This process requires the heat transfer equipment to be thermally very efficient and, more importantly, should be easily cleaned (which had to be conducted daily). It was indeed difficult to meet these operational requirements in most of the early heat transfer equipment that was used for pasteurisation of milk, and which in time led to the development of PHEs. The first commercially operational PHE was invented in 1923 by Dr Richard Seligman, the founder of APV International in England (Seligman, 1964). The device was called a plate pasteuriser, which was destined to revolutionise the 'thermal curing' work in dairies to much the same extent as had the separator in its time. Although the basic concept and operation of a PHE have changed little since then, the overall design and construction have progressed significantly, to accommodate larger throughput capacities, higher working temperatures, and larger working pressures, among other factors. These changes have helped expand the applications from the original milk pasteurisation to a very broad range of industrial heat exchanger needs.

The traditional concept of a PHE is the plate-and-frame heat exchanger, which consists of plates, gaskets, frames and some additional devices, such as the carrying and guiding bars, the support column, etc. (see Fig. 11.1). As shown in Fig. 11.1, the two streams flow into alternate channels between plates, entering and leaving via ports at the corner of the plates. On each plate there is usually a gasket around the edge, and around the port. A stream exchanges heat with the streams in adjacent channels. This is the basic operation principle for PHEs. Because of their structure, PHEs offer a number of advantages over shell-and-tube heat exchangers. These advantages include compactness, high heat transfer coefficient, simple construction, low cost, low fouling formation rate, easy cleaning, flexibility, etc. The main disadvantages are that PHEs are restricted to applications with relatively low pressures and temperatures. The traditional plate-and-frame heat exchanger suffers leakage problems because the gasket cannot withstand the high pressure or temperature or the corrosive fluid. These problems certainly restrict PHEs' applications. To date, the standard operating pressures for plate-and-frame heat exchangers are up to 25 bar, although somewhat higher pressures can be achieved using heavy duty frames.

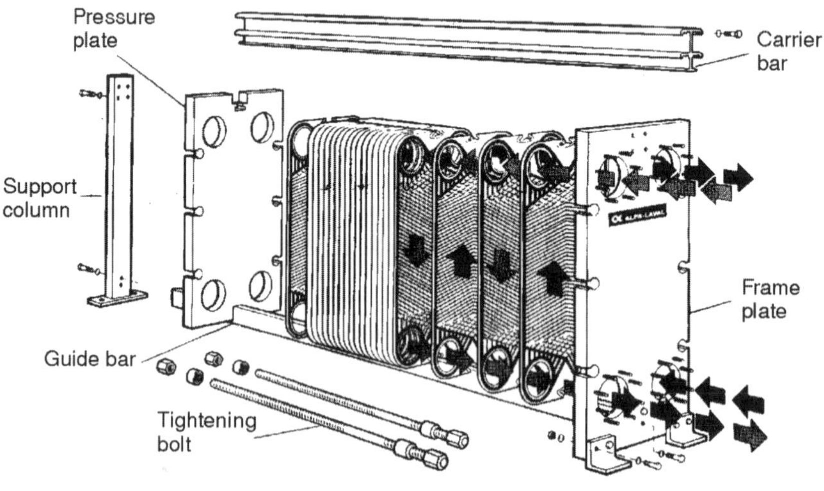

Fig. 11.1 An exploded view of a PHE (courtesy of Alfa Laval).

Upper limits of 160ºC apply to most plate-and-frame heat exchangers. Gaskets with very special materials have been claimed to operate up to temperatures of 400ºC (Wadekar, 1998).

11.2 Types of plate heat exchanger

Although the traditional plate-and-frame heat exchanger is still the most often used in industries, its disadvantages of limited pressure and temperature tricked many efforts, which resulted in variant models of PHEs. The evolution can be either structure construction or plate pattern. In structure construction, brazed, semi-welded and fully-welded PHEs have been developed. In plate pattern, many different surface types have been developed for the food industry, e.g., wide-gap plates, some special plates with minimum contact points, etc. A comprehensive list and description can be found in Wang *et al.* (2003). Here some examples of the new developments for food industry are presented.

11.2.1 Brazed PHEs

A brazed PHE (see Fig. 11.2) is composed of a pack of thin corrugated stainless steel plates brazed together using copper as a brazing material to form a self-contained unit. Brazing eliminates the need for either frames or gaskets, and results in a very compact exchanger. Because the plates are brazed to each other and there are no frames or gaskets, they can handle higher pressure and temperatures than plate-and-frame heat exchangers, e.g., situations with pressures up to 30 bar and temperature up to 400ºC. In addition, brazed PHEs are characterised by very low weight due to the absence of frames. Typical applications in food industry can be found in utilities, e.g., water heating and cooling, CO_2 heating/cooling, refrigeration, etc.

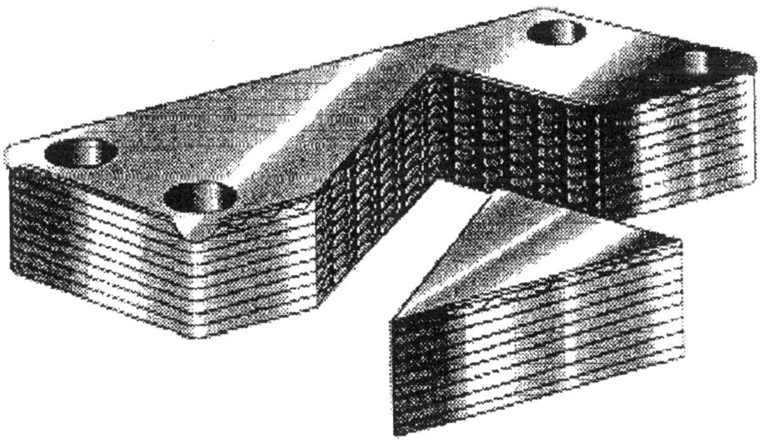

Fig. 11.2 The brazed PHE (courtesy of Alfa Laval).

11.2.2 Semi-welded PHEs

By welding heat exchanger plates in pairs, called twin plates, a semi-welded PHE (see Fig. 11.3) is achieved. It is designed especially for handling aggressive medium. The semi-welded plate pack consists of welded channels which alternate with traditional gasketed channels. The aggressive medium flows in a channel which is formed by welding the reverse side of two plate grooves to each other. The channels containing the non-aggressive secondary medium are sealed by traditional elastomer gaskets. It should be pointed out that frames are still needed in semi-welded PHEs. The semi-welded PHEs can withstand pressures up to 30 bar on the welded side. Typical applications in food industry include beer cooling, utilities (e.g., water heating/cooling, glycol cooling and refrigeration), etc.

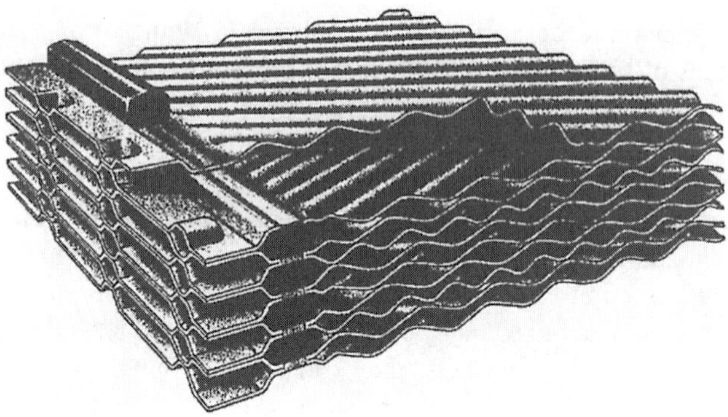

Fig. 11.3 The semi-welded PHE (courtesy of Alfa Laval).

11.2.3 Fully-welded PHEs

The fully-welded PHE (see Fig. 11.4) is a gasket free version of PHEs, where a completely welded plate pack is held in a conventional manner within a frame. The elimination of the gaskets enhances the reliability as well as the temperature and pressure limits of the gasketed PHEs. The laser welds in fully-welded PHEs are applied in two dimensions only, in the plane of the plates. This allows the plate pack to expand and contract along the length of the plate pack as temperature and pressure changes take place, which makes it more fatigue resistant than most other welded solutions. This flexibility of the plate pack allows fully-welded PHEs to handle applications with rapid changes in temperature or pressure, without fatigue problems. The fully-welded PHEs are intended for severe duty processes. Heavily aggressive fluids are permitted on either side. They can withstand temperatures up to 350°C and pressures up to 40 bar. Typical applications in food industry include vegetable oil heater, utilities (CO_2 heating/cooling, refrigeration), etc.

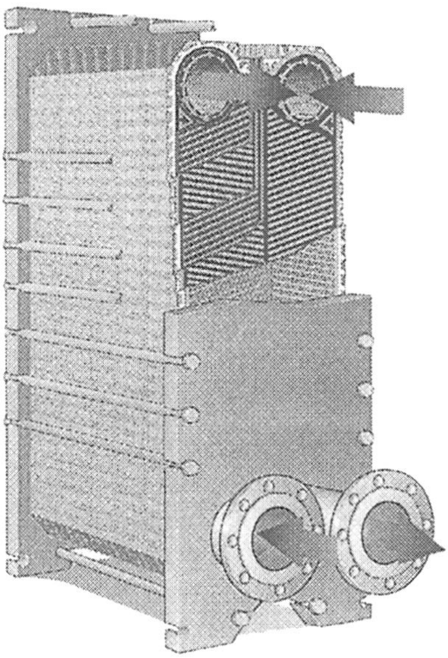

Fig. 11.4 The welded PHE (courtesy of Alfa Laval).

11.2.4 Wide-gap PHEs

The wide-gap PHE (see Fig. 11.5) provides a free-flow channel for fluids containing fibres or coarse particles and high-viscous fluids, which normally clog or cannot be satisfactorily treated in other types of heat exchanger. The free-flow channels are designed with a gap up to 16 mm. The plate corrugation provides high turbulence and high heat transfer coefficients compared to other types of heat exchanger. Typical applications in food industry include the wort boiling for brewery, heating of raw, limed, and mixed juice in sugar mills, sanitisation of fibrous food product slurries, etc.

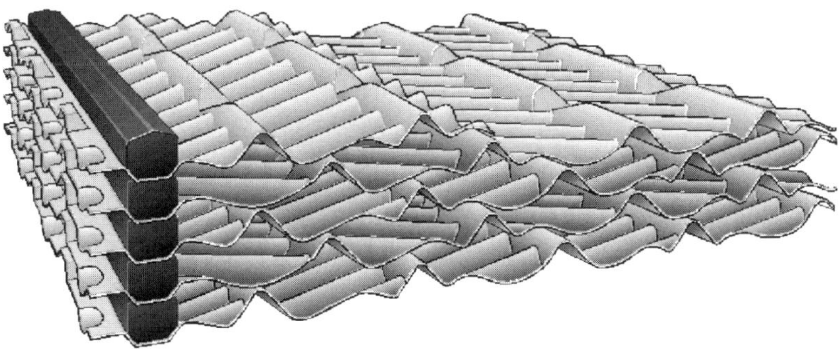

Fig. 11.5 The wide-gap PHE (courtesy of Alfa Laval).

11.3 Applications of plate heat exchangers in food processing: pasteurisation and evaporation

According to the specific characteristics of individual food types, certain processing is required. Because of the large number of food types, many different processes are used. Therefore, different thermal processing approaches are required. It is not possible to list all thermal processing applications where PHEs are involved. But two representative examples are illustrated in this section, to explain why PHEs are employed frequently in the food industry.

11.3.1 Pasteurisation

Pasteurisation is a very important process in the food industry, and is indeed used in almost all sections, e.g., milk, cream, wort, beer, juice, wine, etc. This process was named after Louis Pasteur of France who discovered that spoilage organisms could be inactivated in wine by applying heat at temperature below its boiling point. The process was later applied to milk and remains the most important operation in the processing of milk. The principle of milk pasteurisation is the heating of every particle of milk or milk product to a specific temperature for a specified period of time without allowing recontamination of that milk or milk product during the heat treatment process. During this process, the extent of micro-organism inactivation depends on the combination of temperature and holding time, which is illustrated in Fig. 11.6. On the other hand, heating temperature and holding time are also important to the product quality, i.e., milk taste. Therefore, precise control of temperature and time combinations in the pasteurisation process must be highly regulated, which means the selection of proper heat exchangers becomes very important.

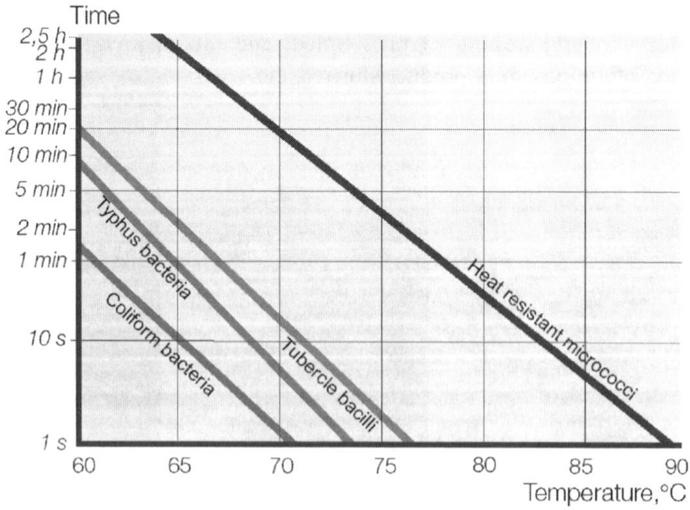

Fig. 11.6 The effect of micro-organism inactivation (Bylund, 1995).

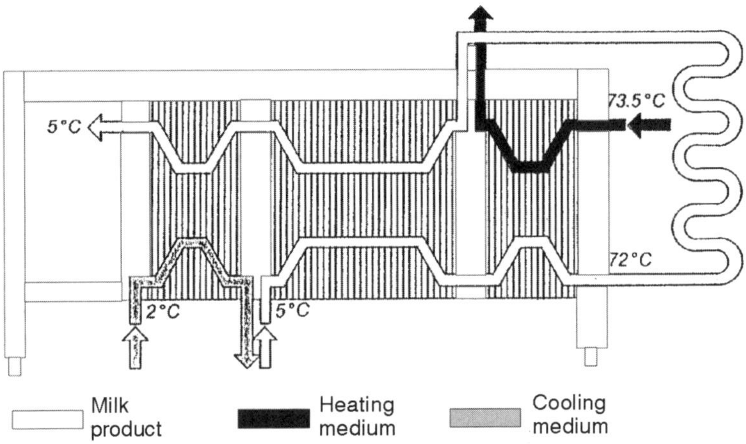

Fig. 11.7 Flow sheet for milk pasteurisation.

For milk pasteurisation, the milk must be kept at 72ºC for not less than 16 seconds. The most common pasteurisation method is the continuous processing, where PHEs are widely used (Bylund, 1995). A typical flow sheet is shown in Fig. 11.7, where a large three-section PHE with dividing frames is used. The cold raw milk at about 5ºC from a constant level tank (not shown in the figure) is drawn into the regenerative part of the PHE (middle section). Here the milk is warmed to approximately 57ºC to 68ºC by the heat given by the hot pasteurised milk flowing in the neighbouring channels. Afterwards, the warm raw milk is continuously heated to at least 72ºC in the right section of the exchanger by either vacuum steam or hot water. The milk, at pasteurisation temperature and under pressure, flows through the holding tube where it is held for at least 16 seconds. Thereafter, the properly pasteurised milk flows through the regenerative part (middle section) where it gives up heat to the raw milk and in turn is cooled to approximately 32ºC to 9ºC. Finally, the warm milk passes through the cooling part of the exchanger (left section) where it is cooled to 5ºC or below by the cooling water. Now, the final product of milk has been produced, and is ready for packaging.

There are many reasons why plate-and-frame heat exchangers have been selected in milk pasteurisation, presently and historically. First, plate-and-frame heat exchangers can be opened and complete cleaning is thus possible, which meets the hygienic requirement of the food industry. Second, high heat transfer coefficients of PHEs permit very close approach temperature difference, down to 1ºC. This makes it easy to maintain the outlet temperature of pasteurised milk with 72ºC. Because the three parts can be assembled in one PHE unit, it is very compact and thus significant reduction in installation space can be achieved. In addition, it is usually much cheaper than other types of heat exchangers, due to the small heat transfer surface required.

11.3.2 Evaporation units

This is another example where PHEs are used in the food industry. Evaporation process is very important to obtain high concentration products, such as milk, whey, sugar, etc. It is also used as a preliminary step to drying, e.g., milk products intended for milk powder are normally concentrated from an initial solids content of 9–13% to a final concentration of 40–50% total solids before product is pumped to the dryer. Common evaporation process is to boil off water from the solution with the supplied heat. The products to be evaporated are normally heat sensitive and can be destroyed by adding heat. To reduce this impact, evaporation often takes place under vacuum. At the same time, the evaporator should be designed for the shortest possible residence time, to have desired product taste. In addition, a multiple-effect evaporator is often used in order to save energy. This means two or more units operate at progressively lower pressures and thus with progressively lower boiling points. In such an arrangement, the vapour produced in the previous step can be used as heating medium in the following step. A typical flow sheet of a multiple-effect evaporator is schematically shown in Fig. 11.8.

Although most of the evaporation systems in use today are based on shell-and-tube heat exchangers, this has been changed during recent decades, and PHEs are good options in this system. There are several reasons for such a system to employ PHEs as evaporators. Firstly, a plate evaporator is much less bulky than traditional tubular evaporators. A tubular evaporator with a height of 10–12 m can be replaced by a plate evaporator less than one-quarter of the size – with lower costs for foundations and piping. The small size reduces hold-up volume and residence time. Therefore, the evaporation system can be started up and shut down quickly, and is also quick to reach a stable state. These factors help to improve the product quality, and are especially important for products that are sensitive to heat. Because they take up so little space, plate evaporators can be used very effectively as boosters, working in parallel with existing evaporation systems to increase production. They are also very flexible – they

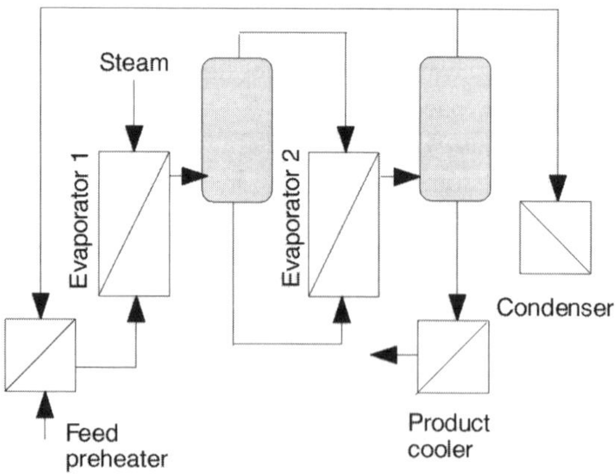

Fig. 11.8 A two-stage evaporation unit.

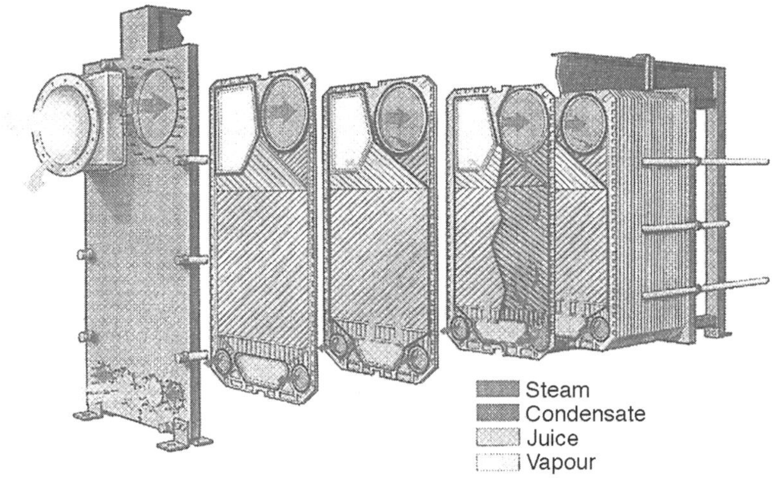

Fig. 11.9 Plate evaporator (courtesy of Alfa Laval).

can be extended, for instance, simply by adding more plates. A further benefit of employing plate evaporators is much greater heat transfer efficiency of the evaporation system. While tubular evaporators can only manage a difference of 5–10°C between the two media, plate evaporators achieve a difference of just 3–5°C. This means more evaporator steps can be used in series, achieving higher concentrations of the end product, and using less steam, and hence less energy.

Due to the density differences at the inlet and outlet, plate evaporators usually have specially designed ports. An example is shown in Fig. 11.9, where big ports are provided for low-density fluids, steam and vapour. This design is to avoid high velocity in ports, which contributes to better fluid distribution: it also improves the operation itself, helps to prolong the exchanger's lifetime, and reduces the pressure inside the port.

It must be pointed out that the applications of PHEs in food industry are too numerous to be listed fully in this chapter. The pasteurisation and evaporation units outlined in this section are quintessentially representative, and serve as an introduction to readers. Many other processes in the food industry, e.g., various heating and cooling processes in pretreatment and fermentation, also involve numerous PHEs, which should not be ignored.

11.4 Improving the design of plate heat exchangers: modelling pressure and heat transfer

The prediction of thermal-hydraulic performance is very important in the design of PHEs. For single-phase applications, most efforts have been conducted in experiments. Although some work has also been carried out by computational fluid dynamics (CFD), this method still suffers from difficulties in treating complex geometries, deficiencies in turbulence modelling, long computational

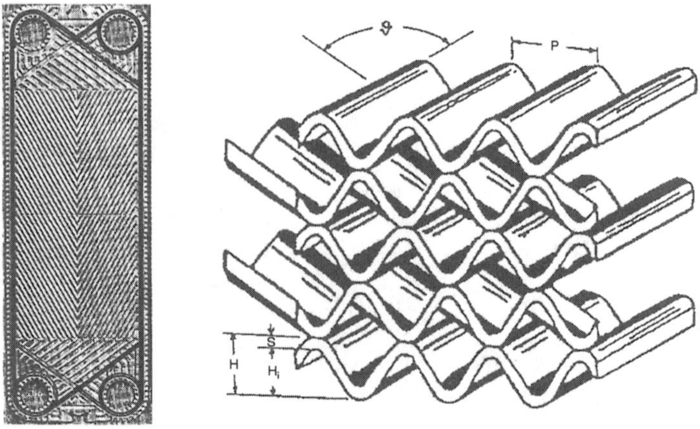

Fig. 11.10 Chevron wave pattern.

time, and others. For two-phase applications, all previous effects are experimental, and the aim has been to develop correlations for accurate prediction of thermal-hydraulic performance. In this section, the most common correlations for predicting thermal-hydraulic performance are summarised, for both single- and two-phase applications. During the past decades, chevron wave pattern (see Fig. 11.10) has proved to be the most successful design offered by the majority of manufacturers (Martin, 1996, 1999). The summarised correlations are mostly applied to this type of plate pattern.

11.4.1 Single-phase applications

A number of detailed experimental studies, using model corrugation patterns and systematically varying parameters like amplitude, wavelength, inclination angle and flow rate, have generated a relatively large amount of interesting facts about heat transfer and pressure drop in PHEs (Okada *et al.*, 1972; Focke *et al.*, 1985; Muley and Manglik, 1999). Based on the individual results, many different correlations for heat transfer and pressure drop have been proposed (Manglik, 1996), and they usually take the following form (for single-phase flows)

$$\mathrm{Nu} = a\,\mathrm{Re}^b\,\mathrm{Pr}^{c_1}(\mu/\mu_w)^{c_2} \tag{11.1}$$

$$\Delta P = 2f\frac{L}{D_e}\frac{G^2}{\rho} \tag{11.2}$$

$$f = f(\mathrm{Re}) = c\,\mathrm{Re}^d \tag{11.3}$$

where D_e is the equivalent diameter, and is defined as

$$D_e = \frac{4WH_i}{2(W + H_i)} \approx 2H_i \tag{11.4}$$

where W is the plate width and is normally very large compared to the internal height H_i.

The corresponding coefficients in Eq. (11.1) and Eq. (11.3) have specific values for specific plates, although in most correlations the values of c_1 and c_2 are taken as 0.33 and 0.17, respectively. These coefficients are usually obtained through individual experiments. However, as the general forms are already determined, the specific experiment does not require a huge database to determine these coefficients. As a matter of fact, this is the way most of the commercial PHE manufacturers are determining their own correlations.

It is of interest to have a general correlation for different plates. Clearly, such correlations should incorporate characteristic parameters of the plates. Muley and Manglik (1999) have made such attempts, and acclaimed that the following equations are suitable for different chevron-type PHEs:

$$\begin{aligned}\mathrm{Nu} = {} & \left(0.2668 - 0.006967\theta + 7.244 \times 10^{-5}\right) \\ & \times \left(20.78 - 50.94\varphi + 41.16\varphi^2 - 10.51\varphi^3\right) \\ & \times \mathrm{Re}^{[0.728+0.054\sin(\pi\theta/45+3.7)]}\mathrm{Pr}^{1/3}\left(\frac{\mu}{\mu_w}\right)^{0.14}\end{aligned} \tag{11.5}$$

$$\begin{aligned}f = {} & \left(2.917 - 0.1277\theta + 2.016 \times 10^{-3}\theta^2\right) \\ & \times \left(5.474 - 19.02\varphi + 18.93\varphi^2 - 5.341\varphi^3\right) \\ & \times \mathrm{Re}^{-[0.2+0.0577\sin(\pi\theta/45)+2.1]}\end{aligned} \tag{11.6}$$

where θ and φ are the corrugation angle and plate surface enlargement factor (ratio of effective corrugated surface area to its projected area), respectively. These equations are valid for the range of $2 \le \mathrm{Pr} \le 6$, $\mathrm{Re} \ge 1000$, $30^{\circ} \le \theta \le 60^{\circ}$ and $1 \le \varphi \le 1.5$. The claimed accuracies for Nusselt number Nu and friction factor f are 10% and 5%, respectively. Because these correlations were based on only three different PHEs, more tests are necessary in order to confirm them.

In addition, Martin (1996, 1999) developed general correlations for chevron-type PHEs. The equations are based on a model of the flow pattern in the channels of some compact heat exchangers. The numerical constants in these equations have been fitted to experimental data from the literature. The Fanning friction factor is:

$$\frac{1}{\sqrt{f}} = \frac{\cos\theta}{\sqrt{0.045\tan\theta + 0.09\sin\theta + f_0/\cos\theta}} + \frac{1-\cos\theta}{\sqrt{3.8 f_1}} \tag{11.7}$$

where $f_0 = 16/\mathrm{Re}$ for $\mathrm{Re} < 2000$ and $f_0 = (1.56\ln\mathrm{Re} - 3.0)^{-2}$ for $\mathrm{Re} \ge 2000$

$$f_1 = \frac{149}{\mathrm{Re}} + 0.9625 \text{ for } \mathrm{Re} < 2000 \text{ and } f_1 = \frac{9.75}{\mathrm{Re}^{0.289}} \text{ for } \mathrm{Re} \ge 2000$$

The correlation for the heat transfer is given as:

$$\mathrm{Nu} = 0.205\,\mathrm{Pr}^{1/3}\left(\frac{\mu_m}{\mu_w}\right)^{1/6}\left(f\,\mathrm{Re}^2\sin 2\theta\right)^{0.374} \tag{11.8}$$

It is claimed by Martin that the friction correlation of Eq. (11.7) is valid for the corrugation angle within 0–80°, and is accurate within –50% and +100%. The Nusselt correlation of Eq. (11.8) is valid for the corrugation angle within 10–80°, and is accurate within ±20%.

11.4.2 Two-phase applications

The main discussion here is limited to evaporation and condensation. The ideal correlations should be able to predict the phase-change heat transfer and pressure drop for different plate patterns and different working fluids. However, a literature survey shows this seems impossible, and in practice specific correlations were developed for specific situations. This is because the thermal-hydraulic performances are dependent on many factors, such as fluid property, plate geometry, system pressure, mass flow rate, vapour quality, etc. The large number of influence factors means that accurate prediction is somewhat difficult. Therefore, most correlations were developed experimentally for specific situations. For the hydraulic performance, the total pressure drop in PHEs consists of several parts: frictional pressure drop, gravity pressure drop, acceleration pressure drop and some additional pressure drops. The frictional pressure drop is usually the main part, and will be discussed below. The details of other individual parts are referred to Wang *et al.* (2003).

Evaporation

The common working fluids in plate evaporators include refrigerants and pure or mixtures of hydrocarbons. A literature survey shows that most of previous work was conducted for refrigerants, and came out with qualitative conclusions instead of refined correlations for predicting the heat transfer coefficient and pressure drop. The reason for this is probably due to the fact that the evaporation process is very complicated and geometry dependent.

The evaporation process is usually divided into the nucleate boiling dominated regime, two-phase convective boiling dominated regime and a regime where both nucleate and two-phase convective boiling contributes. Because the vapour quality changes inside plate channels, it is believed that both nucleate and two-phase convective boiling contributes to the heat transfer in plate evaporators. In previous results, the influences from various factors are not conclusive and dependent on various processes (Palm and Thonon, 1999). Most correlations proposed have a similar format with Chen (1963), Shah (1976), Steiner and Taborek (1992) and others. These correlations take the form:

$$h = h_{nb}S_{nb} + h_{cb}F_{cb} \tag{11.9}$$

where h_{nb} is the heat transfer coefficient for pool boiling; h_{cb} is the heat transfer coefficient for single-phase liquid flow; S_{nb} the nucleate boiling suppression factor; and F_{cb} is a two-phase multiplier. Obviously, Eq. (11.9) is based on the assumption that the two mechanisms (nucleate boiling and convective two-phase boiling) are additive. Various investigators have proposed different correlations for respective heat transfer contributions h_{nb}, S_{nb} and h_{cb}, F_{cb} in their specific cases. The details can be found in Palm and Thonon (1999).

However, Thonon *et al.* (1995) suggested that the evaporation heat transfer coefficient should take the greater value of the nucleate boiling term and the two-phase convective boiling term. They gave the following criteria to distinguish whether the flow was nucleate boiling dominated or two-phase convective boiling dominated.

$$Bo \cdot X_{tt} > 0.00015 \qquad \text{nucleate boiling} \tag{11.10}$$

$$Bo \cdot X_{tt} < 0.00015 \qquad \text{convective boiling} \tag{11.11}$$

where Bo is the boiling number defined in Eq. (11.12) and X_{tt} is the Lockhart-Martinelli parameter defined in Eq. (11.13). Thonon *et al.* used the Gorenflo (1993) correlation for the nucleate boiling dominated region, and his own correlation for two-phase convective boiling region.

$$Bo = \frac{\dot{q}}{G\gamma} \tag{11.12}$$

For evaporation pressure drops, previous investigators have indicated that the frictional pressure drop increases by increasing vapour quality and mass velocity, but decreases by increasing system pressure. For the prediction, the most widely accepted correlation is probably the Lockhart-Martinelli model (Lockhart and Martinelli, 1949; Chisholm, 1967), which was originally developed from a series of tests of isothermal two-phase, two-component flows in horizontal tubes. The Lockhart-Martinelli parameter X and the two-phase friction multiplier ϕ_l for the liquid are expressed as:

$$X^2 = \frac{\Delta P_l}{\Delta P_v} \tag{11.13}$$

$$\phi_l^2 = \frac{\Delta P_f}{\Delta P_l} \tag{11.14}$$

where ΔP_l is the liquid phase frictional pressure drop, ΔP_v is the vapour phase pressure drop and ΔP_f is the two-phase pressure drop. ΔP_l and ΔP_v are calculated from single-phase equations for the prediction of pressure drop in PHEs, assuming liquid and vapour flow alone. The relationship between ΔP_l and ΔP_v has been established for different applications. For horizontal tubes, Chisholm (1967) recommended the following correlation:

$$\phi_l^2 = 1 + \frac{C}{X} + \frac{1}{X^2} \tag{11.15}$$

For two-phase flows in tubes, the value of constant C is chosen according to:

liquid	*vapour flow condition*	*C*
turbulent	turbulent T-T	20
viscous	turbulent V-T	12
turbulent	viscous T-V	10
viscous	viscous V-V	5

For evaporation in PHEs, the selection of the value is quite contradictory in the literature. A value of 8 was suggested by Thonon *et al.* (1995), but a value of 3 was suggested by the same group, based on their own experimental work (Margat *et al.*, 1997). In the experimental work conducted by Sterner (1999), this value is a function of the Reynolds number, and may vary from 1 to 200, well beyond the value recommended for plain tubes (between 5 and 20). A clarification on this issue is certainly needed, and further discussion will be made in the condensation part of this chapter.

Condensation

In plate condensers, the condensing vapour usually has a very high speed, thus shear-controlled condensation usually prevails in most applications. Condensation heat transfer in PHEs is influenced by flow rate, vapour quality, vapour pressure, medium properties as well as plate pattern. It has been demonstrated by respective experiments (Yan and Lin, 1999; Wang *et al.*, 2000) that the condensation heat transfer coefficient increases by increasing flow rate and vapour quality, but decreases by increasing system pressure. The previous correlations developed for shear-controlled condensation in tubes can be employed as a rough estimation for plate condensers. However, verification must be carried out, and further development is certainly needed.

The classical Nusselt equation (Nusselt, 1916) was derived for gravity-controlled conditions. The equation reads:

$$h = \left[\frac{\rho_l(\rho_l - \rho_g) g \gamma_{lg} k_l^3}{4\mu_l z (T_{gi} - T_w)} \right]^{1/4} \tag{11.16}$$

where z indicates the local position, and γ is the latent heat. This equation sets a bottom limit for prediction of the condensation heat transfer in PHEs.

The equations developed by Boyko and Kruzhilin (1967) and Shah (1979) are used frequently. These equations were developed for shear-controlled condensation in horizontal tubes, and perform better than the classical Nusselt equation for condensation in PHEs. However, the equation developed by Wang *et al.* (2000) is recommended, which is a modified version of the Boyko and Kruzhilin equation. This equation was developed from steam condensation in several different PHEs, and good accuracy was claimed. It reads:

$$h = h_l \left(\frac{\rho_l}{\rho_m} \right)^{(a + b \cdot \mathrm{Re}_l^c)} \tag{11.17}$$

where the constants a, b and c are in ranges 0.3 to 0.37, 5.0 to 6.0 and -0.6 to -0.64, respectively. This equation can probably be extended to condensation of other fluids, e.g., refrigerants, but respective thermal properties should be applied.

For condensation pressure drop, frictional pressure drop of condensation is a function of flow rate, vapour quality, system pressure, etc. It has been demonstrated by respective experiments (Yan and Lin, 1999; Wang *et al.*, 2000) that frictional pressure drop increases with increasing flow rate and vapour quality, but decreases with increasing system pressure. For the prediction, the widely accepted method is to apply the Lockhart-Martinelli model, as the same as plate evaporators. The main issue here is still to determine the appropriate value of C in Eq. (11.15). For steam condensation in PHEs, a value of 16 was recommended for variable C by Wang *et al.* (2000), to predict the average frictional pressure drop.

In consideration of the prediction of evaporation pressure drop in PHEs, it is obvious that a big deviation of the C value exists for different applications. One interesting finding by Holt *et al.* (1997) is that this value is a function of the hydraulic diameter. This was brought up from their experimental results in ducts with different cross-sections (circular, trapezoidal and rectangular) for different mixtures (air/water, helium/water and air/60% aqueous glycerol). However, bringing the findings from other researchers in the same figure shows that the deviation is still big (see Fig. 11.11). This may suggest that the value of C is probably a function of geometry, fluid properties, flow condition, etc. Further work must be carried out to clarify this issue.

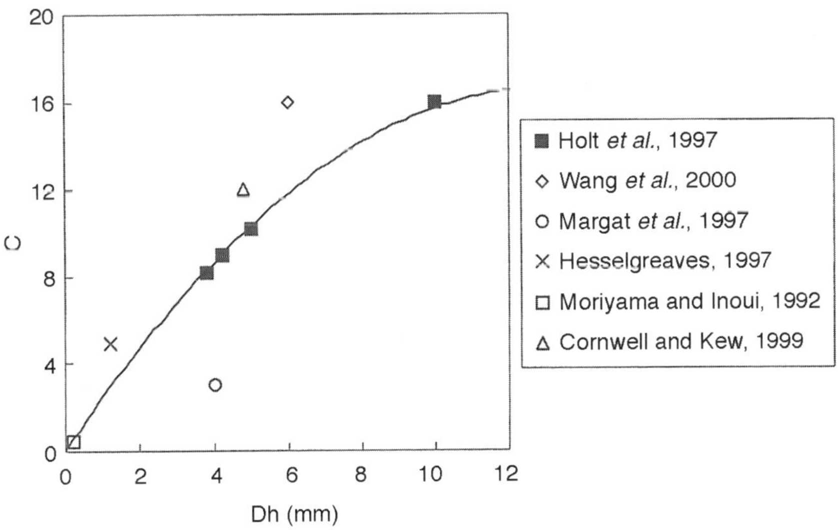

Fig. 11.11 *C* value with hydraulic diameters (Hesselgreaves, 2000).

11.5 Future trends

In order to maintain and expand their industrial applications, PHEs have to be continuously developed. The future developments can be generally divided into two categories: construction and performance. The issue of construction covers the development of new PHE models, the use of new materials, etc. These will result in increased operating limits in both temperature and pressure. New materials can also diminish the threat of corrosion, and permit a wider range of working fluids. Due to the scope of this chapter, no further details will be given on this subject, and the following discussion will concentrate on the general issue of both thermal and hydraulic performance, from the view of all applications including food industry. The following three aspects are believed to be the main subjects in future research and development: more accurate prediction for single-phase and two-phase flows, consideration of maldistribution and mitigation of fouling.

In single-phase flows, one issue is the lack of universal correlations for predicting the thermal and hydraulic performance. Hence, future efforts must be conducted to develop correlations, which can reflect the influences from pitch, wavelength, corrugation angle, plate length and width, etc. The other issue is to develop more accurate correlations for non-Newtonian fluids. Non-Newtonian fluids have significantly different behaviour from Newtonian fluids (Cho and Hartnett, 1985), and to cover this influence is not an easy task. With regard to phase-change processes, the challenge is much more ahead. In the previous sections, it was already demonstrated that the current knowledge is unable to predict accurately either heat transfer coefficients or pressure drops for both evaporation and condensation. More research must be conducted with regard to different working fluids, different operation conditions and different PHE models. In addition, it would be of great interest to visualise the flow, in order to better understand the mechanisms of condensation and evaporation processes in PHEs. Different measurement techniques could be employed, such as liquid crystal thermography for detailed plate surface temperature measurement, laser techniques and high speed camera together with transparent plates for velocity measurement, etc. With such efforts, the fundamental mechanisms of this phase-change process can be revealed, and the correlations for predicting local heat transfer and pressure drop can be developed. For both single-phase and phase-change applications, the CFD method is another promising approach in addition to the experimental work. With the rapid development of computer technology and improved models for the physical transport processes, this method will become more attractive and effective. However, the complex geometry and turbulence modelling are certainly obstacles for the CFD calculations. For phase-change processes, the tracking of the vapour-liquid interface is another obstacle. This means that effective computation algorithms must be developed, which is certainly a challenge for future research.

Research in the future must also be conducted for the reduction of maldistribution in PHEs. Large PHEs working in phase-change applications are more prone to this problem due to the differences in density and velocity of the fluids in the exchanger. The prediction of maldistribution in single-phase applications has been discussed by Heggs and Scheidat (1992), and the appropriate flow arrangement has been suggested for reducing this phenomenon. However, for phase-change processes, there is no effective method to predict this effect, and most efforts were carried out on the development of effective inlet flow distributors to achieve better distribution. The current methods for reducing the maldistribution in phase-change applications include multiple and smaller units, feed from two sides, restrictions at inlet to each individual plate channel, manifold pipe at the inlet, etc. (Holm *et al.*, 2001). However, these methods are either too expensive, or too inconvenient, or associated with some other disadvantages. Therefore, the model for predicting the maldistribution in phase-change applications should be established, and the methods for reducing or preventing maldistribution should be developed in the future.

Due to the high cost of the fouling effect, the research in fouling formation and prevention will continue to be important in the future. It has already been recognised that the traditional concept of fouling resistance is not enough for the accurate design and operation of PHEs. Efforts must be carried out to understand the fundamental mechanisms of fouling formation, and its influence on the heat transfer and pressure drop performance. Various types of fouling mechanisms should be identified, and the influence factors, e.g., temperature, pressure, velocity and concentration, must be investigated. It should be noticed that the maldistribution on the plate surface is an important factor for the local fouling formation as well as the local corrosion formation. Preventing local velocity maldistribution is beneficial to minimise the local hot or cold spots, which is good in reducing both fouling and corrosion. After knowing the fundamental mechanism of fouling formation, design methods for reducing or preventing fouling can be developed. On the other hand, fouling is an inevitable process, and all heat exchangers will be fouled eventually. Therefore, work should also be conducted to develop cleaning methods as well as cleaning strategies.

The above discussion has presented big challenges for the future research for PHEs. If these challenging problems can be solved it will result in less costly but more reliable PHEs. This will certainly improve PHEs competitiveness in the market and widen their industrial applications. Moreover, this is significant to the modern society in respect of conservation of limited resources and development of sustainable energy systems.

11.6 Conclusions

The development of PHEs in food industry has been summarised in terms of application and performance. Thanks to their characteristics, PHEs are widely applied in food processing applications, from pasteurisation to heat recovery systems. During the past decades, the applications have been significantly enlarged in food industry because of new developed models from original plate-and-frame heat exchangers. The correlations for predicting thermal and hydraulic performance in single-phase flows are well established, but a general correlation for handling different plate patterns is still under development. For evaporation and condensation processes, further considerable efforts must be carried out to develop correlations for more accurate predictions of thermal and hydraulic performance. In order to improve overall performance of PHEs, future research should be directed to develop methods for considering and diminishing maldistribution and mitigation of fouling. All these call for considerable theoretical work as well as experimental investigations.

11.7 Sources of further information and advice

(1) WANG L, MANGLIK R M and SUNDÉN B (2003), *Plate Heat Exchangers: Design, Applications and Performance*, WIT Press, Southampton, UK.
This is a comprehensive monograph which gives a state-of-the-art review of knowledge on PHEs. Almost all the related subjects on PHEs are covered in this monograph, which consists of ten chapters: Basic features and development of PHEs; Construction and operations; Industrial applications; Materials and manufacturing; Basic design methods; Multi-pass flow arrangement; Thermal-hydraulic performance in single-phase flow; Thermal-hydraulic performance in condensers and evaporations; Fouling, corrosion and erosion; Extended design and operation issues.

(2) BYLUND G (1995), *Dairy Processing Handbook*, Tetra Pak Processing System AB, Lund, Sweden.
This handbook is published by a big international company, Tetra Pak, which provides integrated processing, packaging and distribution line, and plant solutions for liquid food manufacturing. This is a good resource for obtaining the details of dairy processing, and understand why PHEs are frequently employed in such environments.

(3) HESSELGREAVES J E (2001), *Compact Heat Exchangers: Selection, Design and Operation*, Elsevier, Amsterdam.
This is a good information resource for obtaining knowledge about other types of compact heat exchangers. There is a comprehensive list of major manufacturers and some research organisations of compact heat exchangers.

(4) Some companies' websites provide very fruitful information on PHEs and relevant thermal processing systems. It is also good to contact these companies directly for further information. Some examples are:

Alfa Laval Lund AB	http://www.alfalaval.com
APV	http://www.apv.com
SWEP	http://www.swepphe.com
Tranter	http://www.tranterphe.com
Sondex A/S	http://www.sondex.com
Hisaka Works	http://www.hisaka.co.jp
GEA Ecoflex GmbH	http://www.gea-ecoflex.de

11.8 List of symbols

Bo	boiling number
C	constant in Chisholm correlation
D_e	equivalent diameter, m
D_h	hydraulic diameter, m
f	Fanning friction factor
F	two-phase multiplier
g	gravity acceleration, m^2/s
G	mass velocity, $kg/(m^2 \cdot s)$
h	heat transfer coefficient, $W/(m^2 \cdot K)$
H_i	internal height, m
k	thermal conductivity, $W/(m \cdot K)$
L	plate length, m
Nu	Nusselt number
Pr	Prandtl number
$\dot{q}$	heat flux, W/m^2
Re	Reynolds number
S_{nb}	nucleate boiling suppression factor
T	temperature, K
W	plate width, m
X	Lockhart-Martinelli parameter
ΔP	pressure drop, Pa
ΔP_f	two-phase frictional pressure drop, Pa
ΔP_l	frictional pressure drop of liquid phase, Pa
ΔP_v	frictional pressure drop of vapour phase, Pa

Greek symbols

ϕ	two-phase friction multiplier
φ	surface enlargement factor
μ	dynamic fluid viscosity, $kg/(m \cdot s)$
γ	latent heat, J/kg
θ	corrugation angle, degree
ρ	fluid density, kg/m^3

Subscript

cb	two-phase convective boiling
g	gas or vapour
i	interface
l	liquid
m	mean value
nb	nucleate boiling
tt	turbulent-turbulent

v vapour
w wall

11.9 References

BOYKO L D and KRUZHILIN G N (1967), 'Heat transfer and hydraulic resistance during condensation of steam in a horizontal tube and in a bundle of tubes', *Int. J. Heat Mass Transfer*, 10 (2), 361–373.

BYLUND G (1995), *Dairy Processing Handbook*, Tetra Pak Processing System AB, Lund, Sweden.

CHEN J C (1963), 'A correlation for boiling heat transfer to saturated fluids in convective flow', *6th National Heat Transfer Conference*, Boston, August 11–14.

CHISHOLM D (1967), 'A theoretical basis for the Lockhart-Martinelli correlation for two-phase flow', *Int. J. Heat Mass Transfer*, 10, 1767–1778.

CHO Y I and HARTNETT J P (1985), 'Nonnewtonian fluids', in Rohsenow W M, Hartnett J P and Ganic E N, *Handbook of Heat Transfer Applications* (2nd edition), McGraw-Hill, New York, Chapter 2.

CORNWELL K and KEW P A (1999), 'Compact evaporators', in Convective flow and pool boiling, *Proceeding of the International Engineering Foundation 3rd Conference*, Irsee, Germany: Taylor & Francis, London.

FOCKE W W, ZACHARIADES J and OLIVIER I (1985), 'The effect of the corrugation inclination angle on the thermohydraulic performance of plate heat exchangers', *Int. J. Heat Mass Transfer*, 28 (8), 469–479.

GORENFLO D (1993), *Pool boiling*, VDI-Verlag, Düsseldorf, Germany.

HEGGS P J and SCHEIDAT H J (1992), 'Thermal performance of plate heat exchangers with flow maldistribution', in Shah R K, Rudy T M, Robertson J M and Hostetler K M, *Compact Heat Exchangers for Power and Process Industries*, HTD-Vol. 201, 87–93.

HESSELGREAVES J E (1997), 'Single phase and boiling performance of a novel highly compact heat exchanger surfaces', *5th UK National Heat Transfer Conference*, Session D, IChemE, London.

HESSELGREAVES J E (2000), *Compact Heat Exchangers: Selection, Design and Operation*, Elsevier, Amsterdam.

HOLM M, NILSSON I and WILHELMSSON B (2001), 'Reduction of maldistribution in large rising film plate evaporators', in Shah R K, Deakin A W, Honda H and Rudy T M, *Compact Heat Exchangers and Enhanced Technologies for the Process Industries*, Begell House, New York, pp. 519–524.

HOLT A J, AZZOPARDI B J and BIDDULPH M W (1997), 'Two-phase pressure drop and void fraction in channels', *5th UK National Heat Transfer Conference on Heat Transfer*, Session D, IChemE, London.

LOCKHART R W and MARTINELLI R C (1949), 'Proposed correlation of data for isothermal two-phase, two-component flow in pipes', *Chem. Eng. Prog.*, 45 (1), 39–48.

MAGNUSSON B (1985), *The Origins and Evolution of the Alfa Laval Plate Heat Exchanger*, Norstedts Tryckeri, Stockholm, Sweden.

MANGLIK R M (1996), 'Plate heat exchangers for process industry applications: enhanced thermal-hydraulic characteristics of chevron plates', in Manglik R M and Kraus A D, *Process, Enhanced and Multiphase Heat Transfer*, Begell House, New York, 267–276.

MARGAT L, THONON B and TADRIST L (1997), 'Heat transfer and two-phase flow characteristics during convective boiling in a corrugated channel', in Shah R K, Mochizuki S and Wadekar V V, *Compact Heat Exchangers for the Process Industries*, Begell House, New York, 323–330.

MARTIN H (1996), 'A theoretical approach to predict the performance of chevron-type plate heat exchangers', *Chem. Eng. Process.*, 35, 301–310.

MARTIN H (1999), 'Economic optimisation of compact heat exchangers', in Shah R K, Bell K J, Honda H and Thonon B, *Compact Heat Exchangers and Enhancement Technology for the Process Industries*, Begell House, pp. 75–80.

MORIYAMA K and INOUE A (1992), 'The thermohydraulic characteristic of two-phase flow in extremely narrow channels', *Heat Transfer – Japanese Research*, 21 (8).

MULEY A and MANGLIK R M (1999), 'Experimental study of turbulent flow heat transfer and pressure drop in a plate heat exchanger with chevron plates', *ASME J. Heat Transfer*, 121 (1), 110–117.

NUSSELT W (1916), 'The condensation of steam on cooled surfaces', *Z. Ver. Deutsch. Ing.*, 60, 541–546 and 569–575. (Translated into English by Fullarton D (1982), *Chem. Eng. Fund.*, 1 (2), 6—19.)

OKADA K, ONO M, TOMINURA T, OKUMA T, KONNO H and OHTANI S (1972), 'Design and heat transfer characteristics of new plate heat exchangers', *Heat Transfer – Japanese Research*, 1 (1), 90–95.

PALM B and THONON B (1999), 'Thermal and hydraulic performance of compact heat exchangers for refrigeration systems', in Shah R K, Bell K J, Honda H and Thonon B, *Compact Heat Exchangers and Enhancement Technology for the Process Industries*, Begell House, pp. 455–462.

SELIGMAN R G S (1964), 'The plate heat exchanger in the food industry, part I – Genesis of an idea', *Chemistry and Industry*, Sept., 1602–1603.

SHAH M M (1976), 'A new correlation for heat transfer during boiling flow through pipes', *ASHRAE Trans.*, 82 (2), 66–96.

SHAH M M (1979), 'A general correlation for heat transfer during film condensation inside pipes', *Int. J. Heat Mass Transfer*, 22 (3), 547–556.

STEINER D and TABOREK J (1992), 'Flow boiling heat transfer in vertical tubes correlated by an asymptotic model', *Heat Transfer Engineering*, 13 (2), 43–69.

STERNER D (1999), 'Experiments and analysis of performance of plate heat exchangers as evaporators in refrigerator systems', *Engineering Licentiate Thesis*, Publ. 99/2007, Division of Heat Transfer, Department of Heat and Power Engineering, Lund Institute of Technology, Lund, Sweden.

THONON B, VIDIL R and MARVILLET C (1995), 'Recent research and development in plate heat exchangers', *J. Enhanced Heat Transfer*, 2 (1-2), 149–155.

VALENTAS K J, LEVINE L and CLARK J P (1990), *Food Processing Operations and Scale-up*, Marcel Dekker, New York.

WADEKAR V V (1998), 'Improving industrial heat transfer – compact and not-so-compact heat exchangers', *J. Enhanced Heat Transfer*, 5, 53–69.

WANG L, CHRISTENSEN R and SUNDÉN B (2000), 'An experimental investigation of steam condensation in plate heat exchangers', *Int. J. Heat Exchangers*, 1 (2), 125–150.

WANG L, MANGLIK R M and SUNDÉN B (2003), *Plate Heat Exchangers: Design, Applications and Performance*, WIT Press, Southampton, UK.

YAN Y Y and LIN T F (1999), 'Evaporation heat transfer and pressure drop of refrigerant R-134a in a plate heat exchanger', *ASME J. Heat Transfer*, 121 (1), 118–127.

12

Developments in ohmic heating

R. Ruan, X. Ye and P. Chen, University of Minnesota, USA, and C. Doona and T. Yang, US Army Natick Soldier System Center, USA

12.1 Introduction: ohmic heating principles and technology

12.1.1 Background

Ohmic heating is a thermal process in which heat is internally generated by the passage of alternating electrical current (AC) through a body such as a food system that serves as an electrical resistance. Ohmic heating is alternatively called resistance heating or direct resistance heating, Joule heating, or Electroheating^{TM} (Raztek Corp., Sunnyvale, CA). Extensive research has been done during the past two decades to exploit ohmic heating for food processing, mainly as a result of the increasing shift from batch thermal operations to continuous High Temperature Short Time (HTST) operations. In HTST processes, food is pumped continuously through plate or scraped surface heat exchangers that reach temperatures as high as 140°C. At this temperature, only a few seconds are needed to achieve sterilization, while the products suffer only slight deterioration in quality attributes and retain higher levels of nutrients. HTST processes rely on heat transfer by rapid convection, and are therefore well suited for the preservation of liquid foods. They are limited in their application to particulates, since a processing time of a few seconds is insufficient for heat to transfer to the center of particles more than a couple of millimeters thick and effect sterility. In ohmic heating processes, the food components are parts of the electric circuit through which the alternating current flows and generates heat in the foods based on their intrinsic properties of electrical resistance. For a food item consisting of a liquid-particulate mixture (e.g., beef stew, meat and vegetables in gravy or sauce), heat could be generated at the same or comparable rates in both the liquid and particulate phases using ohmic heating, if the electrical conductivities of the two phases are the same. Ohmic heating thereby provides a technology for processing particulate foods at the rate of an HTST process without the limitations of heat transfer to particulates found in conventional HTST processes.

Applying the concept of ohmic heating of foods is not new. In the nineteenth century, several processes using electrical current for heating flowable materials were patented. In the early twentieth century, 'electric' pasteurization of milk was achieved by passing milk between parallel plates with an applied voltage difference between them, and six states in the US had commercial electrical pasteurizers in operation.[1] In the design of McConnel and Olsson,[2] frankfurter sandwiches were cooked by passing an electric current through them for a predetermined time. Schade in 1951 described a blanching method for preventing the enzymatic discoloration of potato using ohmic heating. This technology virtually disappeared in succeeding years apparently due to a lack of suitable inert electrode materials and suitable control systems. Since that time, the technology has received limited interest, except for electroconductive thawing.[3]

Within the past two decades, new and improved electrode materials and designs for ohmic heating have become available. The Electricity Council of Great Britain has patented a continuous-flow ohmic heater and licensed the technology to APV Baker.[4] The particular interest in this technology emerges from the food industry's ongoing interest in improving aseptic processing of liquid-particulate foods. Conventional aseptic processing systems for particulates rely on heating the liquid phase, and heat transfers from the liquid to the solid phase. Ohmic heating offers an attractive alternative to conventional thermal processing, because of its ability to heat materials through internal heat generation.

Ohmic heating features unique characteristics with associated advantages, which will significantly impact the thermal pasteurization/sterilization processes and the nutritional values of products treated with ohmic heating. Briefly, these characteristics and advantages are:[5,6]

- Heating food materials volumetrically by internal heat generation compared to the limitations of heat transfer in conventional thermal processes or the non-uniformities commonly associated with microwave heating due to the dielectric penetration limit.
- Heating rates in particulate phase similar to or higher than heating rates of the liquid phase can be achieved. Reaching higher temperatures in the particulate center compared to the liquid phase is impossible with conventional heating.
- Reduced risks of fouling on the heat transfer surface and burning of the food product, resulting in minimal mechanical damage of the system and better nutrient and vitamin retention in the food product.
- High-energy efficiency (90% of the electrical energy is converted to heat).
- Optimization of capital investment and product safety as a result of a high solids loading capacity.
- Ease of process control with instant switch-on and shutdown.

- Reduced maintenance cost (no moving parts), and the system is quiet and environmentally friendly.

12.1.2 Principles of ohmic heating

The principles of ohmic heating are illustrated in Fig. 12.1. Ohmic heating is based on the passage of alternating electrical current (AC) through a body such as a liquid-particulate food system that serves as an electrical resistance. AC voltage is applied to the electrodes at both ends of the product body. The rate of heating is directly proportional to the square of the electric field strength, E, and the electrical conductivity. The electric field strength can be varied by adjusting the electrode gap or the applied voltage. The most important factor is the electrical conductivity of the product and its temperature dependence. If the product consists of more than one phase such as in the case of a mixture of liquid and particulates, the electrical conductivity of all the phases has to be considered. The electrical conductivity increases with rising temperature for most food materials, suggesting that ohmic heating becomes more effective as temperature increases, but this could result in runaway heating. Difference in the electrical resistance and its temperature dependence between the two phases can complicate the heating characteristics of the system. Electrical conductivity is influenced by ion content; it is possible to adjust the electrical conductivity of product adding electrolytes (e.g. salts) to manipulate the heating patterns and improve the effectiveness of ohmic heating.

In ohmic heating, microbes are thought thermally inactivated. Other kill mechanisms are also possible. Some suggest that a mild electroporation mechanism may occur during ohmic heating because ohmic heating operated at low frequency (50–60 Hz) allows electrical charges to build up, resulting in pores across cell walls.

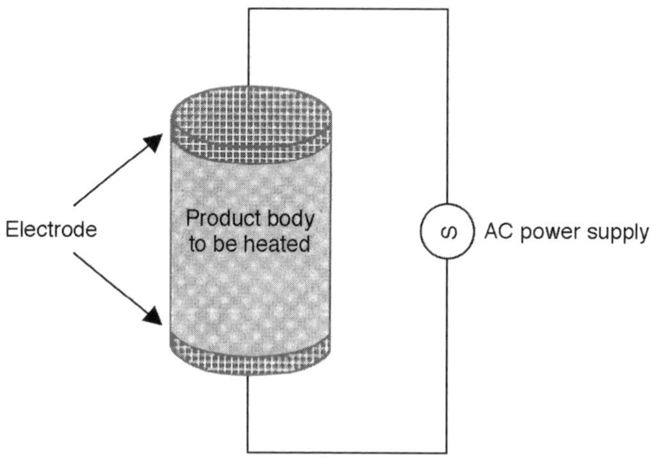

Fig. 12.1 Schematic diagram showing the principle of ahmic heating.

12.1.3 Current status of the technology

Currently, at least 18 ohmic heating operations have been supplied to customers in Europe, Japan, and the US. The most successful of these systems have been exploited for the processing of whole strawberries and other fruits for yogurt in Japan, and low acid ready meals and liquid egg pasteurization in the US.[5] Currently there are two commercial manufacturers of ohmic heating equipment: APV Baker, Ltd., Crawley, UK, and Raztek Corp, Sunnyvale, CA, US. The commercial systems of various capacities of APV Baker have been installed in the food industry, processing a wide variety of high quality, value-added pumpable particulate food products, such as meals and fruit preparations. Raztek's Electroheating™ technology was implemented on an industrial scale at Papetti's high-grade Egg Products in New Jersey, resulting in extended shelf life of their liquid egg products and minimizing production downtime due to necessary cleaning of production equipment.

In the US, a consortium of 25 partners from industry (food processors, equipment manufacturers, and ingredient suppliers), academia (food science, engineering, microbiology and economics) and government was formed in 1992 to develop products and evaluate the capabilities of the ohmic heating system. Research efforts to exploit ohmic heating technology for commercial food processing culminated at that time. A 5–kW pilot-scale continuous-flow ohmic system manufactured by APV Baker, Ltd., Crawley, UK, was evaluated by the consortium at Land-O' Lakes, Arden Hills, Minnesota, from 1992 to 1994. A wide variety of shelf-stable low- and high-acid products, as well as refrigerated extended-shelf-life products was developed. These products featured quality attributes of texture, color, flavor, and nutrient retention that were comparable to or exceeded those characteristic of foods processed by traditional methods such as freezing, retorting, and aseptic processing. The consortium concluded that the technology was a viable alternative food processing technology. In addition to the technical evaluation, an economic study was initiated. Ohmic operational costs were found to be comparable to those for freezing and retorting of low-acid products.[7] While analysis of the operating costs and of the feasibility of the technology for producing value-added food products for the commercial marketplace appear favorable, there remain business risks associated with startup costs and the unknown market potential. For this reason, the consortium did not pursue commercialization of ohmic heating process.

The technological equipment associated with aseptic food processing (i.e., pumps, fillers, heater electrode, etc.) has developed significantly in recent years to afford options for process design. Despite these improvements, the identification, control, and validation of all the critical control points required to demonstrate that a multi-phase food product treated with ohmic processing has been rendered commercially sterile is more difficult than it is for conventional heating (such as canned food products sterilized by retorting). Consequently, the Food and Drug Administration (FDA) has no current filing of continuous ohmically processed multi-phase food products.

Ohmic heating processing has the promise to provide food processors with the opportunity to produce new, high-value-added, shelf-stable products with a quality previously unrealized with current sterilization techniques. Applications which have been developed include aseptic processing of high-value-added ready-prepared meals for storage and distribution at ambient temperature; pre-heating of food product prior to in-can sterilization; and the hygienic production of high-value-added ready-prepared foods for storage and distribution at chilled temperatures. Ohmic heating can also be used for heating high-acid food products such as tomato-based sauces prior to hot-filling, with considerable benefits in product quality. Other potential applications include rapid heating of liquid food products, which are difficult to heat by conventional technologies.[8]

There are three major challenges hindering the commercialization of ohmic heating processing. First, differences in the electrical conductivities of the liquid and solid phases, and variations in the responses of the two phases to increasing temperature, which can cause irregular, complex heating patterns and difficulty in modeling or predicting the heating characteristics of the particulates and carrying media comprising the system, Second, there is a general lack of data regarding critical factors affecting the heating (residence time, particulate size, geometry, and orientation, the relative ratios of electrical conductivity between phases, loading rates, etc.). Third, there is a lack of temperature monitoring techniques for profiling heat distribution and locating potential cold or hot spots during the ohmic heating process.

12.2 Ohmic heating engineering: design and process control

12.2.1 Flow chart and key equipment design

Figure 12.2 is a schematic diagram of a continuous-flow ohmic heating process. A viscous food product containing particulates enters the continuous-flow ohmic heating system via a feed pump hopper. The product then flows past a series of electrodes in the ohmic column, where it is heated to process temperature. Subsequently, the product enters a holding tube for a fixed time sufficient to achieve commercial sterility. The product flows from the holding tube, through tubular coolers, and eventually into hold tanks, where it is stored until filling and aseptic packaging.

Most ohmic heating system configurations consist of three modules: heater assembly, power supply, and control panel. Equipment design is a critical factor that should be considered. The reason for the early failure of ohmic heating was the unavailability of inert electrode materials and sufficiently accurate control equipment to maintain the temperature within the appropriate range that was also sufficiently robust to withstand the conditions of commercial production. Currently, commercially available designs feature electrodes that are located at various positions along the length of the product flow path (in-line field), or located perpendicular to the flow (cross-field). The principal difference in these arrangements is the distribution of E.

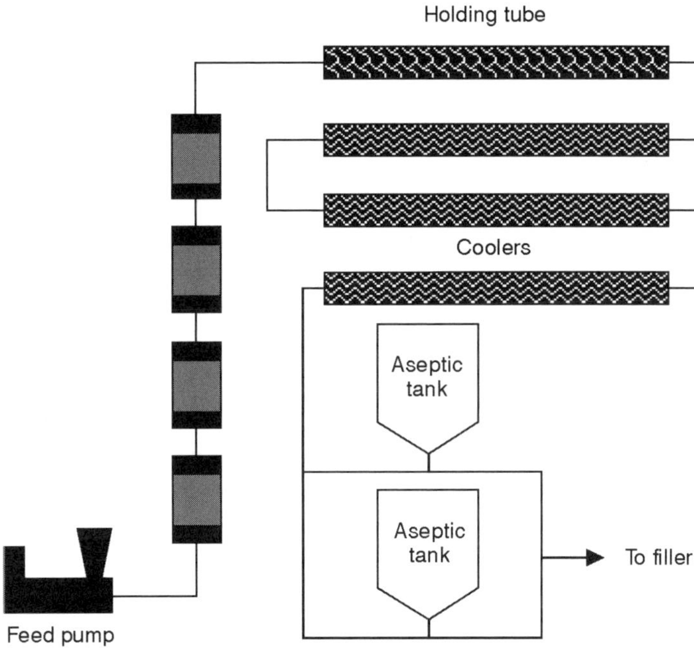

Fig. 12.2 Schematic of a continuous-flow ohmic heating process.

Previous designs attempted to construct electrodes from materials ranging from graphite to aluminum or stainless steel. In food processing, high standards of hygiene are required, and electrodes must be designed carefully. In the early designs, the electrolytic effect that causes the dissolution of the metallic electrodes was completely neglected, and material technology had not progressed to the stage that a suitable inert electrode material was available. For contemporary technologies, such as the APV ohmic process, the use of a food-compatible electrode material and the correct electrical current density has eliminated problems of electrode contamination. Other ways to overcome this problem include utilizing high power frequency, since at alternating frequencies above 100 kHz, there is no apparent metal dissolution. Raztek's current third generation Electroheating™ systems feature the use of common high voltage AC power available worldwide at 50 or 60 Hz. The Raztek Electroheater is constructed of non-conductive FDA-approved materials such as GE Ultemä. These specially treated, pure-carbon electrodes are employed to avoid metal dissolution by electrolysis. Raztek's full-scale industrial Electroheating™ systems are multistage units, configured to take advantage of cost effective three-phase AC power delivery systems.

Designing an ohmic heater for a particular application is somewhat more specific than designing or choosing a heat exchanger. To optimize the ohmic heating process and benefit from its potential advantages over conventional heating, the ohmic heater should be tailored to the specifications of the application. The major design considerations of an ohmic system should include the following.

- *Product characteristics*. The electrical conductivity of the specific product and its change over the range of the temperature rise are the major parameters for the design. The electrical resistance of the product will determine the current density. The actual electrical resistance of an ohmic heating device is a function of the specific resistance of the product and the geometry of the device.
- *Flow rate*. The maximum flow rate of the food product through the ohmic system will determine the power requirement to affect the appropriate temperature increase. In particular, ohmic heating systems heat at a very fast rate, and even small delays in the flow of some of the food product could lead to large differences in temperature. This results when the velocity of the product is not uniform in the cross-section, and the dwell time of the slower moving fluid in the ohmic heater is longer. It is therefore important to avoid even small differences in flow velocity in the cross-section.
- *Temperature rise*. The temperature of the product at the heater entrance and exit determines the power requirement.
- *Holding time*. Using ohmic heating may offer an opportunity to raise the temperature to much higher levels than used in conventional heat exchangers. Pasteurization and sterilization are a function of temperature and time. Higher temperatures require shorter holding times. Possibly generating unprecedented high temperatures with ohmic heating would require re-establishing the commensurately shorter holding times to achieve pasteurization or sterilization.

12.2.2 Process control

Getchell[6] was the first author to emphasize the importance of controlling the ohmic heating process. Since that time, significant developments in semiconductor technology have increased the sophistication of possible control equipment and strategies. In continuous processing, for example, problems can result if a single electrode pair is used to heat food materials through large changes in temperature. Substantial changes in the conductivity of the liquid, and thus in heating rate, may result along the length of the electrode. Employing multiple sets of electrodes imparts greater control to the process. However, Biss *et al.*[9] report that the approach of pure feedback control was not suitable for the APV Baker ohmic process due to the requirement for large time constants, and they described the development of a feed-forward control scheme. For feed-forward processes, some understanding of the characteristics of the system is needed, and control depends as much on process knowledge as on the design of the control loops. Figure 12.3 depicts the Raztek's full-scale multistage Electroheating™ systems for fluid foods that employ a feedback control using a silicon-controlled rectifier (SCR).

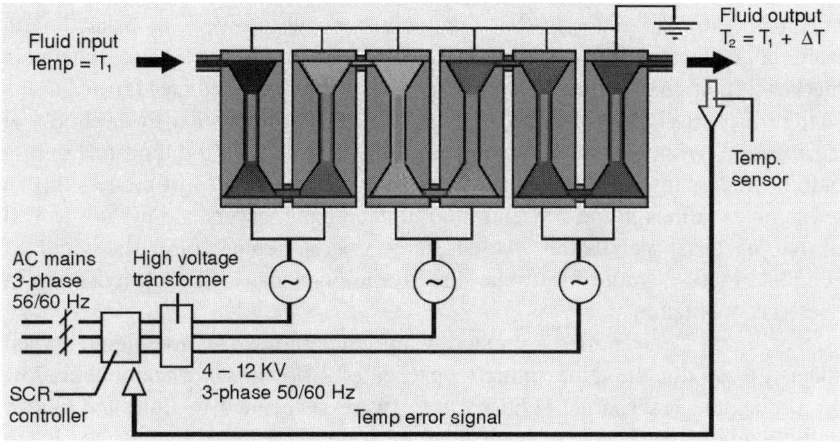

Fig. 12.3 Raztek's mult-stage Electroheating™ system block diagram (courtesy of Raztek Corp.).

12.3 Invasive and non-invasive methods of monitoring ohmic heating

12.3.1 Invasive

Locating the coldest spot within an ohmically heated food system is of paramount concern to food engineers and processors. For a HTST thermal process such as ohmic heating, the determination of the spatial and temporal distribution of temperature within the particulate is necessary and important for ensuring adequate treatment and food safety. The temperature profile (temporally and spatially) not only provides information for calculating lethality and cook value, but it also provides a guide to mathematical modeling of the ohmic process and improved process control. Mapping the dynamic changes in temperature in food materials undergoing ohmic heating, including at intra-particulate locations to ensure food safety, is a complex and difficult task. Teflon coated thermocouples have been used for verification of modeling and for monitoring the process. In this arrangement, temperature could be obtained only from select points and the integrity of the process could be disturbed by the presence of the thermocouples. Moreover, it is difficult to monitor accurately the temperature of a flowing particle. Alternative methods for measuring the temperature profiles of ohmically-heated particulates were sought to circumvent the problems of using thermocouples for this purpose.

Kim *et al.*[10] developed an intrinsic chemical marker approach for mapping temperature distribution and calculating lethality, and applied this method in conjunction with direct microbiological measurements to particulate foods aseptically processed with ohmic heating to obtain valuable information regarding process validation. Chemical markers can be viewed as time-temperature integrators over an HTST domain relevant to the thermal processing of foods. Particulate foods (beef and chicken cubes) were inoculated with bacterial cells and infused with precursor compounds to enhance chemical marker formation. Chemical markers are observable intermediates formed in the Maillard reaction of sugars and lysine residues of proteins that form rapidly at sterilization temperatures. Temperature gradients arising in the foodstuff during ohmic processing were demonstrated in the particulates (post-processing) in terms of variations in the concentration of chemical markers and in the survival of the bacterial population. At constant process temperature, the yield of chemical marker would ideally be directly proportional to the log reduction in bacterial population.

Sastry and Li[11] reported a method using temperature-sensitive liquid crystal sheets to monitor the temperature of particulates flowing in a continuous-flow ohmic heater. Transparent solid particles were suspended in fluid and coated with liquid crystal sheets that changed color from black to red to green to blue to black in response to temperature over a specified range. The sheet was placed parallel to the electric field so that interference with the electric field was minimized and the entire temperature profile inside a solid object could be visualized. This method provided useful information for the temperature distribution and for the model verification.

12.3.2 Non-invasive

Magnetic resonance imaging

Magnetic resonance imaging (MRI) has contributed greatly to a diversity of scientific disciplines in recent years. Applications in the food area have increased significantly as researchers have discovered the power and flexibility of this technique.[12] Food-related MRI research has gone beyond static imaging experiments to experiments involving dynamic processes, diffusion, water mobility, flow, water and lipids distribution, and temperature distribution.[12] The major advantages of MRI are:

- it is non-destructive and non-invasive to the material being imaged and the complexity of foods can be probed without perturbing the sensitive balance
- MRI provides high spatial resolution
- MRI can probe diverse information such as the proton density which is related to moisture or fat concentration, internal structure, chemical shift, diffusion, temperature and flow.

Among all of the food-related MRI techniques, MRI temperature thermometry is emerging as an attractive and promising temperature mapping method.[13–17] As pointed out by Hills[18] the true potential of MRI lies not in static structural determinations, but rather in the non-invasive, real-time, and dynamic changes in foods as they are processed, stored, packaged, or distributed. When using MRI to map the temperature distribution in a food material undergoing a dynamic process like ohmic heating, the data acquisition time should be as fast as possible so that a real-time measurement can be accomplished. If temporal resolution cannot be achieved, the measurement will be inaccurate or perhaps even meaningless. Significant progress has been made in the application of MRI techniques to mapping the temperature distribution in ohmically heated food systems.

Principles

MRI is an extension of nuclear magnetic resonance (NMR) spectroscopy. In brief description, NMR spectroscopy is based on the magnetic behavior of certain nuclei in a sample (such as the protons in water and fat molecules) when placed in an external bulk magnetic field and subjected to a radio frequency (RF) pulse. Atomic nuclei possess a net positive charge, and are considered as spinning about an axis. The spinning of the nucleus generates a magnetic field, which is similar to that generated by a simple bar magnet, and is termed the nuclear magnetic dipole. When placed in a magnetic field B_0 generated by a permanent magnet, for instance, the nucleus will interact with the applied field via its magnetic dipole, and tend to precess about the direction of the applied field at a specific frequency known as *Larmor* frequency (ω), which is proportional to the strength of the applied field B_0, and governed by equation [12.1]

$$\omega = \gamma B_0, \tag{12.1}$$

in which γ is the magnetogyric ratio, a fundamental property of the nucleus. If a second magnetic field generated by using a radio frequency (RF) coil, for instance, whose frequency exactly matches the *Larmor* frequency ω, then resonant absorption of energy occurs. This resonance effect is hence termed nuclear magnetic resonance. In the time following the excitation, the excited spins give up energy and return to their equilibrium state. The energy is released in the form of RF wave, characterized by its *Larmor* frequency, and discharged into the environment through two mechanisms: spin-lattice and spin-spin relaxation processes, each of which is characterized by an associated time constant. The time constant for spin-lattice relaxation is called T_1 and that for spin-spin relaxation is called T_2. In general, T_1 and T_2 can be functions of structure, molecular mobility, temperature, solute and water concentrations, and possibly other physical and chemical properties of the samples. The RF coil also functions as an antenna to receive the RF signals emitted from the nucleus during the relaxation processes, by which the signals decay, and the relaxation behavior of the spins with characteristic T_1 and T_2 can be recorded.

MRI is based on a function of spatial position of magnetic fields instead of using a uniform static field. If a linear field gradient G_x is superimposed on the main magnetic field along the x direction, the resonance frequency at which the spin processes is a function of spatial position along x:

$$\omega(x) = \gamma(B_0 + G_x x) \tag{12.2}$$

Transformation of the data then yields not only the magnitude of the signal but also spatial information. A similar principle can be extended to achieve spatial information in two or three dimensions, which can be used to construct 2- or 3-D magnetic resonance images.

MRI thermometry methods are non-destructive and non-invasive and exploit the temperature dependence of MR properties whose spatial distribution can be visualized. There are several MR properties that exhibit characteristic temperature dependences, and they have been used in MRI thermometry, such as T1 relaxation time, the self-diffusion coefficient of water, and the proton resonance frequency (PRF).[19,20]

Methodologies

Hulbert *et al.*[14] made a temperature map of carrot during heating using T1 weighted MR images. In this instance, the imaging time was about 20 seconds. The data acquisition time was long and unsuitable for a dynamic heating process like ohmic heating. Sun *et al.*[21] used the self-diffusion coefficient of water molecule to map temperature in a potato. To reduce the scan time, the authors adopted the half Fourier transform spin-echo sequence with 16 phase encoding steps. The data acquisition time for a temperature map was claimed to be about 10 seconds. The work also showed that the self-diffusion coefficient of water was more sensitive to temperature than T1 relaxation time. Chang *et al.*[22] improved the T1 relaxation method by utilizing a snapshot FLASH sequence. The method significantly shortened the data acquisition time to $\approx$3 seconds. The image matrix was 128 $\times$ 64. The T1 method requires an enormous amount of post-acquisition data processing to do non-linear regression analysis and determine the T1 value, which is the temperature indication parameter. The complicated temperature and material dependence of the T1 method hinder its application.

The application of MRI temperature mapping techniques to the ohmic heating process raises two principal concerns: First whether the relatively long data acquisition time makes the techniques unsuitable for a dynamic process like ohmic heating, since changes in the system may occur more rapidly than data can be collected; whether the electrical heating power has to be interrupted during the data acquisition to prevent interference with the magnetic field, and thereby producing inaccurate measurements of the temperature maps. The recent work of Ye *et al.*[23,24] employing Proton Resonance Frequency (PRF) Shift MRI Thermometry addressed these two concerns directly.

Temperature mapping during ohmic heating using PRF method

The advantages of the PRF method include its good temperature sensitivity, the linearity of its response to temperature, its reversibility with temperature during both heating and cooling, its effectiveness is independent of the nature of the material, and the ease of measurement and interpretation.[25,26]

Theoretical aspects

The temperature sensitivity of the PRF was first observed by Hindman in 1966.[27] It gained importance for imaging purposes when the idea was introduced by Ishihara *et al.*[25] to measure the temperature related frequency shift from the phase images of gradient-echo sequence. The fractional change of water proton resonance frequency ($\Delta\omega/\omega$) with temperature is defined as δ. It is also referred to as proton chemical shift or PRF shift. Mathematically,

$$\frac{\Delta\omega}{\omega} = \delta \cdot \Delta \mathrm{T} \qquad [12.3]$$

where ΔT is the temperature change. Experimentally, δ has been determined as -0.01 ppm/°C in water.[27] It is generally assumed that the PRF shift of water to lower frequency with higher temperature is caused by rupture, stretching, or a small amount of bending of the hydrogen bonds.[27,28] This indicates a reduction in the average degree of association of water molecules, and hence that these shifts are evidence of an increased average shielding constant of the protons[21] (Ishihara and others 1995). The proton resonance frequency shift imaged in a static field strength B_0 after having undergone a temperature change of ΔT is:

$$\Delta\omega = \delta \cdot \gamma \cdot B_0 \cdot \Delta T \qquad [12.4]$$

where γ is the gyromagnetic ratio of proton. This frequency change manifests as a phase change when imaged with a gradient-echo sequence having an echo time *TE*. This phase change $\Delta\Phi$ can be expressed as:

$$\Delta\Phi = \delta \cdot \gamma \cdot B_0 \cdot \Delta T \cdot TE \qquad [12.5]$$

The phase image $\Phi(x,y)$ can be calculated from the acquired complex data of the MR image using equation [12.6]:

$$\Phi(x,y) = \arctan\left[\frac{Im(x,y)}{Re(x,y)}\right] \qquad [12.6]$$

where $Im(x,y)$ and $Re(x,y)$ are respectively the imaginary and real part of the complex MRI data, and x and y denote the image pixel numbers.

To use the PRF shift technique to map temperature, a reference phase image is first acquired at a known temperature, and then subtracted from subsequent phase images taken at different temperatures. Temperature maps can therefore be obtained based on the reference temperature and the echo time TE of the image sequence according to equation [12.5].

Example: Ohmic heating of brine-potato particulate mixture

An experimental ohmic system was configured as diagrammed in Fig. 12.4. The system consists of a cylindrical heating device, a cylindrical potato particle, and a salt solution with sodium carboxymethylcellulose (CMC). The ohmic heating device consists of a Plexiglas vessel with an inner diameter of 43 mm and a nylon stopper at each end. A 40 mm diameter stainless steel electrode was fixed to each of the stoppers and connected to the power supply. The distance between the two electrodes was 305 mm. A small hole was drilled in the vessel for injection of the fluid and for pressure release during heating. The concentration of NaCl and CMC was 0.2% (w/w) and 1.2% (w/w), respectively. Cylindrical potato particles were freshly cut from Russet potatoes immediately before the

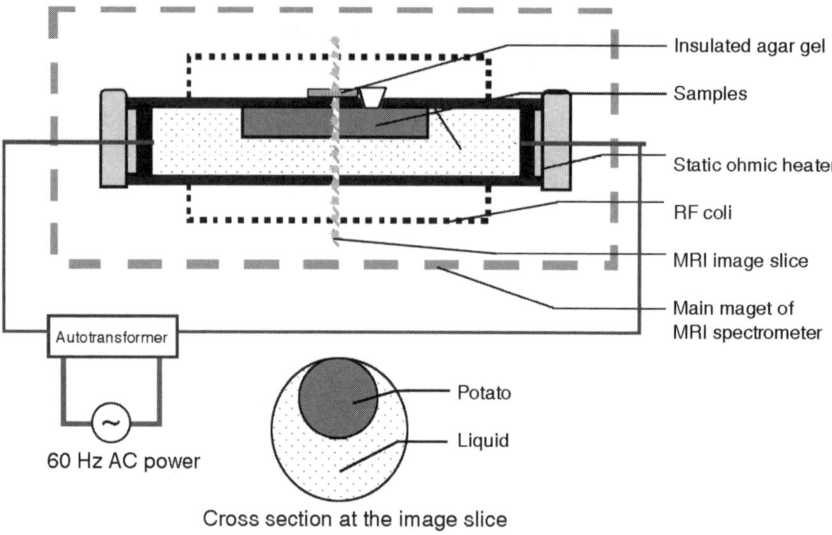

Fig. 12.4 Schematic diagram of experimental set-up.

experiment and put in the middle of each heating vessel. The dimension of the potato particle was 70 mm (length) × 22 mm (diameter). Fluid was injected through the small hole to fill the vessel. Since the solutions featured a higher density than the solids, the potato particulates floated on the surface of the fluids. The application of an AC power supply with a constant voltage of 120 V and a frequency of 60 Hz ohmically heated the food systems.

A volume birdcage radio frequency (RF) coil that was fitted to the configuration of the heating vessel was also used as both an RF transmitter and MR signal receiver. A thin tube containing an agar gel was inserted between the ohmic heating vessel and the RF coil. The tube was thermally insulated by Styrofoam so that the phase change of MR images induced by sources other than the temperature change will be revealed by the phase images in the agar gel. Two-dimensional MR imaging was performed using a 4.7 Tesla SISCO scanner (Varian, Palo Alto, CA) with a 400 mm diameter bore at the Center for Magnetic Resonance Research at the University of Minnesota. The PRF method was incorporated into a Fast Low Angle SHot (FLASH) sequence to achieve a rapid image acquisition. FLASH was first suggested by Haase *et al.*[29] It uses a small flip angle and short repetition time (TR) to obtain a fast scan. The most significant advantage of this sequence is that a high signal-to-noise ratio (SNR) at a short repetition time can be achieved.[30] Transverse images were obtained using the following imaging parameters: repetition time TR = 10 msec; echo time TE = 4.6 msec; flip angle FA ≈12°; slice thickness = 3 mm; field of view FOV = 60 × 60 mm; image matrix size = 64 × 64. The data acquisition time of one image thus was 0.64 sec. Therefore the temporal resolution (data acquisition time for one temperature map) was 0.64 seconds and the spatial resolution of the temperature maps was 0.94 mm.

The built-in filter of the imager successfully eliminated most of the noise due to the electrical power source of ohmic heating, but there were some stripes of noise that appeared in the regular MR images. After the image processing, it turned out that no obvious noise could be discerned in the phase difference images and the temperature maps. The PRF method calculates the relative temperature from the phase difference, and the noise was cancelled out when the subsequent phase image was subtracted. The amount of electrical current passing through the food systems increased as the electrical conductivities of the food materials increased with temperature. This change might have induced a shift in the local magnetic field (B0), which in turn would affect phase images. The agar gel was used as a reference to compensate for the phase change due to the B0 shift. Since the tube was thermally and electrically insulated, any change of phase induced by a source other than temperature change would be reflected by the phase change in the agar gel. Therefore, the phase change in the agar gel was subtracted from the corresponding phase difference image to map the temperature accurately.

The MRI temperature maps of the sample at selected times during ohmic heating are shown in Fig. 12.5. The MRI slice is two–dimensional in the x-y plane and the temperature is projected in the z-axis to generate a 3–D image. The 64×64 pixels of the temperature maps show the details of the temperature distribution since every pixel represents a temperature value at that location. The sample had an NaCl concentration of 0.2% and the maps generally showed that the fluid phase heated up slightly faster than the particulate phase. Most importantly, the MRI maps show that during the initial stages of heating, the cold spot was located at the center of the potato, and it eventually shifted to a location between the liquid-particulate interface and the center of the potato. Temperature mapping using MRI thermometry provided a powerful tool for the development of ohmic heating model.

12.4 Modeling ohmic heating

The future utilization of ohmic heating by the industry will depend on the development of adequate safety and quality assurance protocols. A crucial component in understanding the ohmic heating process is the development of mathematical models, which can then be used to simulate various effects of critical factors and to guide the process design and quality control. Two different modeling approaches currently have been published and are described below

12.4.1 de Alwis-Fryer model

de Alwis and Fryer[31] use the solution to Laplace's equation to calculate heat generation rate together with transient energy balance equations to model a single particulate immersed in a fluid medium without convection. This model has been extended by Zhang and Fryer[32] to include multiple spheres uniformly

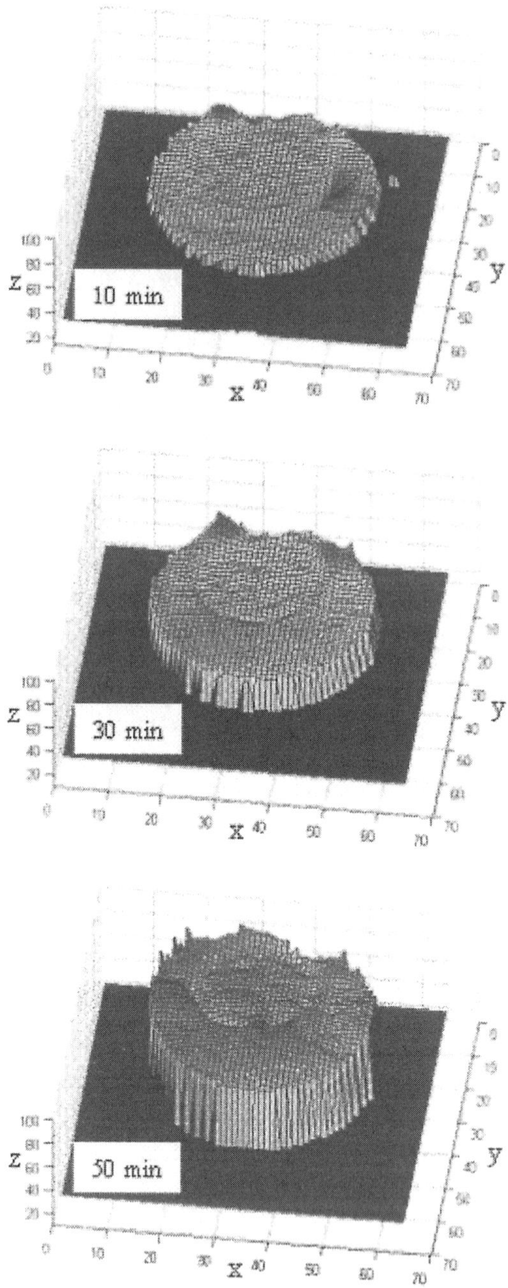

Fig. 12.5 Temperature maps of brine-potato during ohmic heating (10, 30, and 50 min).

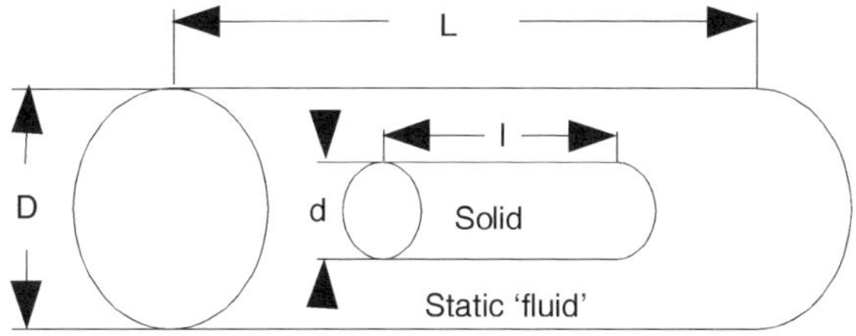

Fig. 12.6 Situation simulated in de Alwis-Fryer model (redrawn from Sastry and Salengke[33]).

distributed on a lattice within a non-convective fluid. The typical situation is illustrated in Fig. 12.6, in which a cylindrical particle is positioned in the middle of a tube filled with stationary, non-convective fluid. An electric field is applied along the length of the tube.

The electric field or voltage distribution can be developed from Maxwell's equations, or by combining Ohm's law and the continuity equation for electrical current [12.7]:

$$\nabla \cdot (\sigma_i \nabla V) = 0 \qquad [12.7]$$

where V = voltage,
∇ = gradient,
σ_i = electrical conductivity of phase I which can take on different values for the particles and liquid.

Ignoring the effect of convection, the heat transfer problem is one of pure conduction with internal energy generation [12.8]:

$$\nabla \cdot (k_i \nabla T) + \dot{u}_i = \rho_i C_{pi} \frac{\partial T}{\partial t} \qquad [12.8]$$

where i again represents the phase,
k = thermal conductivity,
$\dot{u}_i$ = specific internal energy generation rate,
ρ = density,
C_p = specific heat capacity,
T = temperature,
t = time.

The external boundary condition is one of convection to the surroundings [12.9]:

$$-k_{iS} \nabla T \cdot \vec{n} = U(T_{iS} - T_{\infty}) \qquad [12.9]$$

where
k_{iS} = thermal conductivity of phase i at surface,
$\vec{n}$ = unit normal vector,
U = overall heat transfer coefficient,
T_{iS} = surface temperature of phase i,
T_{∞} = surrounding temperature.

The internal energy generation term in equation [12.9] is given by:

$$\dot{u}_i = |\nabla V|^2 \sigma_{0i}(1 + m_i T) \tag{12.10}$$

where
∇V = voltage gradient,
σ_{0i} = initial electrical conductivity,
m_i = temperature compensation constant,
T = temperature.

The system of equations [12.7–12.10] can be solved by the Galerkin-Crank-Nicolson algorithm, a hybrid-spatially finite element, temporally finite difference scheme.

12.4.2 Sastry-Palaniappan model

Sastry and Palaniappan[34] used circuit analogy to approximate electrical conductivity and thus the heat generation for a static heater with a particle immersed in a well-mixed fluid (assuming infinite convective heat transfer within the fluid). A typical situation is as illustrated in Fig. 12.7.

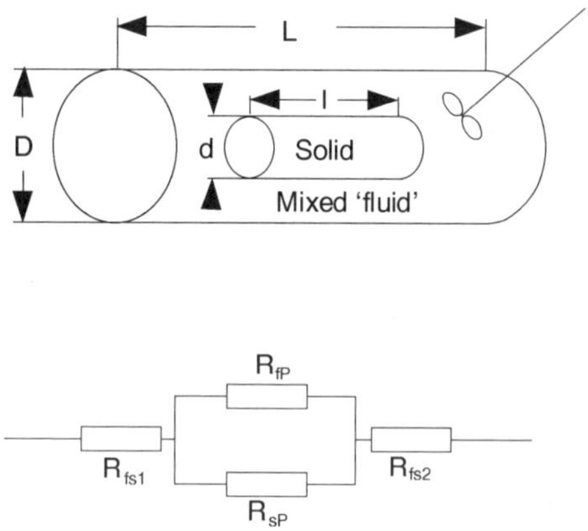

Fig. 12.7 Situation simulated in de Sastry-Palanippan model. Also shown is the circuit analogy (redrawn from Sastry and Salengke[33]).

Sastry[35] extended this approach to a continuous flow ohmic heater for a liquid-particulate mixture containing a high concentration of solids. The effective electrical resistance can be determined using the circuit analogy. The effective resistance of the cell is calculated as [12.11]:

$$R = R_{fs1} + R_{fs2} + \frac{R_{fp}R_{sp}}{R_{fp} + R_{sp}} \tag{12.11}$$

where the resistance R's can be calculated from the electrical conductivity and the geometry. The average voltage gradients (V) in the liquid and particulate can simply be calculated as [12.12]:

$$\nabla V = \frac{V}{L} \text{ or } \nabla V = \frac{IR}{L} \text{ and } I = \frac{V}{R} \qquad [12.12]$$

where V is the applied voltage and L the distance between the two electrodes.

The energy balance on the well-mixed fluid phase is [12.13]:

$$M_f C_{pf} \frac{dT_f}{dt} = \dot{u}_f v_f + n_p h_{fp} A_p (T_{sSm} - T_f) - UA_w (T_f - T_\infty) \qquad [12.13]$$

where the subscript f represents liquid phase and p particulate phase,

M_f = mass of fluid,
C_p = specific heat capacity,
$\dot{u}_f$ = specific internal heat generation rate,
v = volume,
n_p = number of particulates,
h_{fp} = liquid-particulate convective heat transfer coefficient,
A_p = surface area of one particle,
T_{sSm} = mean particulate surface temperature,
U = overall heat transfer coefficient to surroundings,
A_w = area of heater wall.

The specific internal heat generation rate for the liquid phase $\dot{u}_f$ is given by [12.14]:

$$\dot{u}_f = |\nabla V|^2 \sigma_{0f} (1 + m_f T) \qquad [12.14]$$

The particulate heat according to the conduction heat transfer equation with internal energy generation [12.15]:

$$\nabla \cdot (k_s \nabla T_s) + \dot{u}_s = \rho_s C_{ps} \frac{dT_s}{dt} \qquad [12.15]$$

where

$$\dot{u}_s = |\nabla V|^2 \sigma_{0s} (1 + m_s T_s) \qquad [12.16]$$

Temperatures of the two phases are linked by the convective boundary condition [12.17]:

$$-k \nabla T \cdot \vec{n_s} = h_{fp} (T_{sS} - T_f) \qquad [12.17]$$

This system of equations can be solved by forward differences for the liquid phase, and the Galerkin-Crank-Nicolson method for the solid phase.

For any aseptic process, safety deserves the highest priority. The limitations of these two models and the lack of understanding of the temperature distribution mandate the use of the most conservative approach. Regions of low-temperature in the substrate will extend the required processing time needed to ensure satisfactory commercial sterility, and consequently increase the likelihood that other regions or components of the food will be overcooked. Ohmic heating thus will lose its most attractive advantage. The effects of convection were oversimplified in the two models. In reality, the convection heat transfer rate is neither zero nor infinity. A more general model is required that realistically depicts the effects of convection, in order to model the temperature profile during ohmic heating.

12.4.3 Improvements in ohmic model

It is unrealistic to assume that the electrical conductivities for the liquid and particulate phases are the same. When electrical conductivities are significantly different from one another, Fryer *et al.*[36] suggested that local 'shadow regions' of low electric field strength can arise, resulting in large differences in temperature between the liquid and the solid. Experiments and models[37] have been compared in studies to determine the extent to which these shadows might interfere with safe processing in a real industrial situation. The study results showed that the key is matching the conductivities of the two phases, as shadow regions occur with the same order of magnitude as that of the included particles. A centimeter-scale inclusion can create inhomogeneity which cannot be removed by thermal conduction over the timescales typical for ohmic processes. Care must be taken with the model formulations by including the convection effects into the model.

MRI temperature mapping and determination of liquid-particulate heat transfer coefficient

Ye *et al.*[23,24] investigated the differential heating between liquid and particulate phases using MRI temperature mapping. The interface heat transfer coefficients were also estimated using MRI temperature maps during holding period. The experimental set-up is similar to Fig. 12.4. The ohmic heating device is made of a clear PVC tube with an inner diameter of 63.5 mm and a nylon stopper at each end. A 60 mm diameter stainless steel electrode was fixed to each of the stoppers and connected to the power supply. The distance between the two electrodes was 305 mm. The concentration of NaCl and CMC was 0.5% (w/w) and 0.7% (w/w), respectively. Cylindrical potato particles were freshly cut from Russet potatoes immediately before the experiment and placed in the middle of each heating vessel. The electrical conductivity of the liquid is thus substantially higher than that of the solid. The dimension of the potato particle was 75 mm (length) × 25 mm (diameter). The application of an AC power supply with a constant voltage of 120 V and a frequency of 60 Hz ohmically heated the food systems.

The PRF technique was used to obtain temperature maps at 5, 10, 15, 20, 23, 25, 27, and 29 minutes after switching on the electrical power. The heating was stopped at the last point to simulate a holding period. During the holding period, sample #2 was imaged at 5 minute intervals for 50 minutes (while the heating was stopped). The temperature maps from selected times during the heating and holding periods are shown in Fig. 12.8.

Due to the higher electrical conductivity of the fluid phase, in accordance with the principles of parallel electrical circuits, the liquid heated much faster

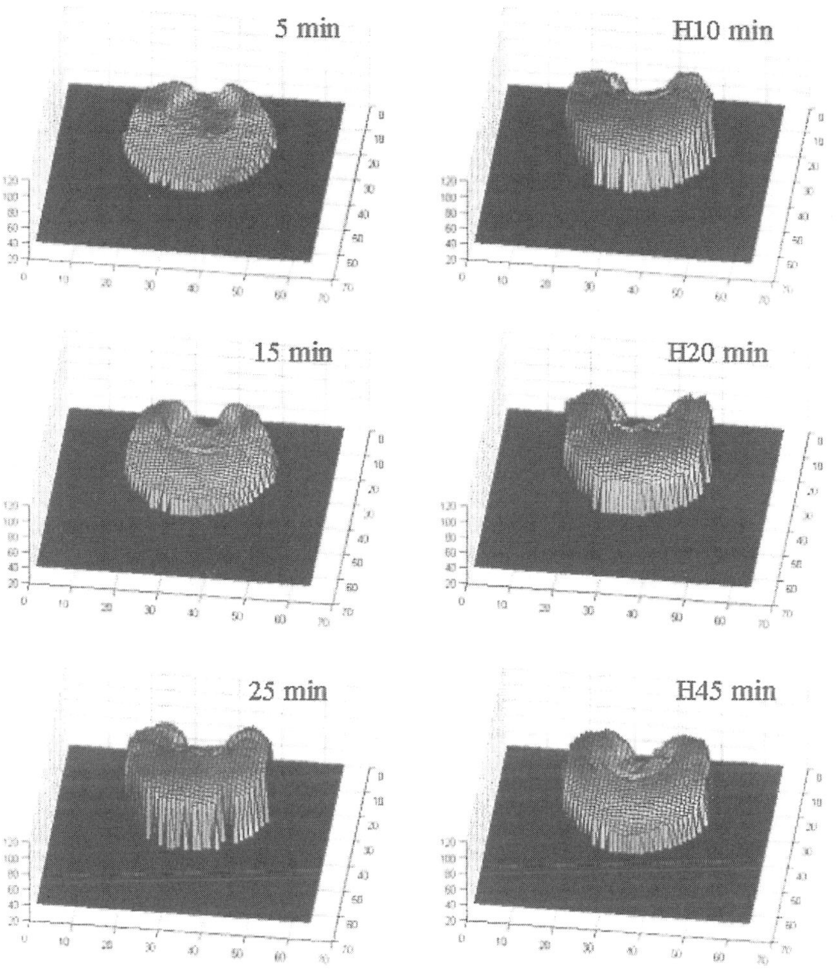

Fig. 12.8 Temperature maps at selected time points. 'H' denotes holding period.

than the potato particle, and the average temperature of the fluid was about 17°C higher than that of the potato particle at 25 minutes of heating. An important aspect demonstrated with the MRI maps is that the coldest region of the potato was always located at its center. As mentioned earlier, the cylindrical potato particles had a lower density than the fluid and floated on the liquid. The top edge of the particles was in contact with the heating vessel wall, and this location heated more slowly because it was not in contact with the more conductive liquid. We can postulate that the electrical current density in the liquid surrounding the potato particle (Zone A, Fig. 12.9) was higher than in regions (Zones B and C). The electrical current was 'diverted' through the liquid phase, which featured a higher electrical conductivity than the solid phase, and

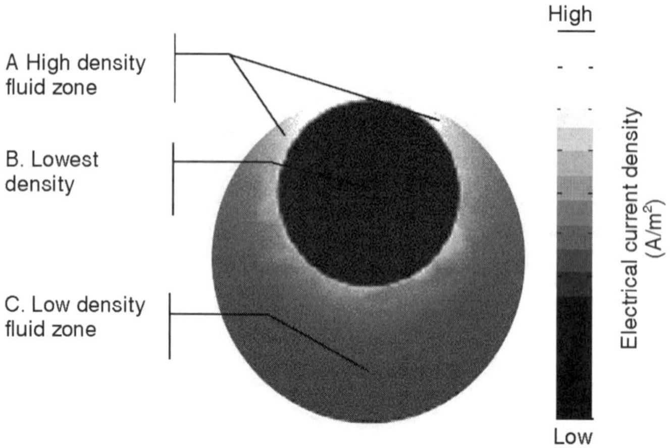

Fig. 12.9 Simulation of electric current density distribution using Laplace equation.

resulted in faster heating around the particle, as is evident from the above temperature maps. This observation supports the application of the Laplace equation to model the electric field during ohmic heating, as suggested by the de Alwis-Fryer model. The Laplace equation was derived from Ohm's law and the equation of continuity, and it models the electrical current flow as curved lines according to the electrical conductivity distribution of the food materials. The electrical current is therefore 'diverted' from the potato to the fluid, if fluid phase has a higher electrical conductivity than the solid phase, and results in a non-uniform distribution of the current density and differential relative heating rates of the two phases. A numerical simulation using Laplace equation was conducted in MATLAB for the same geometry as the experiment in the study and the result is shown in gray scale in Fig. 12.9. The constant electrical conductivity values used in the simulation were 1.3 S/m for the fluid and 0.037 S/m for the potato. These values were taken from Kim and others.[38] The simulation shows that when the potato particle has a lower electrical conductivity than the fluid, there exists a high electrical density zone around the particle, consistent with the observations of the MRI temperature maps.

A modeling procedure with a numerical solution was used to match the calculated temperature distribution with that measured using MRI. Such a non-invasive procedure together with a finite-element method is especially useful for modeling the heating of a liquid-particulate mixture with irregular particulate shapes.

The governing equation is the Fourier's second law without internal heat generation applied to the cylindrical potato particle:

$$\rho_p C_p \frac{\partial T_p}{\partial t} = \nabla \cdot (k_p \nabla T_p) \qquad [12.18]$$

The convective heat transfer between the fluid phase and the solid phase is the required boundary condition and is given by:

$$-k_p \nabla T_p \cdot \overrightarrow{n_s} = h_{fp}(T_{ps} - T_f)|_s \qquad [12.19]$$

where ρ_p is the density of the potato particle, C_P is specific heat of the particle, T_p is the particle temperature that is both time and location dependent, t is time, k_p is the particle thermal conductivity, and T_f the fluid temperature. The subscript 's' in equation [12.19] represents the surface. In most cases reported in the literature, a constant fluid temperature was considered.[34] In this study, the actual time-temperature profile obtained from the MRI temperature mapping was used as a time-dependent boundary condition. By varying h_{fp}, a computationally intensive optimization algorithm was employed to minimize a function $f(h_{fp})$ defined as the sum of the squared differences between the temperatures calculated by the model and those obtained with MRI temperature mapping:

$$f(h_{fp}) = \sum_{i}^{n} (T_{i_{MRI}} - T_{i_{MODEL}})^2 \qquad [12.20]$$

where i is an index corresponding to the pixels in the potato image. The average h_{fp} over a time period was determined as the h_{fp} value that minimized the function $f(h_{fp})$.

Figure 12.10 presents the calculated heat transfer coefficients (h_{fp}) over 5–minute intervals calculated during the holding period for the sample. The results are consistent with the range of values reported in the literature[39] for natural convection conditions.

A trend that can be observed from the figure is that the value of h_{fp} decreased sharply during the first few minutes of the holding period, and then leveled off around 40 (W/m^2 K). One possible physical explanation for this observation may be that fluid flow continued to occur immediately after the ohmic heating power was disconnected, because the fluids were heated to nearly boiling

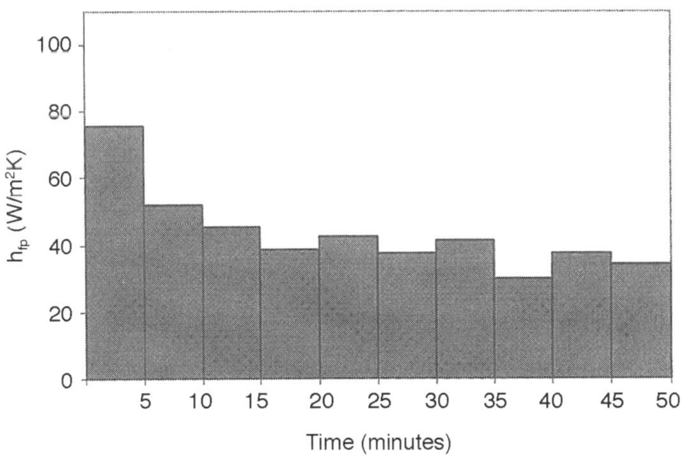

Fig. 12.10 Calculated heat transfer coefficients during holding period.

temperatures (T $\approx$95°C). As time passed during the holding period, the effect of the fluid flow gradually diminished, which consequently resulted in smaller values of h_{fp}. The h_{fp} values are estimated as the average heat transfer coefficients over the entire 5–minute interval. In most cases reported in the literature, temperature measurements at the center of a particle were used to calculate h_{fp}, and ideal conductive heat transfer within the particle was assumed to occur. Presently, the temperature profiles of nearly all of the domains in the particles were taken into consideration, presumably producing a more reasonable estimation of h_{fp}.

Simulation and verification using MRI temperature mapping
The ohmic heating process described in the previous section was simulated by Ye *et al.*[40] using finite element analysis with the commercial software FEMLAB. The simulation improved the de Alwis-Fryer model by considering the interface heat transfer, and the model predictions were verified against temperature maps obtained using MRI. A factor ignored in previous modeling efforts was the electricity-to-heat conversion efficiency. Including this factor in the present model improved the overall performance of the model. The electrical conductivity and its temperature dependence for the liquid and potato particle used in the simulation were determined under constant electric field strength as the simulated ohmic heating processes. Other factors affecting the model predictions (e.g., the heater wall boundary conditions) were also determined *in situ* to increase the accuracy of the estimated parameter values. The temperature profiles predicted by the model and the corresponding MRI temperature maps at the middle cross-section of the ohmic heating systems for select heating times are presented in Fig. 12.11. The model predictions generally yielded good agreement with the MRI temperature maps. Some discrepancies occurring between the MRI maps and model predictions should be pointed out. For example, liquid leaking from the hole on the top of the heater vessel complicated the boundary condition in that area, and caused larger disagreement between the MRI maps and model predictions in that region. It is noteworthy to point out that the formation of bubbles in the ohmic heater may contribute electrically insulated zones, and their impact on the heating procees should be given serious consideration for heater design and process control. An additional factor not included so far is differences in localized heating due to the heterogeneous nature of certain food materials (e.g., fatty meats).

12.5 Future trends

12.5.1 Advancing modeling development and verification

Current modeling approaches tend to address mainly the worst-case scenarios of ohmic heating to ensure product and process safety. In order to further exploit the potential benefits of ohmic heating for producing value-added foods with higher sensory and quality attributes with energy-efficiency and environ-

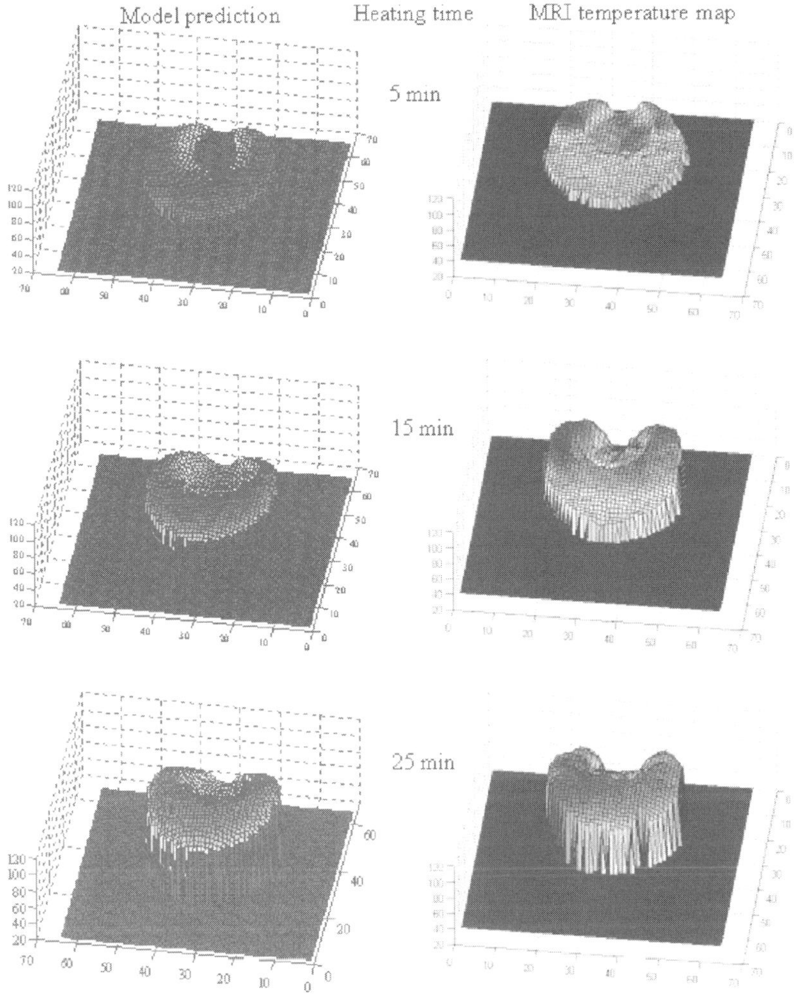

Fig. 12.11 Temperature profile predicted by the model VS. MRI temperature map.

mentally-friendly equipment, it is necessary to model the effects of flow and heat generation occurring concomitantly in a real liquid-particulate mixture food system. This approach requires the solution of the Navier-Stokes equation together with the Laplace equation and the heat transfer equation. This requirement for modeling seems formidable, even with advancements in the techniques of computational fluid dynamics, because of the lack of information on the basic physics involved and the overall complexity of the calculations. Extensive experimental work combined with modeling simplifications will be useful in a number of ways. It can be used to investigate local heating behaviors and predict the behavior of certain food formulations in commercial systems. It will also help identify the types of problems and their effects on real systems in actual practice. In summary, developments in modeling are needed to maximize the benefits and advantages of the ohmic heating process by providing deeper insight and a better understanding of the process how to optimize for achieving high-quality products.

Exploiting the non-invasive technique of MRI to monitor the spatial distribution of temperature in such a complex system is crucial to understanding and controlling the ohmic heating process for enhancing food product quality. Additionally, non-invasive MRI temperature mapping provides vital information for the development of appropriate and accurate models and is essential for the validation of this novel HTST process. The versatility of MRI provides the capabilities of making such non-invasive measurements, but there is a need to further improve these techniques for collecting spatial temperature data under flow conditions, and to apply this technique to validate advanced mathematical models.

12.5.2 Process and product development

Ohmic heating features unique characteristics that offer particular benefits in food processing to the sterilization of liquid-particulate mixtures. In these food-types, particulates are the centerpiece around which the product is formulated. Contrary to conventional heating, where we would expect no difference in heating characteristics when particle orientations are changed, the heating pattern of an ohmically heated food system can be substantially affected by particle orientation in the electric field. de Alwis *et al.*[31] showed that the heating rate of a potato slab depended on its orientation relative to the electrical field, and this unique phenomenon was due to changes in the electrical field associated with the potato slab orientation.

One might assume that there is no limit to the size of particles that can be heated effectively in an electrically uniform mixture. However, both the heating and the cooling stages should be considered, and practical limitations should be expected. The cooling of particulates will always be thermal conduction controlled, and the cooling rates possible may impose an upper limit on the particle size. This is important in HTST processes, where rapid cooling is desired. The center of large particles may cool too slowly to avoid being overprocessed during prolonged cooling. Unlike conventional heating, where the outer surface of the particulate might be overcooked, in ohmic heating (due to the heating inversion phenomenon), the interior of the particulate might be overprocessed. Particulate size is typically limited to 1 in^3. Various combinations of particulates and liquids can be successfully processed when accompanied by suitable product and process controls. The primary considerations on particulates include their size, shape, concentration, density, conductivity, and specific heat capacity. Optimizing the combination of these variables may ensure a uniform and appropriate heating and produce excellent sensory attributes (e.g., texture). The fluid phase cannot be neglected, especially with respect to overcooking. The liquid viscosity should be determined at various temperatures to assure adequate suspension of the particulates over the temperature range in consideration. Moreover, the liquid viscosity may affect heat transfer at the liquid/particle interface and influence the heating rate and process control. More research is needed to address and understand many aspects of the product and process design and their impact on the quality of the product.

Product specifications include information that define the properties of the product and the physical/chemical features of the product that play a critical role in determining the amount of lethal treatment that is delivered during the process. Critical factors may include particle size and shape, liquid viscosity, pH, specific heat, thermal conductivity, solid:liquid ratio, and the electrical conductivities of the phases. As the processor identifies the effects of these individual factors, he will also need to gain an understanding of their interactions and how they subsequently influence the process. Currently, there is only very limited information available on these issues.

Process design is a complete description of the critical processing conditions used in the manufacture of the product and the procedures used in establishing these conditions. It should include batch formulation procedures, initial temperatures, flow rates or particle residence times, exit temperatures, solid loading rates, and the control of them. It is imperative to select process parameters that ensure product safety while simultaneously maximizing product quality, and these parameters will be specific for individual systems and formulations. Understanding the roles of these factors is needed for predictions of the heating behavior. They require rigorous testing and reevaluation when pertinent changes occur.

12.5.3 Other applications

While the research on ohmic heating during the past two decades was focused on the aspect of heat transfer, especially for the aseptic processing of liquid-particulate mixtures, issues involving mass transfer of components in foods during ohmic heating have drawn attention recently. Potential applications for ohmic heating include blanching, evaporation, dehydration, fermentation and extraction.

Lima *et al.*[41] has shown that ohmic heating enhances diffusion of beet dye from beetroot tissue into a fluid at certain thermal conditions. Cho *et al.*[42] investigated the growth kinetics of *Lactobacillus acidophilus* in ohmically heated culture and showed that low-voltage ohmic heating reduces the lag phase of the bacterial cultures and thus is potentially useful in food fermentation. Although the results of these studies are inconclusive, they showed a new pass toward further exploitation of this novel technology.

12.6 Sources of further information

Information on ohmic heating technology and research can be found in the literature listed below. The following research institutes have major research programs on ohmic heating:

Department of Food, Agricultural, and Biological Engineering, The Ohio State University, Expertise: ohmic heating in general, and mathematical modeling, contact: Professor S.K. Sastry.

Department of Chemical Engineering, University of Cambridge, expertise: general and mathematical modeling, contact: Professor P. Fryer.

Department of Biosystems and Agricultural Engineering, University of Minnesota, expertise: MRI temperature mapping and mathematical modeling, contact: Professor R. Ruan.

US Army Soldier Command, Natick RD&E Center, expertise: general, temperature mapping, modeling, contact: Dr. Christopher Doona, Dr. Tom Yang.

12.7 References

1. SASTRY, S.K. and S. PALANIAPPAN, Ohmic heating of liquid-particle mixtures. *Food Technology*, 1992. **46**(12): 64–67.
2. MCCONNEL, S.V. and R.P. OLSSON, Wiener vending machine. 1938: US Patent 2,139,690.
3. DE ALWIS, A.A.P. and P.J. FRYER, The use of direct resistance heating in the food industry. *J Food Engineering*, 1990. **11**: 3–27.
4. SKUDDER, P.J., Ohmic heating: new alternative for aseptic processing of viscous foods. *J. Food Engineering*, 1988. **60**(1): 99–101.
5. ANON, Ohmic heating garners 1996 industrial acheivement award. *Food Technology*, 1996. **20**(10): 114–115.
6. GETCHELL, B.E., Electric pasteurization of milk. *Agric. Engng.*, 1935. **16**(10): 408.
7. ZOLTAI, P.T. and M.A. SWEARINGEN, Product development considerations for ohmic processing, in *IFT Annual Meeting*, 1995. **49**: 49.
8. PARROTT, D.L., Use of ohmic heating for aseptic processing of food particulates. *Food Technology*, 1992. **46**(12): 68–72.
9. BISS, C.H., S.A. COOMBES and P.J. SKUDDER The development and application of ohmic heating for the continuous heating of particulate foodstuffs. In *Engineering Innovation in the Food Industry*. 1987. Bath, UK: Institution of Chemical Engineers.
10. KIM, H.J., Y.M. CHOI, B. HESKITT, S.K. SASTRY and Q. LI, Chemical and microbiological investigation of ohmic heating of particulate foods using a static ohmic heater., in IFT Annual Meeting 1995.
11. SASTRY, S.K. and Q. LI, Modeling the ohmic heating of foods. *Food Technology*, 1996. **50**(5): 246–248.
12. RUAN, R.R. and P.L. CHEN, *Water in Foods and Biological Materials: A Nuclear Magnetic Resonance Approach*. 1998, Lanaster: Technomic Publishing Co.
13. CHANG, K., R.R. RUAN, P.L. CHEN, R.G. FULCHER and E. BASTIAN, Moisture, fat and temperature mapping of cheese block during cooling using MRI. Paper No. 956535, 1995. presented at ASAE Summer meeting (Chicago, Illinois).

14. HULBERT, G.J., J.B. LITCHFIELD and S.J SCHMIDT, Temperature mapping in carrot using T1 weighted magnetic resonance imaging. *J. Food Science*, 1995. **60**(4): 780–785.
15. RUAN, R.R., Z. LONG, K. CHANG, P.L. CHEN and I.A. TAUB, Glass transition temperature mapping using magnetic resonance imaging. *Transactions of the ASAE*, 1999. **42**(4): 1055–1059.
16. SUN, X., J.B. LITCHFIELD and S.J. SCHMIDT, Temperature mapping in a model food gel using magnetic resonance imaging. *J. Food. Science*, 1993. **68**(1): 168–172, 181.
17. SUN, X., S.J. SCHMIDT and J.B. LITCHFIELD, Temperature mapping in a potato using half fourer transform MRI of diffusion. *J. Fd. Proc. Engr.*, 1994. **17**: 423–437.
18. HILLS, B., Food processing: an MRI perspective. *Trends in Food Science & Technology*, 1995. **6**(4): 111–117.
19. DE ZWART, J.A., F.C. VIMEUX, C. DELALANDE and P. CANIONI, Fast lipid-suppressed MR temperature mapping with echo-shifted gradient-echo imaging and spectral-spatial excitation. *Magn Reson Med*, 1999. **42**: 53–59.
20. PETERS, R.D., R.S. HINKS and R.M. HENKELMAN, Ex vivo tissue-type independence in proton-resonance frequency shift MR thermometry. *Magn Reson Med*, 1998. **40**: 454–459.
21. SUN, X., S.J. SCHMIDT and J.B. LITCHFIELD, Temperature mapping in a potato using half Fourier transform MRI of diffusion. *J. Fd. Proc. Engr.*, 1994. **17**: 423–437.
22. CHANG, K., R. RUAN, A. NING, P. CHEN and I. TAUB, Fast MRI temperature mapping of ohmically heated foods. ASAE Annual International Meeting, Orlando, FL, USA, July 12–16, 1998. Paper No. 986068.
23. YE, X., R. RUAN, P. CHEN, C. K. NING, I. TAUB and C. DOONA, Accurate and fast temperature mapping during ohmic heating using PRF shift MRI thermometry. *J. Food Engineering*, 2003. **59**(2–3): 143–150.
24. YE, X., R. RUAN, P. CHEN, C. DOONA and I. TAUB, MRI temperature mapping and determination of the liquid-particulate heat transfer coefficient in an ohmically heated food system. *Journal of Food Science*, 2003. **68**(4): 1341–1346.
25. ISHIHARA, Y., A. CALDERON, H. WATANABE, K. OKAMOTO, Y. SUZUKI, K. KURODA and Y. SUZUKI, A precise and fast temperature mapping using water proton chemical shift. *Magn Reson Med.*, 1995. **34**: 814–823.
26. HEISTERKAMP, J., N.A.A. MATHEIJSSEN, R. VAN HILLEGERSBERG, J.J. VAN VAALS, J.S LAMERIS, J. STOKER and J.N.M. IJZERMANS, Accuracy of MR phase mapping for temperature monitoring during interstitial laser coagulation (ILC) in the liver at rest and simulated respiration. *Magn Reson Med.*, 1999. **41**: 919–925.
27. HINDMAN, J.C., Proton resonance shift of water in the gas and liquid states. *J Chem. Phys.*, 1966. **44**: 4582–4592.
28. MULLER, N. and R.C. REITER, Temperature dependence of chemical shifts of protons in hydrogen bonds. *J. Chem. Phys.*, 1965. **42**: 3265–3269.
29. HAASE, A., J. FRAHM, D. MATTHAEI, W. HANICKE and K.-D. MERBOLDT, FLASH imaging. Rapid NMR imaging using low flip-angle pulse. *J. Magn. Reson*, 1986. **67**: 258–266.
30. PIGNOLI, E., R. MARCHESINI, L. CURTI, A.E. SICHIROLLO, S. TOMATIS and R. MUSUMECI, Potential and limitations of magnetic resonance imaging for real-time monitoring of interstitial laser phototherapy. *Acad. Radiol.*, 1995. **2**: 741–747.
31. DE ALWIS, A.A.P. and P.J. FRYER, A finite element analysis of heat generation and transfer during ohmic heating of foods. *Chem. Eng. Sci.*, 1990. **45**(2): 1547–1559.
32. ZHANG, L. and P.J. FRYER, Models for the electrical heating of solid-liquid food

mixtures. *Chem. Eng. Sci.*, 1993. **48**: 633–643.

33. SASTRY, S.K. and S. SALENGKE, Ohmic heating of solid-liquid mixtures: a comparison of mathematical models under worst-case heating conditions. *J. Food Process Engineering*, 1998. **21**(6): 441–458.
34. SASTRY, S.K. and S. PALANIAPPAN, Mathematical modeling and experimental studies on ohmic heating of liquid-particle mixtures in a static heater. *J. Food Process Engineering*, 1992. **15**(4): 241–261.
35. SASTRY, S.K., Application of ohmic heating to continuous sterilization of food. *Indian Food Industry*, 1992. **11**(4): 28–30.
36. FRYER, P.J., A.A.P.D. ALWIS, E. KOURY, A.G.F. STAPLEY and L. ZHANG, Ohmic processing of solid-liquid mixtures: heat generation and convection effects. *J. Food Engineering*, 1993. **18**(2): 101–125.
37. DAVIES, L.J., M.R. KEMP and P.J. FRYER, The geometry of shadows: effects of inhomogeneities in electrical field processing. *J. Food Engineering*, 1999. **40**(4): 245–258.
38. KIM, H.J., Y.M. CHOI, T.C.S. YANG, I.A. TAUB, P. TEMPEST, P.SKUDDER, G. TUCKER and D.L. PARROTT. Validation of ohmic heating for quality enhancement of food products. *Food Technology*, 1996. **50**(5): 253–255, 257–261.
39. MAESMANS, G., M. HENDRICKX, S. DECORDT and A. FRANSIS, Fluid-to-particle heat transfer coefficient determination of heterogeneous foods: a review. *J. Food Processing and Preservation*, 1992. **16**: 29–69.
40. YE, X., R. RUAN, P.L. CHEN and C. DOONA, Simulation and verification of ohmic heating in a static system using MRI temperature mapping. *Lebensmittel-Wissenschaft und Technologie*, 2004. **37**(1): 49–58.
41. LIMA, M., B.F. HESKITT and S.K. SASTRY, Diffusion of beet dye during electrical and conventional heating at steady-state temperature. *J Food Process. Eng.* **24**(5): 331–340.
42. CHO, H.Y., A.E. YOUSEF and S.K. SASTRY, Growth kinetics of *Lactobacillus acidophilus* under ohmic heating. *Biotechnology and Bioengineering*, 1996. **49**(3): 334–340.

13

Air impingement heating

A. Sarkar and R. P. Singh, University of California – Davis, USA

13.1 Introduction: air impingement processing

In the modern food industry, there is an increasing demand for shorter processing times in freezing, thawing, drying, and baking of foods. Research and development to meet these demands have resulted in many new methods that have been commercialized. One of the latest and most promising developments in this area is the air impingement system.

Air impingement systems have been used in various industrial operations involving heat and mass transfer such as textile and paper drying, electronic cooling and glass quenching. More recently, a promising thermal processing technique has evolved from the air impingement system for industrial food processing operations (Ovadia and Walker, 1998; Li and Walker, 1996) with a significant reduction in process times (Wahlby *et al.*, 2000). Air impingement systems involve arrays of jets that impinge air on the surface of a food product. Industrial impingement systems have been applied to food processes such as drying, baking, toasting, and freezing (Midden, 1995; Borquez *et al.*, 1999; LujanAcosta *et al.*, 1997; Nitin and Karwe, 2001; Moreira, 2001). Other potential operations such as impingement thawing and chilling are being considered.

The basic design of an air impingement system is simple (Fig. 13.1). It essentially consists of a blower arrangement that forces air of the desired temperature into a plenum. The plenum has an array of nozzles that can be either circular or rectangular. A rectangular nozzle releases a slot jet when the length of the nozzle is at least ten times its width. Almost all practical rectangular nozzles produce slot jet. The product is placed on conveyors that run below the jets. Often arrays of jets also impinge the product from the bottom. Typical jet

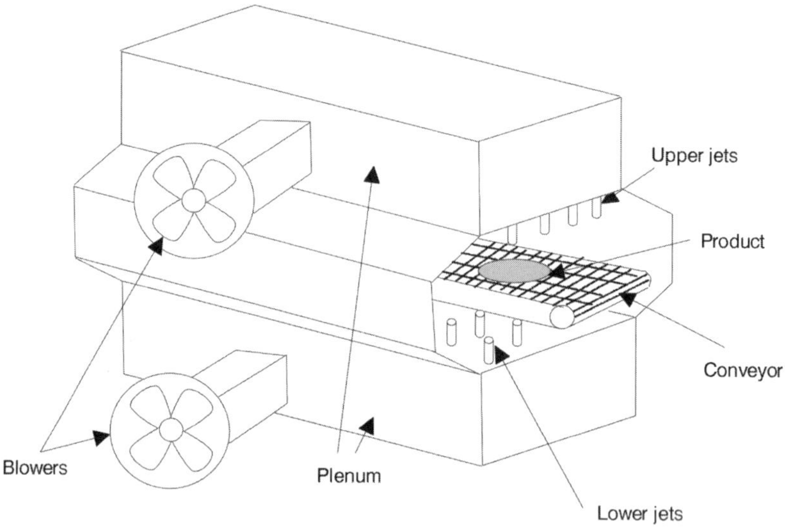

Fig. 13.1 Typical industrial air impingement system.

velocities vary from 20 to 80 m/s. Smaller and lighter products (such as chips during drying or individually quick frozen peas and corn) tend to float between the top and bottom air streams. Such impingement may also be called impingement-fluidization. Larger products such as pizza crusts, cakes, and poultry parts are usually stationary on the conveyor. The conveyor is often shrouded on both sides to conserve energy, and most practical designs have exhausts either on the top or the sides.

In spite of their simple construction, impingement equipments tend to be complicated in practical applications, because the interaction of the airflow with the product varies from product to product. Hence, understanding the underlying physical phenomena in impingement is essential for impingement jets. In this chapter, we will describe the theory behind impingement processing, the design of impingement systems, measurements and measurement techniques that may be needed to design impingement units, and how theoretical understanding may help in improvement of the design for various applications.

13.2 Air impingement processing of food products: principles

When a food product is heated or cooled in a fluid medium such as air or water, a stagnant boundary develops around the product. This boundary layer causes a high thermal resistance and consequently a slow heating and cooling process. The advantage of air impingement systems over traditional systems is that the boundary layer insulation effect between the product and the medium (air in this case) is reduced due to higher air velocity and greater turbulence induced by the jet. Typical jet velocities at the nozzle exit range from 20 to 80 m/s. Note however, that the boundary layer is not eliminated but its characteristics are changed resulting in higher heat transfer rates. This is because the surface heat flux (at $y = 0$) is defined as follows,

$$q_s = -k_f \left.\frac{\partial T}{\partial y}\right|_{y=0} \qquad [13.1]$$

In laminar flows the temperature from the free stream to the wall varies gradually. As a result $(\partial T/\partial y)|_{y=0}$ for laminar flows is small. For turbulent flows the boundary layer has a steep gradient near the wall and the temperature profile is uniform away from the wall. Therefore, in turbulent boundary layer flows $(\partial T/\partial y)|_{y=0}$ is high resulting in an increase in q_s.

Food processing systems such as blast freezers and convection ovens also are developed with the same idea of reducing the insulating effects of boundary layers; however, impingement systems are far more effective considering the low volume of airflow required. Added features of the flow such as stagnation, increase heat transfer further in air impingement systems. Also, unlike traditional forced convection systems, impingement devices target directly onto the product surface, thus conserving energy.

13.2.1 Flow pattern under impingement jets

Heat transfer is implicitly related to fluid flow in impingement and other convective heat transfer processes. Hence, it is important to understand the fluid flow characteristics of impinging jets to understand in turn the physics of heat transfer in air impingement processes. Polat *et al.* (1989) categorized the flow patterns from impinging jets into three characteristic regions: free jet region, stagnation flow region, and radial flow or wall jet region (Fig. 13.2).

The free jet region can be classified into three sub-regions: the potential core region, developing flow region, and developed flow region. The potential core region is the part of the flow where no vorticity exists in the flow. However, the nozzle edges and free shear between the impinging jet and the stagnant air surrounding it causes a mixing boundary layer in the periphery of the jet, resulting in the end of the potential core region and significant energy dissipation from the jet. The rate of dissipation and the length of the potential core are largely dependent on the shape and configuration of the nozzle (Jambunathan *et al.*, 1992; Schlichting, 1979). This region of mixing is called the developing flow region, which eventually manifests as the developed flow region where the potential core is practically non-existent (Fig. 13.2). The characteristics of turbulence in the free jet region have been attributed to the nozzle shape, velocity at nozzle exit, shape factors of the nozzle such as length and diameter, and the sharpness at the nozzle exit by various researchers (Gardon and Akfirat, 1966; Martin, 1977; Polat *et al.*, 1989). The turbulence in the free jet region and associated dissipation of energy is an important factor contributing to heat transfer in the jet further downstream (stagnation and wall

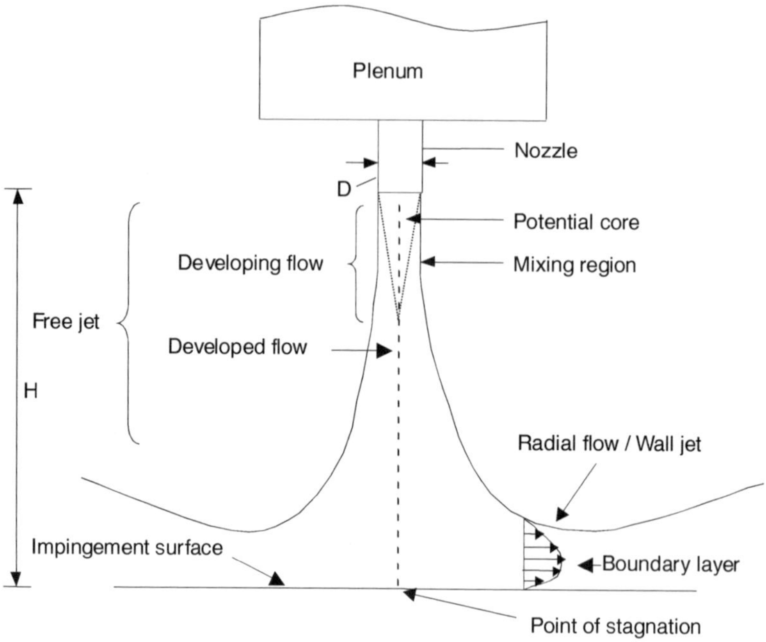

Fig. 13.2 Fluid flow under impinging jets.

jet) where actual heat transfer takes place between the fluid and the product. As a result, understanding the features of turbulence and flow in the free jet region for jets at the temperatures relevant to the application is essential for understanding of the heat transfer downstream.

As the jet approaches the impingement surface, the flow undergoes steep deceleration in the vertical direction, causing a steep velocity and temperature gradient near stagnation. The velocity in the vertical direction eventually comes to zero at stagnation. On the other hand, there is acceleration in the radial direction from the stagnation point with a zero radial velocity at stagnation. Thus, at stagnation point, both vertical and radial velocities are zero (Schlichting, 1979). Stagnation is characterized by very high heat and mass transfer coefficients. This is because the high vertical deceleration makes the gradient $(\partial T/\partial y)|_{y=0}$ very high at stagnation.

From the stagnation point, the fluid velocity increases in the radial direction due to acceleration; this region of the flow is called the radial flow region or wall jet region (Fig. 13.2) because the wall bounds the flow into a boundary layer. The region of the flow very near stagnation may be transitional (changing from laminar to turbulent conditions) if the jet is not fully developed before stagnation occurs (Gardon and Akfirat, 1966). If the jet is fully developed before stagnation, turbulent fluctuations of mean velocities would exist at stagnation though mean velocity would be zero. As a result, the jet would be turbulent even at stagnation and no transitional characteristic would be seen. Transitional wall jets result in secondary, and sometimes tertiary, peaks in heat transfer coefficients causing uneven heat transfer. Radial velocity is zero on the surface of the impingement plate because of no-slip conditions, resulting in formation of a boundary layer, and typical boundary layer characteristics are noticed. The radial flow region eventually develops into a recirculation region if the flow is confined (Jambunathan *et al.*, 1992). Confinement and recirculation, which may cause disturbances in the original jet, can be prevented by proper exhaust arrangements.

From a food processing point of view, the main factor of interest is the heat and mass transfer at the interface of the product surface and the fluid flow. However, features such as stagnation and boundary layer formation result in the uneven heat and mass transfer coefficients on the surface under the jets. Measurements of heat and mass transfer for impingement jets have been made over the last 40 years by various researchers for non-food commodities which have been reviewed in detail by Polat *et al.* (1989). For food applications, two approaches to the determination of heat and mass transfer can be found in the literature: either by using empirical correlations, or by doing actual measurements. Both of these methods will be discussed in the following sections.

13.3 Heat transfer measurements and characteristics in impingement systems

13.3.1 Heat transfer measurements under impinging jets

As described in the previous section, there is a considerable variation of heat transfer coefficients at various locations under impinging jets that may be studied experimentally. Various researchers have studied the spatial variations of heat transfer characteristics of circular and slot jets impinging on flat and curved surfaces (Gardon and Akfirat, 1966; Polat *et al.*, 1989; Sparrow and Lee, 1975; Amano and Sugiyama, 1985; Jambunathan *et al.*, 1992). In these studies, experimental techniques have been developed to study the spatial variation of convective heat transfer, and attempts have been made to develop conditions such that this variation is minimized while maximizing the average heat transfer coefficients. More recently, later researchers have extended the research on heat transfer measurements to food processing (Nitin and Karwe, 2001; Sarkar and Singh, 2003).

Traditionally, the lumped capacitance technique has been used to determine average convective heat transfer coefficients for food processing applications at high Reynolds number. In this technique, a transducer made of high thermal conductivity material is used to represent the product shape, and a temperature sensor is placed at its center to determine the change in temperature. The transducer should be at thermal equilibrium at any point in time, that is, the temperature at any point within the transducer is the same at any given time. The transient heat transfer curve is then used to evaluate the convective heat transfer coefficient. Such a method has been applied to impingement baking systems

(Nitin and Karwe, 2001). This method is simple to use and is good as an average estimate, but use of this technique for air impingement applications will not provide details of the spatial variations of heat transfer coefficients. Therefore, there is a need to study the effect of spatial variation of heat transfer on localized hot and cold spots.

Thermochromic liquid crystals (TLC) provide a way of studying spatial variation of heat transfer coefficients under impinging jets and have been used in various situations (Baughn and Shimizu, 1989; Baughn, 1995; Mesbah *et al.*, 1996; Lee and Lee, 2000). The use of liquid crystals is restricted by the narrow temperature range in which liquid crystals work (−10 to 110ºC). This makes it difficult to apply liquid crystals to freezing or baking applications, since freezing temperatures may reach −40ºC and temperatures in baking processes may reach 400ºC. Also, while considering larger temperature ranges the temperature resolution of TLC systems decrease.

The method of using micro-calorimeters (Westknemper, 1961; Donaldson *et al.*, 1971) has been applied to various heat transfer applications where spatial variation of heat transfer is significant. This technique is an extension of the lumped capacitance technique. Small transducers with thermocouples attached to them (micro-calorimeters) are used at various spatial locations instead of using a single transducer. Donaldson *et al.* (1971) developed a method for studying heat transfer under heated impinging jets using this technique. Their setup consisted of thin copper disks (0.0381 mm thickness) that acted as micro-calorimeters flush mounted on polyurethane foam insulation at various spatial locations. The circumference of the micro-calorimeter was insulated and the exposed surface of the insulation was covered with a copper plate that had holes punched to accommodate the calorimetric disks. The setup was placed under a heated impinging jet, and the heat transfer coefficients were determined by analyzing the transient heat transfer curve assuming that the disks behaved as lumped capacitances. A modified version of this technique has been applied for heat transfer measurements in air impingement applications in the freeze-thaw range (Sarkar and Singh, 2003). However, the use of a lumped-capacitance technique, such as this, may have significant uncertainties due to changing boundary conditions (Butler and Baughn, 1996; Sarkar and Singh, 2003). This is because the change in the temperature of the surface causes a change of thermal boundary layer profile over time (Fig. 13.3(a) and (b)). As a result, the slope $(\partial T/\partial y)|_{y=0}$ changes over time and causes a variation of surface heat flux (q_s). During the beginning of the process, T_s is initial plate temperature, and toward the end of the process, T_s approaches final temperature which is free stream temperature (T_∞). Consequently, the convective heat transfer coefficient (h) varies over time, where h is defined as in Eq. 13.2.

$$h = \frac{q_s}{(T_s - T_\infty)} \qquad [13.2]$$

Since the variation of surface temperatures with time depends on the material properties, using a copper plate introduces errors. On the other hand, using a

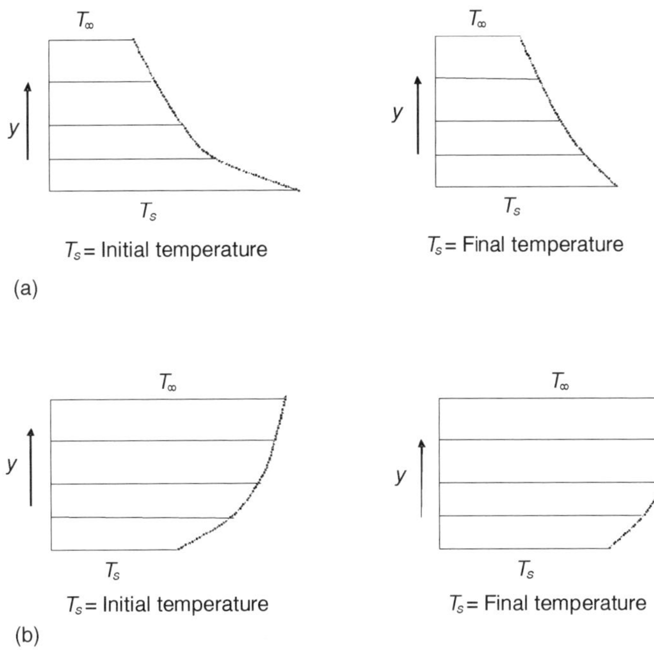

Fig. 13.3 (a) Thermal boundary layer chacteristics during when $T_s > T_\infty$; (b) Thermal boundary layer chacteristics during when $T_s < T_\infty$.

non-metallic surface, such as actual foods, makes the lumped capacitance assumption invalid. An alternative suggested by Butler and Baughn (1996) and used by Anderson and Singh (2002) is to use actual food materials and the solution of the transient heat transfer equation instead of lumped capacitance. The problem with such a technique is to account for mass transfer, thermal properties and phase change during the process. This makes the transient technique difficult to apply for food processing applications such as freezing, baking, and drying. As a result, the lumped capacitance technique with metallic surfaces has been used for impingement (Sarkar and Singh, 2003). The initial part of the transient temperature curve is used to obtain an accurate estimate.

13.3.2 Empirical correlations for estimation of heat transfer under impinging jets

From the discussions presented in the previous section, none of the heat transfer measurement techniques are perfect, and all of them have their own limitations. But with careful experimental design they can be applied with sufficient accuracy and heat transfer coefficients can be determined. It is useful to express the experimental data in terms of non-dimensional correlations to allow extension and interpolation. Certain basic equations used widely in fluid flow and heat transfer studies in general and air impingement research in particular are expressed in Eqs. 13.3 to 13.8.

Nusselt number: $$Nu = hD/k_f \qquad [13.3]$$

Reynolds numbers: $$Re = \rho_f v D/\mu_f \qquad [13.4]$$

All the thermal properties of the fluid are evaluated at the location where the characteristic dimension is defined. Characteristic dimensions for jets are defined as:

Circular jets: $$D = \mathit{diameter\ at\ nozzle\ exit} \qquad [13.5a]$$

Slot jets: $$D = 2 \times \mathit{width\ of\ slot\ jet\ exit} \qquad [13.5b]$$

Nozzle and system dimensions of length (*L*), nozzle exit to product surface distance (*H*) and distance from stagnation (z) are expressed non-dimensionally as:

$$\mathit{Non\text{-}dimensional\ length\ of\ nozzle} = L/D \qquad [13.6]$$

$$\mathit{Non\text{-}dimensional\ distance\ from\ nozzle\ exit\ to\ productsurface} = H/D \qquad [13.7]$$

$$\mathit{Non\text{-}dimensional\ distance\ from\ stagnation} = z/D \qquad [13.8]$$

Gardon and Akfirat (1966) and Polat *et al.* (1989) gave empirical correlations based on their data for estimating *Nu* under turbulent conditions for various *H/D* ratios and *Re* that are valid at their stagnation region. Donaldson *et al.* (1971) compared boundary layer *Nu* to *Re* in various types of heated jets and gave empirical correlations that are valid at the stagnation point and the wall jet region. These studies included free and impinging, and confined and unconfined jets. Martin (1977) has used some of the previously published research and added mass transfer studies to formulate empirical heat and mass transfer correlations along with nomograms. His studies included factors such as *H/D* ratio and spatial variation of heat transfer coefficient for both slot and circular jets. Borquez *et al.* (1999) and Moreira (2001) have used similar empirical relationships for estimation of heat and mass transfer in their studies on air impingement food drying applications. Sarkar (2002) developed correlations for spatial variation of *Nu* for two typical *H/D* ratios for impingement in freezing-thawing applications under both circular and slot jets. The details of the heat transfer characteristics in impingement situations have been discussed in Section 13.3.

Due to their empirical nature, none of these relationships for heat transfer hold good for all temperature regimes. Moreover, because of the presence of turbulence in flow, these models often show considerable deviation from the experimental data, which may be attributed to the difference in transition characteristics, and wall roughness causing differences in the boundary layer characteristics. Martin (1977) noted deviations of up to 30–35% in some of his mass transfer estimates. The extension of these relationships to other spatial locations or product situations is inaccurate because the nature of turbulence associated in the boundary layer is extremely sensitive to the location and surface characteristics.

Thus, ignoring the flow and its effects on heat transfer for impingement applications may cause considerable errors. The latest approach in impingement applications has been to experimentally study or numerically model the flow field in an air impingement situation and relate it to the heat transfer for an improved understanding of the process.

13.3.3 Heat transfer characteristics under impinging jets

Now that we have examined the flow characteristics of air impingement systems and procedures to measure heat transfer, we will consider some of the heat transfer characteristics of air impingement systems and their effects on food processing. The jet velocities at the nozzle exit in typical impingement systems designed for food processing range from 20 to 80 m/s, which correspond to exit *Re* of 10,000 to 100,000, depending on *D*. The *L/D* ratio of the jets influences the jet turbulence at exit. Shorter *L/D* ratios (less than 1) result in jets that have high exit turbulence but require less pressure in the plenum. Long jets ($L/D > 10$) give a fully developed flow with low turbulence at exit. A detailed study of jet length on exit characteristics has been done by Gardon and Akfirat (1966).

The jet loses energy after exiting the nozzle. A higher *H/D* ratio indicates substantial loss of energy in the free steam region while shorter *H/D* ratios cause transitional regimes after impingement, as explained earlier. Figure 13.4 shows the variation of non-dimensional heat transfer coefficients (*Nu*) for various off-

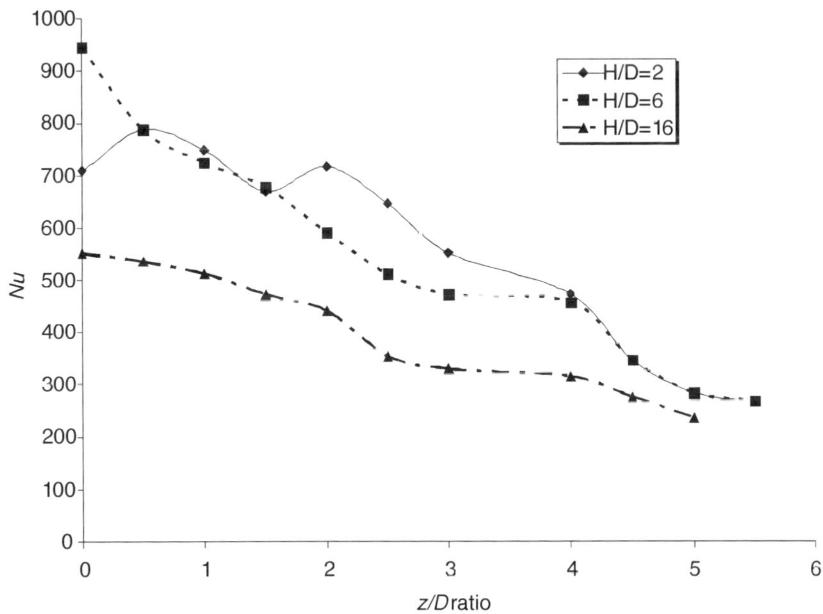

Fig. 13.4 Typical Nusselt number (*Nu*) variations under heated single impinging circular jets for various *H/D* ratios (using data from Gardon and Akfirat, 1966; adapted from Sarkar *et al.*, 2004).

stagnation locations (*z/D*), for a heated jet impinging on a room temperature surface for *H/D* ratios of 2, 6, and 16. The data for *H/D* of 2, 6, and 16 were obtained by Gardon and Akrifat (1965) for *Re* of 28,000 were plotted in terms of *Nu* for various *z/D* ratios. Previous research showed that *H/D* ratios of 6 to 8 result in highest stagnation *Nu,* as seen in Fig. 13.4. Longer *H/D* ratios result in lower *Nu* due to energy dissipation, while shorter *H/D* ratios result in lower *Nu* due to multiple peaks in transitional regimes. These *H/D* ranges were developed for heated conditions and appear to be valid even in cold and freezing temperature regimes (Sarkar and Singh, 2003). Mass transfer coefficients show a direct correlation to heat transfer coefficients due to direct correlations between dimensionless heat and mass transfer coefficients.

Stagnation point *Nu* values have been observed to be as high as 400 to 500 for heated jets (Polat *et al.*, 1989). The corresponding *Nu* are small for jets in freeze thaw conditions (200 to 300) due to frost formation and differences in characteristics of the thermal boundary layer. Circular jets tend to give higher *Nu* at stagnation, though average *Nu* for slot and circular jets are relatively similar. Slot jets tend to have a more uniform heat transfer profile (Sarkar and Singh, 2003) but usually require higher air-flow rates or small nozzle size to compensate for the flow rate and consequent scaling down of the dimensions of the equipment (such as *H* and *L*) to maintain desirable *H/D* and *L/D* ratios.

13.3.4 Extension to fluidized impingement

For smaller products such as chips, crackers and cut vegetables, impingement causes lift of the product (Fig. 13.5a). Use of double impingement, which is common for most practical applications, further increases the lift effect (Fig. 13.6a). These features have been described by Ovadia and Walker (1998). The amount of lift depends on product density, aerodynamic lift based on product shape, and jet velocities. Actual prediction of the amount of lift requires advanced simulation since it is product specific and has not yet been researched in great detail. However, once the product has been lifted, the characteristics of heat and mass transfer would be essentially like impingement over a porous surface and has been studied widely, especially for air impingement drying of textiles and paper (Chen and Douglas, 1995; Martin, 1977). The theory in such a case remains the same but there are added terms for suction in the boundary layer, and the heat and mass transfer in the product works like a porous medium.

13.3.5 Flow visualization and quantitative measurements under impinging jets

Flow measurements in impingement situations have been complicated with traditional intrusive techniques such as hot air anemometry. Hence, most of the earlier experimental flow research was restricted to visualization experiments as in the work of Popiel and Tras (1991) and Cornaro *et al.* (1999). Similar techniques have been applied to impingement studies for food processing application

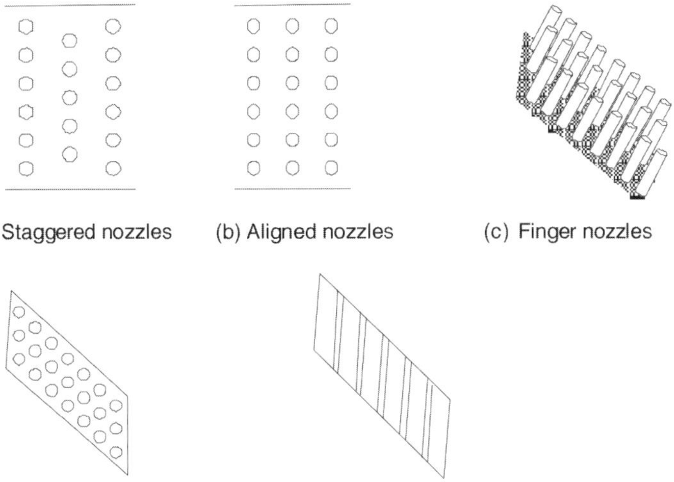

(a) Staggered nozzles (b) Aligned nozzles (c) Finger nozzles

(d) Orifice nozzles (e) Slot jets

Fig. 13.5 Nozzle designs for circular and slot jets (adapted from Sarkar *et al.*, 2004).

by Sarkar and Singh (2003). Recently, researchers have been able to use optical techniques for flow field studies in air impingement systems. The high velocities involved in air impingement require high temporal and spatial resolutions. As a result, most optical techniques applied to air impingement are LASER based. Two of the most popular laser-based technologies are LASER Doppler Anemometry (LDA) and particle imaging velocimetry (PIV). Extensive literature is available on both of these techniques (Adrian, 1991). While the LDA technique is a point measurement technique, PIV techniques can measure the entire flow field. The latter, however, requires more expensive instrumentation. Considerable research on the application of LDA techniques for impingement baking has been done by Marcroft *et al.* (1999) and Marcroft and Karwe (1999). The high degree of sophistication and accuracy of these techniques involves high cost and complicated instrumentation.

13.4 Design and use of air impingement systems in the food industry

Practical application of air impingement systems requires proper design of the equipment. There are few papers in published journals that describe various types of impingement equipment for food processing. Ovadia and Walker (1998) have described various impingement systems mostly with round jets for heating, cooling, baking and drying processes. In the following sections, various practical issues related to the design of air impingement food processing systems are presented.

13.4.1 Design of air impingement systems

Common jet configurations that are used in impingement devices are rectangular-shaped nozzles and arrays of circular nozzles (Fig. 13.5(a)–(e)). Individual slot jets and circular jets emerging from such nozzles can be considered axisymmetric for most practical applications, but, the axisymmetric assumption may not apply when there is a system of double or multiple circular jets. Yet measurements obtained from single jets are always useful, since they can be used to optimize various parameters such as *H/D* ratio, *L/D* ratio, and jet *Re* at nozzle exit. Multiple circular jet designs can be either staggered (Fig. 13.5(a)) or aligned (Fig. 13.5(b)). In this type of multiple-jet configuration, another important factor to be considered is jet-to-jet distance (*S*) expressed non-dimensionally as *S/H* ratio. As seen from heat transfer measurement studies cited in previous sections, slot jets tend to produce more uniform jet flow and heat transfer but need higher flow rates. Hence when the product is sensitive to heat transfer variations or high flow rates are feasible, slot jets are preferred. Recirculation of spent air reduces flow rate requirements and tends to improve applicability of slot jets. Longer jets (Fig. 13.5(c)) with higher *L/D* tend to produce a more uniform flow at exit and may prove beneficial for maintaining uniform heat transfer, but, cleaning and maintenance can be problematic. Longer jets (known as fingers) may sometimes be bent and a few bent fingers can cause considerable flow irregularities compared with shorter jets (known as orifices, as shown in Fig. 13.5(d)). Nozzles are also classified as sharp or bell-shaped (Martin, 1977; Moriera, 2001). The sharpness at nozzle exit significantly affects free jet turbulence.

To maintain uniformity of air flow through the nozzles, usually a plenum is used to maintain constant static pressure. It is important to ascertain that the design of the plenum and nozzles allow equal jet exit velocities from all of the jets, so that the jets do not interfere with each other, and all the jets behave similarly. This allows uniform heat transfer to the product. Industrial designs are sometimes plagued with uneven jet exit velocities at nozzle exit due to improper plenum design.

Another significant factor of concern in impingement design is entrapment and exhaust of air. If the air that has already impinged on the surface is not removed, it may become entrapped in the flow field, as seen in Fig. 13.6(a) with flow field visualization of a confined impinging circular jet. In the figure, the region A has a higher pressure than region B. As a result the air from region B tends to move back to region A (recirculates), thus disturbing the free jet (region C). This can be mitigated by using proper exhaust. An example of proper exhaust design shown in Fig. 13.6(b), consisting of two circular jets emerging from two neighboring nozzle exits into the exhaust in between. The air exhaust prevents disturbance to the original jet and exhausted air can also be recycled, thus improving energy efficiency of the system.

Common exhaust systems used in impingement involve exhaust on the side and on top. One-sided impingement devices (whose application is limited for practical cases) may sometimes have exhaust from the bottom. Exhaust from top is the most efficient design since it causes the least disturbance to the actual jets.

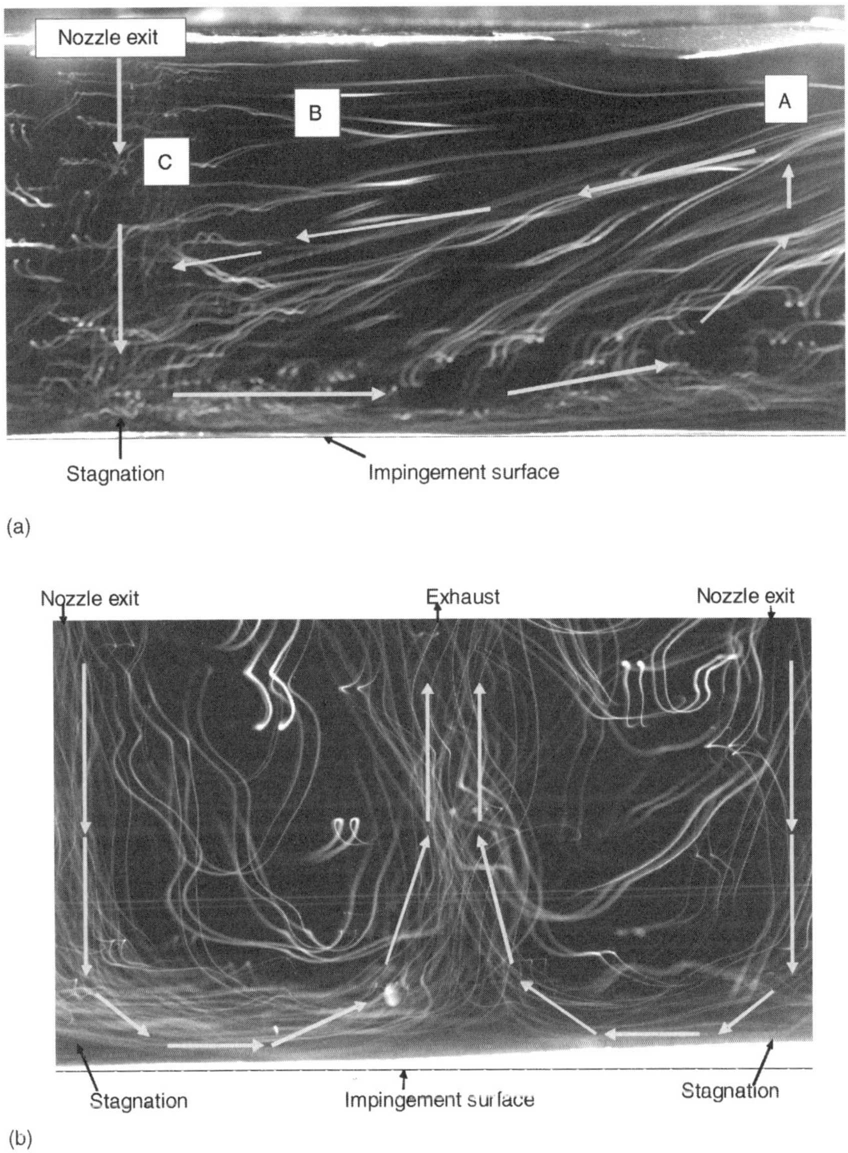

Fig. 13.6 (a) Confinement and recirculation in impinging jet from circular nozzle; (b) exhaust arrangement in impingement system (adapted from Sarkar *et al.*, 2004).

13.4.2 Impingement systems for high-temperature applications (above room temperature)

With the basic principles and design aspects described thus far, there are certain special considerations associated with high-temperature applications. For situations involving baking and toasting, radiation can play a major role in heat transfer. In such cases, the process is more of a combination of radiation and convection. For baking with soft dough, stagnation may cause pressure effects on the surface of the product, resulting in undesirable indentations on the product. As a result, multiple stage-impingement systems may be used where the first stage has low velocities to allow initial crust formation without indentation. Toasting and drying of smaller fragile products such as chips are mostly done in fluidized impingement equipment. In such cases, vigorous flow can cause cracking if the flow is not properly controlled. In drying operations, mass transfer plays a very important role, and exhaust designs involve greater concern as entrapment of spent air without proper exhaust may cause moisture build up. These specific considerations are essential for successful application of air impingement. The information provided here is general, and it is necessary to be aware of product-specific considerations.

13.4.3 Impingement systems for low-temperature applications (freezing and thawing)

For low-temperature applications, there are also special considerations for impingement systems. Impingement applications in thawing will require consideration of frost formation on the product surface and moisture loss (Hosoda and Uzuhashi, 1967). In most freezing applications, moisture loss is not an important issue, because formation of frozen crust prevents moisture loss from the inside. But if the product being frozen is initially at room temperature, there is a significant chilling time until freezing begins. This chilling stage may involve mass loss. However, the impingement technique is still being developed, and most of the factors related to particular food processing applications mentioned here are not yet fully understood. Research is currently in progress on these topics.

13.4.4 Combination of impingement systems with other processing techniques (hybrid designs)

The main advantage of air impingement food processing is the increase in the convective heat transfer coefficient. Hence, if the heat transfer process becomes more dependent on the internal heat transfer, impingement systems may lose their effectiveness. On the other hand, in processes such as microwave heating, the addition of air impingement can aid in the formation of surface crusts, thus increasing versatility of the device.

These considerations have recently sparked interest among commercial equipment makers and researchers to develop hybrid impingement systems. Such systems are a combination of air impingement systems with other rapid heating systems, such as microwave heating and radiofrequency (RF) devices. These devices are still in the development stage and may soon become commercially viable.

13.5 Modeling and optimizing air impingement systems

Heat and mass transfer in air impingement systems are dependent on a wide array of factors such as *H/D* ratio, *S/H* ratio, jet length (expressed non-dimensionally as *L/D* ratio), jet type (slot or circular), turbulence level, surface roughness of the product, and other system-specific factors. As a result, the design and optimization of the equipment for a particular processing operation is easier to do if the system can be modeled and all the variables are optimized. However, the complication of the flow phenomena in air impingement processing requires a combined heat transfer and fluid flow approach for modeling air impingement processes.

Two approaches can be used to model the situation. The first is to model internal heat transfer in a food as a conduction problem, with convective heat transfer coefficient measured experimentally on the food surface as a function of position and time. This requires empirical correlations for convective heat transfer coefficients similar to the ones describe in Sections 13.3.2 and 13.3.3. The second approach is to model the entire flow and heat transfer into the product using simultaneous flow modeling and heat transfer. These two approaches will be described in the following sections.

13.5.1 Modeling and optimization using heat transfer measurements

In this approach, the *Nu* is taken as a function of one or more of the following variables: *Re*, *H/D*, *z/D*, *L/D* and *S/H*, specific jet shape (slot or circular), and temperature regime. The heat transfer in the product is modeled using appropriate mathematical equations and solution techniques (Ozisik, 1994). For a simple case of heating or cooling without significant mass loss or phase change, the model for heat transfer inside the food involves the simple unsteady Fourier conduction equation (Eq. 13.9) with appropriate boundary conditions. In case of air impingement phenomena, the boundary condition is convective boundary (Eq. 13.10) where the convective heat transfer coefficient h is determined as a function of location using *Nu* correlations. The Fourier equation for unsteady state heat conduction in a product in tensor notation is given by:

$$\rho_{prod} C_p \frac{\partial T}{\partial t} = -\frac{\partial}{\partial x_j}\left[k_{prod}\frac{\partial T}{\partial x_i}\right] \qquad [13.9]$$

And the heat flux (q_s) at a convective boundary is defined as:

$$q_s = h(T_s - T_\infty) \qquad [13.10]$$

Depending on the specific situation, modifications to the above equations may be needed. For instance, in case of freezing, considerations for phase change will have to be included in heat transfer calculations. Similarly, in the case of baking, the boundary conditions may have to be modified to incorporate radiation.

This approach to air impingement modeling has been used quite successfully by various researchers because of its simplicity, although it has certain limitations. The limitations include:

1. The *Nu* correlations may have considerable experimental errors (up to 35% according to Martin, 1977).
2. The *Nu* correlations for a given condition may not be applied to another condition. This is true even for similar temperature regimes, since *Nu* is extremely sensitive to boundary conditions (Sections 13.3.1 and 13.3.2). Most of these correlations are developed using non-food materials, where boundary temperature variations are not similar to food materials.
3. This approach cannot take into account heat transfer variation depending on the surface roughness. Also it cannot accurately predict heat transfer if there is a possibility of entrapment and recirculation.
4. Flow modeling is essential for proper exhaust and design of uninhibited flow to ensure proper processing. This is not possible using this approach.

In spite of the above limitations, such an approach may be considered favorable, since the alternative approach would be to simultaneously solve heat transfer in the product and fluid flow around an object. Until recently, this alternative was impractical due to limitations of numerical solution procedures for turbulent flow situations and finite capabilities of computers in terms of memory and speed. But now, with the advancement in computational techniques and the arrival of sophisticated computational fluid dynamics (CFD) solver packages on high-end desktop computers for solution of flow, this alternative has become increasingly possible. This type of approach is discussed in Section 13.5.2.

For optimization of impingement equipment, one can use *Nu* as the objective function and maximize the surface integral of *Nu* while minimizing the surface variation of *Nu*. Another approach is to minimize process time. The constraint in this case will be the variation of temperature over the product. Research regarding optimization of impingement is still in its initial stage.

13.5.2 Modeling simultaneous fluid flow and heat transfer

There are three equations that describe flow and heat transfer in fluids. Known as the Navier-Stokes equations, they describe conservation of mass (continuity equation), momentum, and energy in fluid systems. A complete description of these equations can be found in many textbooks on fluid mechanics (White, 1999; Kundu and Cohen, 2002). The flow variables in the equations are velocity components in the three directional coordinates, temperature and pressure.

In turbulent flows, the flow variables fluctuate with time at a particular location and the variables may be assumed to be constituted of mean flow and turbulent components. For example, the x-velocity (V_x) at any location is broken down into the mean velocity (U_x) and turbulent fluctuations (u_x) as in equation 13.11. The time averaging of equation 13.11 results in 13.12, since the time average of u_x is zero. A similar argument also holds good for velocity components in the y and z directions, temperature and pressure.

$$V_x = U_x + u_x \qquad [13.11]$$

$$\bar{V}_x = \bar{U}_x;\ since,\ \bar{u}_x = 0 \qquad [13.12]$$

Time averaging the variables in the original Navier-Stokes equations yields the turbulent time-average Navier-Stokes equations (Eqs. 13.13–13.15). The basic structure of the Navier-Stokes and time-averaged turbulent Navier-Stokes equations are similar, because time averaging of the turbulent fluctuations of the flow components is zero. However, the process results in certain product terms of the turbulent fluctuations which do not become zero on time averaging. The resulting equations are known as the turbulent Navier-Stokes equations (Eqs. 13.13–13.15), and the fluctuation in product terms is known as the turbulent viscosity term, since it resembles a viscous or frictional loss. A complete description of the mathematical development of the time-averaged turbulent Navier-Stokes equation can be found in the book by Tennekes and Lumley (1972).

Continuity:
$$\frac{\partial \bar{U}_i}{\partial x_j} = 0 \qquad [13.13]$$

Momentum:
$$\frac{\partial \bar{U}_i}{\partial t} + \frac{\partial \overline{U_i U_j}}{\partial x_j} = -\frac{1}{\rho_f}\frac{\partial \bar{P}}{\partial X_i} + \frac{\partial}{\partial x_j}\left[v_f\left(\frac{\partial \bar{U}_i}{\partial x_j} + \frac{\partial \bar{U}_j}{\partial x_i}\right) - \overline{u_i u_j}\right] \qquad [13.14]$$

Energy:
$$\rho_f C_v \frac{\partial \bar{T}}{\partial t} + \rho_f \bar{U}_j C_p \frac{\partial \bar{T}}{\partial x_j} = -\frac{\partial}{\partial x_j}\left[k_f \frac{\partial \bar{T}}{\partial x_i} + \rho_f C_p \overline{u_j \theta}\right] \qquad [13.15]$$

The Navier-Stokes equations in their complete forms are difficult to solve even for two dimensions and laminar flow regimes. The presence of turbulence adds more complications, because there are more unknowns than there are equations. The extra unknowns are the turbulent viscosity terms (also known as viscous shear terms) $\overline{u_i u_j}$ and $\overline{u_j \theta}$. This is known as the closure problem in turbulence, and it has been investigated over the last few decades (Tennekes and Lumley, 1972). Many methods have been suggested to find solutions to these equations by proposing suitable turbulence models (Pope, 2000), that introduce additional equations for evaluating the viscous product terms. One of the most popular turbulence models is the two equation k-ϵ models that suggests two more equations for the turbulent kinetic energy (ke) and the turbulent dissipation rate (ϵ) and some closure terms for these equations (Launder and Spalding, 1972). From these equations, turbulent viscosity terms can be estimated which can then replace the time-averaged turbulent viscosity terms. The closure coefficients for the k-ϵ coefficients were developed on the basis of free shear turbulent length scales, but in the case of impingement, there is a wall-bounded flow in the radial flow region, which introduces smaller turbulent length scales that are different from the free shear length scales. Using the renormalization group theory (RNG), various researchers have suggested different modifications to the standard k-ϵ model. The k-ϵ-RNG model gives reasonably good results for impingement applications in cases of rough walls (Laschefski *et al.*, 1996; Lee and Lee, 2000). Recent research (So *et al.*, 1997) shows that the k-ϵ-RNG model has certain limitations for smooth wall situations where the effect of the viscous sublayer is significant or when transitional domains are involved. However, some simple calculations done in our preliminary studies indicate that food impingement systems can be assumed to be completely rough on the basis of criteria outlined by Schlichting (1979) (roughness Reynolds number, based on roughness dimension > 80). Hence, the k-ϵ-RNG can lead to closure to the turbulent Navier-Stokes equations, though further research needs to be done to verify how well the solution will work.

Hu and Sun (2000), Verboven *et al.* (2000a,b), Hu and Sun (2001), and Mirade *et al.* (2002) have used the standard k-ϵ model for solving problems related to meat chilling, blast freezing and convection ovens using various computational fluid dynamics (CFD). As noted earlier the k-ϵ–RNG is a better formulation for a case when there is a complex wall-bounded flow with pressure gradients as in impingement. However, unlike the above-mentioned applications, there are a few additional factors working in certain impingement operations such as freezing, thawing, and baking. These operations include the phase change and mass transfer in the product, which requires a more complex solution routine for heat transfer inside the product and more challenging boundary conditions in the interface between the solid food and the flow. This makes application of CFD in impingement an interesting challenge and is an area of wide research at present. This shows the possibility of applying CFD in air impingement applications, although the presence of turbulence, the presence of a singularity at stagnation, the possibility of transitional flows, and phase change and mass transport in the product pose interesting challenges that have yet to be overcome.

13.6 Future trends

Commercial food processing equipment using impingement technology has been developed for several scales of applications including industrial processing, restaurants, vending and domestic use. A number of manufacturers offer equipment that incorporates air impingement either as a sole method of processing or in conjunction with other processing systems such as microwave, and infrared radiation. Selected manufacturers of impingement systems for food processing are listed in Table 13.1. Figures 13.7(a–c) show three commercially used designs of air impingement systems. Individual manufacturers provide detailed application notes for each of their models for product specific applications.

The industrial literature provided by these companies claims that impingement systems offer three or four-fold increase of heat transfer coefficients when compared with forced convection systems. When air impingement is incorporated with microwave heating, there is a better control of food characteristics

Table 13.1 Common manufacturers and designs of commercial air impingement systems

Manufacturer	Equipment	Design(s)
APV Baker	Ovens	Enerjet™
Enersyst Development Center	Design and development of impingement and hybrid systems	-
Firgoscandia (FMC FoodTech)	Freezers	FPF Advanted™
Fujimak Corporation	Impingement and microwave hybrid system	SuperJet™
Heat and Control, Inc.	Ovens	AirForce®
KRh Thermal Systems, Inc.	Impingement and microwave hybrid system	The JIM™ (Jet Impingement Microwave)
Lincoln Foodservice Products, Inc	Baking and cooking	Lincoln Impinger™
Mycom	Freezers	Mycom Thermo-Jack™ Freezer
Stein-DSI (FMC FoodTech)	Ovens	JSO Jet Stream™
Thermador	Residential cooker	JetDirect™
Wolverine Proctor and Schwartz	Cooler and dryer/puffer/ toaster	Jetzone™ double impingement cooler Jetzone™ dryer/puffer/ toaster

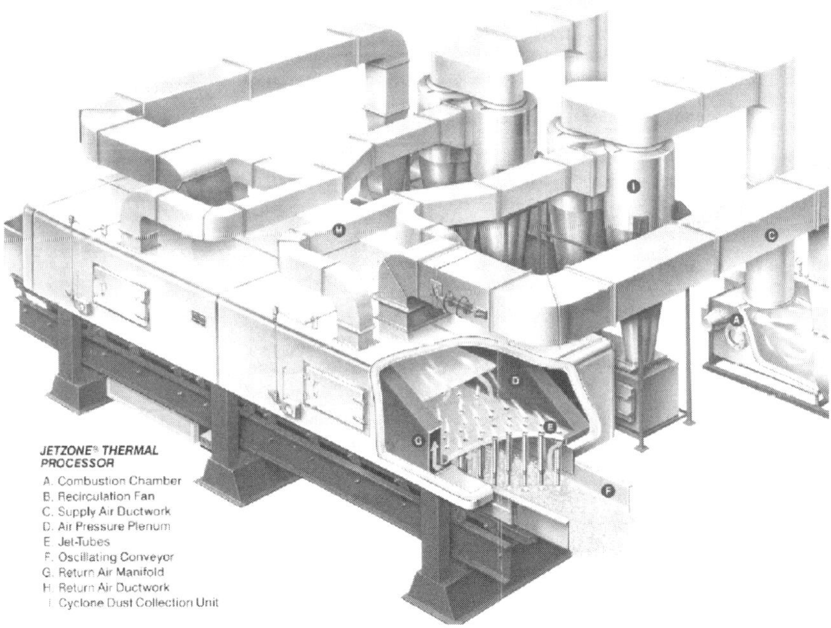

Fig. 13.7 (a) Jetzone impingement cereal toasting oven (courtesy Wolverine Proctor and Schwartz).

Fig. 13.7 (continued) (b) Impingement oven used in baking pizza; (c) Carter Hoffman high speed oven (courtesy Carter Hoffman).

at the exterior surface, and microwaves further enhance the internal heat transfer. Significant reduction in cook times has been claimed, such as 61% reduction in cooking whole turkeys and 85% reduction in baking potatoes. The impingement-microwave ovens are also claimed to be more energy efficient, reducing energy requirement in a significant manner. For example, 43% reduction in energy requirement for cooking pre-frozen pizza when compared with regular domestic ovens. Based on the significant advantages offered by this technology, it is expected that the use of impingement systems in industrial use for food processing will continue to grow at an impressive rate in the future.

13.7 List of symbols

C	specific heat
D	characteristic dimension
h	convective heat transfer coefficient
H	distance from nozzle exit to product surface
k	thermal conductivity
ke	turbulent kinetic energy
L	length of nozzle
Nu	Nusselt number
P	pressure
q	heat flux
Re	Reynolds number
S	jet to jet distance
T	temperature
t	time
U	mean velocity
u	fluctuating turbulent velocity
v	jet velocity at nozzle exit
V	total velocity of turbulent flow
x	arbitrary coordinate axis
y	vertical coordinate
z	distance on impingement surface from stagnation point
ϵ	turbulent dissipation rate
μ	viscosity
ν	kinematic viscosity
θ	fluctuating temperature
ρ	density

Subscript notations

i	coordinate index
j	coordinate index
f	fluid property

$_s$ at the surface
$_\infty$ conditions at free stream
$_{prod}$ product property
$_p$ at constant pressure
$_x$ x coordinate

13.8 References

ADRIAN RJ. 1991. Particle-imaging techniques for experimental fluid mechanics. *Annual Review of Fluid Mechanics*. 23: 261–304.

AMANO RS, S SUGIYAMA. 1985. Investigation of turbulent heat transfer of an axisymetric jet impinging on a flat plate. *Bulletin of the Japanese Society of Mechanical Engineering*. 28(235): 74–79.

ANDERSON, BA, RP SINGH 2002. Air impingement heat transfer on a cylindrically shaped object. *IFT Annual Meeting*, 2002, Anaheim, CA. June 15–19.

BAUGHN JW. 1995. Liquid crystal methods for studying turbulent heat transfer. *International Journal of Heat and Fluid Flow*. 16(5): 365–375.

BAUGHN JW, S SHIMIZU. 1989. Heat transfer measurements from a surface with uniform heat flux and an impinging jet. *Journal of Heat Transfer*. 111: 1096–1098.

BORQUEZ R., W WOLF, WD KOLLER, WEL SPIEB. 1999. Impinging jet drying of pressed fish cake. *Journal of Food Engineering*. 40(1–2): 113–120.

BUTLER RJ, JW BAUGHN. 1996. The effect of the thermal boundary condition on transient method heat transfer measurements on a flat plate with a laminar boundary. *Journal of Heat Transfer* 118(4): 831–837.

CHEN G, WJM DOUGLAS. 1995. Impingement drying of paper. *Drying Technology*. 13(5–7): 1331–1344.

CORNARO C, AS FLEISCHER, RJ GOLDSTEIN. 1999. Flow visualization of a round jet impinging on cylindrical surfaces. *Experimental Thermal and Fluid Science*. 20(2): 66–78.

DONALDSON C DUP, RS SNEDEKER, DP MARGOLIS. 1971. A study of free jet impingement. Part 2. Free jet turbulent structure and impingement heat transfer. *Journal of Fluid Mechanics* 45(3): 477–512.

GARDON R, JC AKFIRAT. 1966. Heat transfer characteristics of impinging two dimensional air jets. *Journal of Heat Transfer*. 88: 101–108.

HOSODA T, H UZUHASHI. 1967. Effects of frost on heat transfer coefficient. *Hitachi Review* 16(6): 254–259.

HU Z, DW SUN. 2000. CFD simulation of heat and moisture transfer for predicting cooling rate and weight loss of cooked ham during air-blast chilling process. *Journal of Food Engineering*. 46(3): 189–198.

HU Z, DW SUN. 2001. Effect of fluctuation in inlet airflow temperature on CFD simulation of air blast chilling process. *Journal of Food Engineering*. 48(4): 311–316.

JAMBUNATHAN K, E LAI, MA MOSS, BL BUTTON. 1992. Review of heat transfer data for single circular jet impingement. *International Journal of Heat and Fluid Flow*. 13(2): 106–115.

KUNDU PK, IR COHEN. 2002. *Fluid Mechanics*. 2nd ed. San Diego: Academic Press.

LASCHEFSKI H, T CZIESLA, G BISWAS, NK MITRA. 1996. Numerical investigation of heat transfer by rows of rectangular impinging jets. *Numerical Heat Transfer*. 30(A):

87–101.

LAUNDER BE, DB SPALDING. 1972. *Mathematical Models of Turbulence*. New York: Academic Press.

LEE J, SJ LEE. 2000. Effect of nozzle configuration on stagnation region heat transfer of axisymmetric jet impingement. *International Journal of Heat and Mass Transfer*. 43(18): 3497–3509.

LI A, CE WALKER. 1996. Cake baking in conventional, impingement and hybrid ovens. *Journal of Food Science*. 61(1): 188–191, 197.

LUJANACOSTA J, RG MOREIRA, J SEYEDYAGOOBI. 1997. Air-impingement drying of tortilla chips. *Drying Technology*. 15(3–4): 881–897.

MARCROFT HE, M CHANDRASEKARAN, MV KARWE. 1999. Flow field in a hot air jet impingement oven- Part II. Multiple impingement jets. *Journal of Food Process Preservation*. 23(3): 2335–2348.

MARCROFT HE, MV KARWE. 1999. Flow field in a hot air jet impingement oven – Part I. A single impinging jet. *Journal of Food Process Preservation*. 23(3): 217–233.

MARTIN H. 1977. Heat and mass transfer between impinging gas jets and solid surfaces. *Advanced Heat Transfer*. 13(1): 1–60.

MESBAH M., JW BAUGHN, CW YAP. 1996. The effect of curvature on the local heat transfer to an impinging jet on a hemispherically concave surface. *Proceedings of the Ninth International Symposium on Transport Phenomena in Thermal-Fluids*.

MIDDEN TM. 1995. Impingement air baking for snack foods. *Cereal Foods World* 40(8): 532–535.

MIRADE PS, A KONDJOYAN, JD DAUDIN. 2002. Three-dimensional CFD calculations for designing large food chillers. *Computers and Electronics in Agriculture*. 34(1–3): 67–88.

MOREIRA RG. 2001. Impingement drying of foods using hot air and superheated steam. *Journal of Food Engineering*. 49(4): 291–295.

NITIN N, MV KARWE. 2001. Heat transfer coefficient for cookie shaped objects in a hot air jet impingement oven. *Journal of Food Process Engineering*. 24(1): 51–69.

OVADIA DZ, CE WALKER. 1998. Impingement in food processing. *Food Technology*. 52(4): 46–50.

OZISIK MN. 1994. *Finite Difference Methods in Heat Transfer*. London: CRC Press Inc. 412 pp.

PATANKAR SV. 1980. *Numerical Heat Transfer and Fluid Flow*. Seattle: Hemisphere Pub. Corp. 197 pp.

POLAT S, B HUANG, AD MAJUMDAR, WJM DOUGLAS. 1989. Numerical flow and heat transfer under impinging jets: A review. *Annual Review of Numerical Fluid Mechanics Heat Transfer*, Vol. 2: 157–197. C. L. Tien (Ed.). Washington, DC: Hemisphere Pub. Corp.

POPE SB. 2000. *Turbulent Flows*. New York: Cambridge Univ. Press. 771 pp.

POPIEL CO, O TRAS. 1991. Visualization of a free and impinging round jet. *Experimental Thermal and Fluid Science*. 4(3): 253–264.

SARKAR A. 2002. *Study of Heat Transfer under Impinging Jets for Freeze-Thaw Conditions*. [MS thesis]. Davis, Calif.: University of California, Davis. 104p. Available from: Library University of California.

SARKAR A, RP SINGH. 2003. Spatial variation of convective heat transfer coefficients in air impingement applications. *Journal of Food Science*. 68(3): 910–916.

SARKAR A, RP SINGH. 2004. Air impingement studies for food processing: visualization studies. *Lebensmittel-Wissenschaft und -Technologie*. Article in Press.

SARKAR A, N NITIN, MV KARWE, RP SINGH. 2004. Fluid flow and heat transfer in air jet impingement in food processing. *Journal of Food Science*. Article in Press.

SCHLICHTING H. 1979. *Boundary-layer Theory*. 7th ed. New York: McGraw-Hill. 817 pp.

SO RMC, A SARKAR, G GERODIMOS, J ZHANG. 1997. A dissipation rate equation for low-Reynolds-number and near-wall turbulence. *Theoretical and Computational Fluid Dynamics*. 9(1): 47–63.

SPARROW EM, L LEE. 1975. Analysis of flow field and impingement heat and mass transfer due to a non-uniform slot jet. *Transactions of American Society of Mechanical Engineering. Series C. Journal of Heat Transfer*. 191–197.

TENNEKES H, JL LUMLEY. 1972. *A First Course in Turbulence*. 1st ed. Cambridge, MA: MIT Press. 300 pp.

VERBOVEN P, N SCHEERLINCK, J DE BAERDEMAEKER, BM NICOLAI. 2000a. Computational fluid dynamics modelling and validation of the isothermal flow in a forced convection oven. *Journal of Food Engineering*. 43(1): 41–53.

VERBOVEN P, N SCHEERLINCK, J DE BAERDEMAEKER, BM NICOLAI. 2000b. Computational fluid dynamics modelling and validation of the temperature distribution in a forced convection oven. *Journal of Food Engineering*. 43(2): 61–74.

WAHLBY U, C SKJOLDEBRAND, E JUNKER. 2000. Impact of impingement on cooking time and food quality. *Journal of Food Engineering*. 43(3): 179–187.

WESTKNEMPER, JC. 1961. On the error in plug in type calorimeters caused by surface temperature mismatch. *Journal of Aeronautical Sciences*. 28(11): 907–908.

WHITE FM. 1999. *Fluid Mechanics*. 4th ed. Boston: McGraw-Hill.

Disclaimer

Reference to a company or product does not imply approval or recommendation of the product to the exclusion of others that may be suitable.

14

Laser-based packaging sterilisation in aseptic processing

K. Warriner, University of Guelph, Canada and S. Movahedi and W. M. Waites, University of Nottingham, UK

14.1 Introduction: limitations in current sterilisation methods for aseptic carton packaging

Aseptic packaging and processing were regarded as the most significant developments in food technology in the latter half of the twentieth century. Advances in process engineering and packaging have been central for the development of sophisticated technology that now exists. In contrast, the area of carton packaging sterilisation has largely been unaltered since the introduction of commercial aseptic processing in the 1950s. Hydrogen peroxide treatment remains the preferred sterilisation method in the flat board Tetra Pak system. However, Elopak (producing pre-formed cartons) adopted a system combining UV-germicidal lamps and hydrogen peroxide. There has been relatively little research performed on alternative (chemical free) sterilisation systems despite the potential advantages of worker safety, less risk of environmental pollution and on-line monitoring for process control. The following section will provide an overview of studies performed to evaluate the application of UV-lasers to achieve chemical-free sterilisation of pre-formed carton packaging. This will include a brief description on the principle of laser operation and selection of appropriate type. Subsequent sections will describe in detail the protocols used for validating UV-laser treatment, spore inactivation and modes of inactivation. Engineering aspects of developing optical arrangements to enhance the effectiveness of UV-lasers will also be discussed. Finally, an overview of alternative carton sterilisation technologies and the potential commercial application of UV-lasers will be evaluated.

14.1.1 Current sterilisation of aseptic carton packaging

Tetra Pak, Elopak and Combibloc are the present dominant companies within the aseptic market. The different packaging approaches undertaken by these companies have historical foundations. The Elopak and Combibloc packs are based on pre-formed cartons whilst Tetra Pak has adopted a form-fill-seal approach. Although all three companies use hydrogen peroxide based sterilisation techniques there are subtle differences in how the operation is performed (Holdsworth, 1992). The Tetra Pak system is characterised by the packaging material being supplied as a reel of polyethylene paper laminate. The packaging material passes through a bath of 35% v/v hydrogen peroxide/wetting agent, after which it is formed around a heated tube to remove residual H_2O_2. Following this, the tube is filled and a traverse seal made through the liquid that forms the top seal for one package and the bottom for the next (Fig. 14.1a). The Combibloc system uses a hot hydrogen peroxide (35% v/v) spray to sterilise cartons with residues being removed by a stream of hot sterile air. The Elopak system uses a combination of hydrogen peroxide and UV germicidal lamps (Fig. 14.1b). The technique was initially devised by Bayliss and Waites (1979, 1980) who observed that by using a combination of UV light and 2% v/v hydrogen peroxide an equal level of spore kill could be achieved compared to when 35% v/v H_2O_2 was applied. The basis for this synergistic effect is thought to be due to the generation of hydroxyl ions that damage the spore DNA. As the hydroxyl ions are in a fluidic matrix, penetration into crevices and pores is readily achieved (Peel and Waites, 1981).

14.1.2 Limitations of current aseptic packaging sterilisation

Hydrogen peroxide is commonly used for aseptic packaging sterilisation because of its effectiveness and relatively cheap cost (Holdsworth, 1992). However, high temperatures and accumulation of impurities in hydrogen peroxide baths can lead to decomposition thereby decreasing decontamination efficacy. Although stabilised hydrogen peroxide is applied there is still a requirement for frequent laboratory testing to ensure critical limits are maintained. The potential of hydrogen peroxide carry-over through inadequate residual removal is a further potential problem as FDA regulations stipulate a limit of 0.5 ppm. In addition, by using chemical based systems there is always the potential of accidental release during handling representing a significant risk to the environment and workers (Jones, 1996). For example, hydrogen peroxide vapour has been linked to long-term respiratory problems among workers of aseptic packaging facilities (Riihimaki *et al.*, 2002).

All the above factors identify the need to develop chemical-free sterilisation systems. UV-lamps are effective at reducing microbial contamination but are insufficiently powerful to achieve carton sterilisation at current line speeds of 150 cartons per minute. In contrast, light derived from lasers has higher intensity and potentially can overcome the limitations of UV-germicidal lamps.

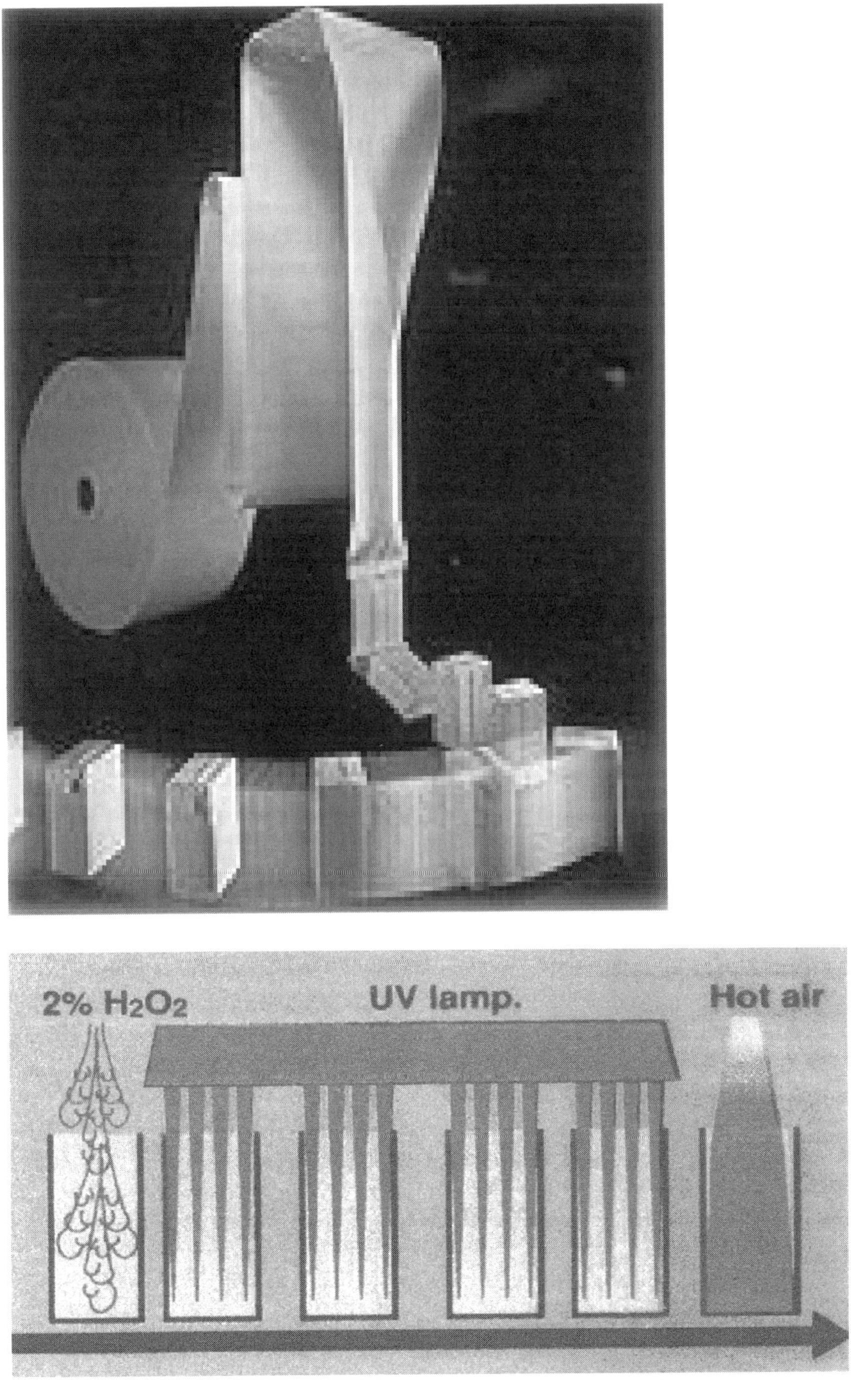

Fig. 14.1 Schematic representation of Tetra-Brik (A) and Elopak pre-formed carton (B) sterilisation process.

14.2 The principles of laser operation

Light is a product of the photons released by electrons returning to ground state. With normal light, such as that derived from lamps, the generation of photons is random, leading to diffuse (low energy) waves consisting of a range of wavelengths. In contrast the formation of Laser (Light Amplification by Stimulated Emission of Radiation) light is organised (stimulated emission) leading to more defined photon beams (Table 14.1). Conceptually lasers are simple in design but development still represents a technical challenge. Essentially a high density active (lasing) medium is excited by an external source (high intensity light or electric discharge). The electrons are moved to a higher excited state and photons are released upon return to ground state. If the photon released should encounter a further atom that has an electron in the same excited state, simulated emission can occur. To further propagate the laser field mirrors at either end of the laser chamber reflect back photons with specific wavelength and phase back into the active medium. The chain reaction continues generating further photons that all have the same phase (i.e. coherent) and wavelength (monochromatic). One of the mirrors within the laser is designed to reflect photons back into the active medium but at the same time allow a proportion through as pulses. It is this light passing through the mirror that constitutes laser beam.

There are many different types of lasers available producing light at different power intensity and wavelength. The active medium used defines the wavelength of the laser light produced. Active or lasing mediums can be solid, gas, liquid or semi-conductor (Table 14.2). Although the majority of lasers have specific wavelengths certain types (for example, Nd:YAG) can be tuned to produce monochromatic light at different values. The ability of tuneable (or mode locking) units provides greater flexibility in terms of application. The parameters used in association with lasers are the wavelength, fluence (energy density) and frequency (pulses per second).

The choice of laser is dependent on the specific application required. For example, for micro-fabrication purposes where high intensity beams are required to etch patterns, solid state lasers such as the Nd:YAG is preferred. For decontamination of packaging surfaces the integrity of the packaging coating needs to be preserved and hence etching effects minimised. Nevertheless, the laser light applied must ensure inactivation of potential biohazards such as vegetative cells and spores. From a historical prospective the application of UV

Table 14.1 Comparative properties of light derived from lasers and lamps

Property	Laser light	Lamp light
Wavelength	Monochromatic	Polychromatic
Intensity	High	Low
Coherent	Yes	No
Directional	Yes	No

Table 14.2 Properties of different laser types

Laser active medium	Wavelength (nm)
Argon fluoride	193
Krypton fluoride	248
Nitrogen	337
Argon (blue)	488
Argon (green)	514
Helium neon	543
Rhodamine 6G dye	570–650
Ruby (CrAlO3) red	694
Nd: YAG	1,064, 532, 355, 266, 213
Carbon dioxide	10,600

lasers was a logical choice due to the established antimicrobial effects of light of this wavelength. Excimer gas lasers are used to generate laser beams in the UV region of the spectrum. The active medium consists of a halide (either chlorine or fluorine) mixed with an inert gas (argon, krypton or xenon) in the presence of a buffer (neon). When a high voltage discharge is applied to the gas mixture the inert gas molecules are excited and are able to react with the halide forming a dimmer (the name excimer is an abbreviation of 'excited dimmer'). These diatomic molecules have very short lifetimes and dissociate releasing the excitation energy through UV photons. KrF excimer lasers (emitting at 248 nm) closely match the wavelength of germicidal lamps and can be assumed to exhibit the optimum antimicrobial property. In addition, the KrF lasers have relatively high efficiency and able to deliver peak powers of the order of 1J at a frequency of 200 Hz (Table 14.3).

Table 14.3 Properties of different excimer lasers

Active medium	Wavelength (nm)	Average power (mW)	Pulse energy (mJ)
KrF	248	550	5.5
XeF	351	200	2.0
ArF	193	350	3.5
KrCl	222	50	0.5
XeCl	308	300	3.0

14.3 Assessing and validating spore inactivation by UV light

Through evolution bacteria have evolved resistance mechanisms to protect them against UV light. From an overview of the relative UV resistance of different types of microorganisms it is evident that vegetative cells are significantly more susceptible to ultraviolet exposure compared to bacterial spores (Table 14.4). This may be unexpected considering that unlike dormant spores, live vegetative

Table 14.4 Relative resistance of microorganisms to UV-light derived from lamps emitting at 254 nm

	UV dose (at 254 nm) mJ/cm^2
	Log 3 (99.9%) Reduction
Bacteria	
Bacillus Cells	11,000[1]
subtilis Spores	58,000[1]
Bacillus Cells	8,700
anthracis Spores	8,250[2]
Bacillus megaterium Cells	2,500[1]
Bacillus megaterium Spores	5,200[1]
Escherichia coli	6,600[1]
Legionella pneumophila	3,800
Pseudomonas aeruginosa	10,500
Salmonella enteritidis	7,600
Staphylococcus aureus	7,000
Mould spores	
Aspergillus niger spores	330,000[1]
Penicillin roqueforti spores	26,400[1]
Rhizopus nigricans	220,000[1]
Viruses	
Bacteriophage (from *E. coli*)	6,600[3]
Hepatitis	8,000[4]
Tobacco Mosaic virus	440,000[5]

1 Legan, 1982.
2 Nicholson and Galeano, 2003.
3 Anon, 1989.
4 Qin and Gerba, 1996.
5 Siegal and Wildman, 1956.

cells have active DNA repair systems (Friedberg *et al.*, 1995). The presence of inner and outer spore coats (Riesenman and Nicholson, 2000) and high dipicolinic acid pyridine 2–6–dicarboxylic acid; DPA (Slieman and Nicholson, 2001; Nicholson *et al.*, 2002) are associated with spore UV resistance (Fig. 14.2a). However, the primary reason for enhanced UV resistance of spores is a combination of DNA photochemistry and nucleic acid repair systems that operate during germination (Setlow, 1988; 2001). In vegetative cells exposure of DNA to UV irradiation typically produces a variety of cyclobutane-type dimers between adjacent pyrimidines (Harm, 1980; Craik *et al.*, 2001). In spores, however, such products rarely occur. Instead the generation of a unique thyminyl-thymine adduct photoproduct (termed the spore photoproduct, SP) is formed (Donnellan and Setlow, 1965; Fig. 14.2b). The different photochemistry is the result of altered DNA conformation (supercoiling) (Setlow, 1992; Griffith *et al.*, 1994) caused by the blanketing of spore DNA with small, acid soluble proteins (α/β SASP). Spores lacking SASP are very UV sensitive and have a

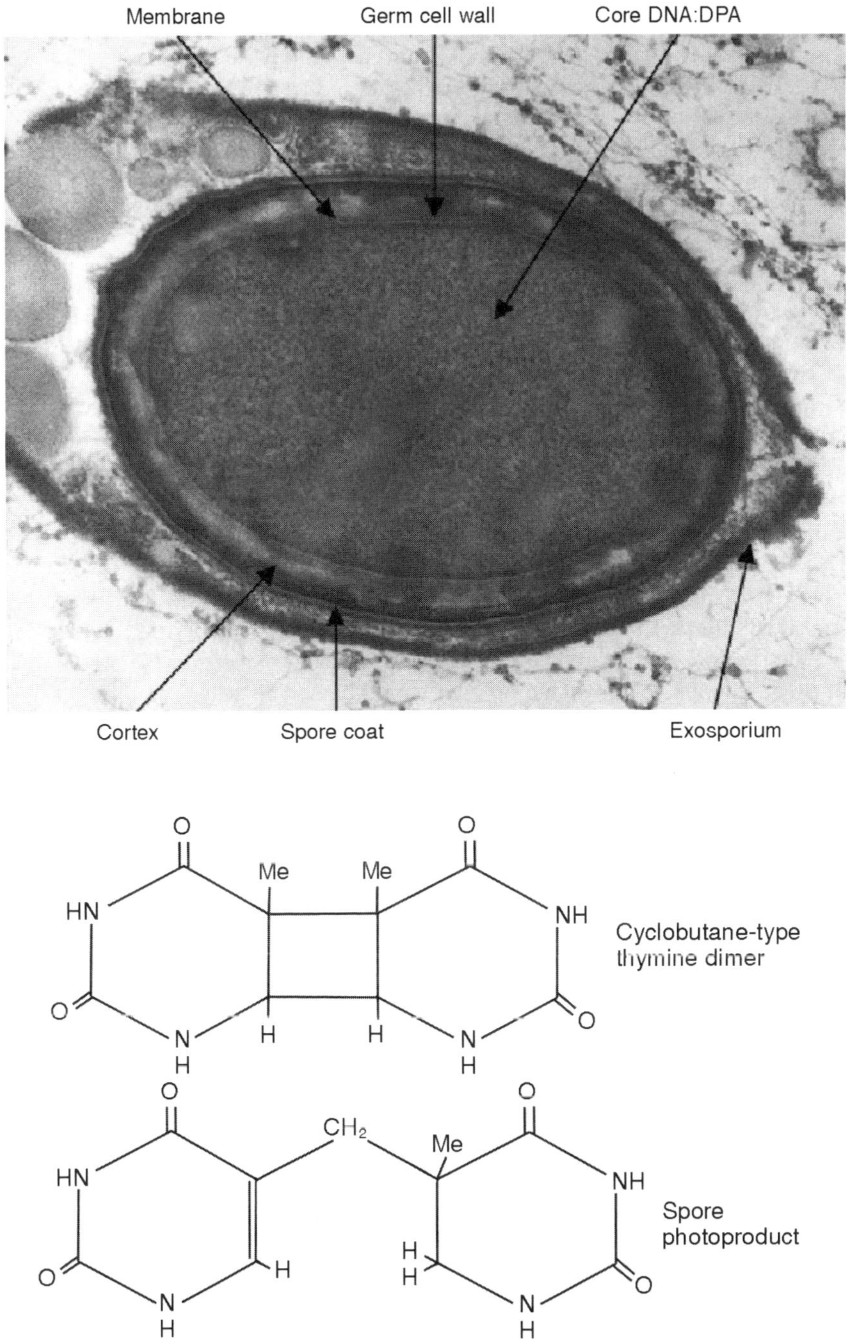

Fig. 14.2 Structure of a Bacillus spore (A) and the different photoproducts produced (B) during UV exposure in vegetative (cyclobutane-type thymine dimer) or spore (spore photoproduct) DNA.

resistance comparable to that of vegetative cells (Fairhead and Setlow, 1992). Therefore, spore DNA is protected from potentially lethal damage by the presence of SASP by ensuring that the photoproduct produced is defined. This enables the efficient repair of damaged DNA via the nucleotide excision repair (NER) and SP lyase systems. Absence of one or both repair systems significantly decreases the observed UV resistance of spores (Setlow, 2001).

14.3.1 Approach to validate the efficacy of UV-laser treatment of packaging material

From the limited studies performed, carton laminates have typically been found to carry a relatively low level of contamination ranging from 1–500 colony forming units per m^2 (Holdsworth, 1992). Nevertheless, all sterilisation standards to attain commercial sterility are based on a 12 log kill of *Clostridium botulinum* spores. Such a 12D reduction is based on thermal processing and cannot be directly related to the decontamination of aseptic packaging. This is the primary reason that no specific regulations currently exist with regard to package decontamination except that the containers used must be sterile. Through various risk analysis studies the general rejection rate for aseptically packaged milk is 1:100,000 (Cerf and Brissende, 1981; Toledo, 1982; Klunge *et al.*, 1988). This infers that by ensuring a 5 log reduction of spores the probability of producing a safe and stable product would be adequate.

Biodosimetry remains the preferred method for assessing the antimicrobial properties of UV based light systems. Here the inactivation kinetics of a test organism by UV treatment can be extrapolated to cover less or equally resistant vegetative cells/spores. *Bacillus subtilis* has been extensively employed due to their high degree of UV resistance (Table 14.5), reproducible inactivation kinetics and robustness (Nicholson and Fajardo-Cavazos, 1997; Nicholson and Law, 1999; Hoyer, 2000).

As with many validation techniques it is critical to establish standardisation in order that true comparisons between different decontamination systems can be evaluated. In the absence of specific regulations commercial packaging companies have developed and applied different validation criteria. For this reason it is informative to describe the experimental methods of spore production, deposition, UV-laser treatment and spore recovery in some detail.

Spore production and deposition

The primary aim of spore production is to obtain homogeneous crops with negligible inter-batch variability in resistance. In this regard by introducing an equal concentration of glucose and ribose into the sporulation medium the yield of *B. subtilis* spores can be improved tenfold (Warriner and Waites, 1999). Repeated washing steps in sterile distilled water further purify spore crops and minimise the risk of vegetative cell carry-over. The final spore crops can be stored at 4ºC for 2–3 months without any significant loss in homogeneity although sporadic mutations can occur over longer storage periods.

Table 14.5 Cumulative dose measured during carton UV-laser treatment. Total dose measured (mJ)[1]

Rig speed[2]	Aluminium coated carton	Polyethylene coated carton
Static, 0 mm/s	608±7	538±1
1-Way Move[3], 1.48 cm/s	1712±3	ND
2-Way Move[4], 2.96 cm/s	1570±13	1329±1
Step change[5]	1831±1	1082±4
Phase 1: 2.2 cm/s		
Phase 2: 0.89 cm/s		

ND: Not determined.
1 Represents the cumulative dose delivered to the interior of the carton as measured at four 4cm^2 points of the carton (pack base, lower side, middle side and upper side).
2 Total of 1800 pulses delivered @35mJ/pulse.
3 The carton was moved in a single motion way from the laser beam output lens.
4 The carton was withdrawn from the output lens and returned after half the number of laser pulses had been delivered.
5 A third of the laser pulses were delivered to the lower half of the carton (Phase 1) with the remaining two-thirds being delivered to the upper part of the pack (Phase 2). (Reproduced with permission from Warriner *et al.* 2000b).

The packaging laminates tested in validation studies were Elopak polyethylene and aluminium cartons (Fig. 14.3). The method of spore deposition is a critical step to minimise clumping effects that leads to shading against UV photons and hence leading to greater data variability. By distributing the spores

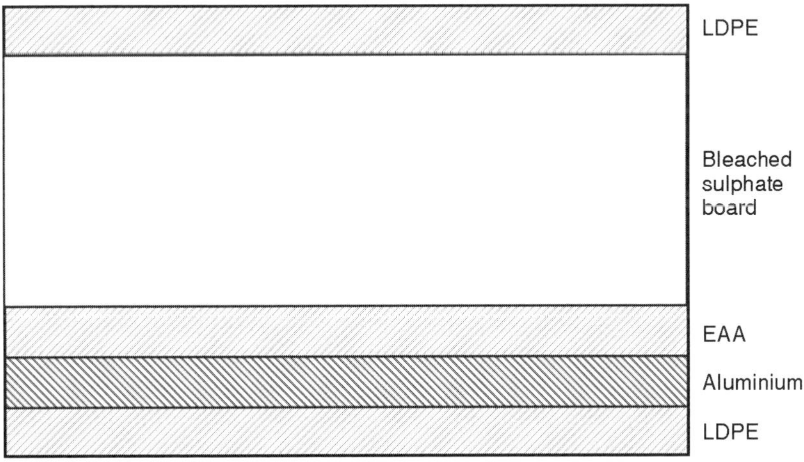

Fig. 14.3 Composition of a typical carton packaging laminate. A layer of low density polyethylene (LDPE) protects the external surface of the carton. The rigidity of the pack is provided by bleached sulphate board. The inner layers consist of ethyl methylacrylic, aluminium barrier (optional depending on product type) and an inner LDPE in contact with the product.

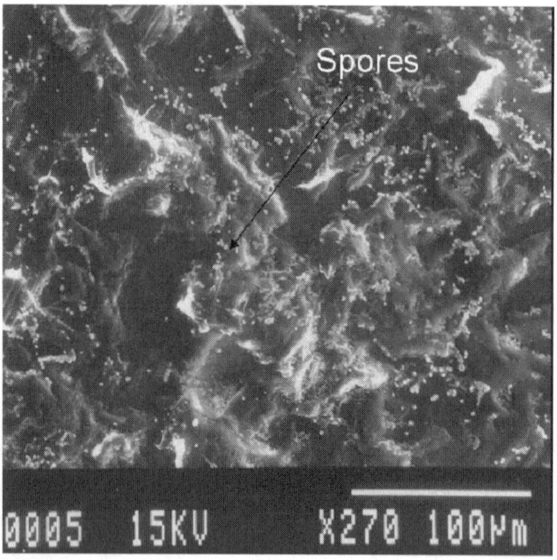

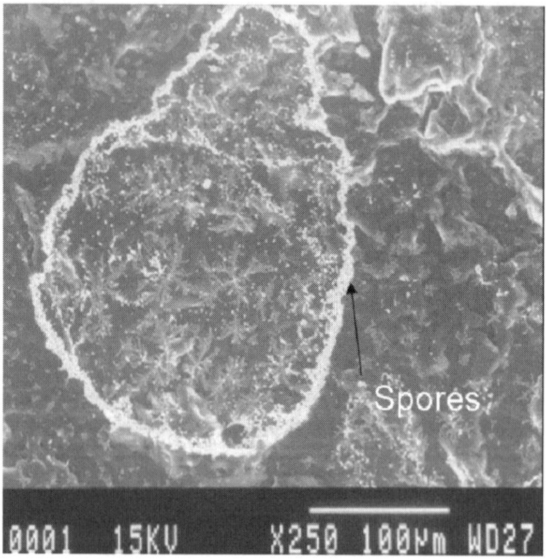

Fig. 14.4 Comparison between the distribution of *B. subtilis* spores deposited onto packaging board using (A) KCl/Tween 80 carrier solution and (B) distilled water. Spore density in both cases was 10^6 cfu/4cm^2 (Reproduced from Warriner *et al.*,unpublished).

in a fine mist using a specially designed chamber it is possible to achieve homogeneous distribution. However, the hydrophobic base polyethylene carton material results in spore clumping during drying (Fig. 14.4a). In addition, the recovery of spores from packaging surfaces is typically 12–40%, presumably due to strong attachment to the pack surface. Both problems can be minimised through the inclusion of Tween-80/KCl in the carrier solution to give even distribution of spores (Fig. 14.4b) and significantly greater spore recovery.

However, the crystals of KCl formed could potentially shield spores from UV photons, thereby giving an underestimation of decontaminating efficacy.

UV-laser treatment and spore recovery

For the studies to be described an LPX 210I model scientific KrF eximer laser (Lambda Physik, Germany) operating at 248 nm was used to irradiate spore laden packaging board samples. To achieve correct uniformity and energy density (fluence) of the beam on the sample the irradiated field was expanded (5 cm^2) on one axis using a simple cylinder lens. The dose applied to each 4 cm^2 sample can be constantly measured using a joule meter pulse energy detector that detects a fraction (18–26%) of the input beam (Warriner *et al.*, 2000a).

Recovery of spores from UV-laser irradiated board is achieved using manual agitation in combination with an aqueous rinse solution of 0.1% (w/v) Tween-80 containing 0.5 mol l^{-1} KCl. Spore survivors are enumerated by plating onto nutrient agar containing 0.15% (w/v) starch and incubating plates at 30ºC for 3 days. With low numbers of spore survivors the rinse solution can be passed through a membrane filter (0.45μm pore size) that is subsequently incubated for 3 d at 30ºC on the surface of a nutrient agar plate containing 0.15% (w/v) starch. The extent of spore kill is expressed in terms of log count reductions (LCR).

$$\text{LCR} = \log_{10}\text{N}_o - \log_{10}(\text{N}_i/\text{n})$$

where N_o was the spore loading of untreated squares, N_i = number of survivors and n was the number of squares irradiated. For each dose, frequency or fluence level a total of 16 × 4 cm^2 squares are used (4 controls and 12 irradiated).

14.3.2 Effect of spore density, UV dose, pulse energy (fluence) and pulse frequency on inactivation kinetics

On flat board samples the extent of spore inactivation by UV-laser treatment is independent on the base packaging material (Fig. 14.5). In the planner format the reflective properties of polyethylene and aluminium is negligible. The diphasic inactivation kinetics of spores treated with UV-laser light are similar to those obtained using germicidal lamps (Soloshenko *et al.*, 1999). Interestingly the transition dose between the two phases is dependent on spore deposition density. This is most evident with boards treated with a total dose of 15 J which give a relatively low LCR, with spore density of 10^4 cfu/4 cm^2 , but significantly increases by three orders of magnitude with a loading of 10^6 cfu/4 cm^2. A possible theory to explain the enhancement of UV lethality could be associated with the accumulation of photosensitisers resulting from the interaction of high-energy photons with spores. It is known that dipicolinic acid (DPA), located within the core of spores, is a strong photosensitiser (Setlow, 1995) and could have been released into the local environment during irradiation. It is unlikely, however, that DPA release occurred from irradiated spores as very few were collapsed or visibly damaged when viewed under scanning electron microscopy (Fig. 14.6).

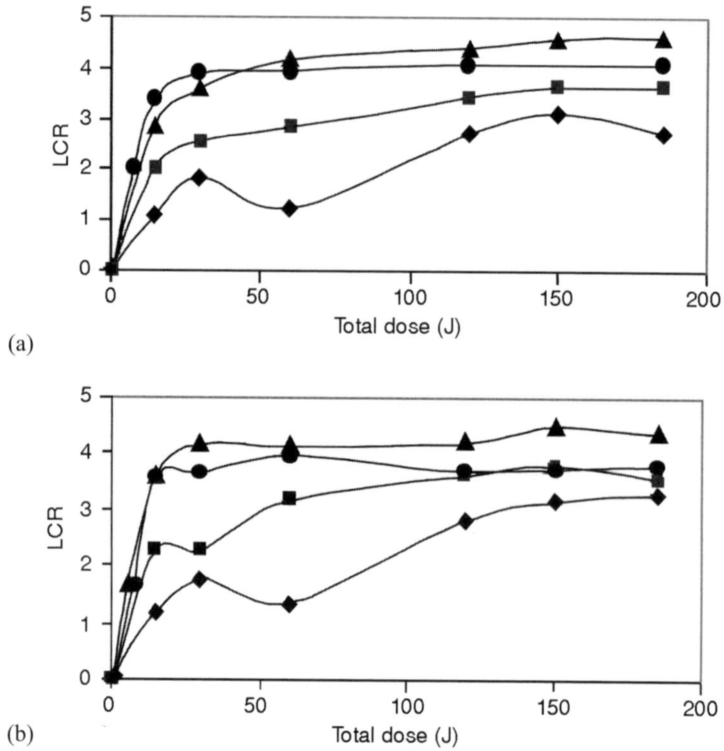

Fig. 14.5 Inactivation of *Bacillus subtilis* spores on (A) aluminium and (B) polyethylene coated packaging boards exposed to UV-laser irradiation. Cards (4 cm^2) carrying 10^4 (◆), 10^5 (■), 10^6 (•) or 10^7 (▲) cfu were irradiated with the appropriate UV dose and survivors subsequently enumerated. The frequency of the laser beam was 10 Hz for dose levels of 15 J, 20 Hz for 30 and 60 J, 40 Hz for 120–185 J dose levels. The fluence of the laser beam was 12 Jm^{-2} as measured at the work piece. (Reproduced with permission from Warriner *et al.*, 2000b).

Extensive tailing is observed in the majority of the dose curves (Fig. 14.5). Indeed, even at the very high dose levels applied (185 J) very few squares were sterile and a number of survivors were present. It has been proposed that tailing on inactivation curves is due to the presence of resistant sub-populations (Sugaware *et al.*, 1981). However, this is unlikely, as the percentage of survivors recovered from the UV-laser irradiated boards is independent of the initial spore loading. The actual numbers of survivors at each spore loading level do not differ significantly, except for boards loaded with 10^7 cfu/4 cm^2. This would indicate that a number of sites (crevices and pores) on the packaging board material existed that may afford protection from the lethal action of UV-laser treatment (Warriner *et al.*, 2000a).

The photon density (fluence) of the laser pulses has a negligible effect on spore inactivation kinetics deposited on packaging material. Therefore, similar to spores in aqueous suspension (Rice and Ewell, 2001) the unit dose delivered

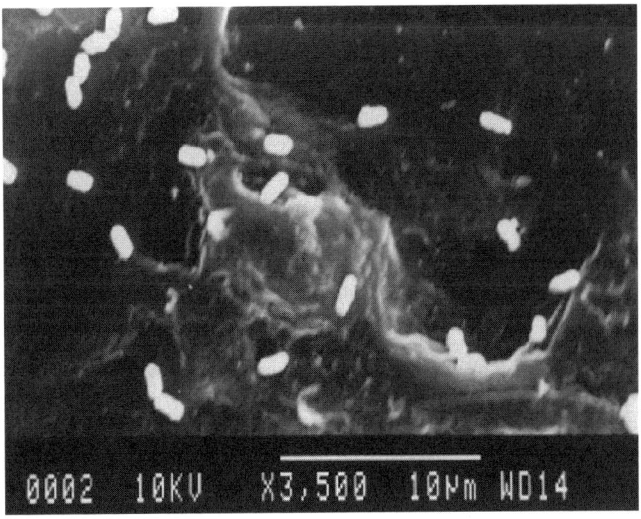

Fig. 14.6 SEM of UV treated B. *subtilis* spores (10^6 cfu/4cm^2) deposited onto polyethylene coated board. The board sample was exposed to a UV dose of 185J. (Reproduced from Warriner *et al.*, unpublished).

is more critical than the laser beam power. In this respect the only advantage of laser light would be to deliver a lethal UV dose more rapidly.

The frequency by which UV-laser pulses are delivered has a negligible effect on spore inactivation kinetics. This is likely to be due to the absence of active DNA repair systems in dormant spores. In contrast, *Escherichia coli* exposed to intermittent UV photons has more resistance than cells continuously exposed (Harm, 1980, Dunn *et al.*, 1995). The theory is that during the dark period, between UV pulses, DNA repair systems are active thereby leading to apparent enhanced resistance (Harm, 1980).

14.4 Application of UV-laser light in package sterilisation

14.4.1 Treatment of pre-formed cartons with UV-excimer laser light

The illumination of carton interior by laser light is problematic compared to treatment with germicidal lamps. With the latter, the diffuse nature of the light enables the pack interior to be fully illuminated. As laser light is coherent the beam has to be physically projected to different areas of the carton interior. This can be achieved through moving the carton into the laser beam. The programmed movement of the rig has a significant effect on the UV distribution within the carton interior (Warriner *et al.*, 2000b). When the rig is held static at the furthest point away from the beam output lens the dose is predominantly delivered to the carton base. The upper part of the pack receives a minor dose by comparison, whilst no light was measured on the sides of the carton interior regardless of the packaging type. However, by moving the carton during UV-

Table 14.6 Log count reduction for *Bacillus subtilis* spores deposited on pre-formed cartons and treated with different doses of UV-laser light

UV-laser dose delivered (J)[1]	Log count reduction	
	Polyethylene cartons	Aluminium cartons
63	3.62	4.14
97	3.90	4.56
126	4.04	4.70
158	4.28	5.98
209	4.94	5.88

1, Total UV dose delivered during treatment. Spore loading on cartons was $9.5 \times 10^5 \pm 2.4 \times 10^5$ cfu. (Adapted from Warriner *et al.*, 2000b).

laser treatment a more even dose is achieved. The most even distribution is obtained when the carton is initially withdrawn in a rapid motion away from the laser output lens and subsequently moved at a slower rate. In this step change motion a third of the UV dose is delivered to the lower half of the carton and the remainder on the upper part of the pack. Although the laser output is constant for the various rig set-ups used the cumulative dose measured differs (Table 14.5). In a static arrangement the dose is relatively low due the majority of the UV laser light being lost to the surroundings. When the carton is moved during the irradiation treatment the majority of the laser beam is contained within the carton and this leads to higher doses being recorded. The cumulative dose measured in the various rig set-ups is higher with aluminium cartons than with polyethylene (Table 14.5). Aluminium films efficiently reflect KrF excimer laser light with negligible loss of photon energy (Matsuura and Miyagi, 1998). Therefore, the internal reflection of UV photons within the aluminium pack augments homogeneous field distribution. With polyethylene packs UV photons are absorbed and hence fewer internal reflections occur. The importance of photon internal reflections during UV-laser treatment of cartons is most evident from spore inactivation data. Here a greater spore kill is obtained when aluminium packs are used compared to polyethylene cartons (Table 14.6). Interestingly, when UV-lamp light is used to treat aluminium and polyethylene cartons a greater spore kill is obtained for the latter. For example, Stannard *et al.* (1985) reported that the log reduction in *B. subtilis* spores exposed to UV-lamp irradiation was only 1.21 on aluminium cartons compared to 3.32 on polyethylene. Although higher internal photon reflections were observed with aluminium packs it is thought that this was mainly composed of wavelengths in the 325–550 nm region which cause a degree of DNA repair (photoreactivation).

14.4.2 Mode of action of UV-laser light

The main mode of spore inactivation by UV-germicidal lamps is through cumulative DNA damage (Mason and Setlow, 1986; Setlow 1988). In terms of

Table 14.7 Occurrence of mutations amongst colonies derived from survivors of UV-laser irradiated *Bacillus subtilis* wild type and SASP deficient spores

Total dose (J)[1]	LCR	Total survivor colonies screened	Mutant numbers (% of Survivors)[2]		
			Aux	*spo*	*aux:spo*
(α^-/β^- SASP)					
Control		200	2 (1%)	0	0
1.0	3.82	100	26 (26%)	27 (27%)	12 (12%)
2.0	3.75	200	61 (31%)	60 (30%)	34 (17%)
3.0	3.96	200	32 (16%)	57 (29%)	24 (12%)
4.0	4.03	175	48 (27%)	30 (17%)	13 (7%)
5.0	3.91	200	101 (51%)	56 (28%)	15 (8%)
Wild type					
Control		225	4 (2%)	1 (>1%)	1 (>1%)
1.0	1.51	501	14 (3%)	19 (4%)	12 (2%)
7.5	2.98	1443	54 (4%)	56 (4%)	50 (4%)
30.0	3.91	836	278 (33%)	338 (40%)	209 (25%)

1 Total UV dose delivered to 4 cm^2 polyethylene packaging flat squares loaded with 10^6 cfu *Bacillus subtilis* PS 346 or PS 361 spores. (Reproduced with permission Warriner *et al.*, 2000a).
2 *aux*, auxotrophic; *spo*, asporogenous.

UV-laser light the lower resistance of *Bacillus* spores lacking SASP (α/β) proteins would also confirm DNA as the primary target (Warriner *et al.*, 2000a). Additional evidence of DNA damage is obtained by screening survivors of UV-laser treatment for mutations in DNA. This is readily achieved through identifying survivors with increased nutritional demands (auxotrophic) or that fail to form spores (asporogeneous) (Table 14.7). It has been proposed that high intensity light can result in the etching and vaporisation of spores (Moisan *et al.*, 2001). However, UV-laser treated spores remained on packaging surfaces (Fig 14.6) and retain phase brightness. Hence, vaporisation or etching effects do not contribute to inactivation of spores.

A further potential, but minor, mode of inactivation of spores by UV-laser light is through the disruption of spore germination systems. In the presence of an activator (for example, L-alanine) spores rapidly (<2 min) become hydrated and lose phase brightness. After cortex hydrolysis the spore initiates elongation and outgrowth within 40 min (Gould, 1969; Vary and Halvoson 1995; Billion *et al.*, 1997). In a typical population over 90% of spores will follow the germination process although a proportion will remain dormant (super-dormant spores). Spores receiving a sub-lethal UV-laser dose exhibit a range of germination responses. Here the majority of spores lose phase brightness but only after an extended lag period of 90 min (Warriner *et al.*, 2002). If the spores are heat activated (at 70°C) prior to germinant addition this lag can be decreased to 40 min. Interestingly, by the sixth hour of germination the majority of spores become phase dark but fail to elongate, a proportion undergo partial phase darkening (grey spores) and some elongate but fail to replicate (Warriner *et al.*, 2002). Considering that no DNA expression occurs until late into germination (Setlow and Johnson, 1997) it is expected that spores should at least reach the elongation stage even with extensive nucleic acid damage. Although not confirmed there is believed to be a set of so-called 'out' genes that are required for successful outgrowth (Scotti *et al.*, 1993; Nessi *et al.*, 1995). It is possible that damage to these genes, if they exist, could block germination. However, a more likely event is the damage of spore cortex material by UV-laser photons. This is supported by the fact that heat injured *B. cereus* and *B. subtilis* with defective cortex hydrolysing enzymes undergo delayed and incomplete germination (Hashimoto *et al.*, 1971; Ishikawa *et al.*, 1998). *Bacillus* spores have three types of hydrolytic enzymes (amidase, endopeptidase and transglycosylase) that act sequentially during cortex degradation (Foster, 1997). Therefore, damage to the enzyme or alteration in the cortex structure by the action of UV photons would affect spore germination.

14.4.3 Recovery of UV-laser treated spores in food matrices

Spore resistance to inimical processes is determined by the colony forming units formed from survivors, and the recovery conditions applied will affect the inactivation data obtained. The effect of post-treatment recovery of injured spores has primarily focused on those exposed to sub-lethal levels of biocidal agents (Williams and Russell 1993) or heat (Lopez *et al.*, 1997). For example, sub-optimal incubation temperature for growth (Condon *et al.*, 1996) and neutral pH (Mallidis and Scholefield, 1986) are known to enhance the apparent thermal resistance of spores. For unknown reasons, supplementing nutrient media with starch also enhances the recovery of thermally damaged spores (Adams, 1978; Lopez *et al.*, 1997). Lysozyme is an additional agent that enhances the recovery of sub-lethally injured spores (Duncan *et al.*, 1972; Alderton *et al.*, 1974; Lund and Peck, 1994; Faille *et al.*, 1997). It is thought that lysozyme when included in the recovery medium results in the cleavage of the cortex peptidoglycan where the endogenous lytic enzymes have been inactivated (Lund and Peck, 1994). Although egg white is an abundant source of lysozyme, similar lytic enzymes can also be found in vegetable extracts (Lund and Peck, 1993). Such enzymes are thought to enhance the recovery of damaged *Clostridium botulium* spores (Stringer and Peck, 1996; Stringer *et al.*, 1999). However, with UV-laser treated spores no enhanced recovery in the presence of lysozyme or vegetable extracts occurs even when spore coats were removed to enhance permeation (Warriner *et al.*, 2002). This would support the theory that UV-laser treatment does indeed alter the cortex structure of spores.

14.4.4 Limitation of UV-lasers in package sterilisation

A key limitation of UV-lamps is the inability of photons to penetrate into the protective crevices on food surfaces (Cerny, 1977; Wong *et al.*, 1998, Gardner and Shama, 1998). In many respects the influence of surface topography on spore inactivation kinetics is greater for UV-lasers compared to lamps. With lamps the diffuse nature of light gives a greater probability of achieving a lethal photon strike. In contrast, laser light is coherent and the beam has to be physically directed to each part of the pack to ensure spore inactivation. This in turn increases the required dose applied resulting in over-exposure of some parts of the carton (Warriner *et al.*, 2000b). Although the use of aluminium packs contributes to enhancing the distribution of UV photons through internal reflections any commercial system would require the flexibility to sterilise different packaging types. In addition, with heterogeneous UV distribution it is likely that sub-lethally injured spores could recover in food matrices.

From the inactivation curves (Fig. 14.5) a relatively high level of spore kill (LCR 4 $\log_{10}$) can be achieved using a minor dose (10J). In effect the higher dose is required to inactivate the low number of residual spores presumably located within protective pores and crevices. Therefore, the route to overcoming the high dose requirement and packaging effects on spore inactivation is to produce a homogeneous UV-photon field similar to that produced by lamps but only with greater energy. As previously stated the coherent nature of laser light is a key advantage over that derived from lamps and enables precise projection of photons onto surfaces via optical lens arrangements.

14.5 Optimising UV-laser sterilisation of cartons: optical and other novel systems

14.5.1 Modelling virtual optical delivery units

Design of laser optics is primarily applied for micro-machine applications where the light beam has to be highly focused to create intricate patterns. In contrast, when applying UV-laser light for carton packaging sterilisation the beam has to be expanded to illuminate the interior of the pack homogenously.

Previously the design of optical systems involved constructing physical optics with optimisation being performed on a trial and error basis. This made optical design a specialised, laborious and high cost process. However, modelling software packages now enable optical designs to be evaluated without the need for physical systems. Not only does this improve precision of the final optics but also drastically reduces development costs. The fundamental concepts of all optical design software are essentially the same but packages have specific strengths depending on the intended application. In terms of constructing an optical device for carton sterilisation the modelling programme was required to enable the import of data (i.e. CAD data) describing the carton dimensions and standard/non-standard optical elements. As laser light was going to be converted from coherent to non-coherent an ability to trace non-sequential rays is also a prerequisite. Essentially non-sequential rays are where the light during dispersion becomes non-coherent in the sense that the path a ray takes during propagation is not affected by the presence or absence of other rays.

OptiCAD (OptiCAD Corperation, USA) is the software of choice for such applications and enables the intensity distribution profiles to be calculated at the walls of the carton as well as ray tracing plots showing the effects of scattered light.

The initial design of the optics still requires a high degree of creativity by the optical designer. The base materials, geometry and arrangement of elements have to be selected to achieve the required dispersion of light in a defined manner. By importing the optical configurations into the modelling programme the distribution of light within carton interiors can be assessed and optimised through changing parameters. The final design chosen for further development was based on a cone-shaped lens fabricated from a single block of shaped fused silica (Fig. 14.7). Four different annular regions of the silica block processed the beam in different ways, as shown in Fig. 14.7(a). The inner central on axis zone (zone A) acts as a diverging lens. UV radiation in the next annular region (zone B) is totally internally reflected on the cone interior wall and caused to impinge on the outer diffusive surface. The beam impinging on the next zone (zone C) is totally internally reflected on the cone face and diffused on the annular exit face of the block while the beam passing through the outer most zone (zone D) passes directly through the silica block and diffused on the end face. By using the modelling programme the appropriate values for the block dimensions, radius of inner zone, etc. can be readily determined (Fig. 14.7).

14.5.2 Lens construction and diagnostics for laser beam uniformity

There is a limit to the information that can be derived from modelling in terms of predicting its behaviour on optical devices. Therefore, there is a need to construct a physical optical device to confirm the modelling data in terms of ray distribution and spore inactivation. The laser light/dose distribution within cartons can be measured in one of two ways. The first is achieved through using a carton containing a series of small apertures cut into the base and side. A high sensitivity pyroelectric energy detector is coupled to a compatible readout unit to allow individual pulses of the cumulative dose to be recorded. Alternatively, beam profiling can be performed using a carton fabricated from lime silica. Here, the fluorescence caused by UV-photons striking the walls of the glass carton is detected by silicon array sensors linked to computer driven software. Nevertheless, spore inactivation data still represents a measure of the true effectiveness of any optical designs developed. By applying the cone designed lens a high LCR of *Bacillus* spores can be achieved within 5 second treatment times (Table 14.8). Times could be further reduced potentially by developing more rapid systems to move either the carton or laser optics during treatment. In the present case the surviving spores were randomly distributed on the carton surface suggesting no 'cold spots' were present. By using the optical head delivery system the effect of carton packaging on spore inactivation kinetics had been eliminated, presumably due to enhanced dispersion of UV-photons in polyethylene packs. In addition, the prolonged incubation of UV-irradiated spores does not result in post-treatment recovery, confirming that sub-lethal injury to spores does not occur to any significant extent.

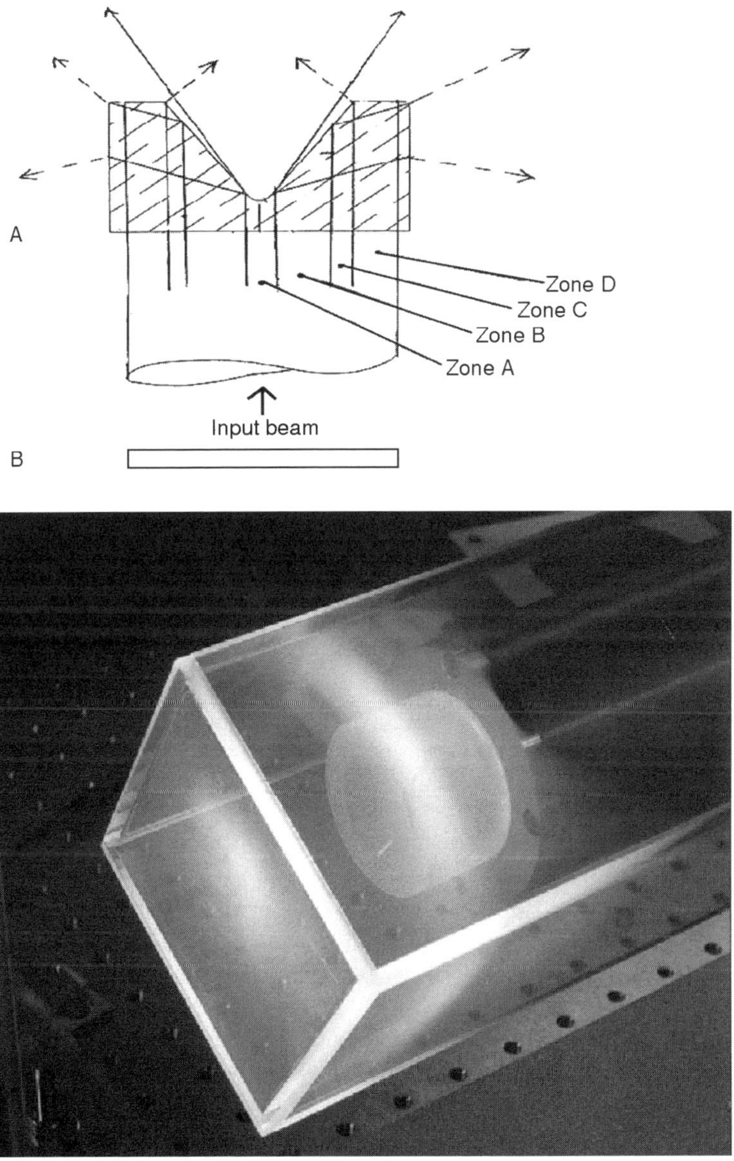

Fig. 14.7 Schematic representation of the conical lens (A) used to deliver UV-laser light during carton sterilisation (B). The final design values were outer diameter, 40 mm, height, 26.5 mm; cone angle, 60 degrees; cone aperture 25 mm; cone apex radius, 1.5. (Reproduced from Warriner *et al.*, unpublished).

Table 14.8 Log count reduction for *Bacillus subtilis* spores deposited on pre-formed cartons and treated with different doses of UV-laser light delivered via a conical silica lens

Irradiation time (seconds)	Log count reduction		
	Total UV dose	Aluminium cartons	PE cartons
5	60	4.04	4.21
10	120	3.91	4.02
30	360	3.72	4.18
60	720	3.58	4.39
120	1440	4.15	4.52

Spore loading on cartons was 1×10^5 cfu.

14.5.3 Alternative chemical-free aseptic carton sterilisation technologies

Other lasers used for surface sterilisation have included the Nd:YAG lasers operating at 1064, 532 and 355 nm (Sadoudi *et al.*, 1997). Here *E. coli* inoculated onto the surface of stainless steel were effectively vaporised by the laser action. However, the high power required to vaporise cells/spores using Nd:YAG lasers would likely result in damage to packaging surfaces.

Pulsed UV light sources have received attention in non-thermal sterilisation of surfaces. The system uses a pulsed power source to drive a xenon lamp which emits intense, broad band white light (200–1000 nm; 20–30% in the UV spectrum). It has been demonstrated that 20×1 J sq.cm^{-1} flashes of 0.3 s duration can result in >6 $\log_{10}$ reduction of *B. pumilus* spores in aqueous suspensions held within a polyethylene container (Dunn *et al.* 1997). Multiple wavelengths in the UV spectrum provide a combination of DNA and macromolecular (e.g. enzyme) damage (Harm, 1980; Hiromoto, 1984; Dunn *et al.*, 1995; Dunn *et al.*, 1997; Peterson *et al.*, 2000; Craik *et al.*, 2001; Kalisvaart, 2001; Oppezo and Pizarro, 2001; Oguma *et al.*, 2002). In addition, non-UV-C wavelengths result in the formation of unique photoproducts that cannot be repaired by DNA repair systems (Munakata *et al.*, 1991). Although primary spore/cell inactivation is the result of UV-C induced damage (MacGregor *et al.*, 1998; Rowan *et al.*, 1999) there is some localised heating (Wekhof, 2000; Wekhof *et al.*, 2001). Pulsed light sources have been commercialised through the Xenon Corporation, WEK-Tec and Purepulse Inc, although the latter has ceased trading. In terms of aseptic packaging sterilisation the problems of photon penetration into protective crevices will be a major technical barrier to overcome.

A further promising technique for aseptic packaging sterilisation is through the application of atmospheric (cold) gas plasmas. Known as the fourth state-of-matter, gas plasmas are formed by an electrical discharge in a gas atmosphere (typically composed of oxygen and helium). The plasma produced is composed of both UV photons and short-lived antimicrobial free radicals (Moisan *et al.*,

2001; Philip *et al.*, 2002). The combination of UV to inactivate surface located spores and the penetration of free radicals into pack crevices can potentially overcome the limitations of light based sterilisation techniques. The process is relatively inexpensive and compares well with UV:hydrogen peroxide based treatments. A major technical barrier is the high temperature (70–100°C) typically reached within the gas plasma that potentially could damage heat sensitive materials (Park *et al.*, 2001). In addition, from the limited studies performed a 6 $\log_{10}$ reduction of *Bacillus* spores is achieved with treatment times in excess of 30 s (Hermann *et al.*, 1999) which is not compatible with high speed lines. However, further developments in electrode configurations and gas compositions have enabled a 4.5 LCR of *Bacillus* spores deposited on flat polyethylene packaging with a treatment time of 1 s (Trompeter *et al.*, 2002).

Ion electron beams (including x-rays) or 'cold pasteurization' are generated by accelerators that operate in both a pulse and a continuous-beam mode (Mittendorfer *et al.*, 2002). Although commonly linked with nuclear sources (for example, cobalt 60) electron beams are easily controlled and contained. Electron beam technology for packaging sterilisation was introduced in the 1970s but poor reliability and the emergence of alternative techniques limited applications. Nevertheless, sustained developments in electron beam technology have made commercialisation a reality. To date the majority of trials have focused on high energy beams to treat pre-packed medical devices for terminal sterilisation. However, trials have been performed using lower-energy (less costly) electron beams. Here, *B. pumilus* spores inoculated on strips and positioned within High Density Polyethylene (HDPE) bottles and treated with electron beams have shown promising results. A 6 $\log_{10}$ reduction of spores can be achieved at 10 bottles per second using a beam power of 5 kW (Cleghorn *et al.*, 2002; Weiss *et al.*, 2002). In addition, electron beam treatment enables real-time performance monitoring of the sterilisation process and is therefore readily integrated into a Hazard Analysis Critical Control Point scheme. However, the adverse public perception of 'irradiation' based processes and high relative cost may limit widespread commercialization in aseptic food packaging sterilisation.

14.6 Future trends

The combination of UV-laser light and sophisticated optical delivery systems have the potential to provide an alternative to current UV-lamp:hydrogen peroxide based systems. For the present application a typical laser unit would cost in the order of £200,000 although running costs would be comparable to those of UV lamp:hydrogen peroxide based systems (£1.25/1000 cartons). Through further advances in technology it is likely that unit costs will be decreased and reliability increased making laser technology a commercially viable option (Endert *et al.*, 1999). However, further developments are still required to achieve the required 5 $\log_{10}$ reduction which, as yet, is still to be obtained with the current optical delivery system (Fig. 14.7).

A further barrier to UV-laser carton sterilisation is the continuing shift from paper board to HDPE containers. In practical terms HDPE packaging offers more flexibility in design, innovation and production. Although environmental issues associated with the use of plastic containers have been raised numerous recycling programmes have sufficiently addressed consumer concerns. It is unlikely that UV-lasers would find utility for sterilising plastic bottles due to the need to fabricate specific optical arrangements for each design used.

The market for aseptic cartons may be declining in US and Europe but in countries such as Brazil, Mexico and China the market is rapidly expanding. Whether UV-laser sterilisation will find utility in this new market is debatable, especially considering recent advances in packaging technology where heat resistant barrier materials can replace polyethylene (Robertson, 2002). This would logically lead the way to thermal sterilisation of cartons thereby negating the need for chemicals and advanced technologies.

In conclusion, it is possible that UV-laser surface sterilisation will find application in the food and medical industries. However, based on the lack of movement away from chemical based techniques, coupled with advances in packaging technology this is unlikely to be in aseptic carton sterilisation.

14.7 Sources of other information and advice

Packaging companies

Elopak Ltd: Elopak, PO Box 24, N-3431 Spikkestad, Norway (website: http://www.Elopak.com).

Packaging organisations

The Aseptic Packaging Council, 2111 Wilson Boulevard, Suite 700, Arlington, VA 22201.

Aseptic packaging books

Aseptic Processing and Packaging of Food Products. S. D. Holdsworth. Kluwer Academic Publishers (1994). ASIN 185166775X.

Aseptic Processing and Packaging of Foods: A Food Industry Prospective. J. David, R. Graves and V. Carlson (1996), CRC Press Inc., Boca Raton, FL.

General laser information

Exitech Ltd, Oxford Industrial Park, Yarnton, Oxford OX5 1QU, UK (www.exitech.co.uk).

Factory Automations (www.factoryautomationhq.com/keyword_Lasers.html; http://uvc.start4all.com).

Laser optics design
Laser Tools Incorporated, 3520 West 69th Street, Suite 401, Little Rock, AR 72209 (http://www.lasertoolsco.com).

Lasers for decontaminating surfaces
Laser and Optical Systems Engineering, University of Glasgow, Glasgow G12 8QQ, UK (http://www.mech.gla.ac.uk/Postgrad/colloq/Abstract.html?AbstractID=142).

14.8 References

ADAMS D M (1978), 'Heat injury of bacterial spores', *Advn. Appl Microbiol*, 23, 245–261.

ANON (1989), 'The use of ultraviolet light for microbial control', *Ultrapure Water* (Apr).

ALDERTON G, CHEN J K and ITO K A (1974), 'Effect of lysozyme on the recovery of heated *Clostridium botulinum* spores', *Appl Microbiol*, 27, 613–615.

BAYLISS C E and WAITES W M (1979), 'The combined effect of hydrogen peroxide and ultraviolet irradiation on bacterial spores', *J Appl Bacteriol*, 47, 263–269.

BAYLISS C E and WAITES W M (1980), 'The effect of hydrogen peroxide and ultraviolet irradiation on non-sporing bacteria', *J. Appl. Bacteriol*, 48, 417–422.

BILLION C M-P, MCKIRGAN C J, MCCLURE, P J and ADAIR C (1997), 'The effect of temperature on the germination of single spores of *Clostridium botulinum* 62A', *J Appl Microbiol*, 82, 48–56.

CERF O and BRISSENDE C H (1981), 'Aseptic packaging. New monograph on UHT milk', Bulletin No. 133 pp. 93–104. *Int Dairy Fed*, Brussels, Belgium.

CERNY G (1977), 'Sterilisation of packaging materials for aseptic packaging 2. – Investigations of the germicidal effects of u.v.-C rays', *Verpackungs Rundschau*, 28, 77–82.

CLEGHORN D A, DUNN J and NABIO S V (2002), 'Sterilisation of plastic containers using electron beam irradiation directed through the opening', *J. Appl. Microbiol*, 93, 937–943.

CONDON S, PALOP A., RASO J and SALA F J (1996), 'Influence of the incubation temperature after heat treatment upon the estimated heat resistance values of spores of *Bacillus subtilis*', *Lett Appl Microbiol*, 22, 149–152.

CRAIK S A, WELDON D, FINCH G R, BOLTON J R and BELOSEVIC M (2001), 'Inactivation of *Cryptosporidium parvum* oocysts using medium and low pressure ultraviolet radiation', *Water Res*, 35, 1387–1398.

DONNELLAN, J E, JR and SETLOW, R B (1965), 'Thymine photoproducts but not thymine dimers are found in ultraviolet irradiated bacterial spores', *Science* 149, 308–310.

DUNN J, OTT T and CLARK W (1995), 'Pulsed light treatment of food and packaging', *Food Technol*, 49, 95–98.

DUNN J, BURGESS D and LEO F (1997), 'Investigation of pulsed light for terminal sterilisation of WFI filled blow/fill/seal polyethylene containers', *J Pharm Sci Technol*, 51, 111–115.

DUNCAN C L, LABBE R G and REICH R R (1972), 'Germination of heat-and alkali-altered spores of *Clostridium perfringens* type A by lysozyme and an initiation protein', *J Bacteriol* 109, 550–559.

ENDERT H, SCAGGS M., BASTING D and STAMM U (1999), 'New ultraviolet lasers for material processing in industrial applications', *J Laser Appl*, 11, 1–6.

FAILLE C, LEBREST V, GAVINI, F and MAINGONNAT, J-F (1997), 'Injury and lethality of heat treatment of *Bacillus cereus* spores suspended in buffer and in poultry meat', *J Food Prot*, 60, 544–547.

FAIRHEAD H and SETLOW P (1992), 'Binding of DNA to α/β-type small, acid-soluble proteins from spores of *Bacillus* or *Clostridium* species prevents formation of cytosine dimers, cytosine-thymine dimers, and bipyrimidine photoadducts after UV irradiation', *J Bacteriol*, 174, 2874–2880.

FOSTER S J (1997), 'The role of cortex hydrolysis during sporulation', In *Abstracts of the 9th International Conference on Bacilli*. p. 42, Universite de Lausanne, Lausanne, Switzerland.

FRIEDBERG E C, WALKER G C and SIEDE W (1995), 'DNA Repair and Mutagenesis, pp. 92–107. ASM Press, Washington DC.

GARDNER D W M and SHAMA G (1998), 'The kinetics of *Bacillus subtilis* spore inactivation on filter paper by u.v. light and u.v. light in combination with hydrogen peroxide', *J Appl Microbiol*, 84, 633–641.

GOULD G W (1969), 'Germination', In *The Bacterial Spore* ed. Gould, G W and Hurst, A. pp. 397–444. Academic Press, London.

GRIFFITH J, MAKHOV A, SANTIAGO-LARA L and SETLOW P (1994), 'Electron microscopic studies of the interaction between a *Bacillus subtilis* α/β small, acid soluble spore protein with DNA: Protein binding is co-operative, stiffens the DNA, and induces negative supercoiling', *Proc Nat Acad Sci*, 91, 8224–8228.

HARM W (1980), 'Repair related phenomena', In *Biological Effects of Ultraviolet Radiation*. ed. Harm, W. pp. 124–134. Cambridge University Press.

HASHIMOTO T, FRIEBEN W R and CONTI S F (1971), 'Kinetics of germination of heat injured *Bacillus cereus* spores', In *Spores V* ed. Halvorson, H. O., Hanson, R. and Campbell, L. L. pp. 406–415. American Society of Microbiology, Washington, USA.

HERMANN H W, HENINS I., PARK J and SELWYN G S (1999), 'Decontamination of chemical and biological warfare agents using atmospheric pressure plasma jet', *Phys, Plasmas*, 6, 2284–2289.

HIROMOTO A (1984), 'Method of sterilization', USA Patent 4464336.

HOLDSWORTH S D (1992), 'Packaging operations', In *Aseptic Processing and Packaging of Food Products*, ed. Holdsworth, S. D. pp. 253–306. Elsevier Applied Sciences.

HOYER O (2000), 'The status of UV technology in Europe', *IUVA News* 2, 22–27.

ISHIKAWA S, YAMANE K and SEKIGUCHI J (1998), 'Regulation and characterisation of a newly deduced cell wall hydrolase gene (*cwlJ*) which affects germination of *Bacillus subtilis* spores', *J Bacteriol* 180,1375–1380.

JONES R L (1996), 'Excimer's pulsed beams sterilise packaging', *Opto and Laser Europe*, 32, 7.

KALISVAART B F (2001), 'Photobiological effects of polychromatic medium pressure UV lamps', *Water Sci Technol*, 43, 191–197.

KLUNGE C H, KUHNET S, LIEBERS H and LINKE L (1988), 'Continuous sterilization and aseptic filling of pulp foods', *Proc Int Symp Progress in Food Preservation Processes*, Vol. 1 pp. 213–225. CERIA, Brussels, Belgium.

LEGAN R W (1982), 'Alternative disinfection methods: A comparison of UV and Ozone', *Industrial Water Engineering*, Mar/Apr. 15–19.

LOPEZ M, GONZALEZ I, MAZAS M, GONZALEZ J, MARTIN R and BERNARDO A (1997), 'Influence of recovery conditions on apparent heat resistance of *Bacillus stearothermophilus* spores', *Int J Food Sci Technol*, 32, 305–311.

LUND B M and PECK M W (1994), 'Heat resistance and recovery of spores of non-proteolytic *Clostridium botulinum* in relation to refrigerated, processed foods with an extended shelf-life', *J Appl Bacterio Symp Suppl*, 76, 115s–128s.

MACGREGOR S L, ROWAN N J, MCLIVANEY L, ANDERSON J G, FOURACRE R A and FARISH O (1998), 'Light inactivation of food-related pathogenic bacteria using a pulsed power source', *Lett Appl Microbiol*, 27, 67–70.

MALLIDIS C G and SCHOLEFIELD J (1986), 'Evaluation of recovery media for heated spores of *Bacillus stearothermophilus*', *J Appl Bacteriol* 61, 517–523.

MASON J M and SETLOW P (1986), 'Essential role of small, acid-soluble spore proteins in resistance of *Bacillus subtilis* spores to uv. Light', *J Bacteriol*, 167, 174–178.

MATSUURA Y and M MIYAGI (1998), 'Flexible hollow waveguides for delivery of excimer-laser light', *Optics Lett*, 23, 1226–1228.

MITTENDORFER J, BIERBAUMER H P, GRATZL F and KELLAUER E (2002), 'Decontamination of food packaging using electron beam-status and prospects', *Rad Phys Chem* 63, 833–836.

MOISAN M, BARBEAU J, MOREAU S, PELLETIER J, TABRIZIAN M and YAHIA L H (2001), 'Low-temperature sterilization using gas plasmas: a review of experiments and an analysis of the inactivation mechanisms', *Int J. Pharm*, 226, 1–21.

MUNAKATA N and RUPERT C S (1972), 'Genetically controlled removal of "spore photoproduct" from deoxyribonucleic acid of ultraviolet-irradiated *Bacillus subtilis* spores', *J Bacteriol*, 111, 192–198.

MUNAKATA N, SAITO M and KOTARO H (1991), 'Inactivation spectra of *Bacillus subtilis* spores in extended ultraviolet wavelengths (50–300 nm) obtained with synchrotron radiation', *Phytochem. Phytobiol*, 54, 761–768.

NESSI C, ALBERTINI A M, SPERANZA M L and GALIZZI A (1995), 'The *outB* gene of *Bacillus subtilis* codes for NAD synthetase', *J Biol Chem*, 270, 6181–6185.

NICHOLSON W L and FAJARDO-CAVAZOS P (1997), 'DNA repair and the ultraviolet radiation resistance of bacterial spores: from the laboratory to the environment', *Recent Res. Dev. Microbiol*, 1,125–140.

NICHOLSON W L and LAW J F (1999), 'Method for purification of bacterial endospores from soils: UV resistance of natural Sonoran desert soil populations of *Bacillus* spp With reference to *B. subtilis* 168', *J Microbiol Methods*, 35, 13–21.

NICHOLSON W L and GALEANO B (2003), 'UV resistance of *Bacillus anthracis* spores revisited: Validation of *Bacillus subtilis* spores as UV surrogates of spores of *B. anthracis* Sterne', *Appl Environ Microbiol*, 69, 1327 1330.

NICHOLSON W L, FAJARDO-CAVAZOS P, REBEIL P. SLIEMAN, T A, RIESENMAN P J, LAW J F and XUE Y M (2002), 'Bacterial endospores and their significance in stress resistance', *Int. J. Gen. Mol. Microbiol*, 81, 27–32.

OGUMA K, KATAYAMA H and OHGAKI S (2002), 'Photoreactivation of *Escherichia coli* after low- or medium-pressure UV disinfection determined by an endonuclease sensitive site assay', *Appl. Environ. Microbiol*, 68, 6029–6035.

OPPEZZO O J and PIZARRO R A (2001), 'Sub-lethal effects of ultraviolet A radiation on *Enterobacter cloacae'*, *J. Photochem Photobiol B Biol*, 62, 158–165.

PARK, J, HENINS, I, HERRMANN, H W, SELWYN, G. S and HICKS, R F (2001), 'An atmospheric pressure plasma source', *J Appl Phys*, 89, 20–28.

PECK M W, FAIRBAIRN, D A and LUND B M (1992), 'The effect of recovery medium on the estimated heat-inactivated of spores of non-protolytic *Clostridium botulinum*', *Lett*

Appl Microbiol, 15, 146–151.

PEEL J L and WAITES W M (1981), 'Improvements in the methods of sterilisation', United States Patent 4, 289, 728.

PETERSEN A B, GNIADEKI R, VICANOVA J, THORN T and WULF H C (2000), 'Hydrogen peroxide is responsible for UV-a induced DNA damage measured by alkaline comet assay in HaCaT keratinocytes', *J Photochem. Photobiol. B Biol*, 59, 123–131.

PHILIP N, SAOUDI B, CREVIER M C, MOISAN M, BARBEAU J and PELLETIER J (2002), 'The respective roles of UV photons and oxygen atoms in plasma sterilization at reduced gas pressure. The case of N-2–O-2 mixtures', *IEEE Trans Plasma Sci*, 30, 1429–1436.

QIN S and GERBA C P (1996), 'Comparative inactivation of enteric adenoviruses, poliovirus and coliphages by ultraviolet irradiation', *Water Res*, 30, 2665–2668.

RICE J K and EWELL M (2001), 'Examination of peak power dependency in the UV inactivation of bacterial spores', *Appl Environ Microbiol*, 67, 5830–5832.

RIESENMAN P J and W L NICHOLSON (2000), 'Role of the spore coat layers in *Bacillus subtilis* spore resistance to hydrogen peroxide, artificial UV-C, UV-B, and solar UV radiation', *Appl. Environ. Microbiol*, 66, 620–626.

RIIHIMAKI V, TOPPILA A, PIIRILA P, KUOSMA E, PFAFFLI P and TUOMELA P (2002), 'Respiratory health in aseptic packaging with hydrogen peroxide: a report of two cases', *J Occup Health*, 44, 433–438.

ROBERTSON G L (2002), 'The paper beverage carton: Past and future', *Food Technol*, 56, 46–52.

ROWAN N J, MACGREGOR S J, ANDERSON J G, FOURACRE R A, MCLLVANEY L and FARISH O (1999), 'Pulsed light inactivation of food-related microorganisms', *Appl Environ Microbiol*, 65, 1312–1315.

SADOUDI A-K, HERRY, J-M and CERF O (1997), 'Elimination of adhering bacteria from surfaces by pulsed laser beams', *Lett Appl Microbiol*, 24, 177–179.

SCOTTI C, PIATTI M, CUZZONI A, PERANI P, TOGNONI A, GRANDI G, GALIZZI A and ALBERTINI A M (1993), 'A *Bacillus subtilis* large ORF coding for a polypeptide highly similar to polyketide synthesis', *Gene* 130, 65–71.

SETLOW P (1988), 'Resistance of bacterial spores to ultraviolet light', *Com Mol Cell Biophy*, 5, 253–264.

SETLOW P (1992), 'I will survive: protecting and repairing spore DNA', *J Bacteriol*, 174, 2737–2741.

SETLOW P (1995), 'Mechanisms for the prevention of damage to DNA in spores of *Bacillus* species', *Ann Rev Microbiol*, 49, 29–54.

SETLOW P (2001), 'Resistance of spores of *Bacillus* species to ultraviolet light', *Environ. Mol. Mutagenesis*, 38, 97–104.

SETLOW P and JOHNSON E A (1997), 'Spores and their significance'. In *Food Microbiology: Fundamentals and Frontiers*, ed. Doyle, M. P., Beuchat, L. R. and Montville, T. J. pp. 30–62 ASM Press, Washington, DC.

SIEGEL A and WILDMAN S G (1956), 'The inactivation of the infectious centres of tobacco mosaic virus by ultraviolet light', *Virology*, 2, 69–82.

SLIEMAN TA and NICHOLSON W L (2001), 'Role of dipicolinic acid in survival of *Bacillus subtilis* spores exposed to artificial and solar UV radiation', *Appl Environ Microbiol*, 67, 1274–1279.

STRINGER S C and PECK M W (1996), 'Vegetable juice aids the recovery of heated spores of non-proteolytic *Clostridium botulinium*', *Lett Appl Microbiol*, 23, 407–411.

STRINGER, S. C., HAQUE, N and PECK, M. W. (1999). 'Growth from spores of nonproteolytic

Clostridium botulinum in heat-treated vegetable juice'. *Appl. Environ. Microbiol.* 65, 2136–2142.

SOLOSHENKO I O, KHOMICH V A, TSIOLKO V V, MIKHNO I L, SHCHERDIN A I, RYABSTEV A V *et al.* (1999), 'Experimental and theoretical investigation of cold sterilization of medical instruments by plasma DC glow discharge', In *Proceedings of the 14th International Symposium on Plasma Chemistry*, Prague, p. 2551.

STANNARD C J, ABBIS J S and WOOD J M (1985), 'Efficiency of treatments involving ultraviolet irradiation for decontaminating packaging board of different surface composition', *J Food Prot*, 48, 786–789.

SUGAWARA T, YONETA M, OHASHI H and TAMAGAWA S (1981), 'Inactivation rate of *Bacillus subtilis* spores irradiated by ultraviolet rays', *J Chem Enginee Jap*, 14, 400–405.

TOLEDO R T (1982), 'Update on container and equipment sterilization techniques in aseptic packaging', *ALChE Symposium Series*, 78, 31–41.

TROMPETER F-J, NEFF W J, FRANKEN O, HEISE M, NEIGER M, LI S and PIETSCH G J (2002), 'Reduction of *Bacillus subtilis* and *Aspergillus niger* spores using non-thermal atmospheric gas discharge', *IEEE Trans. Plasma Sci*, 30, 1416–1423.

VARY J C and HALVORSON H O (1965), 'Kinetics of germination of *Bacillus* spores', *J Bacteriol*, 89, 1340–1347.

WARRINER K and WAITES W M (1999), 'Enhanced sporulation in *Bacillus subtilis* grown in medium containing glucose:ribose', *Lett Appl Microbiol*, 29, 97–102.

WARRINER K, RYSSTAD G, MURDEN A, RUMSBY P, THOMAS D and WAITES W M (2000a), 'Inactivation of *Bacillus subtilis* spores on packaging surfaces by UV-excimer laser irradiation', *J Appl Microbiol*, 88, 678–685.

WARRINER K, RYSSTAD G, MURDEN A, RUMSBY P, THOMAS D and WAITES W M (2000b), 'Inactivation of *Bacillus subtilis* spores on aluminium and polyethylene pre-formed cartons by UV-excimer laser irradiation', *J Food Prot*, 63, 753–757.

WARRINER K, KOLSTAD J, RUMSBY P and WAITES W M (2002), 'Carton sterilisation by U.V.-C excimer laser light: recovery of *Bacillus subtilis* spores on vegetable extracts and food simulation matrices', *J Appl Microbiol*, 92, 1051–1057.

WILLIAMS N D and RUSSELL A D (1993), 'Revival of *Bacillus subtilis* spores from biocide-induced injury in the germination process', *J Appl Bacteriol*, 75, 76–81.

WEISS D E, CLEGHORN D A and NOBLO S V (2002), 'Electron beam process validation for sterilization of complex geometries', *Radiation Phy Chem*, 63, 581 586.

WEKHOF A (2000), 'Disinfection with flashlamps', *J Pharm Sci Technol*, 54, 264–267.

WEKHOF A, TROMPTER F-J and FRANKEN O (2001), 'Pulsed UV-disintegration, a new sterilization mechanism for board packaging and medical-hospital applications', *Proc. First Int. Congress on UV technologies*, Washington, DC, June 14–16.

WONG E, LINTON R H and GERRARD D E (1998), 'Reduction of *Escherichia coli* and *Salmonella senftenberg* on pork skin and pork muscle using ultraviolet light', *Food Microbiol*, 15, 415–423.

Part IV

Improving validation of thermal processes

15

Modelling heat penetration curves in thermal processes

F. Eszes and R. Rajkó, University of Szeged, Hungary

15.1 Introduction: assessing boundary conditions for heat treatment

The most widely used heat treatment method is the constant temperature programme containing three phases: the come up, holding and cooling down phases (Fig. 15.1). There are three basic equations for representing heat transfer during this process (Table 15.1). The first can be used for the determination of the heat transfer coefficient (Fig. 15.2). The second one characterises the heat treatment processes of foods and the heat transfer influenced by both the surface heat transfer coefficient and thermal conductivity of foods. The third one represents internal heat transfer, but it is only true for steam-air autoclaves and for pure steam condensation.

15.1.1 Overview of the initial and boundary conditions of heat treatment

In the course of the evaluation of heat penetration curves, the most important task is to survey the initial and boundary conditions of non-steady state heat transfer. This is needed because, if the calculations can be simplified, the evaluation time can be reduced. The most frequently used assumptions are listed in Table 15.2.

Many heat treated products are mostly solid in which the heat is transferred only by conduction. In these cases, convection heat transfer can be expected only in the case of ready meals (gravy) and with vegetable products in sugar or salt solution, for example. Special additives such as starch and modified starch take up water at about 60–70°C and convective heat turns into conduction heat transfer. Fat melts at 40°C, at which point fat-containing products become

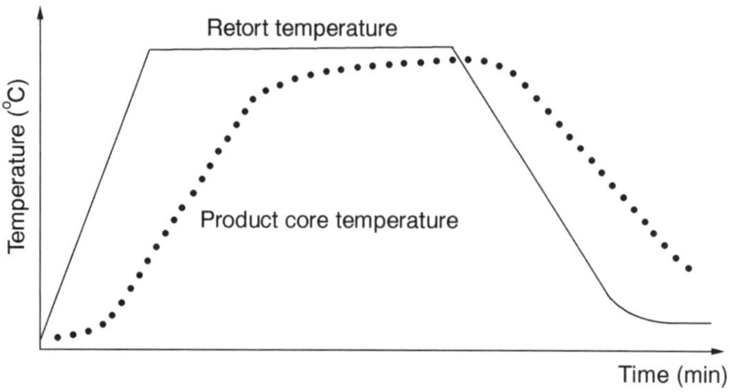

Fig. 15.1 Schematic picture of the traditional heat treatment.

capable of partial convection heat transfer. This convection phenomenon can be seen in a breaking point in the infinite series solution curves and Ball curves of transient conduction heat transfer, because convection is 3–4 times quicker than conduction. This difference decreases the correlation coefficient of the slope index and increases the sum of squares of the fitted lines/curves. The ratio of conduction and convection in heat transfer can be assessed by measuring the heat penetration of two different size cans according to Eq. 15.1 (Ball and Olson, 1957, Reichert, 1985). If K_{va} is near to 1 the heat transfer carries out mainly by conduction; if K_{va} is near to 0 then the heat transfer carries out mainly by convection.

$$K_{va} = \frac{K_v f_h - f_{h'}}{f_h K_v - f_h K_a} \qquad [15.1]$$

K_{va} Conduction-convection heat transfer index [-]
K_v Can conduction index [-]

Table 15.1 Base cases of the heat treatment

1. $\alpha >> \lambda$ and $Bi < 0,1$	2. Finite α and λ and $1 < Bi < 200$	3. $\alpha >> \lambda$ and $Bi = \infty$
$Y = \frac{T - T_a}{T_o - T_a} = e^{-\frac{\alpha A}{\rho c_p V} t}$	$Y = \frac{T - T_a}{T_o - T_a} = A_1 e^{-B_1 F_o} C_1$	$Y = \frac{T - T_a}{T_o - T_a} = A_2 e^{-B_2 F_o} C_2$

Y	Dimensionless temperature [-]	T	Time [s]
T	Measured temperature [°C]	A_1, A_2, B_1, B_2	Constants $f(Bi)$
T_0	Initial temperature [°C]	Bi	hX/λ [-]
T_a	Ambient temperature [°C]	X	Characteristic length [m]
h	Surface heat transfer coefficient [W/m^{2}K]	λ	Thermal conductivity [W/mK]
A	Surface [m^2]	Fo	Fourier number [at/X/X] [-]
ρ	Density [kg/m^3]	α	Thermal diffusivity [m^2/s]
c_p	Specific heat [J/kgK]	C_1, C_2	Constants $f(Bi,x/X)$
V	Volume [m^3]	x	Distance from the core [m]

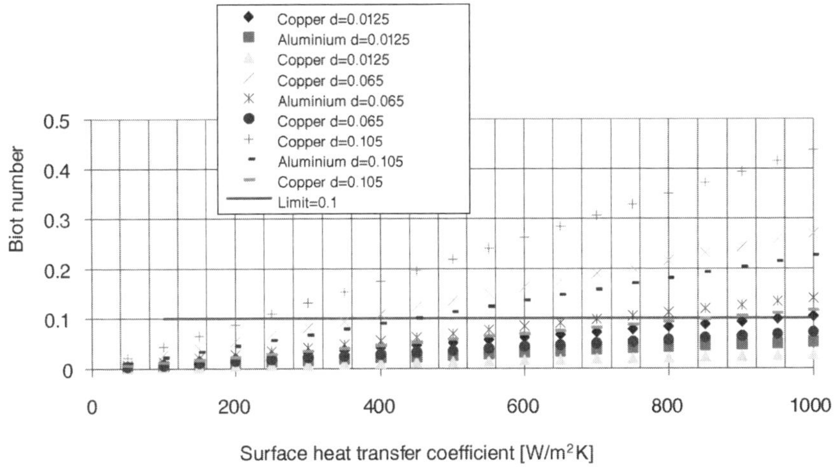

Fig. 15.2 Biot number development of 1 dimensional bodies vs. surface heat transfer coefficient and length.

K_a Can convection index [-]
f_h Slope index of the first can [min]
$f_{h'}$ Slope index of the second can [min]

Most heat treated products are cylindrical or rectangular in shape. In the case of special packagings, there is some deviation from these basic shapes (e.g. special glasses, pear shaped cans, tubes, semi rigid retortable pouches). In most cases these can be approximated with geometrical indices or with one dimensional heat transfer through the shortest side. Leonhardt (1976a and 1976b) found only small deviations in temperature development of different groups of container. These are shown in Table 15.3 and 15.4 for oblong and Pullman cans and

Table 15.2 Fulfilling the simplifying assumptions of the Fourier heat conduction equations: fulfilled (+++), approximately fulfilled (++), not fulfilled (0) assumption

Assumption	Fulfilling
1. Heat is transfered only by conduction	+++
2. Simple geometry (slab, cylinder, sphere or bodies can be approximated with them)	+++
3. There is no phase change	++
4. The product is homogeneous	++
5. The thermal parameters (thermal conductivity, thermal diffusivity specific heat) are constant	++
6. The product initial temperature is constant	++
7. Ambient temperature is constant and there is no come up time	++
8. Infinite heat transfer coefficient	0

Table 15.3 Sizes and dimension neglections of large rectangular packagings

Packaging	Sizes (mm)			Packaging	Sizes (mm)		
12 lb oblong	105	169	323*	16 lb Pullman	115	115	545**
12 lb oblong LANGEN	103	164	305	11 lb Pullman	115	115	385*
Ham mould	98	105	318	8 lb Pullman	100	100	400**
Roll form	105	105	287	6 lb Pullman	100	100	303
Ham mould (long)	82	82	400**	4 lb Pullman	100	100	207

* 3 times difference; ** four to five times difference.

traditional cans. The head space in the cans and the air pockets can decrease further the amount of heat penetrating into the can. The effects of different height/diameter (H/D) ratios are shown in Table 15.4.

On the base of experimental analysis we can expect standard deviations of ±1% for water and protein content and ±2% for fat content. The distribution of constituents may be homogeneous and random (e.g comminuted meat batters), heterogeneous random (cold cuts, luncheon meat) or heterogeneous with definite distributions (fat covered cooked hams, hot smoked bacon). The first two have similar thermal properties, but the third one must be treated separately with different thermal properties (Gaffney *et al.*, 1980). When we investigate initial

Table 15.4 H/D ratios for traditional cylindrical cans

USA Code	Diameter (mm)	Height (mm)	H/D	USA Code	Diameter (mm)	Height (mm)	H/D
202X109	52	40	0,769	300X402	73	105	1,438
211X102	65	28	0,431	300X404	73	107	1,466
202X204	52	57	1,096	307X306	83	86	1,036
211X109	65	40	0,615	401X208	99	64	0,646
300X108	73	38	0,521	211X510	65	143	2,200
202X214	52	73	1,404	300X407	73	113	1,548
211X200	65	51	0,785	307X309	83	91	1,096
202X306	52	85	1,635	401X211	99	68	0,687
202X314	52	98	1,885	303X406	78	111	1,423
211X212	65	69	1,062	307X409	83	116	1,398
300X205	73	58	0,795	307X509	83	140	1,687
307X113	83	46	0,554	401X400	99	102	1,030
300X208	73	64	0,877	401X411	99	119	1,202
307X113	83	51	0,614	401X512	99	146	1,475
211X300	65	76	1,169	404X502	105	129	1,229
307X201,25	83	53	0,639	404X700	105	178	1,695
307X206	83	60	0,723	502X512	127	156	1,228
211X400	65	102	1,569	603X400	153	101	0,660
300X304	73	82	1,123	603X600	153	152	0,993
300X308	73	88	1,205	603X700	153	178	1,163
211X414	65	124	1,908	603X908	153	241	1,575
401X205	99	58	0,586				

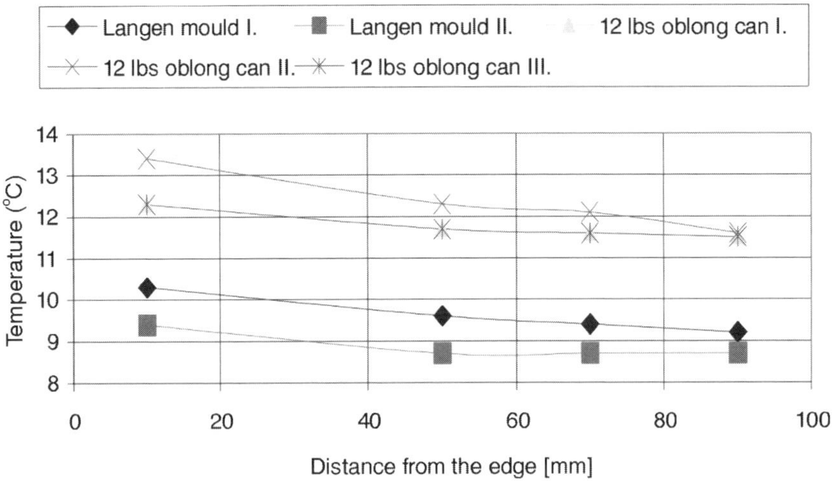

Fig. 15.3 Development of the initial temperature in 12 lb oblong cans along the longest length after 10 and 30 minutes.

product temperature conditions, products can be divided into two groups: cold (e.g. minced meats and potted meats) and hot filled products (e.g. ready to eat products, blanched raw materials and hot gravies). In the first group the waiting period before retorting can cause warming up (Fig. 15.3), whilst in the second group cooling down can occur. In the first case there is no problem in large cans but problems can arise in small cans in reaching a target temperature for a defined pasteurising/sterilising time. In case of the hot filled product, for example, the initial temperatue of 70°C can decrease to 50°C (Stumbo, 1973). These fluctuating initial temperature problems can be avoided by control in processing technology. In the case of semi-preserved hams we experienced a region of 6–14°C and a 9.99 average and ±1.98 standard deviation The lower region occurred when the raw materials were cold stored for long time (especially at weekend), and the upper region reached when the product was waiting for heat treatment.

As has been noted, the constant ambient temperature heat treatment is separated into two phases: the come up and holding stages. In normal technical conditions, the holding temperature can be reached within 5–10 minutes. There is no problem in holding constant temperature in modern autoclaves equipped with automatic process control. According to our experience about 0.3–1°C difference can be expected in new and old units respectively. An experienced autoclave operator can reach 1°C difference in manual operations. Deviant temperature drops were very rare in our industrial experiments. During cooling the constant cooling water temperature cannot be reached immediately. The time required depends on the cooling water supply. Sometimes the co-running of heat treating units exhausts the supply, and other demands on the water supply can also cause serious deviances (Figs 15.4 and 15.5). We have also observed a 10°C difference is water temperature between summer and winter. These problems

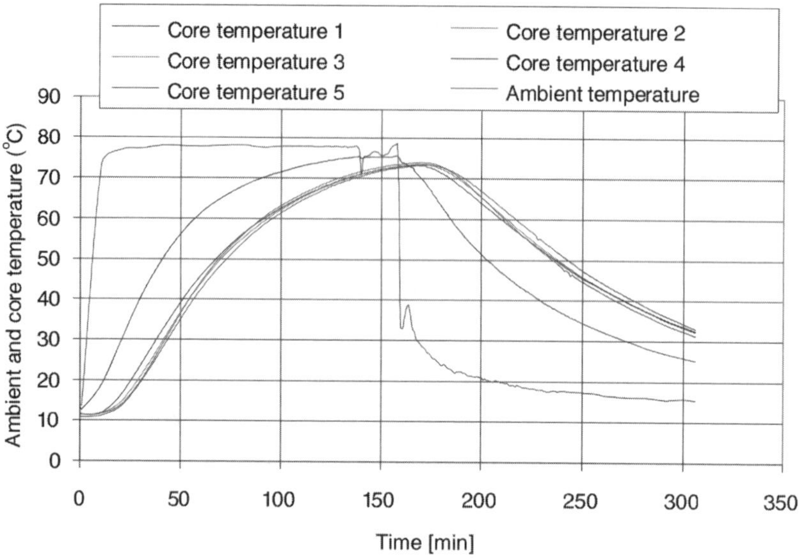

Fig. 15.4 Warming up of cooling water during the overshooting.

can be avoided only by separate water source tanks operating only for the heat treating units.

The infinite surface heat transfer coefficient and Biot number can only be true in very few cases (Fig. 15.6 and Table 15.5). Although several large numbers, for surface heat transfer coefficient can be found in the literature (Table 15.6)

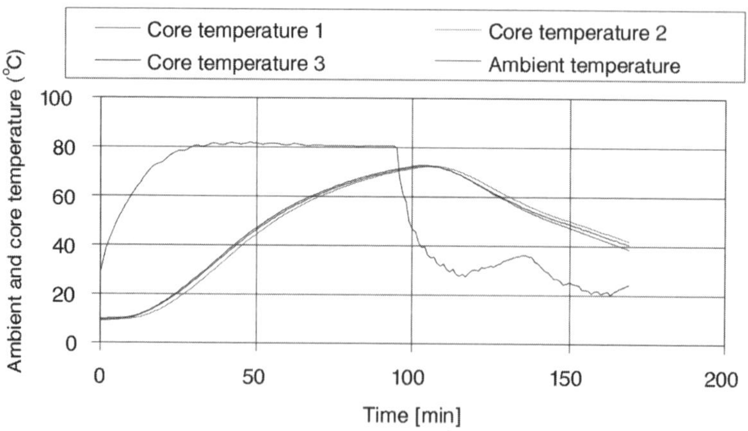

Fig. 15.5 Warming up of cooling water during the remaining part of cooling.

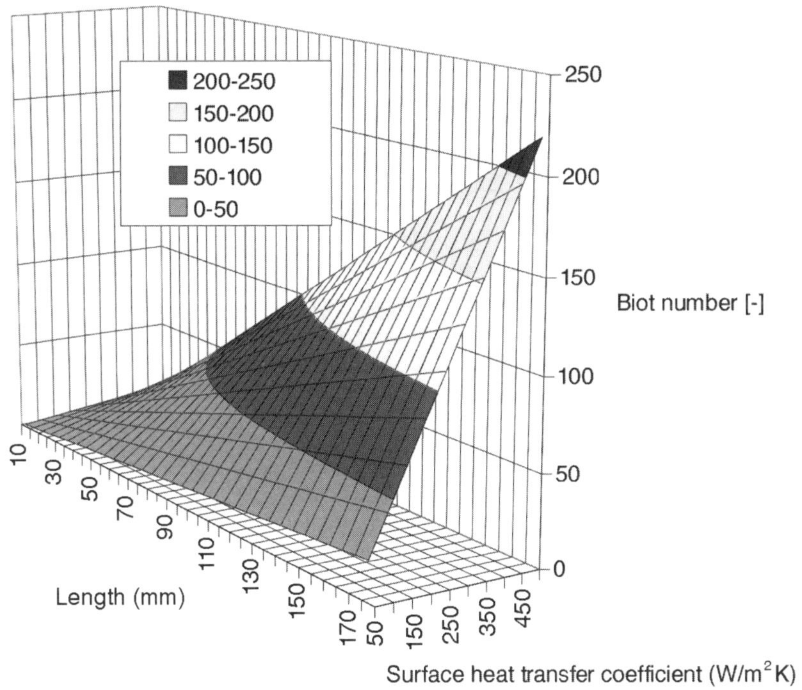

Fig. 15.6 The development of the Biot numbers vs. surface heat transfer coefficient and length (thermal conductivity = 0.5 W/mK).

the expected average value is 200–400W/m^{2}K. These give Bi>200 only in very few cases, even with the largest cans (Ramaswamy *et al.*, 1982). The Biot numbers of the largest cans fall into the region of 20–60 in most cases (Fig. 15.7). It is 1–20 for standard size cans.

Table 15.5 Order of the magnitude of surface heat transfer coefficient in different types of medium (Adams and Rogers, 1973)

Heating medium	Surface heat transfer coefficient [W/m^{2}K]
Still air	5–10
Still water	40–60
Circulated water	280–1200
Moist air stream	200–400
Stream air mixture	25–300
Pure steam condensation	15000

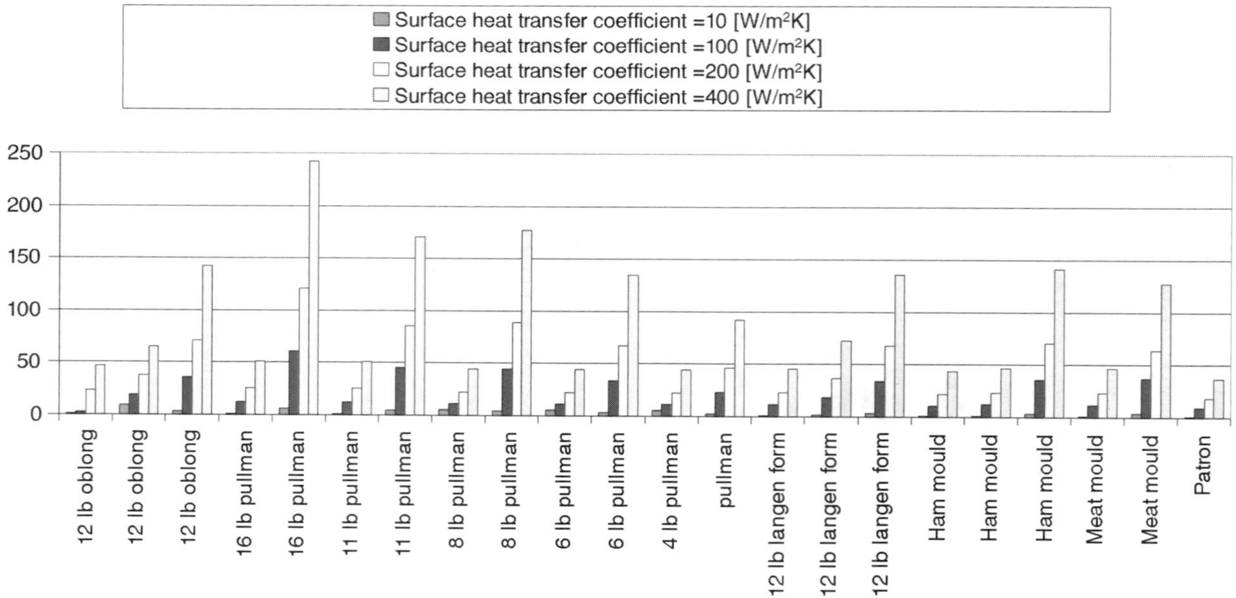

Fig. 15.7 Biot numbers of large cans used for ham at surface heat transfer coefficient of 10; 100; 200; 400 W/m^2K and thermal conductivity of 0.45 W/mK.

Table 15.6 Surface heat transfer coefficient (h) in circulating heat treating units

Author and publication year	h [W/m^{2}K]	Notice
Hayes (1987)	280–1200	Forced convection of water
Burfoot and Self (1988)	830–1550	Circulating water bath
Peterson and Adams (1983)	187–278	Circulating water in rectangular duct
Chang and Toledo (1989)	222–317	Streaming water around a cube
Pátkai *et al.* (1990)	650	Horizontal autoclave
	55–370	Horizontal autoclave
	143–287	Laboratory ultra thermostat
Lebowitz and Bhowmik (1990)	186±54	Water circulating autoclave
Bhowmik and Tandon (1987)	295,94	Water circulating autoclave
Bhowmik and Shin (1991)	202–278	Surface heat transfer coefficient range
	174±35	During come up time
	243±117	During holding time
	186±54	Come up and holding time together
	222±128	During cooling
	178±27	Average for plastic body

15.2 Determining thermal diffusivity

The thermal diffusivity is an important parameter in heat treatment calculation. Methods for its determination are summarised in Table 15.7. The values from existing studies are difficult to use because it is not easy to find the same product, and experimental conditions are not always fully documented. If we want to determine it from analysis of food constituents, we can also produce errors: see Fig. 15.8 (Miles *et al.*, 1983). Instrumentation for measuring thermal diffusivity is expensive and gives the same order of error as heat penetration measurements, e.g. the Line Heat Source Technique has about 5–10% error according to the equipment producers. Determination from chemical analytical values is useful because the main constitutents such as water, fat and protein content are regularly controlled and measured to meet regulatory and consumer requirements.

The simplest and most frequently used method for determining thermal diffusivity is the evaluation of heat penetration curves either with the infinite series solution using Fourier differential equations or with the Ball method. The former method uses the Y vs. Fo number, the latter one uses the $\log(T_k - T_{mag})$ vs. time coordinate sytem. The infinite series method requires the constants of the FDE VSM solution (Eq. 15.2) determined from temperature measurements, expressed as lag factor (j) and slope index (f_h) which can be determined graphically or mathematically. The two methods have both advantages and disadvantages (Table 15.8). The infinite series solution is used in non-linearised form while the Ball method is used in linearised form.

$$Y = \frac{T - T_a}{T_o - T_a} = Ae^{-BFo}C \qquad [15.2]$$

$$Y = \frac{T - T_k}{T_o - T_k} = j10^{-t/f_h} \quad [15.3]$$

$$f_h = -BFo = \frac{R^2 \ln(10)}{\alpha\beta^2} \quad [15.4]$$

$$f_h = -BFo = \frac{R^2 \ln(10)}{\alpha\beta^2} \quad [15.5]$$

$$j = AC \quad [15.6]$$

Y	Dimensionless temperature [-]	R	Characteristics radius [m]
T	Measured temperature [°C]	Bi	$\alpha X/\lambda$
T_0	Initial temperature [°C]	λ	Thermal conductivity [W/mK]
T_a	Ambient temperature [°C]	Fo	Fourier number $[a\tau/X/X]$ [-]
h	Surface heat transfer coefficient [W/m^{2}K]	α	Thermal diffusivity [m^2/s]
		τ	Time [s]
A	Surface [m^2]	C	Constant $f(Bi,x/X)$
ρ	Density [kg/m^3]	x	Distance from the core [m]
c_p	Specific heat [J/kgK]	j	Lag factor [-]
V	Volume [m^3]	f	Slope index [min]
T	Time [s]	t	Time [min]
A,B	Constant $f(Bi)$	β	Root of the characteristic equation [-]
X	Characteristic length [m]		

Hayakawa and Bakal (1971) used the infinite series solution to measure thermal diffusivity. Larkin and Steffe (1982, 1987) compared the Ball method and infinite series solutions and gave 2–4% differences between them. Flambert (1973) calculated the thermal diffusivity from the gradient of the core

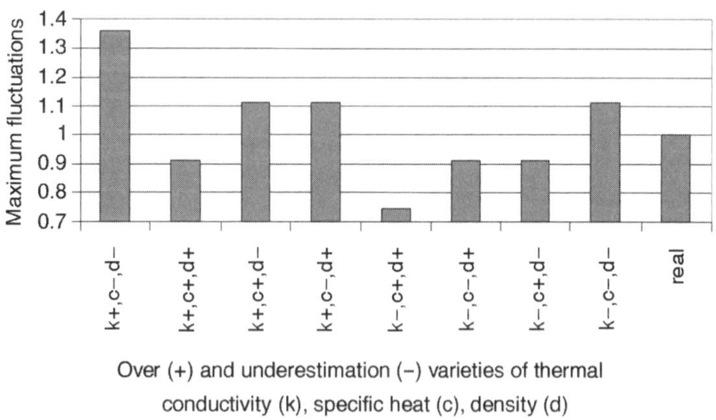

Fig. 15.8 A thermal diffusivity estimation from its elements assuming 10% error in thermal conductivity (k), specific heat (c) and density (d) - = underestimation, + = overestimation.

Table 15.7 Classification of the methods for determinations of thermal diffusivity

A. From literature	B. From instrumental measurements	C. From chemical compositions
• Directly	• Line Heat Source technique	• From water and dry matter content
• From its constituents (thermal conductivity, specific heat, density)	• Guarded Hot Plate method • Differential Scanning Calorimeter	• Only from the water content • From water-, fat-, and protein content • From all constituents

D. Temperature Matching

Surface heat transfer coefficient $= \infty$		Surface heat transfer coefficient $\neq \infty$	
Linearised	Non-linearised	Linearised	Non-linearised
• 1st term • limited number of terms • many terms	• 1st term • limited number of terms • many terms	• 1st term • limited number of terms • many terms	• 1st tag • limited number of terms • many terms

E. From heat penetration curves

Constant ambient temperature			Variable ambient temperature
One measuring point		More measuring points	
Surface heat transfer coefficient $= \infty$	Surface heat transfer coefficient $\neq \infty$	Surface heat transfer coefficient $\alpha \neq \infty$	Dickerson (1965), Dickerson and Read (1968), Hayakawa and Bakal (1971)
Stumbo (1973), Flambert (1973), Poulsen (1982)	Pflug *et al.* (1965) Ramaswamy *et al.* (1983)	Uno-Hayakawa (1980), Bhowmik-Hayakawa (1980), Nesvadba (1982)	Dickerson and Read (1975), Rizvi *et al.* (1980)

temperature increase. He determined the maximum of this gradient vs. time and give a constant value for Fo_{max} number for different height/diameter ratios. The advantage of the Flambert method is that we need only measure the core temperature. The disadvantage of this method is that it assumes infinite surface heat transfer coefficient.

$$\frac{\alpha \tau_{max}}{X^2} = Fo_{max} \qquad [15.7]$$

In the ratio method more points in the body are measured. These can be the surface and the core temperature (Bhowmik and Hayakawa, 1979) or several

Table 15.8 Characterisation of the Ball method and and infinite series solution relating to determination of thermal diffusivity

FDE VSM	Ball method
More sensitive to changes in parameters	More robust relating to changes in parameters
Does not treat convection	Applicable for convection
Can be derived for changing tempertures	Applicable only for constant temperature
More sensitive in case of convection	Slope index proportional with V/A and $1/X^2$ for convection and conduction respectively
More sensitive on time measurements	Not sensitive for time measurements because it measures temperature difference
More sensitive for sensor placement error	Not so sensitive for sensor placement error
More sensitive for the come up time	Not so sensitive for come up time
More sensitive for the temperature measurements	Not so sensitive (Gaffney *et al.*, 1980).

points around the core (Uno and Hayakawa, 1980). The infinite series solutions described for these points are divided by each other. After division only the factors C depending on the relative distance and Biot numbers (Eq. 15.2) remain constant for a period of time, from which the constant can be calculated. Awuah *et al.* (1995) compared the ratio method with Ball methods and concluded that more errors arise with ratio methods.

Nesvadba (1982) divided the time and place derivative of the Fourier differential equation to get the thermal diffusivity. This ratio was evaluated each time steps at the extremum. It needed small time steps and more sensors (10) built in a row. Pflug *et al.* (1965) suggested Eq. 15.8 for a one-dimensional case. If we had more dimensions we would have to sum the reciprocals of the f_h values (Eq. 15.9). In the case of the most frequently used cylindrical can shape, this has the form of Eq. 15.10 (Ramaswamy *et al.*, 1982)

$$\alpha = \frac{X^2 \ln(10)}{\beta^2 f_h} \qquad [15.8]$$

$$f_h = \sum_{i=1}^{n} \frac{1}{f_i} \qquad [15.9]$$

$$\alpha = \frac{2,305}{\left(\frac{\beta_{1R}^2}{R^2} + \frac{\beta_{1H}^2}{H^2}\right) f_h} \qquad [15.10]$$

Where

α	Thermal diffusivity [m^2/s]	f_i	Slope index of dimension of i [min]
X	Characteristic length of the body [m]	β_{1R}	First root of the $\beta J_1(\beta) = Bi\, J_0(\beta)$
β	First root of the characteristic equation [-]	β_{1H}	First root of the $\beta \mathrm{tg}(\beta) = Bi$
		R	Radius of the can [m]
f_h	Ball slope index [min]	H	Half height of the can [m]

Poulsen (1982) used only two points for the evaluation of f_h value after reaching the dimensionless temperature of 0.4 for 6 minutes time difference. Rogacsev and Babarin (1983) summarised the characteristic roots and geometrical measures into one constant for the radius and height to give these constants for $Bi < 10$. The determination of the thermal diffusivity from the Ball curve is very insensitive for the long come up time. Rizvi *et al.* (1980) determined a thermal diffusivity from slope index determination during the constant holding temperature and the infinite series solution. Sometimes researchers measure the Ball slope index and use it to calculate thermal diffusivity with an infinite surface heat transfer coefficient. In this case irrational high thermal diffusivity can result (Carciofi *et al.*, 2002). Pátkai *et al.* (1990) ascertained thermal conductivity and thermal diffusivity using heat penetration curves. These were 30% higher than values determined from chemical composition (Choi and Okos, 1986) and the methods of Pflug *et al.* (1965) according to Eq. 15.8.

Although the constant holding temperature is used most frequently today, a variable retort temperature has been proposed (Durance, 1997). Dickerson (1965) gave a very simple equation for thermal diffusivity determination for constantly rising ambient temperature, an infinite cylinder and a well stirred thermostat (Eq. 15.11)

$$\alpha = \frac{4M}{(T_k - T_m)R^2} \qquad [15.11]$$

Dickerson (1968) and Dickerson and Read (1975) have used Eq. 15.11 for determination of thermal diffusivity of meats, Rizvi *et al.* (1980) for meat analogues, and Luna and Bressan (1985) for cooling of cheese. Tung *et al.* (1989) used a sliding evaluation of the linear portion of the Dickerson's curves.

15.3 Determining surface heat transfer coefficients

There are several methods for determining surface heat transfer coefficient. These are the lumped capacity, infinite series solution with known thermal diffusivity, Ball method, and Nusselt functions. In case of lumped capacity, using the surface heat balance, Newtonian cooling law and Fourier's first law gives Eq. 15.12. The surface heat transfer coefficient can be determined from the slope of the linearised form of Eq. 15.12. It gives very accurate estimation of surface heat transfer coefficient but has the limitations of $Bi < 0.1$ and being true only for metals. Food simulants (e.g. bentonite suspensions) and different plastic materials give at least $Bi > 1$ conditions. Another disadvantage is that the experimental data is only valid for calculating the surface heat transfer coefficient during the come up stage. In this way this method is useful when beginning with the stationary phase at ambient temperature. Ramaswamy *et al.* (1983) gave an equivalent solution with slope index of the Ball heat penetration curves (Eq. 15.12).

$$Y = \frac{T - T}{T_o - T_k} = e^{-(hA/\rho c_p V)t} \qquad [15.12]$$

$$h = \frac{\ln(10)Vk}{A\alpha f_h} \qquad [15.13]$$

The infinite series solution (Eq. 15.2) is useful if we know the thermal properties of the materials to be heat treated. Artificial materials (e.g. bentonite suspensions, plastics) can be used to simulate the food. The method has the advantage that thermal properties for common foods can be taken from material properties handbooks, and that only constants depending on Biot numbers remain unknown. The surface heat transfer coefficient can be determined by two techniques. The first uses the least sum of squares. The second is to fit a line based on the measured against one based on calculated values and then changing the calculated values till the slope reaches 1 ± error limit (Lebowitz and Bhowmik, 1989, 1990).

The Ball method can also be used for surface heat transfer coefficient determination (Eq. 15.8) for 1D (Pflug *et al.*, 1965). When we have a high heat transfer coeffcient, the fluctuation of f_h value can result in an irrationally high constant of β depending on the surface heat transfer coefficient (Eq. 15.14).

$$\beta = \sqrt{\frac{2{,}303 R^2}{\alpha f_h}} \qquad [15.14]$$

Peterson and Adams (1983) also used the Ball slope index for the determination of surface heat transfer coefficient (Eq. 15.15).

$$h = \lambda \sqrt{\frac{2{,}303}{\alpha f_h}} \tan \sqrt{\frac{2{,}303 L^2}{\alpha f_h}} \qquad [15.15]$$

Where h Surface heat transfer coefficient [W/m^{2}K]
α Thermal diffusivity [m^2/s]
L Volume/Surface ratio
f_h Ball slope index

The Nusselt function is used in special circumstances. These circumstances include free convection (Eq. 15.16), forced convection (Eq. 15.17) and mixed forced and free convection, and pure condensation. In case of mixed convection the Nu numbers of free and forced convection are summarised and then rooted.

$$\mathrm{Nu} = C_1 (Gr\mathrm{Pr})^m \qquad [15.16]$$

$$\mathrm{Nu} = C_2 \mathrm{Re}^p \mathrm{Pr}^r \frac{\mu^s}{\mu_f^s} \qquad [15.17]$$

Where Nu Nusselt number $\mathrm{Nu} = hD_e/\lambda$ [-]
h surface heat transfer coefficient [W/m^{2}K]
λ Thermal conductivity [W/mK]
C_1, C_2 Constants [-]
Gr Grashoff number $G = \beta g \Delta T X^3/\mu^2$ [-]
β Cubic thermal expansion coefficient [1/K]
ΔT Temperature difference between surface and ambient [C]
X Equivalent length [m]
μ Kinematic viscosity [m^2/s]
Pr Prandtl number [-]
Re Reynolds number $D_e v/\mu$ [-]
D_e Equivalent diameter [m]
v Velocity [m/s]

If the product configuration has no circular cross-sectional area, the equivalent diameter is substituted into Eq. 15.17.

$$D_e = \frac{4A}{K} \qquad [15.18]$$

Where D_e Equivalent diameter [m]
A Free flow cross section [m^2]
K Wetted perimeter [m]

If the product is covered with thin layers (foil, metal can wall, contact resistance between product and a can wall, e.g. air pockets, fat layer) the surface heat transfer coefficient can be lowered given resistance to heat transfer (Eq. 15.19).

$$k = \frac{1}{\frac{1}{\alpha_k} + \sum_{i=1}^{n} \frac{\delta_i}{\lambda_i} + \frac{1}{\alpha_b}} \qquad [15.19]$$

k Overall heat transfer coefficient [W/m^{2}K]
h_k Outside surface heat transfer coefficient [W/m^{2}K]
δ Thickness of the heat resisting thin layer [m]
λ Thermal conductivity of the heat resisting thin layer [W/mK]
h_b Inside surface heat transfer coefficient [W/m^{2}K]

15.4 Increasing the accuracy of the parameters determined from heat treatment penetration curves

15.4.1 Evaluation of the lethality of the process

The first thing to evaluate in the heat treament is the required microbial stability. Since we usually only have discrete values in even time steps, we usually use

Table 15.9 Integral methods for calculation of the heat equivalents of heat treatment

Method	Continuous case	Discrete case	Error
Additional (rectangular)	$\int_a^b y(x)dx = \sum_a^b y(x_i) \int_a^b 1dx = (b-a)y(x_0)$	$F = \int_a^b 10^{-(T-T_r)/z} = dt = \sum_a^b 10^{-(T-T_r)/z}\Delta t$	Quadratic decrease if we decrease the time step
Trapezoidal method	$\int_a^b y(x)dx = \frac{b-a}{2}(y(a) - y(b))$	$\int_a^b y(x)dx = \frac{b-a}{2}\left(\frac{y(a)}{2} + \sum_a^b y(x_i) - \frac{(y(b)}{2}\right)$	
Tangential method		$\int_a^b y(x)dx = \frac{b-a}{n}\sum_b ay(x_k)$	$E(y) = \frac{1}{12}\left(\frac{b-a}{2}\right)^2 y''_{\max}(x)$
Simpson method	$\int_a^b y(x)dx = \frac{b-a}{6}(y(a) + 4y\left(\frac{a+b}{2}\right) + y(b)$	$\int_a^b y(x)dx = \frac{h}{3}[y_0 + y_n + 4(y_1 + y_3 + \ldots y_{n-1}$ $+2(y_2 + y_4 + \ldots y_{n-2})]$	$E(y) = \frac{1}{180}\left(\frac{b-a}{2}\right)^4 IVy^{IV}_{\max}(x)$
Gauss method	$\int_a^b y(x)dx = \int_{-1}^{+1} \frac{b-a}{6}(y(x)d\left(\frac{b-a}{2}\xi + \frac{a+b}{2}\right)$	$\int_a^b y(x)dx = \sum_a^b A_i\xi_i$	

numerical integration. The simplest approach is the so-called addition method (Eq. 15.20).

$$F = \sum_{t_1}^{t_2} 10^{(T-Y_t)/z} \Delta t \qquad [15.20]$$

T measured (core) temperature [ºC]
T_r Reference temperature [ºC]
z Decimal reduction temprature [ºC]
Δt Time step [min]
F Lethality [min]
t_1 Start of time [min]
t_2 End of time [min]

The trapezoidal method means the average between two neighbouring points and is frequently used in computerised modelling because the preceding and actual value can be stored easily. The Simpson methods use a parabolic approximation for three points. In all these case the practical accuracy is about the same. The calculation and the errors are summarised in Table 15.9. The weighed summaries of the Gauss method (Hayakawa, 1968) are rarely used but can give a quick evaluation.

Only occasionally do we have the opportunity for the temperature measurements with larger time steps. In this case we have to be careful because the values can fluctuate around the real value and, because of the time step, constant multiplication can give even higher lethality than the real value (Fig. 15.9). In this case we require trapezoidal or Simpson methods. The Gauss Integration showed a similar fluctuating picture according to the number of points considered. A 4 points formula is needed for accurate calculation.

New research has revealed that the D and z value vary as a result of a long come up time and prolonged heat treatment. Peleg (2000, 2002) demonstrated that the elongated come up time causes a longer shoulder in the surviveal curve, and that the linear portion of the curve has a different D value compared to

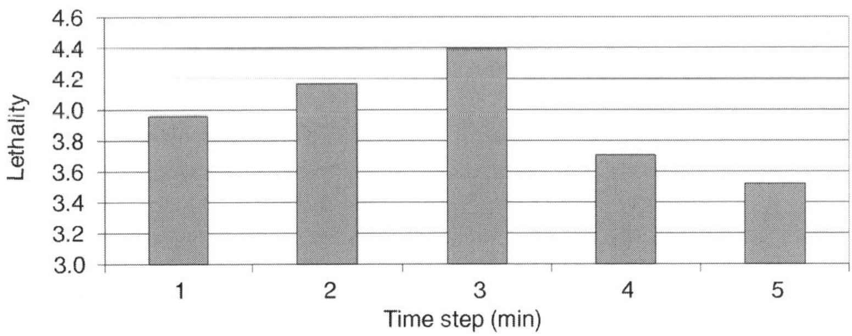

Fig. 15.9 Development of the lethality in case of different time steps (additive method, Cl. Botulinum, sterilisation of 99 × 63 DIN can).

constant ambient temperature measurements. This phenomenon is even more important in the case of large cans, multi-stage cooking and variable retort retort temperature, and in heat treatment of heat sensitive glass and plastic packages. Microbiologists have related variability of D and z value to water activity and pH value. Mafart and Leguernel (1998) have suggested including pH value in the lethality calculation as a quadratic function (Eq. 15.21 and 15.22). Körmendy and Körmendy (1997) have proposed a way of evaluating D and z values for a known time section if they do not follow the linear relationship.

$$\log D = \log D_r - \left(\frac{1}{z}\right)(T - T_r) - \left(\frac{1}{z_{pH}^2}\right)(pH - pH_r)^2 \qquad [15.21]$$

$$F = \int_a^b 10^{\frac{T-T_r}{z_r} + \left[\frac{pH - pH_r}{z_{pH}}\right]^2} dt \qquad [15.22]$$

15.4.2 Increasing the accuracy of thermal parameter estimation from heat penetration curves

The two main parameters in calculating heat penetration are the surface heat transfer coefficient and thermal diffusivity. A key issue is to reduce the degree of error (Taguchi, 1989). We have depicted the characteristic root vs. Biot number in Fig. 15.10. The similar picture can be drawn for slab and plate geometry as well. We can see that at low Biot numbers, relatively large errors in determining the root from the heat penetration curves cause relative little error in surface heat transfer coefficient. With increasing Biot numbers, the degree of error increases, even to 20% error in determination of surface heat transfer coefficient (Uno and Hayakawa, 1980). It can be also seen that very large errors

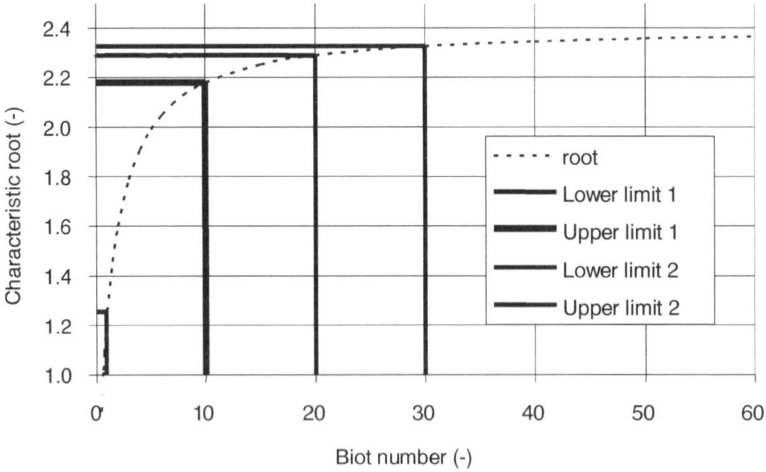

Fig. 15.10 The dependence of the characteristic root from Biot number for cylinder at constant ambient temperature.

in determining the characteristic root cause only 10 Biot number differences at limit 1. As a result, in determining the surface heat transfer coefficient, we have to set parameters resulting in low Biot numbers. In determining thermal diffusivity the lowest error is in β resulted from high Biot number. This is true for the upper part of Fig. 15.10. We therefore need the opposite setting of Biot numbers compared to the surface heat transfer coefficient determination. If feasible, we have to apply the lumped capacity method for surface heat transfer coefficient measurement (come up time). In the case of finite surface heat transfer coefficients and Biot numbers, we have to change the dimension of the body because the thermal conductivity is determined mainly by the composition of the food, whilst the equipment determines the range of surface heat transfer coefficient.

In determining thermal diffusivity accurately, we must also decide whether the Biot number or slope index determination is the more important factor if we use the Ball method. In Fig. 15.11 we show the resulting error in thermal diffusivity if we made a 10% error in the Biot number and 5% in the Ball slope index. We can see that the error in slope index determination results in higher error in thermal diffusivity than error resulting from the surface heat transfer coefficient/biot numbers. Some studies involve an error 5–13% or even 20% in the calculation of thermal diffusivity (Lund, 1978; Singh, 1982; Smout *et al.*, 2000). Larkin and Steffe (1982) have proposed using the $0.15 < Y < 0.85$ values of the curves to decrease the standard deviation in determination of thermal diffusivity. In parallel, we can also use robust regression methods. These methods disclose the points in the curve having the largest deviation. Roughly speaking, robust procedures are sufficiently insensitive to outliers and the slight departures from the assumptions of the applied stochastic model. The exact

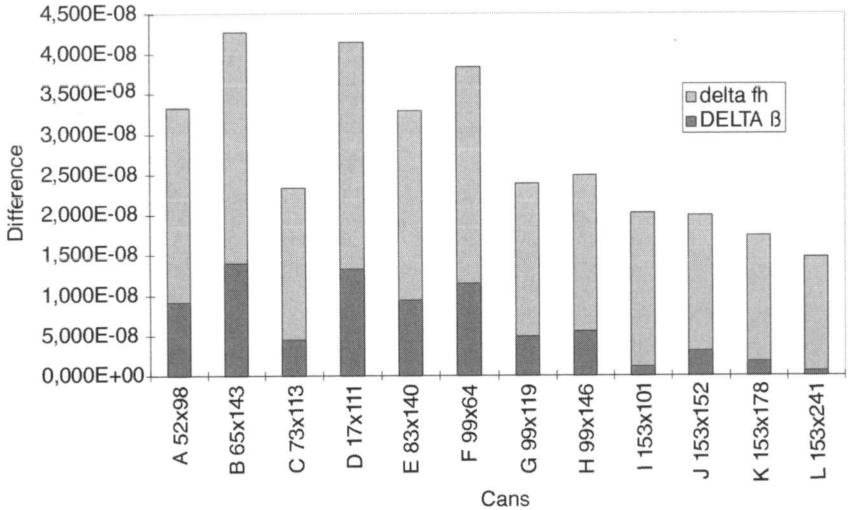

Fig. 15.11 Difference in the determination of thermal diffusivity due to the errors in characteristic root of β and Ball slope index for cans with 400 g filling weight.

definition of *qualitative* robustness for estimates is the following: *the estimation is robust if the distribution of estimates is a uniformly continuous functional of the mother distribution, i.e., near-lying distributions of estimates belong to near-lying mother distributions*. Robustness can be measured quantitatively in two different ways. The overall sensitivity of an estimator is measured by its breakdown point (this is often described as *quantitative* robustness), which is roughly the smallest fraction of outliers in the collection that can take the estimator over all bounds. It describes the overall behaviour of an estimator under large perturbations. The influence function (the *infinitesimal* robustness) is defined to measure robustness locally. It describes how the estimator alters under infinitesimal perturbations at some single point. It is seen that a robust estimator can treat the first-mentioned model error if it has a high breakdown point and a bounded, continuous influence function. The theoretical and practical properties of the used parameter estimators are shown in Table 15.10. Much more about robust methods and related literatures can be found in Rajkó (1994).

The differences between the classical least squares and robust regression estimation are shown in Fig. 15.12. It can be seen that the robust regressions gave smaller standard deviations of slope index. We have done the regressions for the infinite series solution as well for the same 117 curves of Fig. 15.12 and Pullman can. The thermal diffusivity values for the holding and cooling phase are summerised in Table 15.11. Table 15.11 shows that there is a significant difference between the holding and cooling phase relating to thermal diffusivity and can size (Figs 15.13 and 15.14).

Table 15.10 Theoretical and practical properties of the parameter estimators investigated

Methods	Derivations	Breakdown points	Robustness	Recommendations
CLS and ILS	min L_2	0%	Not robust at all	Recommended only for large data (30–50 measurements)
IRLS9	M-estimator	20%	Slightly robust	Recommended only for large data (15–30 measurements)
IRLS6	M-estimator	25%	Robust	
MFV	M-estimator	25%	Robust	
SM	Based on rank correlation	30%	Robust	Recommended for parameter estimation even for small data (8–15 measurements)
RM	Based on modified U-statistic	50%	Very robust	
LMS	Min M_2	50%	Very robust	

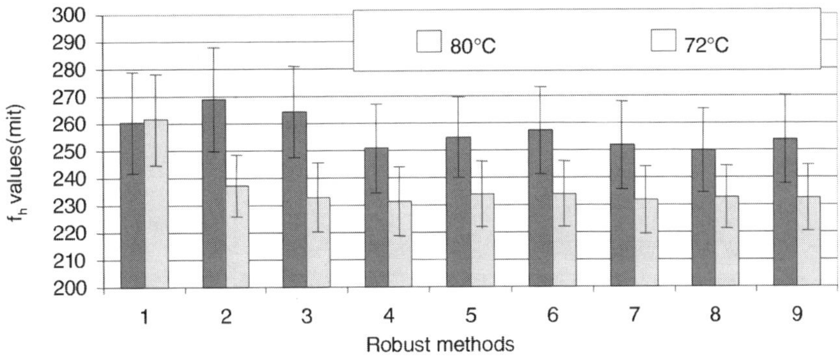

Fig. 15.12 Classical least squares and robust regression of Ball slope index (f_h) for 12 lb oblong cans (1. Subjective exclusion (SE) before CLS, 2. Classical Least squares (CLS), 3. Inverse least squares (ILS), 4. Iteratively reweighted least squares k = 6 (IRLS6), 5. Iteratively reweighted least squares k = 9 (IRLS9), 6. Simple median (SM) 7. Most frequent value (MFV), 8. Least median of squares (LMS), 9. Repeated Median (RM)).

Table 15.11 Thermal diffusivity values obtained with non-linearised least squares method for Pullman and oblong cans [m^2/s]

Packaging	Holding		Cooling	
	Average	Standard deviation	Average	Standard deviation
11 lb Pullman	1.404E-7	5.48E-9	1.161E-07	6.80E-09
12 lb oblong	1.305E-7	7.40E-9	1.130E-7	6.50E-9

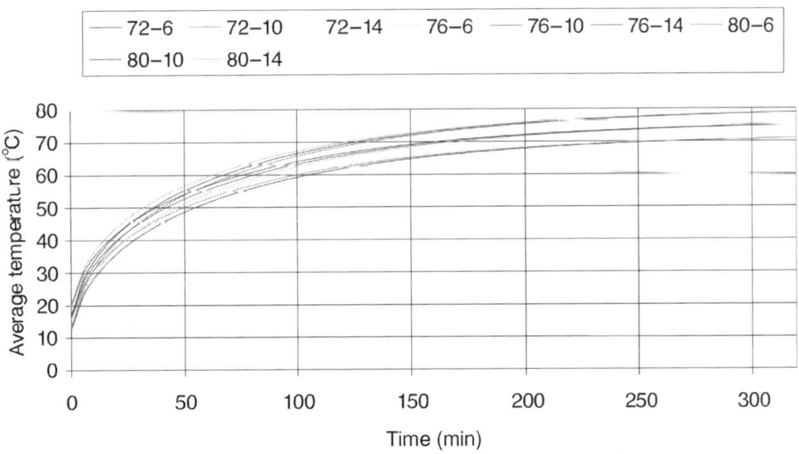

Fig. 15.13 Development of the average temperature in 12 lbs oblong cans for different initial and ambient temperature (thermal diffusivity = 1.3E-7m^2/s).

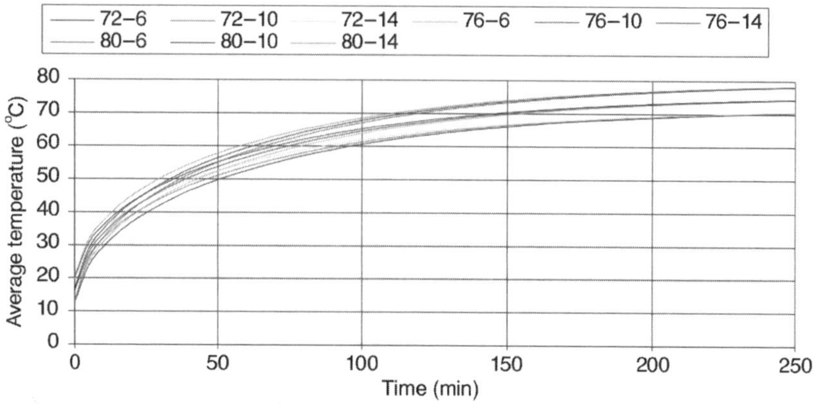

Fig. 15.14 Development of the averge temperature of 11 lb Pullman cans for different ambient and initial temperature (thermal diffusivity = 1.3E-7m^2/s).

We can see from this figure that the average temperature is about 55–60°C after an hour is holding time. The average temperature of the 11 lbs cans is about 10°C higher than the 12 lbs oblong cans. We can state that the average of the average temperature is between 60 and 70°C during the whole holding cycle. These phenomena are in agreement with the temperature depending calculation, based on composition, of thermal diffusivity. On the basis of the figures we can conclude that different ambient and initial temperatures give differing thermal diffusivity values from heat penetration curves, despite measuring the same material. The ambient temperature has a greater influence than the initial temperature.

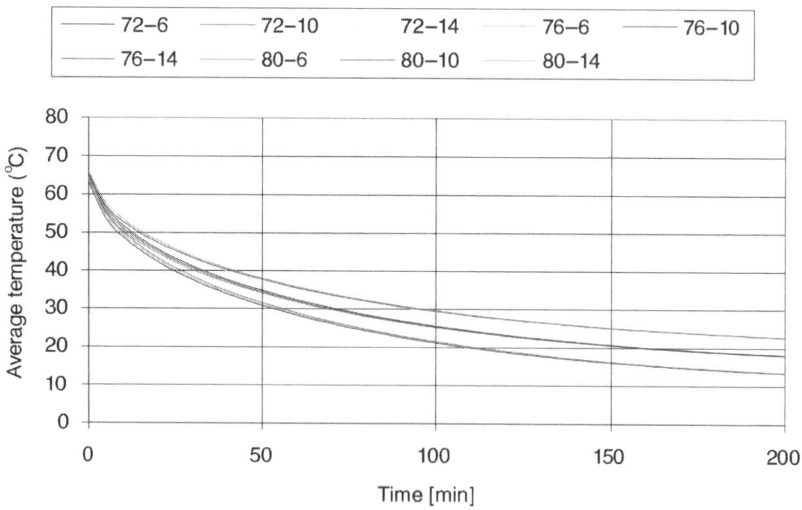

Fig. 15.15 Development of the average temperature in 12 lbs cans during cooling.

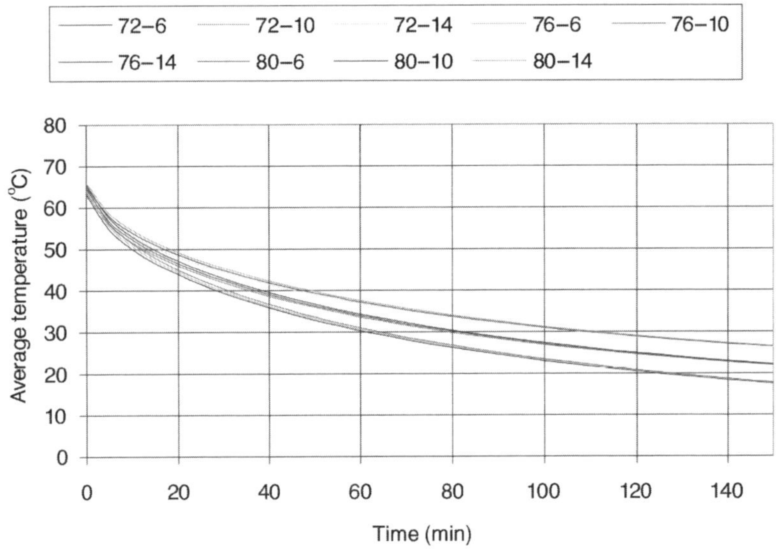

Fig. 15.16 Development of the average temperature in 11 lbs cans during cooling.

We have found a significant difference in thermal diffusivities between the holding and cooling phase. The average temperature develops 25–35°C at the end of cooling in the case of 12 lbs cans, depending on the initial temperature and cooling water temperature. This is lower for 11 lbs cans. Because we stop

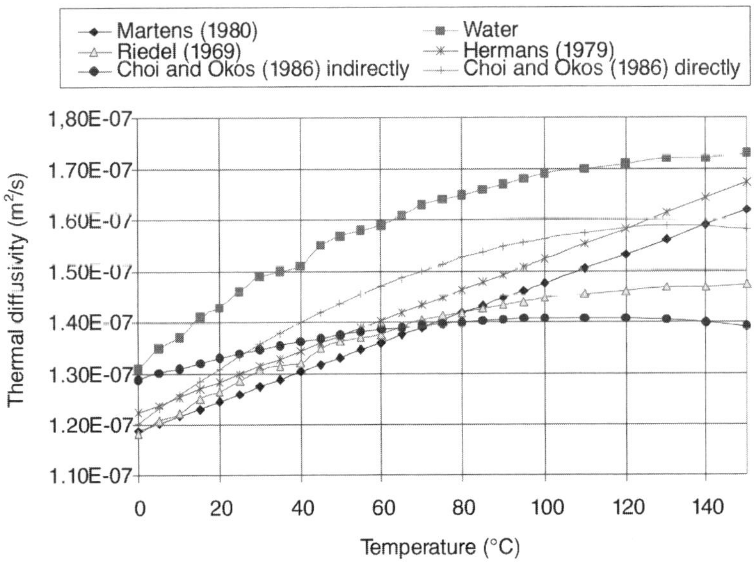

Fig. 15.17 The thermal diffusivity of water and hams in dependence of temperature.

heating at 40°C core temperature, the differences are not so high between the two cans as in the case of the holding phase. These are shown in Figs 15.15 and 15.16. Temperatures above 60°C during holding time and between 15–30°C during cooking time produce differences in thermal diffusivity of 1.3–1.4*10^{-7} m^2/s and 1.1–1.2*10^{-7} m^2/s respectively. If we consider that the water content of some heat-treated products (e.g. meat) is high, one explanation may be the thermal parameter of the water. Figure 15.17 shows the parameters of water. This values can be regarded as possible maximum, because the other constituents of meat products (fat, protein) have lower specific heat and thermal conductivity whilst protein has higher and fat has lower density. Altogether the thermal conductivity and thermal diffusivity is lower as the water content decreases. The changes in specific heat are low. The density decreases as the temperature increase. The agreement between the determination methods of chemical composition and heat penetration is only valid if the average temperatures are considered. If we consider that the curve fitting seeks for the best fit in time and space, and the ambient temperature influences the core temperature development in the body, this average value for thermal diffusivity from heat penetration curves is not surprising.

15.5 Future trends

As we have seen, the heat treatment schedules influence temperature development within the product and thus thermal diffusivity and surface heat transfer coefficient determination from heat penetration curves. We need therefore to turn to finite difference and finite element methods to find the temperature dependence of thermal properties of different materials. If we use the infinite series solution of transient heat conduction with average constant thermal diffusivity for core temperature, we have an acceptable degree of error because we reach or overstep the reference temperature of the equivalent sterilising time calculation only for short time. If we calculate the average and surface temperature, these differences are higher and we will be above the reference temperature for longer so the C values may differ significantly. In case of the numerical methods, the overestimation speeds up and the underestimation slows down the temperature development because the numerical methods calculate the temperature field from the preceding temperature field. Therefore we need to lower the determination error. Sensitivity analysis (e.g. Nicolai and Baerdemaeker, 1996) and the Taguchi method (Taguchi, 1989) is very useful to find the parameter constellation giving the lowest error in the determination. Nevertheless we need to investigate each case separately. The neural network method has been proposed (Chen and Ramaswamy, 2002) for complicated circumstances, e.g. the variable temperature heat treatments, non-regular forms, etc. With this method, we do not need to establish exact relationships for heat penetration but the results will relate only to the applied parameter range.

15.6 Acknowledgement

This work has been partly supported by the Hungarian Science Foundation (OTKA T-035125 and T-037480).

15.7 References

ADAMS J A, ROGERS D F (1973), *Computer Aided Heat Transfer Analysis*, McGraw-Hill, New York.

AWUAH G B, RAMASWAMY H S, SIMPSON B K (1995), 'Comparison of two methods for evaluating fluid-to-surface heat transfer coefficients', *Food Research International*, 28 (3), 261–271.

BALL C O, OLSON F C W (1957), *Sterilization in Food Technology*, McGraw-Hill, New York.

BHOWMIK S R, HAYAKAWA K I (1979), 'A new method for determining the apparent thermal diffusivity of thermally conductive food', *Journal of Food Science*, 44 (2), 469–474.

BHOWMIK S R, SHIN S (1991), 'Thermal sterilization of conduction-heated foods in plastic cylindrical can using convective boundary condition', *Journal of Food Science*, 56 (3), 827–842.

BHOWMIK S R, TANDON S A (1987), 'Method for thermal process evaluation of conduction heated foods in retortable pouches', *Journal of Food Science*, 52 (1), 202–209.

BURFOOT D, SELF K P (1988), 'Prediction of heating times for cubes of beef during water cooking', *International Journal of Food Science and Technology*, 23 (3), 247–257

CARCIOFI B A M, FAISTEL J, ARAGO M F G, LAURINDO J B (2002), 'Determination of thermal diffusivity of mortadella using actual process data', *Journal of Food Engineering*, 55 (1), 89–94.

CHANG S Y, TOLEDO R T (1989), 'Heat transfer and simulated sterilization of particulate solids in a continuously flowing system', *Journal of Food Science*, 54 (4), 1017–1030.

CHEN C R, RAMASWAMY H S (2002), 'Modelling and optimisation of constant retort temperature (CRT) thermal processing using coupled neural networks and genetic algorithms', *Journal of Food Process Engineering*, 25 (5), 351–379.

CHOI Y, OKOS M R (1986), 'Effects of temperature and composition on the thermal properties of foods', in Le Maguer M, Jelen P. *Food Engineering and process application, Volume 1. Transport Phenomena*, Elsevier Applied Science, London, 93–101.

DICKERSON R W (1965), 'An apparatus for the measurement of thermal diffusivity of foods', *Food Technology*, 19 (5), 198–204.

DICKERSON R W, READ R B JR (1968), 'Calculation and measurement of heat transfer in foods', *Food Technology*, 22 (12), 1533–1548.

DICKERSON, R W, READ R B JR (1975), 'Thermal diffusivity of meats', *ASHRAE Paper*, No. 2349. 356–364.

DURANCE T D (1997), 'Improving canned food quality with variable retort temperature processes', *Trends in Food Science and Technology*, 8 (4), 113–118.

FLAMBERT C M F (1973), *La conduction de la chaleur dans les produits alimentaires solides et son application aux calculs de thermobactériologie*, PhD Thesis, Fac. Univ. Sci. Agron., Gembloux, Belgium.

GAFFNEY J J, BAIRD C D, ESHLEMANN W D (1980), 'Review and analysis of the transient method for determining thermal diffusivity of fruits and vegetables', *ASHRAE Transactions*, 86 (2), 261–280.

HAYAKAWA K I (1968), 'A procedure for calculating sterilizing value of a thermal process', *Food Technology*, 22 (7), 905–907.

HAYAKAWA K I, BAKAL A (1971), 'New computational procedure for determining the apparent thermal diffusivity of a solid body approximated with an infinite slab', *Journal of Food Science*, 38 (3), 623–629.

HAYES G D (1987), *Food Engineering Data Handbook*, Wiley, New York.

HERMANS F (1979), 'The thermal diffusivity of foods', cited in Jowitt R, Escher F, Hallström B, Meffert H F Th, Spiess W E L, Vos G (1983), *Physical Properties of Foods*, Applied Science Publishers, London, 269–312.

KÖRMENDY I, KÖRMENDY L (1997), 'Considerations for Calculating heat inactivation processes when semilogaritnómic thermal inactivation models are non-linear', *Journal of Food Enginering*, 34 (1), 33–40.

LARKIN B J, STEFFE J F (1982), 'Error analysis of heat penetration data for determination of thermal diffusivity', *Journal of Food Processing and Preservation*, 6, 135.

LARKIN B J, STEFFE J F (1987), 'Experimental errors associated with the estimation of thermal diffusivity from thermal process data', *Journal of Food Science*, 52 (2), 419–428.

LEBOWITZ S F, BHOWMIK S R (1989), 'Determination of retortable pouch heat transfer coefficients by optimization method', *Journal of Food Science*, 54 (6), 1407–1412.

LEBOWITZ S F, BHOWMIK S R (1990), 'Effect on retortable pouch heat transfer coefficients of different thermal processing stages and pouch material', *Journal of Food Science*, 55 (5), 1421–1424, 1434.

LEONHARDT G F (1976a), 'Estimation of the central temperature of thermally conductive food in cylindrical and rectangular cans during heat processing', *Journal of Food Science*, 41, 685–690.

LEONHARDT G F (1976b), 'Estimation of the average temperature of thermally conductive food in cylindrical and rectangular cans during heat processing', *Journal of Food Science*, 41, 691–695.

LUNA J A, BRESSAN J A (1987), 'Mass transfer during ripening of Cuartilo Argentino cheese', *Journal of Food Science*, 52 (2), 308–311.

LUND D B (1978), 'Statistical analysis of thermal process calculations', *Food Technology*, 32 (3), 76–78

MAFART P, LEGUERNEL I (1998), 'Modeling combined effects of temperature and pH on heat resistance of spores by a linear-Bigelow Equation', *Journal of Food Science*, 63 (1), 6–8.

MARTENS T (1980), *Mathematical model of heat processing in flat containers*, Ph.D. Thesis, Katholieke Universiteit te Leuven, Faculty of Agricultural Sciences, Laboratory of Food Technology, Leuven, Belgium.

MILES C A, VAN BEEK G, VEERKAMP C H (1983), 'Calculation of thermophisical properties of Foods', in Jowitt R, Escher F, Hallström B, Meffert H F Th, Spiess W E L, Vos G (1983), *Physical Properties of Foods*, Applied Science Publishers, London and New York, 269–312.

NESVADBA P (1982), 'A new transient method for measurement of temperature dependent thermal diffusivity', *Journal of Physics D. Applied Physics*, 15, 725–738.

NICOLAI B M, DE BAERDEMAEKER J (1996), 'Sensitivity analysis with respect to the surface heat transfer coefficient as applied to thermal process calculations', *Journal of*

Food Engineering, 28 (1), 21–33.

PÁTKAI GY, KÖRMENDY M, ERDÉLYI M (1990), 'Outline of a system for the selection of the optimum sterilization process for canned food II', *Acta Alimentaria*, 19 (4), 305–320.

PELEG M (2000), 'Microbial survival curves the reality of flat shoulders and absolute thermal death times', *Food Research International*, 33, 531–538.

PELEG M (2002), 'A model of survival curves having an "Activation shoulder"', *Journal of Food Science*, 67 (7), 2438–2443.

PETERSON W R, ADAMS J P (1983), 'Water velocity effect on heat penetration parameters during institutional retort pouch processing', *Journal of Food Science*, 48 (3), 457.

PFLUG I J, BLAISDELL J L, KOPELMAN J (1965), 'Developing Temperature-Time curves for objects that can be approximated by a sphere, infinite plate or infinite cylinder', *ASHRAE Paper*, No. 1935, 238–247.

POULSEN K P (1982), 'Thermal diffusivities of foods measured by simple equipment', *Journal of Food Engineering*, 1, 115–122

RAJKÓ R (1994), 'Treatment of model error in calibration by robust and fuzzy procedures', *Analytical Letters*, 27(1), 215–228.

RAMASWAMY H S, LO KV, TUNG M A (1982), 'Simplified equations for transient temperatures in conductive foods with convective heat transfer at the surface', *Journal of Food Science*, 47 (6), 2042–2065.

RAMASWAMY H S, TUNG M A, STARK R (1983), 'A method to measure surface heat transfer coefficient from steam air mixtures in batch retorts', *Journal of Food Science*, 48 (4), 900–904.

REICHERT J E (1985), *Wärmebehandlung von Fleischwaren*, Hans Holzman Verlag GmbH & Co., Bad Wörishöfen.

RIEDEL L (1969), 'Temperaturleitfähigkeitsmessungen an wasserreichen Lebensmitteln', *Kältetechnik und Klimatisierung*, 21 (11), 315–316.

RIZVI S S H, BLAISDELL J L, HARPER W J (1980), 'Thermal diffusivity of model meat analog systems', *Journal of Food Science*, 45 (6), 1727–1731.

ROGACSEV V I, BABARIN V P (1983), *A konzervipar folyamatos hőkezelő berendezései [in Hungarian]*. Mezőgazdasági Könyvkiadó, Budapest.

SINGH R P (1982), 'Thermal diffusivity in food processing', *Food Technology*, 2, 87–91.

SMOUT C, VAN LOEY A, HENDRICKX M, BEIRLANT J. (2000), 'Statistical variability of heat penetration parameters in relation to process design', *Journal of Food Science*, 65 (4), 685–693.

STUMBO C R (1973), *Thermobacteriology in Food Processing. 2nd ed.*, Academic Press, New York.

TAGUCHI G (1989), *Introduction to Quality Engineering*, Asian Productivity Organization, Tokyo

TUNG M A, MORELLO G F, RAMASWAMY H S (1989), 'Food properties, heat transfer conditions and sterilisation considerations in retort processes', in Singh R P and Medina A G (eds), *Food Properties and Computer Aided Engineering of Food Processing Systems*, NATO ASI Series E: Applied Sciences, Vol. 168, Kluwer Academic Publishers Dordrecht, The Netherlands, 49–71.

UNO J I, HAYAKAWA K I (1980), 'A method for estimating thermal diffusivity of heat conduction food in a cylindrical can', *Journal of Food Science*, 45 (3), 692–695.

16

Validation of heat processes: an overview

G. S. Tucker, Campden and Chorleywood Food Research Association, UK

16.1 Introduction: the need for better measurement and control

This chapter provides an update to that in the first book on Thermal Processing Technologies. In order that the chapter is a complete guide to heat process validation, the chapter structure remains unchanged and some content is repeated out of necessity. Section 16.2 sets out the aims and objectives of process validation, which includes methods based on temperature measurement, microbiological techniques and other more novel methods such as using time-temperature integrators. These methods are treated in more detail in subsequent sections.

Sections 16.3 and 16.4 of this chapter deal with temperature measurement systems, which should be, and are likely to remain as, the first option for the validation method chosen by most companies. Section 16.3 describes temperature distribution testing where the aim is to prove that the processing system delivers heat to the product in a uniform way. Section 16.4 describes heat penetration testing where probes are inserted into the food as a means of measuring temperatures that are converted to integrated process values. The methodology for temperature distribution and heat penetration testing evolved in the early part of the 1900s for the canned foods sector where full sterilisation processes in metal cans were the norm.[1] In recent years, the temperature measurement hardware and analysis software have developed to allow measurements to be made for the plethora of pasteurisation treatments[2] that have emerged in response to the consumer demands for less processed foods.

Sections 16.5 and 16.6 in this chapter consider microbiological and biochemical methods for process validation, to illustrate where they can be best applied. These methods are important where trailing wires and dataloggers would interfere with the measurements; for example, continuous in-line processes for foods containing particulates or continuous oven and fryer processes (intended for products stored under refrigeration). Alternatives to probes are needed, such as immobilised spores in alginate beads, encapsulated biochemical time-temperature integrators or microbiological challenge tests. Emphasis is placed on time-temperature integrators because they represent an emerging technique with widespread applications.

16.2 Validation methods: objectives and principles

As the number and variety of manufactured food products increases, food companies are faced with the challenge of proving that these products are safely pasteurised or sterilised. This can sometimes be difficult if conventional temperature probe systems cannot be used and other more complex approaches need to be adopted. The main product categories that introduce these complexities include products cooked in continuous ovens or fryers (e.g. poultry joints, chicken nuggets, burgers, bread) and products with discrete pieces cooked in steam-jacketed agitated vessels (e.g. ready meals, soups, cook-in-sauces, fruit preparations) or in heat exchangers (e.g. cook-in-sauces, preserves, dressings). If an alternative to temperature probes is needed, the following approaches to validating microbiological process safety are the options available to prove the microbiological process safety:

- Microbiological methods can be used whereby cells or spores of a non-pathogenic organism, with similar temperature-induced death kinetics to the target pathogen, are embedded into an alginate bead. The beads are made to mimic the food pieces in their thermal and physical behaviour and so pass through the process with the food. Enumeration of the surviving organisms allows the log reduction and process value to be calculated. Section 16.4 deals with microbiological methods.
- Simulated trials are carried out in a laboratory where the heat transfer conditions of the process are replicated and temperature measurement is feasible.
- No validation is attempted, with the process safety being inferred from temperature probing of the bulk product or the environment. Substantial over-processing is allowed, in order that the thermal process delivered to the product thermal centre is sufficient. End product testing for microbiological activity is usual. This approach will not feature in this chapter.
- Process models can be developed that predict, for example, the temperature-time history of the critical food particles as they travel through the heating, holding and cooling zones of the process. Such modelling issues are dealt with elsewhere in this book.

- Time-temperature integrators (TTIs) can be applied to gather similar process data to that from microorganisms. Section 16.5 deals at length with TTIs because this method is one of the most exciting to have emerged in recent years.

One of the core activities involved with establishing a thermal process is the selection of 'worst case conditions' likely to be experienced during normal production. The current thought process used by most food companies is to validate the microbiological process safety under worst case conditions so that, by default, the process will be safe under normal production conditions.

16.2.1 What are 'worst case conditions'?

It has been stated above that the process validation study should be conducted using worst case conditions. Therefore, by inference it should not be possible for a normal production batch to heat more slowly than the combination of factors evaluated as worst case. To determine the worst case conditions it is necessary to consider the product, process and package separately. The following lists suggest the factors that should be addressed in a process validation study, although the lists are not intended to be exhaustive.

Product factors

Formulation; weight variation in ingredients, e.g. high starch levels that could lead to increased viscosity.
Fill weight; percent overfill of the key components, e.g. solids content.
Consistency or viscosity of the liquid components; before and after processing.
Solid components; size, shape and weight before and after processing, potential for matting and clumping.
Preparation methods; e.g. blanching.
Rehydration of dried components.
Heating mode; convection, conduction, mixed or broken heating.

Container factors

Type; metal cans, glass jars, pouches, semi-rigid containers.
Nesting of low profile containers; geometry of the divider plates.
Vacuum and headspace; residual gases with flexible containers.
Orientation.
Fill method; initial temperature and the effects of delays in getting instrumented containers into the retort
Symmetry of rotation.

Retort or processing system factors

Type; steam, steam/air, water immersion, raining water. Venting schedule if steam. Overpressure profile if water or steam/air.
Retort come-up time; this should be as short as possible to minimise the quantity of heat absorbed by the product during this phase.

Racking and dividing systems.
Rotation; slowest heating position usually along the retort axis.

The combination of conditions to arrive at the worst case scenario should be determined by a competent individual, sometimes referred to as a thermal process authority. It does not matter which validation method is used (probes, microbes or biochemical), the thought process for product, package and process conditions should be the same.

16.2.2 Setting the target process value

The next stage will be to measure or validate the thermal process to ensure that the target process value is achieved. Commercial sterility for a thermally processed product is the target condition, which depends on the types and numbers of organisms present before and after the process, and on the intended storage conditions. When the process value (F) is divided by the decimal reduction time (D_T) this gives the number of log reductions of surviving spores (see equation 16.1):

$$F = D_T \cdot \log\left(\frac{N_{\text{initial}}}{N_{\text{final}}}\right) \qquad [16.1]$$

where, N_{final} is the final number of organisms after a specific time-temperature history, N_{initial} is the initial number of organisms, D_T is the decimal reduction time at a fixed temperature (T) to reduce the number of organisms by a factor of ten, min.

For example, a '12D cook' is often quoted as the commercially acceptable minimum for *Clostridium botulinum* spores in ambient stable products, which assumes a starting position of 100 viable spores per container (N_0) and a final container with 10^{-12} viable spores. In fact, this actually equates to a probability of 1 processed container in 10^{12} containing a viable spore (N). The 12D minimum botulinum cook is quoted as F_0 3, which is calculated using a $D_{121.1}$ of 0.21 minutes and rounding up the F-value to the nearest integer. In commercial thermal processing practice, however, it would be common to operate at substantially increased safety margins, which would equate to log reductions upwards of 25. This will result in a fully-sterilised food product.

Pasteurisation processes on the other hand are usually operated to only 6 log reductions of the target organism, presumably because of the less lethal nature of the organisms when compared with *Clostridium botulinum*. Further details on pasteurisation treatments can be found in CCFRA.[2] The method for calculating the process values is similar to that for the sterilised products, and utilises equation 16.1.

Having calculated the target process value from data on the heat resistance of the target pathogenic or spoilage organisms, the thermal process achieved in the food containers is validated using temperature or other appropriate in-process measurements. These methods are discussed in subsequent sections.

16.3 Validation based on temperature measurement

For an in-container process, there are two main stages in a process validation exercise; *temperature distribution or TD tests* to identify the location of the zone of slowest heating in the retort, and *heat penetration or HP tests* to measure the temperature response at the product cold point. For a continuous flow process, a specialised form of the HP test is required, with various assumptions made regarding the positions of temperature measurement that make the TD test redundant. Some of the methods used for continuous flow processes are common with those for in-container processes and are described in this chapter.

16.3.1 Temperature distribution: locating the retort cold point

Any thermal processing system (retort, autoclave or steriliser) will contain regions in which the temperature of the heating medium is lower than that measured by the master temperature indicator (MTI). The location of these cold spots should be determined by performing 'temperature (or heat) distribution' (TD) tests throughout the system. The concepts for TD testing are simple; however the practicalities of making the relevant measurements are fraught with difficulty. For example, regions of low temperature may exist within a retort crate because of the flow restrictions imposed by the close packed containers; but to get at a container near the crate centre it is necessary to re-pack the crate manually and trail thermocouple wires between containers, which may open up the flow channels.

Temperature distribution (TD) tests in the heating medium are the first stage in a heat penetration study. However, a uniform TD throughout the retort does not necessarily imply that the lethalities delivered to the product are also uniform, since uniform temperature does not guarantee uniform heat transfer. Therefore, the uniformity in temperature is the minimum that has to be studied and an additional heat distribution study is advisable if there are concerns about air entrapment or heat transfer coefficient reductions throughout a container load. Rotary retorts using excessive air overpressure are one example where the potential exists for air to collect at the crate centres.

For a steam retort, if the TD is unsatisfactory this can normally be resolved simply by increasing the length of the period of air removal at the start of the process (venting). This contrasts with non-steam retorts where a large temperature range may be attributed to the design/loading of the retort and simple corrective action is not possible, so testing can only provide the relationship between the retort instrumentation and the retort interior but does not necessarily improve it.

The TD within a retort should be tested on its installation, with intermittent re-testing being required as factors change that could affect the retort performance. Retorts require, as a minimum, re-testing in the event of any engineering work likely to affect the TD of the retort, such as:

- Relocation of the retort or installation of another retort that uses the same services.

- Modification to the steam, water or air supply.
- Failure of the key components (for example pumps and valves).
- Repair or modification to water or steam circulation systems within the retort.
- If there are any doubts about the performance of the circulation system.

In addition, if the load to be processed in a retort changes, re-testing is required. Such circumstances include the use of:

- New container sizes and shapes.
- New container loading patterns.
- New crate or layer pad design, or mode of use.

It is also necessary to ensure that a retort's performance does not deteriorate over a period of time, as corrosion or fouling in the steam, water or air supply pipes builds up. Retort instrumentation and process records should be inspected regularly to identify when a TD problem has arisen. Regular re-testing of a retort's TD is good practice to ensure that these faults are not overlooked.

If the retort uses condensing steam as the medium, it is necessary to establish the time of vent in order that the distribution of temperatures across the retort is reduced to an acceptable limit. For venting trials, the following guidance is derived from CCFRA:[8]

- Note precisely the time at which the retort reaches 100ºC.
- Do not close the main vent until all thermocouples reach the same temperature within 0.5 Cº.
- Close the vent and record when the master temperature indicator and chart recorder reach process temperature.
- All thermocouples should indicate the same temperature within 1 minute of the first thermocouple indicating that (process) temperature.
- Record the venting time as the number of minutes for which the main vent was left open after 100ºC was reached on the thermocouple in the thermometer pocket.

The vent test is specific to condensing steam retorts and is required in addition to TD testing. For retort systems utilising water or mixtures of steam and air, the TD tests are unlikely to result in a 1 Cº distribution in temperatures across the crates at the start of the hold phase. This is because of less favourable heat transfer coefficients with these heating media when compared with condensing steam and also the reduced quantity of heat available in for example a raining water system. It is common practice to quote a time into the hold phase by which the temperature distribution has stabilised to within 1 Cº, and to take this into account when establishing the hold time at constant temperature.

Although TD tests basically sample the performance of retort systems, in industry they are frequently used as an opportunity to 'audit' the installation to ensure long-term compliance. There are no UK regulations of general application to retort processing, although there are specific regulations, derived from EC directives, which apply to the heat processing of certain products.

These contain some implications for thermal processing of these products (e.g. milk, milk products, egg products, meat and fish products). Food manufacturers producing products not covered by these specific regulations must comply with the Food Safety Act (General Food Hygiene Regulations) 1995, implementing EC Directive 931431EC.

The Good Manufacturing Practice guidelines for TDs in batch retorts as defined in DH,[7] section 10.3.4, are as follows: 'In steady state operation, the temperature spread across the sterilising vessel should ideally be 1 C° or less. However, when this degree of control is not achievable due to design or characteristics of the equipment, any deviation from the limit should be allowed for in the scheduled process.'

16.3.2 Heat penetration: establishing times and temperatures

The aim of a heat penetration (HP) study is to determine the heating and cooling behaviour of a specific product in order to establish a safe thermal process regime and to provide the data to analyse future process deviations. Design of the study must ensure that all of the critical factors are considered to deliver the thermal process to the product slowest heating point. The numbers of instrumented sample containers and replicate retort runs have been subject to much discussion (CCFRA;[4] NFPA;[5] IFTPS[9]), with the final decision linked to the measured variability between samples and between runs.

Modern datalogging systems can provide the facility for taking multiple temperature measurements, therefore large quantities of data can be taken more easily than was the case when the CCFRA guidelines were written in 1977. These recommendations were for three samples in three replicate runs, providing a total of nine measurements. The more common situation now is to take up to ten samples in two replicate runs, providing that the variability between runs is within acceptable limits. However, there can be limitations on the number of probes that can be inserted through a packaging gland or through the central shaft of a rotating system, and in these situations at least two replicate runs should be completed.

The HP study should be carried out prior to commencing production of a new product, process or package. Changes to any of the criteria that may change the time-temperature response at the product slowest heating point will require a new HP study to be conducted. The conditions determined in the study are referred to as the scheduled heat process and must be followed for every production batch, with appropriate records taken to confirm that this was followed. No further temperature measurement within containers is required in production, although some companies do measure temperatures in single containers at defined frequencies. However, the conditions used in single container testing will not represent the worst case, and it would be expected that the instrumented container would show a process value in excess of that measured from the HP study. Such data are intended to show due diligence and are more of a comfort factor.

A HP test is usually sub-divided into two further stages when conducting the tests, firstly to locate the product cold point in the container, and secondly to establish the process conditions that will lead to the scheduled process.

Locating the product cold point: Within each food container there will be a point or region that heats up more slowly than the rest. This is referred to as the 'slowest heating point' or 'thermal centre' and should be located using thermocouples or some other sensing method positioned at different places in a food container. For foods that heat mainly by conduction, the slowest heating point will be at the container geometric centre. However, for foods that permit movement and can thus convect heat, this point is between the geometric centre and approximately one-tenth up from the base (in a static process). During a thermal process the food viscosity will decrease in response to increasing temperature, and as a result the slowest heating point will move downwards from the container geometric centre. The critical point is that when the lethal effect on the target microbiological species is at its most significant, which will be towards the end of the constant temperature hold phase. If the process utilises rotation or agitation, the slowest heating point will be at the container geometric centre.

Establishing the scheduled process time and temperature: The thermal process is finally established by measuring the temperature at the container slowest heating point for a number of replicates that are placed in the cold spot(s) of the thermal processing system. The data obtained are usually referred to as 'heat penetration' data. A point open to discussion is the number of replicates required for confidence in the data. In general, this depends on the variability between data sets, with CCFRA[4] recommending 3 sensors from each of 3 replicate runs, and NFPA[5] suggesting at least 10 working sensors from a run with replicate runs required where variability is found. Further details on heat penetration testing are given in section 16.4.

16.3.3 Process establishment methods

The need for caution in process establishment can be reinforced by reminding ourselves that the calculation of an integrated process value from heat penetration data requires the use of three mathematical models. Each model introduces an error in the calculated process value, thus the potential clearly exists for these errors to be additive. The models are for:

- Decimal reduction time (or D_T value) for the target organism, which is calculated by using first-order death kinetics to model the spore survivor curves. The D_T value is the time (usually in minutes) which is required at a constant heating temperature to reduce the numbers of surviving spores by a factor of ten.
- Kinetic factor (or z value) for the target organism, which is calculated by using a first-order relationship to model the relationship between the D_T value and heating temperature. The z value is a measure of the relative 'killing

power' of the heating temperature, and is the temperature difference required to effect a ten-fold change in the D_T value.

- Integrated Lethal Rate, usually referred to as the F_0 value (or P-value, Pu), which is calculated by integrating the area beneath the curve obtained when the lethal rates are plotted against time. It is usual to employ the trapezoidal rule to effect this integration, which is one further model with an associated error. The calculation procedure is referred to as the General method and derives from Bigelow *et al.*[6] Process values calculated using the General method should not be considered exact values due to the need for using the three models outlined above, but as estimates and therefore quoted to 1 decimal place (d.p.).
- F-values (sterilisation) or P-values (pasteurisation) can also be calculated by integrating the killing power of a thermal process over the time-temperature history experienced by the product. An F-value calculated using equation 16.1 will be the same as that calculated from the time-temperature integration, provided that first-order kinetics have been followed for the microorganism's destruction by heat (see Equation 16.2).

$$P \text{ or } F = \int_0^t 10^{(T-T_{\text{ref}})/z} \cdot dt = D_T \cdot \log\left(\frac{N_{\text{initial}}}{N_{\text{final}}}\right) \qquad [16.2]$$

where T is the product temperature, °C
T_{ref} is the reference temperature for the D_T value, °C
t is the process time, min
z, the kinetic factor, is the temperature change required to effect a ten-fold change of the D_T value (C°)

This integration is usually done within the datalogger software to allow a thermal process to be operated until the target F- or P-value is reached. This method would be referred to as the General method. Other methods that use the measured times and temperatures within predictive models are also acceptable and used widely. The most common ones being the Ball method and numerical methods using finite differences, for example with CTemp (CCFRA) and NumeriCAL (FMC FoodTech). These give flexibility to evaluate the process for different input variables such as product initial temperature, process temperature and process time.

16.3.4 Instrumentation for TD and HP testing

Modern dataloggers are typically multi-channel systems with digital outputs allowing data to be recorded directly to a laptop PC for display and to maintain permanent records. Thermocouples based on type T (copper/constantan) with PTFE insulation are most common because they are inexpensive, accurate over the desired temperature range, and respond rapidly to changing temperature. Other types, based on a change in electrical resistance with temperature, such as thermistors and platinum resistance thermometers (PT100), are used in dataloggers where the logging unit is remote (e.g. Ball Datatrace and Ellab Tracksense). These are referred to as resistance temperature detectors, or RTDs.

Calibration of a temperature sensor against a traceable instrument is essential each time it is used in a set of TD or HP trials. This can be achieved using the master temperature indicator on the retort (MTI) which must be calibrated at no less than six-monthly intervals.[7]

16.4 Validation based on microbiological methods

Microbiological spore methods are often referred to as direct methods, but they in fact rely on measuring the achieved log reductions for a process using a non-pathogenic microorganism and converting this to a process value for the target pathogen using equation 16.1. For example, in a sterilisation process where the target was *Clostridium botulinum* spores, the marker organism could be spores of *Bacillus stearothermophilus* which are reported to have similar death kinetics, or specifically the kinetic factor or z-value.

It is critical that the z-value is close to that for the target microbial species, otherwise significant errors in the calculated process values can be introduced.[10,11] If the z-value is not the same as the target value, the processing temperature should be close to the reference temperature, otherwise significant errors can arise between values estimated with organisms and probes. Also, the decimal reduction time should allow sufficient log reductions to be measured in order that the process can be correctly calculated from equation 16.2. A test that results in no surviving spores does not allow the process F-value to be calculated, and raises doubt as to where or when the total kill occurred. A microbiological method can be conducted using organisms distributed evenly throughout a food product or concentrated in small beads.

16.4.1 Inoculated containers

This method is also known as the count reduction method and involves inoculating the entire food with organisms of known heat resistance. It is essential that some organisms survive the heat process in order that the containers can be incubated and the surviving organisms counted. The average thermal process received by a container can be calculated using equation 16.1. If the product is liquid it is relatively easy to introduce the organisms, but for solid products it is necessary to first mix the organisms in one of the ingredients to ensure that they are dispersed evenly throughout the container.

Few studies on sterilisation processes use *Clostridium botulinum* spores because of the hazards associated with their handling, but also due to the very low number of surviving spores that would result from a commercial process. The theory of a sterilisation process is that 3 minutes equivalent at 121.1°C will result in at least 12-log reductions in spores. Almost all commercial processes operate to safety margins many times larger than 3 minutes, so it would be impractical to incubate the vast numbers of containers required to find the surviving spores. Incubation and testing of full production runs would be required, with little chance of finding the surviving spore. Hence, a non-pathogenic organism with a high $D_{121.1}$ value is used, such as spores of *Clostridium sporogenes* or *Bacillus stearothermophilus*. Typical levels of the inoculum are between 10^3 and 10^5 spores per container. An alternative is to use a gas-producing organism and estimate the severity of the process by the number of blown cans.

16.4.2 Encapsulated spores or organisms

This method allows the organisms to be placed at precise locations within a container or within the food particulates, by encapsulating known numbers of spores in an alginate bead.[12] The alginate bead can be made up with a high percentage of the food material so that the heating rate of the bead is similar to the food. This method has been used for continuous processes where the food contains particulates that require evaluation at their centres, and conventional temperature sensing methods cannot be used. Large numbers of alginate beads are used to determine the distribution of *F*-values that can occur in continuous processes as a result of the distribution of particle residence times.[13]

Estimating the exact number to use in a test is not straightforward because it depends on the *F*-value distribution; which is not known until after the test is conducted and the results analysed. The number of organisms used will be greater than for an inoculated container test and can be of the order of 10^6 per bead, but it is also important that not all are destroyed by the heat process otherwise it is not possible to estimate a process value using equation 16.1. If there are no surviving organisms then it is only possible to conclude that the process achieved greater than 6 log reductions for the example of a 10^6 initial loading. In this situation, there will be uncertainty as to whether the organisms died as a result of the process, during transportation to or from the factory, or if the spores germinated during the come-up time making them more susceptible to destruction at milder temperatures than for the heat resistant spores. Hence, controlling how these tests are performed is critical and the expertise to conduct a test using encapsulated spores or organisms tends to be restricted to a limited number of microbiology laboratories.

16.5 Validation based on biochemical time-temperature integrators

The use of time and temperature integrators (TTIs) as an alternative means of process evaluation, to either temperature or microbial systems, has received considerable attention recently.[10–11,14–15] A TTI can be an enzyme, such as amylase or peroxidase, that denatures (an unwinding of the structure) as it is heated. If the reaction kinetics of the temperature-induced denaturation match

Table 16.1 Key attributes of the *Bacillus amyloliquefaciens* α-Amylase TTI

Category	Description
Operating principle	Reduction in amylase activity in response to time and temperature
Measurement method	Amylase assay to measure absorbance rate, using a spectrophotometer
Active temperature range	60–100 (°C)
Kinetic factor, or *z*-value	9.7±0.3 (C°)
Decimal reduction time	$D_{80.7}$ = 18.7 (min)
Process value	'pasteurisation-value'
Sample size	0.02 (mL)

those of the microbial death kinetics for the target species, it is possible to use such TTIs as non-biological markers of a process.

Table 16.1 presents the key attributes of one such system, an amylase from *Bacillus amyloliquefaciens*[16] that has suitable kinetics for estimating pasteurisation, or *P*-values, in processes. This TTI has been used recently by Tucker and Wolf[21] for studies on fruit processing and poultry cooking where the target processes were within the measurement range. The heat denaturation rate of this amylase is minimal at ambient temperatures, simplifying transportation of the amylase TTI from the laboratory to the factory.

P-values for TTIs are calculated from the initial and final activities using equation 16.3, which assumes that the first-order thermal death time model applied. This is the same equation for calculating integrated process values when using microbiological spore methods except that the number of organisms is replaced by the amylase activity.

$$P = D_T \cdot \log\left(\frac{A_{\text{initial}}}{A_{\text{final}}}\right) \qquad [16.3]$$

where A_{final} is the final activity after a specific time-temperature history
A_{initial} is the initial activity
D_T is the decimal reduction time to achieve a 1-log reduction in amylase activity (minutes)

Some of the industrial tests reported by Tucker and Wolf[21] are given in this section, to illustrate the scope for using TTIs to validate thermal processes. Encapsulating the TTI is a key step that prevents it coming into contact with the food or processing environment. Their TTI encapsulation method used lengths of silicone tubing of 2.0 or 2.5 mm bore with the liquid TTI sealed by silicone end plugs. Many of the TTI trials required the tubes to be made into simulated food particles to match the thermal characteristics to those of the food particulates. The chosen silicone had similar physical and thermal properties to the high water content food particulates.

Table 16.2 illustrates the product and process types investigated. Two of the examples given in Table 16.2 will be examined in more detail.

Table 16.2 Type of products and processes evaluated with pasteurisation TTIs. (BAA is *B. amyloliquefaciens* amylase and BLA is *B. licheniformis* amylase)

Product type	Process description	Amylase type (BAA or BLA)	TTI particle method
Fruit preparations with particulates	Ohmic heating	BAA	Silicone particulates moulded to represent fruit pieces
Cook-in-sauces	Sprayed water pasteurisation-cooling tunnel	BAA and BLA	Tubes suspended within sauce
Poultry pieces	Continuous oven cooking-cooling	BAA	Tubes inserted directly into product
Fruit preparations with particulates	Tubular heat exchanger	BAA	Silicone particulates moulded to represent fruit pieces
Fruit products in liqueurs	In-jar pasteurisation	BAA	Tubes inserted directly into larger products (e.g. peaches). Silicone particulates moulded to represent smaller fruit pieces (e.g. blackcurrants)
Ready meals	Sous-vide processing in water baths	BLA	Tubes inserted directly into the meat components
Hot-fill sauces	Sprayed water top-up pasteurisation of internal jar surfaces	BLA and BAA	Tubes attached onto jar surfaces

- *Sprayed water pasteurisation-cooling tunnel for cook-in-sauces.* The aim of this trial was to map the pasteurisation achieved at various locations within a jar of tomato cook-in-sauce during processing in a sprayed water pasteuriser. Amylase TTIs were used to provide P-values from 16 positions in the jar and comparisons made with those measured by probes, where possible. The probes were thin-wire type K thermocouples with soldered junctions, which allowed several temperature measuring points within the single jar. These tests were conducted at CCFRA using a water bath to effect the heating and cooling.
- *Bacillus licheniformis amylase was the TTI.* Its kinetics of destruction by heat were represented by decimal reduction times (D_T value) at 93.3°C of 8.2 minutes and a kinetic factor (z-value) of 9.1 C°. These kinetics were suitable for measuring the butyricum process that was used for these sauces, which was equivalent to 5 minutes at 93.3°C with a z of 8.3 C°.

Figure 16.1 shows the location of paired TTIs and probes in the jar, with estimated P-values against each position. A TTI calibration was introduced into this type of process evaluation trial to ensure that minor changes in amylase

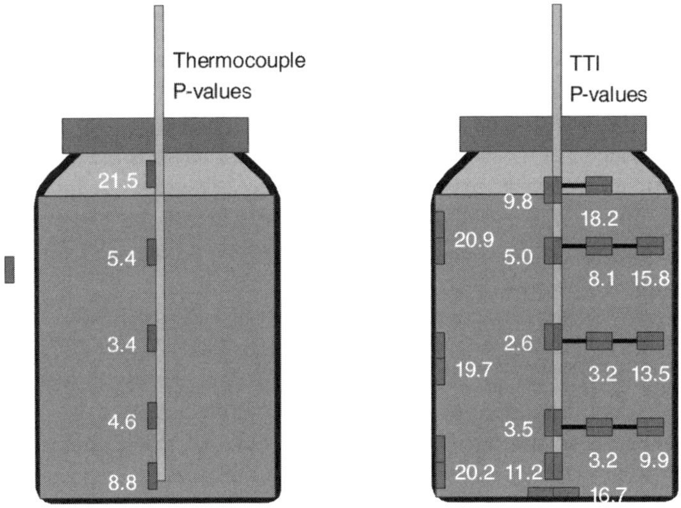

Fig. 16.1 *P*-values measured with thermocouples (left jar) and TTIs (right jar) at various locations within a jar of tomato cook-in-sauce.

$D_{93.3}$ values did not affect the accuracy of the results. This approach is similar to that used with temperature sensors. A TTI calibration test involved heating three to four TTI tubes in water at a temperature close to the reference temperature (93.3°C) to achieve approximately 1-log reduction in activity. A cross-check against the expected probe *P*-value allowed the accuracy of the TTI *P*-value to be assessed.

At all locations other than the headspace, the TTI and thermocouple P-values agreed within the experimental limits of thermocouple and TTI accuracy. The major and undefined source of error was in the precise location of TTIs and thermocouple junctions, partly as a result of their small sizes, but also due to the movement of TTIs or thermocouple junctions during processing. Hence, the P-value results provide a reliable map of the thermal treatment received throughout the jar. The discrepancy in headspace P-values between TTIs and thermocouples was thought to have been caused by the TTIs just penetrating the top surface. Evidence for this was the uptake of red/orange coloration from the sauce. The thermocouple junction was not in contact with the sauce and as such would have measured a *P*-value in the headspace.

TTI results given in this test showed that a wide range of pasteurisation values occurred within a single jar of sauce, ranging from 2.6 minutes at the centre to 20.9 minutes towards the sides. The jar centre *P*-values measured with TTIs and probes showed lower values than the target minimum of 5 minutes at 93.3°C. However, for the purpose of this test, the initial sauce temperature, the pasteuriser zone temperatures and residence times were set at deliberately conservative values.

Continuous oven cooking-cooling of poultry pieces
Reformed meat and poultry products are known to present difficulties for inserting and maintaining the position of temperature sensors. The aim was to use TTIs to measure the processes achieved in chicken fillets cooked in a continuous oven and to assess the effects of taking 2 minutes off the cook time.

TTI particles were silicone tubes of diameter 2.5mm and length 7–10mm, containing approximately 15 μL of *Bacillus amyloliquefaciens* amylase. Kinetics of destruction by heat were represented by a decimal reduction time (D_T value) of 6.8 minutes at 85°C or 268 minutes at 70°C, with a *z*-value of 9.4 C°. To ensure the process achieved at least a 6-log reduction in 'aerobic pathogens' such as *Listeria, Salmonella* and *E. Coli*, the target *P*-value was 2 minutes at 70°C (CCFRA[2]). Although this was a low process to measure with the amylase when compared with the D_{70} value, there was a need to operate with a substantial margin and data from previous trials had resulted in measurable activity loss. TTI tubes were inserted by slicing the fillets to expose the centre and placing the tubes 'end-to-end' in a row along the centreline. This created a good seal when the cut half was folded back. Fillets with TTIs were placed onto a wire mesh tray in order to identify them from the production fillets. They were retrieved immediately after the oven, and taken to the chiller for a few minutes to remove sufficient heat to enable the TTIs to be handled.

The first products tested were 25–40 g chicken fillets marinaded in lime and coriander, and the second were 120–130 g chicken fillets. For each product, 15 TTIs were used to assess the level of pasteurisation at the normal production speed and 15 at a faster throughput. Conventional temperature probing was not used because of the difficulties in positioning probe tips at the fillet centres and holding them in place for the duration of the process. Temperature probing of the fillets after the oven followed standard company practices to ensure the fillet centres had reached a threshold of 85±3°C.

P-values for the 25–40 g and 120–130 g fillets showed high levels of pasteurisation, substantially in excess of the target of 2 minutes at 70°C for achieving 6-log reduction in 'aerobic pathogens'. Typical *P*-values for production conditions were in the range 250–300 minutes for the 25–40 g fillets and 430–500 minutes for the 120–130 g fillets. Reducing the cook time by 2 minutes for each product showed reductions in *P*-value to 35–200 and 400–450 minutes respectively, but the lowest values were still substantially greater than the minimum safety value of 2 minutes. The lowest process measured at *P*-value 33.6 minutes still represented a 100-log reduction in 'aerobic pathogens'.

These poultry processes were shown to operate with a high level of microbiological safety. This is important to allow for variations in processing and product conditions (e.g. fillet size, belt loading, air temperatures, etc). For small products such as the 25–40 g fillets, where temperature probes cannot be used with accuracy, the TTI approach can give commercially valuable results. A 100-log reduction represented a satisfactory safety margin, particularly when the trials did not take account of the additional pasteurisation that occurred between the oven, searer and chiller.

The breakthrough that enabled these complex processes to be evaluated was in encapsulating the amylase solutions in silicone tubes. When stored in chilled water to minimise the rate of amylase degradation, the TTI tubes could be used for up to 14 days, with reports of one company in the project finding minimal loss in activity after 1-month. Chilling the amylase increases the scope for applying TTIs to factories overseas. To further extend their usable shelf life, filled TTIs can be frozen in large numbers, and TTI tubes removed as and when required. Freezing has little impact on amylase structure or on the rate at which its structure degrades by heat. It is conceivable that several hundred TTIs could be made at one time, frozen individually, and used over a period of months. This would be an economical method for producing TTIs and would ensure that the kinetics for each tube would be similar. Studies are ongoing to confirm the usable shelf life for periods held at −12°C for greater than 2 months.

16.6 Future trends

The use of time and temperature measurements to validate the degree of thermal process achieved is likely to remain the most widely used method. However, the advances in microchip technology will provide more computing power to analyse these results and increase the accuracy in defining the calculated process value. Traditional canning processes have been evaluated using lethal rates at time steps of one minute, because of the capabilities of available data recorders, but modern process values can be estimated from temperature measurements taken at much reduced frequencies, for example at every second. The increased data storage capabilities also allow for more temperature probes to be used in each TD or HP test; with multiple loggers linked together. With such systems it is easily possible to exceed the number of suggested working probes to define a HP test, however, the limitation on the number of probes that can go through the packing gland of a retort remains the same. To overcome this limitation, the diameter of the thermocouple cables must be reduced without compromising the strength or measuring accuracy of the system.

The use of more advanced mathematical models to evaluate and predict process times and temperatures will also increase as the computing power available on desk-top PCs increases. Examples of predictive models currently used are the CTemp and NumeriCAL software, both of which utilise a numerical approach to solving complex equations using finite differences. At present, most companies rely on the General method for their scheduled times and temperatures, but this does not allow process deviations to be assessed. A predictive modelling approach would not only help with deciding the fate for batches of product that had undergone a process deviation, but the task of process establishment would be made more straightforward by applying the models to evaluate low initial product temperatures, short come-up times or low retort temperatures.

In the introduction to this chapter it was stated that the scope of the market for thermally processed foods is increasing, as the pasteurised foods sector becomes more prominent. This is a trend that will continue because of the consumer demand for products of high quality that can only be produced by applying minimal thermal processes. There will always be demand for fully sterilised canned food, but it is unlikely that this industry sector will increase in popularity. To service the needs of the pasteurised foods sector, the methods available to the food processor for validating the process must evolve. Many of these process types cannot be validated using conventional wire-based systems, and so alternative methods are required. For food products manufactured in a cook-chill or cook-freeze system, dataloggers are available that can pass with the product as it moves between ovens and chillers or freezers. The types of products manufactured in this way have traditionally been processed to an end of oven product temperature, but are now being evaluated in terms of a pasteurisation process and the data converted to pasteurisation values. Self-contained logging units are also a recent introduction, which can be inserted into food containers that are processed in continuous retorts such as hydrostatic or reel and spiral sterilisers. While these units have enabled continuous processes to be evaluated without recourse to process simulators, improvements can be made to their robustness and longevity. Research is active in the application of TTIs to various pasteurisation and sterilisation processes, particularly to those where the food is heated and cooled in large vessels or heat exchangers. These processes cannot be evaluated using wire-based systems.

In conclusion, the scope of the thermally processed foods sector will increase to include products, packages and processes of increased complexity, and in response to this, the methods used to validate the process severity will also need to develop. The requirements of the traditional canned foods industry are disappearing, with an exciting new thermal processing sector developing with all of its demands on process validation techniques.

16.7 Sources of further information and advice

Many of the useful texts for advice of thermal process evaluation are given in the list of references, so will not be repeated here. However, the guidelines produced by the Campden and Chorleywood Food Research Association[2,4,8,17–19] and the Institute For Thermal Processing Specialists[9,20] are of particular relevance for their use as industry standard documents. The CCFRA guidelines for overpressure retorts[17–19] contained information that updated some aspects of Technical Manual No. 34, and the IFTPS protocols for carrying out temperature distribution[20] and heat penetration[20] studies provided detailed listing of factors to consider for TD and HP testing respectively. Both CCFRA and IFTPS documents were produced in consultation with representatives from the food industry. One further reference text produced from an industrial working party is the Department of Health guidelines for establishing safe thermal processes.[7] This booklet covers all aspects of thermal processing, from handling of raw materials through the processing stages and dispatch of processed containers.

16.8 References

1. BALL C.O. and OLSEN F.C.W., *Sterilization in Food Technology. Theory, Practice and Calculation.* New York, McGraw-Hill Book Co., 1957.
2. CCFRA Pasteurisation heat treatments, *CCFRA Technical Manual No. 38*, CCFRA, Chipping Campden, Glos., GL55 6LD, 1992.
3. MAY, N.S. Thermal underware, *Food Manufacture*, 1992.
4. CCFRA Guidelines to the establishment of scheduled heat processes for low-acid foods, *CCFRA Technical Manual No. 3*, CCFRA, Chipping Campden, Glos., GL55 6LD, 1977.
5. NFPA *Guidelines for Thermal Process Development for Foods Packaged in Flexible Containers*, NFPA, Washington, DC, USA, 1985.
6. BIGELOW, W.D., BOHART, G.S., RICHARDSON, A.L. and BALL, C.O. Heat penetration in processing canned foods, *National Canners Association Bulletin*, 16-L: 128, Washington, DC, USA.
7. DEPARTMENT OF HEALTH *Guidelines for the Safe Production of Heat Preserved Foods*. HMSO, London. 1994.
8. CCFRA Canning retorts and their operation, *Technical Manual No. 2*, CCFRA, Chipping Campden, Glos., GL55 6LD, 1975.
9. IFTPS *Protocol for Carrying Out Heat Penetration Studies*. IFTPS, PO Box 2764, Fairfax, VA, USA, 1995.
10. HENDRICKX, M., MAESMANS, G., DE CORDT, S., NORONHA, J., VAN LOEY, A. and TOBBACK, P. Evaluation of the integrated time-temperature effect in thermal processing of foods. *Critical Reviews in Food Science and Nutrition*, 1995 35 (3), 231–262.
11. VAN LOEY, A.M., HENDRICKX, M. E., DE CORDT, S., HAENTJENS, T.H. and TOBBACK, P. P. Quantitative evaluation of thermal processes using time-temperature integrators. *Trends in Food Science and Technology*, 1996 7, 16–26.
12. BROWN, K.L., AYRES, C.A., GAZE, J.E. and NEWMAN, M.E. Thermal destruction of bacterial spores immobilised in food/alginate particles. *Food Microbiology*, 1984 1, 187–198.
13. TUCKER, G.S. and WITHERS, P.M. Determination of residence time distribution of non-settling food particles in viscous food carrier fluids using Hall effect sensors. *Journal of Food Process Engineering*, 1992 17, 401–422.
14. MAESMANS, G., HENDRICKX, M., DE CORDT, S., VAN LOEY, A., NORONHA, J. and TOBBACK, P. Evaluation of process value distribution with time temperature integrators. *Food Research International*, 1994 27, 413–423.
15. TUCKER, G.S. Comment calculer les valeurs de pasteurisation dans le produits avec morceaux avec l'integrateur temps-temperature amylase. *Symposium Technique International de l'appertise UPPIA/CTCPA 'Securite et appertisation: de nouveaux outils pour la maitrise des treatments thermiques'*, Paris. 1998.
16. ADAMS, J.B. Determination of $D_{80°C}$ for α-amylase inactivation. *CCFRA Internal Project Report, Ref: 12598/1*, CCFRA, Chipping Campden, Glos., GL55 6LD, 1996.
17. CCFRA Guidelines for batch retort systems – full water immersion – raining/spray water – steam/air. *Guideline No. 13*, CCFRA, Chipping Campden, Glos., GL55 6LD. 1997.
18. CCFRA Guidelines for performing heat penetration trials for establishing thermal processes in batch retort systems. *Guideline No. 16*, CCFRA, Chipping Campden,

Glos., GL55 6LD. 1997.

19. CCFRA Guidelines for establishing heat distribution in batch overpressure retort systems. *Guideline No. 17*, CCFRA, Chipping Campden, Glos., GL55 6LD. 1997.
20. IFTPS *Temperature Distribution Protocol for Processing in Steam Still Retorts, Excluding Crateless Retorts*. IFTPS, PO Box 2764, Fairfax, VA, USA, 1992.
21. TUCKER, G.S. and WOLF, D. Time-temperature integrators for food process analysis, modelling and control. *CCFRA R&D Report No. 177*. CCFRA, Chipping Campden, Glos., GL55 6LD, 2003.

17

The use of data loggers to validate thermal processes

G. H. Shaw, Campden and Chorleywood Food Research Association, UK

17.1 Introduction: the role of data loggers in validating thermal processes

The ability to establish a safe thermal process is essential for the continued existence of many sectors of the food industry. The establishment of a process requires the accurate measurement of process temperatures that products are exposed to and the interpolation of the level of microbial destruction at the tested process conditions. Data logging systems and their associated fittings provide a means by which such temperature measurement can be easily undertaken.

A temperature measuring system used for the validation of a thermal process consists of two constituent parts: a temperature sensor and a data logging component. The sensor type that may be connected to a data logger varies from device to device and as such provides flexibility for the individual application. These will be discussed in the next section of this chapter. The type of retort system that is to be validated often determines the type of data logging device that is used. The wide range of commercially available data loggers gives the individual plenty of options in providing solutions to the practical problems presented by each retort type. The final part of this chapter will discuss approaches to be used for process evaluation.

17.2 Types of data logger

The first stage of any process evaluation is the selection of equipment to be used. This decision has an impact on the outcome of the trial, specifically with the accuracy of the results. CCFRA (1977) and Department of Health (2000) Guidelines recommend that measurements need to be taken to an accuracy of greater than ±0.5°C. In addition to the accuracy of the device itself, it is also important that the equipment does not have a detrimental effect to the test object in terms of heating or cooling rates.

Two types of sensor are commonly used with process evaluation equipment: thermocouples and resistance thermometers. This latter type may be sub-divided into platinum resistance thermometers (PRTs) and non-metallic thermistors. Both of these types are based upon the principle of changes of electrical resistance in a sensing element due to temperature. The thermistor has a larger response to changes in temperature, and can be smaller than the platinum resistance thermometer and so closer to the thermocouple in terms of performance.

Both of the main types of sensor, the thermocouple and the resistance thermometer, have advantages and disadvantages. Of the two, the platinum resistance thermometer is regarded as the more accurate for the monitoring of a stable temperature, but the sensor size is large compared with that of a thermocouple junction. When carrying out measurements within food containers, this has two implications. Firstly, point measurements are not possible. If a PRT is placed inside a fairly small particle, it can only give an average temperature rather than provide an accurate reflection of the actual conditions at the centre of that particle. Secondly, the relatively large size of the PRT means that the measured temperatures lag behind the actual temperature when this is changing.

The thermocouple provides a cheaper option than the resistance thermometers, but depending on the processing environment and their type of construction can have a shorter life span. Where the process evaluation work requires a high level of accuracy it is often preferable to use a ready-made thermocouple. These are generally supplied as part of a whole system that provides all the accessories necessary for data acquisition such as glands for enabling access to containers with minimal disruption of container preparation and product heating. The food industry has generally adopted Type T thermocouples (copper/constantan) as standard. This is despite the fact that the copper component is a relatively good conductor of heat, which could provide conduction errors, and also rapidly deteriorates when exposed to retort processing conditions. To allow the thermocouple to withstand such conditions the commercial ones are often coated with silicon rubber that is sealed at both ends. One end of the thermocouple will be connected to a jackplug that can then plug into the data logger. The other end is typically encased in a stainless steel needle with a screw type thread that connects with the gland that has been placed in the container.

When access to the inside of a food container is not necessary, for example when carrying out temperature distribution within a retort system, it is often favourable to use the cheaper form of thermocouple, which are those hand-made from lengths of thermocouple cable. The cable can be single or multicore with the latter being more robust and less susceptible to errors. For most retort processes a Teflon outer coating is used, but cheaper and less temperature resistant options can be used for pasteurisation processes.

Preparing the thermocouples involves cutting the wire to length and stripping short lengths of the outer coating from both ends. One end is connected to a jackplug or directly to the terminals of the data logger, and at the other end the two wires are twisted and soldered together to produce the measurement junction. For some types of cable it is also recommended that the outer casing be cut to prevent pressured water escaping from the retort system leaking onto the data logger. In some cases or to minimise conduction errors, very thin wire thermocouples can be made up for obtaining measurements at the centre of a small particle. Special glands are available to provide entry of such cables into containers.

All the measuring devices so far outlined have focused on the acquisition of temperature measurements, as these are the most common to be taken. Specialist equipment is also available that allows other measurements such as pressure or container deflection to be obtained. These devices connect to a data logger in a similar manner to temperature devices, and can be highly useful in optimising a process, particularly ones for products in flexible packaging.

17.3 Using data loggers to measure thermal processes

The type of data logger that can be used for validating a retort system is often dictated by the design of the retort system itself. The next section of this chapter will consider the selection of the correct logger type for some of the most commonly encountered retort systems.

17.3.1 Batch static retorts

Acquisition of accurate temperature measurements within batch retort systems containing stacked containers is possibly the simplest type of evaluation, although some difficulties may be encountered. Instrumentation for evaluation work on such retorts will typically consist of a data logger and thermocouples due to the low cost of the sensors, which allows for a high resolution of data for temperature distribution and heat penetration (see Fig. 17.1).

Most retort systems will have tappings in the retort wall, often in the instrument pockets, and it is through these that access is gained into the retort for the thermocouples. The tappings are typically of either 1/2″ or 3/4″ British Standard Pipe (BSP) sizes into which commercially available compression glands can be fitted. These glands take either four or six of the commercial

Fig. 17.1 A typical range of commercially available data logging units, including systems for temperature logging, deflection measurement, and pressure recording (Courtesy Ellab A/S)

silicon protected thermocouples or up to 30 of the thin wire thermocouples, although some leakage may be experienced with the latter.

It is important when the crates of product are being prepared with the thermocouples that care is taken to get them to their desired location within the crate without their being damaged. To do this the thermocouples should be threaded in between cans and through the holes in the layer pads to prevent their being crushed. Where possible it is generally easier for the thermocouples to pass through the side of the crate rather than from the top.

One problem that can occur is if the containers are placed into the retort crates using an automatic crate loading system. Thermocoupled containers cannot be loaded using such a system, so consequently the containers must be loaded by hand. When such loading is undertaken, it is important that the same container packing pattern normally applied by the automatic system is adhered to. Consideration must also be given to the additional time needed to load the crate or crates with the test samples in as this may lead to variation of the sample heating rates. Initial temperatures of the product may be affected, whilst some products, such as pasta, may undergo rehydration during the additional stand time.

17.3.2 Batch rotary retorts

In batch rotary retort systems, similar equipment can be used as for static batch retorts, with one key addition, which is that of a slip ring. The major problem with using thermocouple cables for taking measurements within a rotary system is that of allowing for the rotary motion. The use of slip rings removes this problem by using a stationary electrical contact impinging onto a rotating disk to pass the thermocouples' output. The maximum number of channels that can be logged through a slip ring is usually 12, with the internal diameter of the hollow drive shaft precluding the use of any more cables, so the resolution for individual runs in rotary batch retorts is lower than that for static retorts. This is one potential difficulty that should be considered with this type of retort system, as the high temperature, short processes typical of these systems can lead to more variation between runs.

A key factor that can be found with rotary processes is the improved heat transfer imparted by mixing due to the headspace within the container. Consequently, stringent control over can preparation and process pressure conditions is needed to guarantee headspace size and so obtain accurate heat penetration results.A final consideration for rotary batch retorts is to verify that the rotation speed setting is accurate. In the simplest form, this can be carried out using a stopwatch. Alternatively, a commercially available rotation counter data logger may be used. These will be covered in more detail later in this chapter.

17.3.3 Hydrostatic processing systems

The very nature of a hydrostatic processing system with the long conveying system makes them unsuitable for carrying out process validation using thermocouple cables. Traditionally, validation in such systems has been carried out using either expensive and time-consuming microbiological spore inoculation methods, or by using simulators. The latter are pilot scale retort systems that replicate the temperature and agitation of the full-scale counterparts. Due to their smaller size, it is possible for these to be considered in the same manner as batch retorts.

The last decade has seen groundbreaking advances in electronic miniaturisation. This in turn has led to the continued development of temperature measurement devices (data acquisition units) that are able to operate within a retort system independent of any outside connection (see Fig. 17.2). Consequently, such wireless data loggers are ideal for obtaining data from within continuous processing systems such as a hydrostat. Examples of the wireless data loggers are the Ellab Tracksense Pro and the Datatrace Micropack III systems. Both of these types work in a similar way, with the data logger being activated by a reader station connected to a computer. The data logger can then be used within a process with the measured data being saved to a computer style memory. The unit can then be downloaded at the end of the run to allow analysis of the data. Typically, the unit will be externally mounted on the can with just the probe inside the container. This is possible within a hydrostat cooker as the carrier bars used to convey containers through the system can allow for such longitudinal protrusions from the can. To ensure that the measurements are taken at the correct point within the container, the units are available with different length probes. Some, such as the Datatrace systems, have a fixed probe to the body, whilst others, such as the Ellab units, have removable probes of differing lengths that can be fitted to the main body of the logger.

The inability of being able to obtain real-time data for continuous processes is one that needs to be considered. Advances are being made using radio frequency

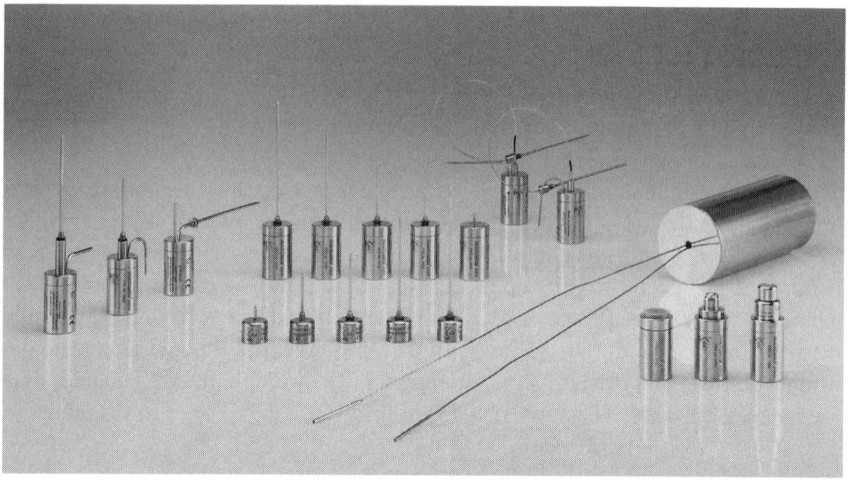

Fig. 17.2 A range of wireless data logging units highlighting the flexibility possible with these units in terms of size and probe type, with temperature, pressure and humidity loggers all being shown (Courtesy Ellab A/S)

to transmit signals from the data logger to a computer outside the retort, such as the TechniCAL RTemp wireless data collection system, but as yet the limiting factor of the radio frequency transmission precludes these from use within all hydrostatic systems.

17.3.4 Reel and spiral processing systems

As with hydrostatic systems, the reel and spiral continuous retort system is unsuitable for carrying out process validation using thermocouple cables. The nature of the reel and spiral retort also prevents using an externally mounted data acquisition unit, as could be the case for the hydrostat. This is because to the transfer valves that maintain pressure within the retort shell will only allow the passage of a single can.

Like hydrostats, the traditional method of carrying out a process validation has been dependent on the use of either microbiological methods or carrying out trials using a simulator. Two types of the latter are typically available. One is in effect a cross-section of the whole system with containers being processed on one single reel in one plane, rather than progressing down the system along the spiral as in an industrial machine. The other type is of an even smaller scale, with individual cans being intermittently rotated using rollers within a steam chamber. Heat penetration measurements can be obtained using a similar method as that used in a rotary batch retort with thermocouples going through a hollow drive shaft to be connected to a slip ring. However, special thermocouples will need to be used, located in the end of the cans, which allow the cans to rotate on the thermocouple and so allowing typical rotation. One disadvantage of using such a thermocouple system is that there is a need to prevent direct access of steam to the thermocouple tip. The presence of a rubber washer to make a seal around the probe and so avoid this, can cause a braking effect which conflicts with the need for free rotation of the cans. An alternative to the rubber washer is to apply silicon grease around the probe, which will maintain the integrity of the thermocouple and allow the correct motion of the can.

The data acquisition units mentioned in the previous section offer an alternative method of acquiring process validation data. The dimensions of these cylindrical units can be as little as 16 mm by 20 mm which is small enough for them to be mounted inside all but the smallest can sizes. Typically, a unit will operate within a temperature range of −50°C to +150°C and at a pressure of 10 bar when fully immersed. The memories will hold up to 30,000 data points and can be programmed to have a recording interval of between one second and 24 hours. The units are started before a process validation run from a reader station connected to a computer. In addition to programming the logging interval, it is possible to input constants such as a reference temperature and z-value for the logger to apply to the acquired data. If necessary, a delayed start time can be configured or the unit can be set to start once the environment temperature increases above a set point and continue logging until the temperature decreases below another set point. To prevent the logger from starting and stopping before the process has commenced, an additional set point is also specified that the logger must record before it can stop. Such a value needs to be one that would only be found within a process environment. Once the run has been completed, the unit is placed in the reader station again and the data transferred to the computer.

An often perceived problem of data acquisition units used inside cans is that their presence could alter the rate of product heating. It is possible that the presence of the units may result in heat sink effects and, for rotary processes, additional mixing. For reel and spiral processes the presence of the logger inside the can is unavoidable, so any potential errors should be minimised by using one of the new mini loggers such as the Datatrace Micropack III or the Ellab Tracksense Pro Mini Logger. For other types of process using these data acquisition units, the logger should wherever possible be mounted with the body outside the can and the probe passing through a suitable compression fitting to the interior of the container.

Research at CCFRA has indicated that as long as a correction is made to the headspace to take into consideration the presence of the data logger, for rotary processes internally mounted data loggers can be used for convective and forced convective products without too much concern. However, for conductive products errors such as texture changes may occur. Similar results were found for static processes, although any forced convective product types should now be considered conductive, whilst externally mounted data loggers showed a similar performance to traditional thermocouples for all typical processes.

17.4 Using data loggers to validate thermal processes

To ensure the microbiological safety and organoleptic quality of in container pasteurised and sterilised food products, it is important that accurate process validation of thermal processes is carried out routinely. When carrying out such validation trials, it is important that the pitfalls associated with product temperature measurements are avoided, as incorrect measurements will result in the overestimation of process lethality and release on to the market of unsafe product. Consequently, the main basis for heat penetration trials should always be that of a cautious approach with care taken to minimise errors and ensure that the correct or even an underestimated process is determined. There are several potential sources of error in the acquisition of product temperature data.

Firstly, the system that is being used to acquire the data should be considered. Individual sensors will have slight variation so calibration is essential to provide compensation for these. For a process validation trial in a steam retort, the easiest way to carry this out is by positioning all the sensors next to the Master Temperature Indicator in a uniform steam environment. From this it is possible to generate a calibration offset or correction factor for each sensor at that temperature. The Master Temperature Indicator itself should be regularly calibrated against another instrument of a higher traceable standard. As an alternative to steam, oil baths may be used for calibration purposes, provided they are of suitable temperature uniformity and if it is proven that the retort heating medium does not give differing sensor performance.

To allow for confidence in the process validation measurements, it is important that the measurements are taken at the worst case conditions which will be the coldest point within a pack and also within a retort. Such conditions must be determined experimentally for all product and pack configurations. In a properly vented steam retort, the variation in temperature distribution should be at the very worst 0.5°C with no significant cold points existing within the retort during the hold phase of the process. However, during the come up and cooling phases of the process there will be greater variation with low temperature zones present. Steam/air and water retorts may have less consistent temperature distribution, so it is important that accurate cold point determination is carried out within such systems.

Once the cold point within a retort has been determined, it is then necessary to consider the product container. Cold points within containers are dependent upon the geometry of the container and the characteristics of the food product. In a canned product, the cold point determination is simply a matter of taking measurements at differing heights along the central axis of the can. Products that heat by conduction will have a cold point located at the geometric centre, although there might be slight displacement due to the headspace or lidding materials of differing heat transfer properties. Conductive products will have a cold point that has been shifted downwards to a position that is determined by the amount of product movement that takes place within the container. Other packs of more complex geometry such as plastic trays or pouches, will have a less predictable cold point, so widely dispersed measurements will need to be taken to determine the cold point. Such packs may also have an additional difficulty presented by the flex of the pack during the process contributing to movement of the sensor. It is therefore important that steps are taken to ensure that the correct sensor location is maintained at all stages of the process.

A final potential source of error in the acquisition of temperature measurements for process validation is that of the time basis for the measurements. The lethality of a thermal process is dependent on both the temperature to which the product is exposed and the length of time at which it is held at that temperature. The ready availability of data loggers ensures that readings can be easily taken at an accurate time interval, eliminating the problems that could occur if manual readings were used to validate a process. Wherever possible it is important that the data logger is used at the greatest resolution or smallest time interval to give as accurate a set of data as possible: 30 seconds should be considered the maximum logging interval for most processes.

The potential sources of errors in acquiring accurate temperature measurements considered so far have been linked to either the data logger itself or its application. It is important when carrying out a process validation to also consider the containers and the product within them as these too are a potential source of error in the readings obtained by a data logger.

Process evaluation should always be carried out with a conservative approach with consideration to the worst case. To keep this in mind, fill weights of product in the test containers should always correspond to the highest anticipated from the production line. If production records are available, then an overfill of at least two standard deviations above the mean weight is common practice when making up a heat penetration sample. This will account for at least 95% of variation in production. Alternatively, a 10% overfill is usually considered to be adequate. An exception to this would be if a 10% overfill is not feasible within the constraints of the container capacity. In such cases a 10% overfill of product components which result in a slower heat transfer should be implemented.

For a process validation trial in a rotary cooker, it is important that the size of the headspace is largely maintained. This is due to the discrepancy in heat transfer rates that can occur with variation in headspace movements within a container during the process. Consequently, there is a responsibility on the part of the food manufacturer not to deviate from headspaces used in heat penetration trials when in production.

It is important to define an initial temperature of the product used during heat penetration trials, as this is an issue of direct commercial interest to the manufacturer. The temperature tested should correspond to the lowest that may occur in production at the start of a retort cycle. For hot fill products, it is therefore necessary to determine a level of cooling that is permissible between filling and the start of processing, and so define a maximum production delay for this phase of production that is adhered to. Any product that fails to meet this specification for initial product temperature should be safely disposed of. By using a range of initial temperatures during the validation work, either practically or using a modelling package such as the CCFRA CTemp software, an appropriate process duration can be specified and wastage minimised.

A variation in initial product temperature is only one factor that can change within a product before processing. A delay between filling and processing can also have an effect on the heat penetration characteristics of a product that is independent of temperature. Products such as pasta and other dehydrated products absorb surrounding sauces with a resultant thickening and reduction in heat transfer rates. To control this for such products, both time and temperature between filling and processing should be specified within the heat penetration trial conditions. If a product has particulates within it, such as carrot, potato or meat chunks, then the thermocouple needs to be located within these particles. For smaller particles such as peas or beans, probe location within a particle is not considered practical. CCFRA (1997a) states that only a particle with a smallest dimension of greater than 10 mm should be mounted on a probe. Generally measurements will be taken at the centre of a particle. Some vegetable products will be an exception to this if harvesting practices and sound physiology can ensure low levels of contamination at the centre of the tuber. For such products, measurements should be taken at the expected maximum depth of surface damage. When selecting a particle from raw material, the largest particle size must be selected. Such a practice can unavoidably lead to variation in temperature measurements dependent on size of selected tubers.

Wherever possible, product formulations used for heat penetration trials should match as closely as possible those intended for production. Sometimes it will be desirable during product development to carry out a 'look see' trial to establish important process parameters and test quality characteristics. Such trials should match the final production formulation as much as possible, particularly with reference to those that most have an effect on temperature measurements. Ingredients that need to be considered include starches that can cause considerable variation to the equivalent process received by a product. If there is a doubt over the effect that a specific ingredient might have on the heat penetration trial, it is wise to allow for any potential variation by using additional quantities during the trial.

Finally, one of the most important factors that can ensure the accuracy of results during heat penetration work is the good communication between those carrying out the trial and production personnel. This can help to ensure that the correct process parameters are used, particularly conditions such as time/temperature combinations, rotational speeds, retort loading patterns and container specifications, which can all affect the rate of heating of the product.

17.5 Future trends

The equipment used for process validation has, with some exceptions, followed a fairly standard format for the last couple of decades, whether it be the data logger box with thermocouples attached or the wireless data logger units. That is not to say the industry is standing still. On the contrary, there is a constant process of evolution and refinement that greatly assists those carrying out process validation work. The most positive advances are undoubtedly those made over the last couple of years with wireless data acquisition units. Both Ellab and Mesa Medical have replaced their existing models of DAU with new versions, these being the Ellab Tracksense Pro and the Datatrace Micropack III respectively. Both systems show much greater reliability than previously seen. The Tracksense system provides greater flexibility for the user with over 100 sensor configurations for use with the standard body. These feature a reduced probe diameter of 2 mm and can also feature a second probe that can log the temperature outside the container at the same time as the first logs the internal temperature. For systems such as a hydrostat or a rotary batch retort, this can allow temperature distribution within the retort environment and heat penetration within the product containers to be carried out simultaneously with the same loggers.

The new Datatrace Micropack III system is ideal for use within reel and spiral retorts, with its small size of only 20 mm in length, ensuring the effect on the process due to the location of the logger with the container is minimised. The Ellab Tracksense Pro mini logger provides competition for this system with dimensions of 16 mm in length by 20 mm in width.

However, as previously discussed, both these systems are restricted by a lack of real-time information. Systems to avoid this restriction are being developed such as the TechniCAL RTemp wireless data collection system. This system uses radio frequency to transmit from inside a retort system to a receiver on the outside. Limitations on signal transmission from within the retort can affect the functionality of such systems. It is the metallic structure of the retort itself that provides these limitations, and whilst data logger manufacturers will continue to research ways of getting real-time data for wireless systems, it is not certain when or if it may be achieved for all types of processing system.

17.6 Acknowledgements

The input and assistance of Nick May (CCFRA), Mark Seiboth (Ellab UK Ltd) and Jeremy Rouse (Wessex Power Ltd) in the preparation of this chapter is gratefully acknowledged.

17.7 References and further reading

CAMPDEN AND CHORLEYWOOD FOOD RESEARCH ASSOCIATION (1975), *Technical Manual No. 2, Canning Retorts and Their Operations.*

CAMPDEN AND CHORLEYWOOD FOOD RESEARCH ASSOCIATION (1977), *Technical Manual No. 3, Guidelines for the Establishment of Scheduled Heat Processes for Low Acid Canned Foods.*

CAMPDEN AND CHORLEYWOOD FOOD RESEARCH ASSOCIATION (1989), *Technical Memorandum No. 567.* Review of Data Acquisition Units for In Container Temperature Measurement and Initial Evaluation of the Ball Datatrace System.

CAMPDEN AND CHORLEYWOOD FOOD RESEARCH ASSOCIATION (1997a), *Guideline No. 16.* Guidelines for Performing Heat Penetration Trials for Establishing Thermal Processes in Batch Retort Systems.

CAMPDEN AND CHORLEYWOOD FOOD RESEARCH ASSOCIATION (1997b), *Guideline No. 17.* Guidelines for Establishing Heat Distribution in Batch Overpressure Retort Systems.

DEPARTMENT OF HEALTH (2000), *Guidelines for the Safe Production of Heat Preserved Foods* ISBN 0-11-321801-X.

INSTITUTE FOR THERMAL PROCESSING SPECIALISTS. Guideline Protocols (1992–2002).

18

The use of time-temperature integrators (TTIs) to validate thermal processes

A. Van Loey, Y. Guiavarc'h, W. Claeys and M. Hendrickx, Katholieke Universiteit Leuven, Belgium

18.1 Introduction: the importance of TTIs

Thermal processing has been and is still one of the most widely used physical methods of food preservation. The changing consumer demands for high quality convenience products have resulted in the optimisation of existing heating technologies and the development and application of new heating technologies, such as continuous processing in rotary retorts, aseptic processing, ohmic and microwave heating and combined processes. Quantification of the impact of a thermal process in terms of safety and quality (consumer acceptability, nutritional value) is a prerequisite for thermal process design, optimisation, evaluation and control.

The impact of a thermal process on a food attribute is usually quantified using the concept of an 'equivalent time at reference temperature', commonly referred to as process-value F. This concept translates the time-temperature variable product profile into an equivalent time at a chosen constant reference temperature that will affect the target attribute in the same way as the actual time-temperature variable profile to which the food (i.e. the attribute of interest) was subjected. Mathematically, the processing-value can be written in terms of the temperature history of the product (eqn. 18.1 and 18.2) or alternatively in terms of a response status before and after processing (eqn. 18.3 and 18.4).

$$^{z}F_{T_{ref}} = \int_0^t 10^{(T-T_{ref})/z}dt \tag{18.1}$$

$$^{E_a}F_{T_{ref}} = \int_0^t \exp\left(\frac{E_a}{R_g}\left(\frac{1}{T_{ref}} - \frac{1}{T}\right)\right)dt \tag{18.2}$$

Hence, based on the temperature history of the product aspect, either recorded or simulated by mathematical modelling, combined with knowledge on the kinetics of the target attribute, the thermal process impact on that aspect can be calculated, an approach that is commonly referred to as the physical-mathematical method. However, the need for time-temperature data can seriously limit the applicability of this method. Direct temperature registration inside the product using temperature sensors can be inappropriate (e.g. continuous processes, particulated foods). The accuracy of mathematical models for simulation of the time-temperature profile depends on the accuracy of the physical input parameters required. Often a lack of accurate estimates of the values for the model parameters (e.g. data on viscosity or conductivity at high temperature, fluid to particle heat transfer coefficients) restricts the constructive computation of the temperature history.

The impact of a thermal process on a specific product attribute can also be determined relying solely on the initial and final status of the attribute of interest and on its kinetics. Eqn. (18.3) is valid for a target attribute that obeys a first-order decay and eqn. (18.4) for an nth order ($n \neq 1$) decay.

$$^{z(E_a)}F_{T_{ref}} = D_{ref}\log\left(\frac{X_0}{X}\right) = \frac{1}{k_{ref}}\ln\left(\frac{X_0}{X}\right) \tag{18.3}$$

$$^{E_a}F_{T_{ref}} = \frac{1}{k_{ref}}\left(\frac{X^{1-n} - X_0^{1-n}}{n-1}\right) \tag{18.4}$$

Hence, based on the response status of an attribute before (X_0) and after (X) thermal treatment, combined with its kinetics, the process impact can be calculated using eqn. (18.3) or (18.4) depending on the order of the heat induced reaction occurring to this attribute. In case the level of the actual target attribute of interest is measured before and after processing, this approach is referred to as the *in situ* method. However, in practice, the measurement of microbial counts, vitamin content, organoleptic quality, etc., can be laborious, time-consuming or expensive, and in some cases even impossible because of the detection limit of the analytical techniques at hand and/or sampling requirements.

To overcome the above described limitations and inherent disadvantages associated with the physical mathematical approach and the *in situ* method, time-temperature integrators (TTIs) have been and are being developed as an alternative means for process evaluation. In this chapter, the principles and use of TTIs for thermal process evaluation will be discussed, including their strengths and weaknesses. The current state of the art regarding TTI development and use is presented, and future trends are indicated.

18.2 The principles of TTIs

A TTI can be defined as 'a small measuring device that shows a time-temperature dependent, easily, accurately and precisely measurable irreversible change that mimics the changes of a target attribute undergoing the same variable temperature exposure' (Taoukis and Labuza, 1989a,b). The target attribute can be any safety (e.g. micro organisms (spore) inactivation) or any quality attribute (loss of a specific vitamin, texture, colour) of interest. The change occurring during the thermal process, which can either be a decrease/degradation or an increase/formation of a specific component, has to be irreversible in order to be able to quantify the impact after processing.

Next to some convenience (e.g. inexpensive, quickly and easily prepared, easy to recover, accurate and user-friendly read-out), and thermophysical (e.g. no disturbance of heat transfer) requirements, this definition implies specific kinetic requirements: the temperature sensitivity of the rate constant (i.e. z-value or activation energy) of the TTI and the target attribute should be equal and the reaction rate constant of the TTI should be sufficiently (but not too) low in the relevant temperature domain to allow a detectable response to the temperature history.

Theoretically, the use of a TTI characterised by its z-value or activation energy is strictly only justifiable to monitor product aspects with an equal z-value or activation energy. On practical grounds, the allowed difference in z-value between TTI and target attribute to ascertain a pre-set accuracy in process impact determination has been studied theoretically (Van Loey *et al.*, 1995a). In general, the allowed difference in z-value increases the closer the actual processing temperature approaches the reference temperature and the higher the z-value of the target attribute. From a food safety point of view, an overestimation of the actual processing impact by use of a TTI with a deviating z-value should be avoided, because it can lead to understerilisation of food products, which involves a risk for public health.

In general, TTIs can be classified according to their working principle, type of response, origin, application in the food material and location in the food (Fig. 18.1). Depending on the working principle, TTIs can be subdivided into biological (microbiological and enzymatic), chemical and physical systems. A TTI can be a single response system (i.e. one heat sensitive component with equal z-value as the target attribute) or a multi response system (i.e. a set of individual components each characterised by its z-value, deviating from the one of the target attribute). With respect to the origin of the TTI, extrinsic and intrinsic TTIs can be distinguished. An extrinsic TTI is incorporated into the food whereas intrinsic TTIs are intrinsically present in the food and represent the behaviour of another target attribute. With regard to the application of the TTI in the food product three approaches can be distinguished: dispersed, permeable or isolated. Dispersed systems allow the evaluation of the volume average impact while all three approaches can be the basis for single point (specific locations in the food, e.g. coldest spot) evaluations.

18.3 Application of TTIs to measure thermal processes

After preparation of the monitoring system (in case of extrinsic TTIs), the initial level of the TTI response property (X_0) should be accurately measured.

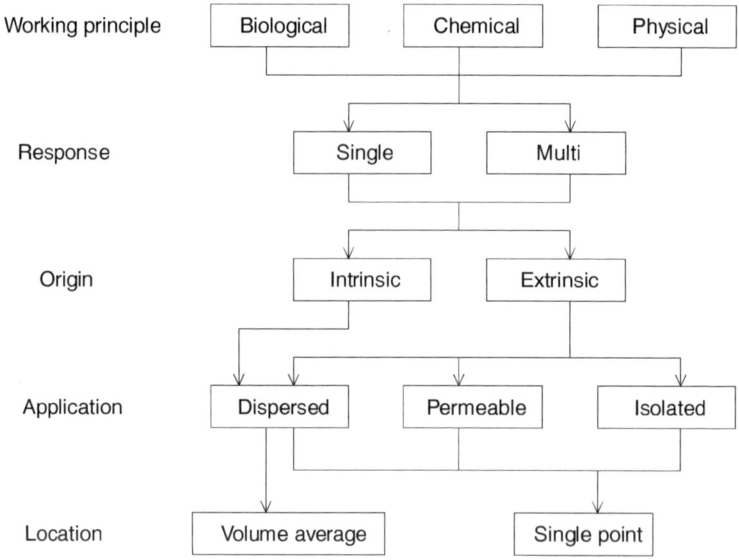

Fig. 18.1 General classification of TTIs.

Depending on the nature of the monitoring system this involves, e.g. the determination of initial microbial count, the measurement of the initial enzymic activity or the determination of the initial concentration of a chemical compound (which can be zero or very low if product formation during thermal processing is the monitoring system). After incorporation of the TTI in a food product at the desired location (in case of extrinsic TTIs), the food can be thermally treated. Post-processing steps involve the removal of the TTI from the food product and the determination of the final level of the response property of the TTI (X). In between the removal of the TTI and the read-out it is advised to store the TTI refrigerated or in ice-water. Based on the measured response status of the TTI before (X_0) and after thermal treatment (X), combined with the decimal reduction time or reaction rate constant at reference temperature of the TTI, the process impact can be calculated using eqn. (18.3) or eqn. (18.4) provided that the TTI obeys a first order or n^{th} order reaction respectively. For each freshly prepared batch of TTIs a calibration in terms of the decimal reduction time at reference temperature is indispensable and the determination of the z-value is highly recommended. An appropriate way to determine these heat inactivation kinetic parameters is by following the decrease/increase in the response property of the TTI as a function of inactivation time at different constant inactivation temperatures.

18.3.1 Current state of the art on TTIs

Based on the classification described above, an overview of existing TTIs is given in Table 18.1, which reports the z-value or the activation energy for each

Table 18.1 State of the art of TTIs for thermal process evaluation

Origin	Application	Principle	E_a (kJ/mol) or z (°C)	Reference
Microbiological				
Extrinsic	Dispersed	Inoculated Pack[a]		Yawger (1978)
		Bacillus stearothermophilus	NR	Burton *et al.* (1977)
		Bacillus coagulans	NR	Jones & Pflug (1981)
		Bacillus subtilis	NR	Pflug & Odlaug (1986)
		Clostridium sporogenes	NR	Pflug & Odlaug (1986)
	Permeable	*Bacillus stearothermophilus* in alginate	8.5°C	Bean *et al.* (1979)
		Bacillus stearothermophilus in alginate	11.4–11.8°C	Brown *et al.* (1984)
		Clostridium sporogenes in alginate	12.5–12.7°C	Brown *et al.* (1984)
		Bacillus stearothermophilus in alginate	9°C	Heppel (1985)
		Bacillus stearothermophilus in alginate	NR	Sastry *et al.* (1988)
		Clostridium sporogenes in a turkey cube	8.5°C	Segner *et al.* (1989)
		Bacillus stearothermophilus in polyacrylamide gel	11.7°C	Rönner (1990a, b)
		Bacillus stearothermophilus in alginate	8.8–9.1°C	Gaze *et al.* (1990)
		Bacillus stearothermophilus in alginate	NR	Ocio *et al.* (1997)
		Bacillus stearothermophilus in alginate	8.6°C	Rodrigo *et al.* (1998)
	Isolated	*Bacillus anthracis* in perspex	60°C	Hunter (1972)
		Bacillus stearothermophilus in plastic	7.8–10°C	Pflug *et al.* (1980a, b)
		Bacillus stearothermophilus in glass bulb	10°C	Hersom & Shore (1981)
		Bacillus stearothermophilus in aluminium	NR	Rodriguez & Teixeira (1988)
		Talaromyces flavus in pastic spheres	9.3–14.7°C	Rönner (1996)
Protein based/enzymic				
Intrinsic	Dispersed	Enzyme-linked immunosorbent assay for lactate dehydrogenase	NR	Smith (1995)
		Acid phosphatase (beef)	7.4°C	Orta-Ramirez *et al.* (1997)
		Lactate dehydrogenase (beef)	4.0°C	Orta-Ramirez *et al.* (1997)
		Peroxidase (beef)	7.8°C	Orta-Ramirez *et al.* (1997)
		Phosphoglycerate mutase (beef)	5.2°C	Orta-Ramirez *et al.* (1997)

Table 18.1 (continued)

Origin	Application	Principle	E_a (kJ/mol) or z (°C)	Reference
		Glyceraldehyde-3-phosphate dehydrogenase (beef)	4.7°C	Orta-Ramirez *et al.* (1997)
		Triose phosphate isomerase (beef)	5.6°C	Orta-Ramirez *et al.* (1997)
		Acid phosphatase (turkey)	6.3°C	Veeramuthu *et al.* (1998)
		Lactate dehydrogenase (turkey)	3.8°C	Veeramuthu *et al.* (1998)
		Creatine kinase (turkey)	4.8°C	Veeramuthu *et al.* (1998)
		Immunoglobulin G (turkey)	8.6°C	Veeramuthu *et al.* (1998)
		Glyceraldehyde-3-phosphate dehydrogenase (turkey)	4.3°C	Veeramuthu *et al.* (1998)
		Triose phosphate isomerase (turkey)	5.8°C	Veeramuthu *et al.* (1998)
		Triose phosphate isomerase (roast beef)	NR	Hsu *et al.* (2000)
		Alkaline phosphatase (milk)	4.8°C	Claeys *et al.* (2001a)
		Lactoperoxidase (milk)	3.3°C	Claeys *et al.* (2001a)
		Hydroxymethylfurfural (milk)	90.2 kJ/mol	Claeys *et al.* (2001a)
		Lactulose (milk)	99.1 kJ/mol	Claeys *et al.* (2001a)
		Furosine (milk)	88.7 kJ/mol	Claeys *et al.* (2001a)
Extrinsic	Permeable	β-galactosidase in alginate	NR	Matthiasson & Gudjonsson (1991)
		Lipase in alginate	NR	Matthiasson & Gudjonsson (1991)
		Nitrate reductase in alginate	NR	Matthiasson & Gudjonsson (1991)
		Immobilised amylase in alginate matrix	NR	Wunderlich (1995)
	Isolated	Immobilised peroxidase in decanol	11.6°C	Weng *et al.* (1991)
		Immobilised peroxidase in dodecane	10.1°C	Weng *et al.* (1991)
		Immobilised *Bac. licheniformis* α-amylase	302 kJ/mol	De Cordt *et al.* (1992)
		Bac. amyloliquefaciens α-amylase + 83 wt% glycerol	701 kJ/mol	De Cordt *et al.* (1994)
		Bac. amyloliquefaciens α-amylase + 49 wt% glycerol +31 wt% sucrose	560 kJ/mol	De Cordt *et al.* (1994)
		Bac. subtilis α-amylase (solution)	6–12°C	Van Loey *et al.* (1995a)
		Bac. amyloliquefaciens α-amylase (solution)	7.9°C	Van Loey *et al.* (1995b)

		Bac. subtilis α-amylase (aw = 0.76)	9.7°C	Van Loey *et al.* (1997a)
		Bac. licheniformis α-amylase (aw = 0.81)	260 kJ/mol	Haentjens *et al.* (1998)
		Horseradish peroxidase	10°C	Lemos *et al.* (2000)
		Fluorescence loss of phycoerythrin at pH 9.0	9.2°C	Orta-Ramirez *et al.* (2000)
		Bac. amyloliquefaciens α-amylase (solution)	9.4°C	Tucker *et al.* (2002)
		Bac subtilis α-amylase (solution)	9.1°C	Tucker *et al.* (2002)
		Bac. licheniformis α-amylase (aw = 0.81)	9.0°C	Guiavarc'h *et al.* (2002a)
		Bac. licheniformis α-amylase (aw = 0.85)	7.9°C	Guiavarc'h *et al.* (2002a)
		Tomato pectinmethylesterase +50% glycerol	5.3°C	Guiavarc'h *et al.* (2003)
		Cucumber pectinmethylesterase +60% glycerol	5.2°C	Guiavarc'h *et al.* (2003)
Chemical				
Intrinsic	Dispersed	Formation of 2,3-dihydro-3,5dihydroxy-6-methyl-(4H)-pyran-4-one	96 kJ/mol	Kim & Taub (1993)
Extrinsic	Dispersed	Thiamine breakdown	26°C	Mulley *et al.* (1975a, b)
	Permeable	Maillard's reaction on paper disc	NR	Favetto *et al.*, (1988, 1989)
	Isolated	Hydrolysis of dissacharides	18°C	Wen Chin (1977)
		Methylmethionine sulfonium breakdown	20–22.8°C	Berry *et al.* (1989)
		Acid hydrolysis of sucrose	94.6 kJ/mol	Sadeghi & Swartzel (1990)
		Acid hydrolysis of sucrose	99 kJ/mol; 31°C	Pinheiro Torres *et al.* (1994)
		Destruction of Blue #2 at pH 11.3	58.2 kJ/mol	Sadeghi & Swartzel (1990)
		Destruction of Blue #2 at pH 9.5	74.5 kJ/mol	Sadeghi & Swartzel (1990)
		Nitrophenylglucoside hydrolysis	21.7°C	Adams & Langley (1998)
		Formation of 2,3-dihydro-3,5dihydroxy-6-methyl-4(H)-pyran-4-one from glucose/serine	65 kJ/mol; 44°C	Eliot-Godéreaux *et al.* (2003)
Physical				
Extrinsic	Isolated	Thermal Memory Cell[b]		Swartzel *et al.* (1991)
		Thermalog S	9.4–9.6°C	Witonsky (1977)

a: In inoculated pack studies, micro-organisms relevant to the type of product and process are added; thus a single *z*-value cannot be given as this depends on the particular micro-organisms used in a given study
b: The thermal memory cell is a multi-component TTI, based on the diffusion of at least two different ions, each with their own activation energy, in the insulator layer of a metal-insulator-semiconductor capacitor. Hence the activation energies that characterise the 'thermal memory cell' depend on the ions used.
1: NR: not reported

TTI system. As mentioned before, when no information is available on the time-temperature history of the product, a TTI can only be used to monitor target attributes with equal *z*-values or activation energies. Depending on the thermostability of the reported TTI systems (i.e. rate constant or decimal reduction time at a selected reference temperature), they can be used either in the pasteurisation or in the sterilisation area. To identify potential target attributes, databases or overview articles on the heat inactivation kinetics of micro-organisms or quality aspects can be consulted (Norwig and Thompson (1986), Villota and Hawkes (1986), Pflug (1987), Betts (1992), Betts and Gaze (1992), Villota and Hawkes (1992)). In general, the *z*-value for thermal inactivation of vegetative cells and spores is in the range of 4°C–12°C and for quality aspects (e.g. colour, texture, flavour, vitamins) in the range of 25°C–45°C (Lund (1977)).

With regard to safety considerations, the design of sterilisation processes for low acid canned foods are directed to the destruction of spores of proteolytic strains of *Clostridium botulinum* ($z = 10$°C). For the evaluation of the safety of pasteurisation processes several micro-organisms with *z*-values ranging from 5°C to 12°C have been advanced for use as reference micro-organism because depending on the secondary preservative barriers other micro-organisms might be the main cause of spoilage and therefore be of particular concern. From Table 18.1, it can be observed that most research on development of TTIs has in the past five years focused on enzymic TTIs (either intrinsic dispersed or extrinsic isolated) and on extrinsic isolated chemical TTIs. The performance and the reliability of several developed TTIs to monitor thermal impacts have been evaluated. Several TTIs have been further validated under semi-industrial thermal processing conditions in a number of pilot plant installations. Van Loey *et al.* (1996a) used a *Bacillus amyloliquefaciens* α-amylase based TTI to monitor in-pack process lethality distribution in a particulated food due to a rotational pasteurisation process in a water cascading retort. The same group of researchers successfully applied a *Bacillus subtilis* α-amylase based TTI to investigate the influence of the end-over-end rotational speed and of the viscosity of the brine on the spatial distribution of process-lethalities in a particulated food model system, sterilised in a water cascading retort, and to evaluate the lethality distribution and hence determine the coldest zone in a water cascading retort (Van Loey *et al.*, 1997b). Ocio *et al.* (1997) investigated the performance of a TTI based on immobilised *Bacillus stearothermophilus* spores in a cylindrical particle consisting of an alginate-starch-mushroom purée in terms of spore distribution, spore leakage during and after the thermal process and frozen storage conditions. They concluded that this artificial particle could be a very reliable TTI for monitoring the thermal impact on micro-organisms during sterilisation processes in continuous aseptic systems. Afterwards, pilot plant studies were carried out in a water immersion static retort to evaluate the accuracy and general performance of this TTI (Rodrigo *et al.*, 1998). Knap and Durance (1998) used calibrated *Bacillus stearothermophilus* spores sealed in differential scanning calorimetry pans to investigate the effect of fixing food particles at the tip of thermocouples for process evaluation of particulated food products, thermally processed in an agitating steam/air overpressure simulator retort. The applicability of triose phosphate isomerase as an intrinsic TTI for roast beef processing has been validated in water bath and pilot oven studies (Hsu *et al.*, 2000). Tucker and co-workers applied successfully a *Bacillus* α-amylase based TTI to monitor the thermal processing impact during pasteurisation of fruits and fruit products thermally treated in a tubular heat exchanger and in an ohmic heater (Tucker, 2000; Tucker *et al.*, 2002). *Bacillus licheniformis* α-amylase at reduced moisture content was used as a TTI to investigate potential differences between process values in freely moving spherical particles as compared to a centrally fixed particle inside cans thermally processed in a rotating water cascading retort. Results showed that the process value inside freely moving particles can be up to 19.7% smaller than the process value inside a centrally fixed sphere, which implies that process impact evaluation in a fixed particle can lead to potential overestimation of the actual process value with possible hazardous quality/safety implications (Guiavarc'h *et al.*, 2002b). Eliot-Godéreaux *et al.* (2003) validated a TTI based on the formation of 2,3-dihydro-3,5dihydroxy-6-methyl-4(H)-pyran-4-one in a glucose/serine mixture in a laboratory-scale ohmic heater to predict the cooking effect. These application studies highlight the potentials of TTI technology to monitor the safety/quality of heat-processed foods in different types of heating technologies.

18.4 Strengths and weaknesses of TTIs

The major strength of TTIs is the ability to quantify the integrated time-temperature impact on a target attribute without information on the actual temperature history of the product. Hence, TTIs can be used as an alternative in process design, for process evaluation and for process optimisation when the physical mathematical method or *in situ* approach is infeasible. It should be mentioned that if temperature is not the only reaction rate determining factor, using only a TTI to monitor food safety or quality loss would result in error, as other factors that change with time can be critical (e.g. moisture content). To evaluate such processes a 'product history integrator' would be necessary to mimic the behaviour of its target attribute exactly by responding in the very same way, to all of the changing intrinsic and extrinsic factors, as the quality attribute that it is designed for.

A major weakness of TTIs is that they are not suitable for on-line monitoring of the efficacy of thermal processes – which is feasible by direct registration and consecutive integration of the product temperature history. Indeed, all TTIs are *post factum* indicators of the impact of a thermal process, because the calculation of this process impact is based on the change in status of the TTI after thermal treatment, as compared with its initial status. If recovery of the TTI at different stages in a multistage treatment is possible, TTIs can be used to evaluate the contribution of each stage to the total process impact.

The relative strengths and weaknesses of different types of TTIs with regard to their working principle, their response type, their origin and application mode and their positioning in the food are briefly discussed in the following section. More details are described in Hendrickx *et al.* (1995) and Van Loey *et al.* (1996b).

18.4.1 Working principle of TTIs

Any system, whether biological (microbiological or enzymatic), chemical or physical, can be used to mimic changes of food safety/quality attributes provided that the temperature sensitivity of the rate constant (*z*- or E_a value) of the TTI and the target attribute is equal in the relevant temperature range, and the reaction rates encountered during heating of the TTI will induce a detectable response to the temperature history. The time after thermal processing needed to calculate the process impact by read-out of a TTI depends upon the nature of the monitoring system: for example microbiological assays are time-consuming (several days), whereas the evaluation of enzyme activity or quantification of a chemical compound is usually much faster (up to minutes). The advantage of a microbiological system to monitor the safety of thermal processes is that the measuring device and target are sensitive to heat in the same temperature range. The major weakness of any microbiological monitoring system is the length of the assay. The long incubation time between process and read-out of the system (up to several days) does not allow for rapid intervention upon any kind of failure or process deviation.

Additional problems arise from the inherent variability of living organisms. Thus, before use as marker to determine the killing power of a given heat treatment, spores should be thoroughly calibrated to determine their heat resistance (Pflug and Odlaug, 1986). Several characteristics of thermostable enzymes are favourable for their use in the context of TTI development: they are small in size, both enzyme activity and denaturation enthalpy, properties that can be measured rapidly and accurately, may serve as a response, and their heat inactivation kinetics can be manipulated in various ways (e.g. enzyme immobilisation, 'solvent engineering').

Chemical systems are based on a purely chemical response towards time and temperature. Flexibility of handling and high analytical precision in the detection of chemical reactions make chemical TTIs promising tools for the evaluation of thermal processes. A major weakness, however, is that no chemical reactions have thus far been identified in the open literature on heat treatments of food, that feature the activation energy required for monitoring food safety in the sterilisation temperature range, and only few are available that can be used to follow the deterioration of quality attributes.

18.4.2 Response types of TTIs

The response of a TTI can be the measurement of the change in the status of a single component or of several individual components. If a single component

TTI that has a z-value/activation energy equal to the one of the target attribute is at hand, a direct relation between the change in status of the TTI and that of the target attribute it monitors can be proposed. However, when the z-value/activation energy of the TTI and target attribute differ, additional information on the time-temperature history is needed to predict the thermal impact on the target quality attribute based on the change in status of the TTI. Mere correlation studies whereby the response of the TTI is statistically coupled to safety/quality changes for a given set of time-temperature conditions without any matching of temperature dependency of rate constants, offer useful but limited information because the correlations found are valid only for the exact conditions tested. Extrapolation of the correlation equations to other temperatures or for fluctuating conditions could lead to errors.

The multicomponent TTI concept involves the reading of the impact of a thermal process on a set of individual components, each responding according to its own z-value/activation energy, and predicting from this multiple responses the change in status of a safety/quality attribute with a different z-value/activation energy. Maesmans *et al.* (1993, 1994, 1995) studied the possibility to combine the use of 'the equivalent point method' (a transformation function that summarises a variable temperature history by an equivalent time at an equivalent temperature, independent of the activation energy (Swartzel *et al.*, 1991; Sadeghi and Swartzel, 1990)) and a multicomponent time-temperature integrator. Because a theoretical basis for this 'equivalent point method' can only be formulated for isothermal heating profiles, and the method becomes an empirical engineering approach for time-variable temperature profiles, the method should be used carefully, and validation is recommended for each time-temperature profile and for each set of temperature sensitive components.

The multicomponent TTI concept has also been studied by Stoforos and Taoukis (1998) and they concluded that the best option, when developing multicomponent TTIs is to include components with z-values below and above the one of the target attribute under concern. In case this option cannot be achieved they advise to use components showing z-values above rather than below the z-value of the target attribute of interest. In spite of the theoretical interest of multicomponent TTIs, they show some potential limitations due to the necessity to measure the thermal impact on at least two components instead of one (with single component TTI). However, multicomponent TTIs remain an interesting tool to monitor process-values in case the measures of the thermal impact on the several (at least two) components is easy and fast to perform. Moreover, with a multicomponent TTI, it is possible to determine thermal impacts on target attributes showing various z-values. In other words, multicomponent TTIs can be multitarget oriented.

18.4.3 Origin and application in the food

When using an intrinsic food component as a TTI, the TTI is more or less homogeneously dispersed in the food, which eliminates heat transfer limitations and allows evaluation of the volume average or a single point impact of the thermal process. A major advantage of intrinsic markers is that no monitoring component or device has to be added and/or placed in position before heating, and that problems related to the selection of an appropriate carrier system, as discussed below, are avoided. A weakness of intrinsic markers, on the other hand, is the need for elaborate kinetic calibration for each TTI/product combination, because kinetic parameters are sensitive to environmental properties such as moisture content, pH, and the concentration of various substances. The same disadvantage holds for extrinsic dispersed (i.e. mixed in with the actual food product) TTIs. To avoid the influence of the food environment on the kinetic behaviour of the TTI, extrinsic TTIs have been encapsulated, either in an inert carrier material (e.g. glass, plastic, metal) completely isolated from the food environment, or in a permeable carrier system (e.g. alginate, polyacrylamide). In isolated systems, TTI kinetics can be determined independently of the type of food and nature of the surrounding medium. For example, incorporating the TTI in a small hermetically sealed and high conductive carrier system guarantees the elimination of any influence of the environment other than temperature and allows its incorporation into a real food particle without significantly disturbing the entire particle's response to given processing conditions. In permeable carrier systems, again the influence of intrinsic food properties on the kinetic behaviour of the TTI should be known. Indeed, when the TTI is incorporated in a permeable carrier, diffusion of food components into the particle containing the TTI might alter its kinetic behaviour. Care also has to be taken to avoid leakage of the temperature sensitive component out of the carrier. In the case of encapsulation, either in an inert or in a permeable carrier material, it is important to verify that the heating rate is determined by the food product and not by the carrier material. Depending on the type of heating technology to be monitored by a TTI, specific criteria regarding the TTI-carrier system have to be imposed. For example, density of the TTI-carrier should be identical to that of the real food particle when the influence of free rotation and translation of the monitored particle is of importance, as is the case for agitated or aseptic thermal processing of particulated foods. Next to real food particles where TTIs have been implanted, alginate particles meet the requirements on thermal diffusivity and density. In cases where a TTI is used, in for example aseptic processing of liquids containing particles or a surface scraped heat exchanger, the mechanical behaviour of the TTI-carrier should be investigated to avoid deformation of the particle, which influences among other things the residence time of the particle. In case where encapsulated TTIs are to be applied in microwave heating, specific dielectric properties of the carrier material are important (e.g. the penetration depth and loss factor of the carrier material).

18.4.4 Positioning the TTI: measuring a volume average or single point effect

A TTI can be homogeneously dispersed in the food system to measure the volume average impact of the process by analysis of the response value of the

TTI over the entire volume before and after processing. The volume average process impact on the food is especially of interest when hidden quality attributes such as nutrient retention are considered. When the option is taken to evaluate the impact of the process at a single 'point', one should locate the TTI where it will experience worst-case conditions. To evaluate the safety of a food product, the TTI should be located preferentially in the slowest heating point (zone) of the product. The location of a TTI to monitor the quality retention of a thermally treated food depends on the nature of the quality attribute to be monitored: for example, appearance will be monitored at the product surface. The positioning of an extrinsic TTI at a specific point in the product is critical in process impact determination, just as the placement of temperature sensors is critical to register the temperature history of the food product for process calculations.

18.5 Future trends

To conclude, potentials and limitations of the use of TTIs and of the physical-mathematical approach are mutually compared. When direct registration of the product temperature history is feasible, the use of the physical-mathematical approach (i.e. integration of the recorded temperature history) is preferably applied for process evaluation because this method allows on-line process control, which is not possible by use of TTIs and because the process impact can be calculated for a variety of target attributes. Neither this integration of the registered temperature profile, nor the use of TTIs has predictive power for processing conditions different from those under which experiments were run. Because of the lack of predictive power and the inability of direct measurement of the time-temperature profile under certain processing conditions, predictive formula methods have been developed based on empirical formulae or theoretical models for heat transfer. Empirical formula methods, that allow one to handle complex heating systems, which, indeed, can be encountered in real food, have been developed and are widely used (Ball, 1923; Ball and Olson, 1957). Major drawbacks are: (i) the need for an accurate estimate of the heat penetration characteristics of the product which again incurs the problems related to temperature registration; and (ii) the inability to evaluate thermal processes with variable boundary conditions. On the contrary, evaluation of thermal processes with a variable heating medium temperature is possible by use of TTIs, independent of the mode of heat transfer involved. With regard to theoretical models, conductive heating/cooling can be adequately predicted from the Fourier equation using analytical or numerical solutions (including finite differences and finite elements). Also, for convection and mixed convection/ conduction heating, theoretical models have been developed, though to a limited extent. However, the lack of accurate input parameters (thermophysical properties of foods, flow characteristics, motion and temperature of particles inside the package) limits the applicability of the existing models for complex heating systems. Hence, modelling of heat transfer in complex heating situations still represents a major challenge in thermal process evaluation.

As in the physical mathematical approach, in case of direct temperature registration, calibrating thermocouples and proper integration routines are essential; or in case of temperature reconstruction from theoretical or empirical heat transfer models, the need for reliable and accurate input parameters (e.g. thermal conductivity, heat capacity, heating rate factor, fluid-to-particle heat transfer coefficient) is a central issue, calibration and evaluation of the TTI monitoring system in terms of kinetic parameters and accurate reading of the system after thermal treatment are critical to the proper use of the developed TTIs.

Most of the developed TTIs described in Table 18.1 are research tools, but at the moment are not commercialised. Progress in alternative procedures, such as the use of TTIs, will complement continuing efforts in improving physical-mathematical procedures (Stoforos *et al.*, 1997). The future of TTIs lies in association with temperature sensors, each with its place in the risk assessment procedure. Thermal process evaluation is a fundamental part of risk assessment and invariably becomes a critical control point in any HACCP (hazard analysis and critical control points) plan. TTIs can be used in monitoring such critical control points, both for conventional thermal processing as well as for new technologies to produce heat-preserved foods. The development of TTIs to measure the safety/quality of heat preserved foods is in line with the needs of companies in the field of heat preservation as well as the consumer demands for microbiologically safe high quality foods and the demands of legislative bodies, such as the US Food and Drug Administration (FDA). TTI developers should continue to produce definitive data on the ability of a TTI to predict the thermal impact on safety/quality aspects if they are to convince the industry and regulatory authorities of their worth. It is expected that TTIs will not only be useful to the food industry but also be valuable to the pharmaceutical and chemical industries where the same problems exist.

In practice the food industry strives to maintain the tightest heating impact distribution. Whilst thermocouples have enabled processors to achieve this in static batch retort systems we are a long way from that in continuous heating processes and unable to set some novel processes with much confidence. The future development and use of TTIs will enhance the further introduction of these new thermal processing technologies for the production of microbiologically safe, high quality convenience foods, where existing approaches are not applicable. The use of TTIs can provide a better approach for establishing thermal processing conditions, leading to possible reductions in production costs and an improved safety–quality balance. TTI's are necessary in these areas and hopefully will provide food manufacturers with a means of controlling these processes.

18.6 Sources of further information and advice

A major development task on TTIs has been undertaken by a coordinated EU project (AIR1-CT92-0746, 'The development of new time-temperature integrators (product history indicators) for the quantification of thermal processes in terms of food safety and quality'), which has been investigating microbial, enzymatic and chemical systems. Several review articles or book chapters on the development and use of TTIs for thermal process evaluation can be found in literature (Tucker, 1999a,b; Van Loey *et al.*, 1996b, 1997c, 1998, 1999). Selman (1995) presents an overview of commercially available indicators for thermal process validation. Often, these commercially available process indicators consist of thermosensitive inks that change colour after exposure to a specific temperature profile and are limited to qualitative indications of the heating medium temperature. Few integrate the full temperature history but can only be placed on the outside of a container and hence respond only to heating medium temperature and provide no relevant information on the in pack thermal efficacy of a process. These indicator systems are used extensively in the drug industry, for example for the sterilisation of medical tools, but are not or less appropriate to monitor thermal processes in the food industry.

Besides TTIs for thermal process evaluation, TTIs have been and are being developed to monitor quality-losses during low temperature (refrigeration/freezing) storage and distribution (e.g. Tsoka *et al.*, 1998). Those monitors are based on reactions occurring at chilling or freezing temperatures and are not applicable in pasteurisation or sterilisation processes. Reviews on TTIs for low temperature applications are available in literature (e.g. Taoukis *et al.*, 1991; Sherlock *et al.*, 1991; Labuza and Fu, 1995).

18.7 References

ADAMS J B, LANGLEY F M (1998), 'Nitrophenyl glucoside hydrolysis as a potential time-temperature integrator reaction', *Food Chemistry*, 62(1), 65–68.

BALL C O (1923), 'Thermal process time for canned food', *Bull.* 7 1(37), Natl. Res. Council, Washington D.C., USA.

BALL C O, OLSON F C W (1957), *Sterilization in Food Technology*, McGraw-Hill, New York.

BEAN P, DALLYN H, RANJITH H (1979), 'The use of alginate spore beads in the investigation of ultra-high temperature processing', *J Food Technol*, 281–294.

BERRY M R, SINGH R K, NELSON P E (1989), 'Kinetics of methylmethionine sulfonium in buffer solutions for estimating thermal treatment of liquids foods', *J Food Proc Preserv*, 13, 475–488.

BETTS G D (1992), 'The microbiological safety of sous-vide processing', *Technical Manual No. 39*, Campden and Chorleywood Food Research Association, pp. 1–58, UK.

BETTS G D, GAZE J E (1992), 'Food pasteurisation treatments', *Technical Manual No. 27*, Campden and Chorleywood Food Research Association, UK.

BROWN K L, AYRES C A, GAZE J E, NEWMAN M E (1984), 'Thermal destruction of bacterial spores immobilized in food/alginate particles', *Food Microbiol* 1, 187–198.

BURTON H, PERKIN A G, DAVIES H L, UNDERWOOD H M (1977), 'Thermal death kinetics of *Bacillus stearothermophilus* spores at ultra high temperatures. III. Relationship between data from capillary tube experiments and from UHT sterilizers', *J Food Technol*, 12, 149–161.

CLAEYS W, LUDIKHUYZE L, VAN LOEY A, HENDRICKX M (2001a), 'Inactivation kinetics of alkaline phosphatase and lactoperoxidase, and denaturation kinetics of β-lactoglobulin in raw milk under isothermal and dynamic temperature conditions', *J Dairy Res*, 68, 95–107.

CLAEYS W, LUDIKHUYZE L, HENDRICKX M (2001b), 'Formation kinetics of hydroxymethylfurfural, lactulose and furosine under isothermal and non-isothermal conditions', *J Dairy Res,* 68, 287–301.

DE CORDT S, HENDRICKX M, MAESMANS G, TOBBACK P (1992), 'Immobilized α-amylase from *Bacillus licheniformis*: a potential enzymic time-temperature integrator for thermal processing', *Int J Food Sci Technol*, 27, 661–673.

DE CORDT S, AVILA I, HENDRICKX M, TOBBACK P (1994), 'DSC and protein-based time-temperature integrators: case study of α-amylase stabilized by polyols and/or sugar', *Biotechnology and Bioengineering*, 44(7), 859–865.

ELIOT-GODÉREAUX S, GOULLIEUX A, QUÉNEUDEC T'KINT M (2003), 'Elaboration and kinetic modelling of the formation of a biochemical marker to quantify HTST processing', *Food Research International*, 36, 131–139.

FAVETTO G J, CHIRIFE J, SCORZA O C, HERMIDA C (1988), 'Color-changing indicator to monitor the time-temperature history during cooking of meats', *J Food Protect*, 51(7), 542–546.

FAVETTO G J, CHIRIFE J, SCORZA O C, HERMIDA C (1989), 'Time-temperature integrating indicator for monitoring the cooking process of packaged meats in the temperature range of 85–100ºC', United States Patent, 4 834 017.

GAZE J E, SPENCE L E, BROWN G D, HOLDSWORTH S D (1990), 'Microbiological assessment of process lethality using food/alginate particles', *Technical Memorandum No. 580*, Campden and Chorleywood Food Research Association, 1–47.

GUIAVARC'H Y P, DELI V, VAN LOEY A M, HENDRICKX M E (2002a), 'Development of an enzymic time temperature integrator for sterilization processes based on *Bacillus licheniformis* alpha-amylase at reduced water content', *J Food Sci*, 67(1), 285–291.

GUIAVARC'H Y P, DINTWA E, VAN LOEY A M, ZUBER F T, HENDRICKX M E (2002b), 'Validation and use of an enzymic time-temperature integrator to monitor thermal impacts inside a solid/liquid model food', *Biotechnology Progress*, 18(5), 1087–1094.

GUIAVARC'H Y, SILA D, DUVETTER T, VAN LOEY A, HENDRICKX M (2003), 'Influence of sugars and polyols on the thermal stability of purified tomato and cucumber pectinmethylesterases: a basis for TTI development', *Enzyme and Microbial Technology*, 33, 544–555.

HAENTJENS T, VAN LOEY A, HENDRICKX M, TOBBACK P (1998), 'The use of α-amylase at reduced water content to develop time temperature integrators for sterilisation processes', *Lebensmittel-Wissenschaft und Technologie,* 31, 467–472.

HENDRICKX M, MAESMANS G, DE CORDT S, NORONHA J, VAN LOEY A, TOBBACK P (1995), 'Evaluation of the integrated time temperature effect in thermal processing of foods', *CRC Crit. Rev. Food Sci. and Nutrition* 35(3), 231–262.

HEPPEL N J (1985), 'Measurement of the liquid-solid heat transfer coefficient during continuous sterilization of foodstuffs containing particles'. Proceedings of IUFOST symposium on aseptic processing and packaging of foods, September 9–12, Tylosand, Sweden, 108–114.

HERSOM A C, SHORE D T (1981), 'Aseptic processing of foods comprising sauce and solids', *Food Technol*, 35(4), 53–62.

HSU Y C, SAIR A I, BOOREN A M, SMITH D M (2000), 'Triose phosphate isomerase as an endogenous time-temperature integrator to verify adequacy of roast beef processing', *J Food Sci*, 65(2), 236–240.

HUNTER G M (1972), 'Continuous sterilization of liquid media containing suspended particles', *Food Technol Aust*, April, 158–165.

JONES A, PFLUG I (1981), '*Bacillus coagulans*, FRR B666, as a potentional biological indicator organism', *J Parenter Sci Technol*, 53(3), 82–87.

KIM H-J, TAUB I A (1993), 'Intrinsic chemical markers for aseptic processing of particulate foods', *Food Technol*, 47(1), 91–99.

KNAP R P, DURANCE T D (1998), 'Thermal processing of suspended food particles in cans with end-over-end agitation', *Food Research International*, 31(9), 635–643.

LABUZA T P, FU B (1995), 'Use of time/temperature integrators, predictive microbiology, and related technologies assessing the extent and impact of temperature abuse on meat and poultry products', *Journal of Food Safety*, 15(3), 201–227.

LEMOS M A, OLIVEIRA J C, SARAIVA J A (2000), 'Influence of pH on the thermal inactivation kinetics of horseradish peroxidase in aqueous solution', *Lebensmittel Wissenschaft und Technologie*, 33(5), 362–368.

LUND D (1977), 'Design of thermal processes for maximizing nutrient retention', *Food Technol*, 31(2), 71–78.

MAESMANS G, HENDRICKX M, DE CORDT S, TOBBACK P (1993), 'Theoretical consideration on design of multi-component time temperature integrators in evaluation of thermal processes', *J Food Proc Preserv*, 17(5), 369–389.

MAESMANS G, HENDRICKX M, DE CORDT S, VAN LOEY A, NORONHA J, TOBBACK P (1994), 'Combined use of the equivalent point method and a multi-component time temperature integrator in thermal process evaluation: influence of kinetic characteristics and reference temperature', *Food Control*, 5(4), 249–256.

MAESMANS G, HENDRICKX M, DE CORDT S, TOBBACK P (1995), 'A critical evaluation of the equivalent point method', *J Food Eng*, 24, 225–248.

MATTHIASSON, E. AND GUDJONSSON, T. (1991). 'Enzymes as thermal-time indicators in food', *ITI*, Reykjavik.

MULLEY A, STUMBO C, HUNTING W (1975a), 'Kinetics of thiamine degradation by heat. A new method for studying reaction rates in model systems and food products at high temperatures', *J Food Sci*, 40(5), 989–992.

MULLEY A, STUMBO C, HUNTING W (1975b), 'Kinetics of thiamine degradation by heat. Effect of pH and form of the vitamin on its rate of destruction', *J Food Sci*, 40(5), 993–996.

NORWIG J F, THOMPSON D R (1986), 'Microbial population, enzyme and protein changes during processing', in Okos M R, *Physical and Chemical Properties of Food*, ASAE, Michigan, USA.

OCIO M J, SALVADOR FERNANDEZ P, RODRIGO M, PERIAGO P, MARTINEZ A (1997), 'A time temperature integrator for particulated foods: thermal process evaluation', *Food Research and Technology*, 205(4), 325–328.

ORTA-RAMIREZ A, PRICE J F, HSU Y C, VEERAMUTHU G J, MERRITT J S, SMITH D M (1997), 'Thermal inactivation of *Escherichia coli* O157:H7, *salmonella senftenberg*, and enzymes with potential as time-temperature indicators in ground beef', *J Food Protect*, 60(5), 471–475.

ORTA-RAMIREZ A, MERRILL J E, SMITH D M (2000), 'pH affects the thermal inactivation parameters of R-phycoerythrin from Porphyra yezoensis', *J Food Sci*, 65(6), 1046–1050.

PFLUG I J (1987), *Textbook for an Introduction Course in the Microbiology and Engineering of Sterilization Processes*, Sixth edition, Environmental Sterilization Laboratory, Minneapolis, USA.

PFLUG I J, ODLAUG T E (1986), 'Biological indicators in the pharmaceutical and medical device industry', *J Parenter Sci Technol*, 40(5), 242–248.

PFLUG I J, JONES A, BLANCHETT R (1980a), 'Performance of bacterial spores in a carrier system in measuring the F_0-value delivered to cans of food heated in a Steritort', *J Food Sci*, 45(4), 940–945.

PFLUG I J, SMITH G, HOLOCOMB R, BLANCHETT R (1980b), 'Measuring sterilizing values in containers of food using thermocouples and biological indicator units', *J Food Protect*, 43(2), 119–123.

PINHEIRO TORRES A, OLIVEIRA F A R, SILVA C L M, FORTUNA S P (1994), 'The influence of pH on the kinetics of acid hydrolysis of sucrose', *Journal of Food Process Engineering*, 17, 191–208.

RODRIGO F, RODRIGO M C, MARTINEZ A (1998), 'Evaluation of a new time-temperature integrator in pilot plant conditions', *Food Research and Technology*, 206(3), 184–188.

RODRIGUEZ A C, TEIXEIRA A A (1988), 'Heat transfer in hollow cylindrical rods used as bioindicator units for thermal process validation', *Trans ASAE,* 31(4), 1233–1236.

RÖNNER U (1990a), 'A new biological indicator for aseptic sterilization', *Food Technology International Europe*, 90, 43–46.

RÖNNER U (1990b), 'Bioindicator for control of sterility', *Food Laboratory News*, 22(6:4), 51–54.

RÖNNER U (1996), 'Microbiological TTIs for pasteurisation processes', Oral presentation at the seminar 'Time Temperature Integrators for thermal process evaluation-from development to industrial application', November 26, 1996, Leuven, Belgium.

SADEGHI F, SWARTZEL K R (1990), 'Time-temperature equivalence of discrete particles during thermal processing', *J Food Sci*, 55(6), 1696–1698, 1739.

SASTRY S K, LI S F, PATEL P, KONANAYAKAM M, BAFNA P, DOORES S, BEELMAN R B (1988), 'A bioindicator for verification of thermal processes for particulate foods', *J Food Sci*, 53(3), 1528–1531.

SEGNER W P, RAGUSA T J, MARCUS C L, SOUTTER E A (1989), 'Biological evaluation of a heat transfer simulation for sterilizing low-acid large particulate foods for aseptic packaging', *J Food Proc Preserv*, 13, 257–274.

SELMAN J D (1995), 'Time-temperature indicators', in Rooney M L, *Active Food Packaging*, Blackie, 215–237.

SHERLOCK M, FU B, TAOUKIS P S, LABUZA T P (1991), 'A systematic evaluation of time-temperature indicators for use as consumer tags', *J Food Protect*, 54(11), 885–889.

SMITH D M (1995), 'Immunoassays in process control and speciation of meats', *Food Technol*, 49(2), 116–119.

STOFOROS N G, NORONHA J, HENDRICKX M, TOBBACK P (1997), 'A critical analysis of mathematical procedures for the evaluation and design of in-container thermal processes for food', *CRC Critical reviews in Food Science and Nutrition*, 37(5), 411–441.

STOFOROS N G, TAOUKIS P S (1998), 'A theoretical procedure for using multiple responses time-temperature integrators for the design and evaluation of thermal processes', *Food Control*, 9(5), 279–287.

SWARTZEL K R, GANESAN S G, KUEHN R T, HAMAKER R W, SADHEGI F (1991), 'Thermal memory cell and thermal system evaluation', US patent 5 021 981.

TAOUKIS P S, LABUZA T P (1989a), 'Applicability of time-temperature indicators as shelf life monitors of food products', *J Food Sci*, 54(4), 783–788.

TAOUKIS P S, LABUZA T P (1989b), 'Reliability of time-temperature indicators as food quality monitors under nonisothermal conditions', *J Food Sci*, 54(4), 789–791.

TAOUKIS P S, FU B, LABUZA T P (1991), 'Time temperature indicators', *Food Technol*, 45(10), 70–82.

TSOKA S, TAOUKIS P S, CHRISTAKOPOULOS P, KEKOS D, MACRIS B J (1998), 'Time temperature integration for chilled food shelf life monitoring using enzyme-substrate systems', *Food Biotechnology*, 12(1/2), 139–155.

TUCKER G (1999a), 'Novel time-temperature integrators', *Food Review*, 26(11), 13–14.

TUCKER G (1999b), 'A novel validation method: application of time-temperature integrators to food pasteurization treatments', *Food and Bioproducts Processing*, 77, 223–231.

TUCKER G S (2000), 'Estimation of pasteurisation values using an enzymic time-temperature integrator', *Food Australia*, 52(4), 131–136.

TUCKER G S, LAMBOURNE T, ADAMS J B, LACH A (2002), 'Application of a biochemical time-temperature integrator to estimate pasteurisation values in continuous food processes', *Innovative Food Science and Emerging Technologies*, 3(2), 165–174.

VAN LOEY A, LUDIKHUYZE L, HENDRICKX M, DE CORDT S, TOBBACK P (1995a), 'Theoretical consideration on the influence of the *z*-value of a single component time temperature integrator on thermal process impact evaluation', *J Food Protect*, 58(1), 39–48.

VAN LOEY A, HENDRICKX M, LUDIKHUYZE L, WEEMAES C, HAENTJENS T, DE CORDT S, TOBBACK P (1995b), 'Potential Bacillus subtilis α-amylase based time temperature integrators to evaluate pasteurisation processes', *J Food Protect*, 59(3), 261–267.

VAN LOEY A, ARTHAWAN A, HENDRICKX M, HAENTJENS T, TOBBACK P (1996a), 'The development and use of an α-amylase based time temperature integrator to evaluate in-pack pasteurisation processes', *Lebensmittel-Wissenschaft und Technologie*, 30(1), 94–100.

VAN LOEY A, HENDRICKX M, DE CORDT S, HAENTJENS T, TOBBACK P (1996b), 'Quantitative evaluation of thermal processes using time-temperature integrators', *Trends in Food Science and Technology*, 7, 16–26.

VAN LOEY A, HAENTJENS T, HENDRICKX M, TOBBACK P (1997a), 'The development of an enzymic time temperature integrator to assess thermal efficacy of sterilization of low-acid canned foods', *Food Biotechnology*, 11(2), 147–168.

VAN LOEY A, HAENTJENS T, SMOUT C, HENDRICKX M, TOBBACK P (1997b), 'The use of an enzymic time temperature integrator to monitor lethal efficacy of sterilization of low-acid canned foods', *Food Biotechnology*, 11(2), 169–188.

VAN LOEY A, SMOUT C, HENDRICKX M (1997c), 'European research results on new approaches to assess the safety of thermal processes', *European Food & Drink Review*, autumn, 43–49.

VAN LOEY A, HAENTJENS T, HENDRICKX M (1998), 'The potential role of time-temperature integrators for process impact evaluation in the cook-chill chain', in Ghazala S, *Sous Vide and Cook-chill Processing for the Food Industry*, Aspen Publishers Inc., Maryland, 89–110.

VAN LOEY A, HAENTJENS T, SMOUT C, HENDRICKX M (1999), 'Enzymic time temperature integrators for the quantification of thermal processes in terms of food safety', in

Oliveira F A R and Oliveira J C, *Processing of Foods: Quality Optimization and Process Assessment*, CRC Press, Boca Raton, 13–40.

VEERAMUTHU G J, PRICE J F, DAVIS C E, BOOREN A M, SMITH D M (1998), 'Thermal inactivation of *Escherichia coli* O157:H7, *salmonella senftenberg*, and enzymes with potential as time-temperature indicators in ground turkey thigh meat', *J Food Protect*, 61(2), 171–175.

VILLOTA R, HAWKES J G (1986), 'Kinetics of nutrients and organoleptic changes in foods during processing', in Okos M R, *Physical and Chemical Properties of Food*, ASAE, Michigan, USA, 266–366.

VILLOTA R, HAWKES J G (1992), 'Reaction kinetics in food systems', in Heldman D R, Lund D B, *Handbook of Food Engineering*, Marcel Dekker inc., NY, USA, 39–144.

WEN CHIN, L. (1977). Disaccharide hydrolysis as a predictive measurement for the efficacy of heat sterilization in canned foods. Ph.D. thesis, Department of Food Science and Nutrition, University of Massachusetts, Amhers, Massachusetts, USA.

WENG Z, HENDRICKX M, MAESMANS G, TOBBACK P (1991), 'Immobilized peroxidase: a potential bioindicator for evaluation of thermal processes', *J Food Sci*, 56(2), 567–570.

WITONSKY R J (1977), 'A new tool for the validation of the sterilization of parenterals', *Bull Parenter Drug Assoc,* 11(6), 274–281.

WUNDERLICH J (1995), 'Temperature bio-indicators based on immobilized enzyme for the controlling of continuous HTST heating of particulate foods and other heating processing'. Proceedings of New Shelf-life Technologies and Safety Assessments, VTT symposium, Helsinki, Finland, (Ahvenainen, R., Mattila-Sandholm, T. and Ohlsson, T., eds), pp. 197.

YAWGER E S (1978), 'Bacteriological evaluation for thermal process design', *Food Technol*, 32(6), 59–62.

19

New techniques for measuring and validating thermal processes

K. P. Nott and L. D. Hall, University of Cambridge, UK

19.1 Introduction: limitations of current temperature measurement

The measurement of temperature is fundamental to most thermal processing operations on food materials, first and foremost for microbiological safety but also on grounds of food quality and economics. Currently there is a need for more comprehensive multi-dimensional temperature measurement techniques because:

1. Heating is often heterogeneous, particularly when the food is a composite of different materials.
2. As a consequence, commercially many foods are systematically overcooked to ensure 100% microbiological safety and longer shelf-life.
3. Most of the commonly used temperature measurement techniques are invasive, destructive, require physical contact for good response times, and provide only point measurements.
4. Most of the multi-dimensional Computational Fluid Dynamics (CFD) models from which process equipment is designed have yet to be rigorously validated.

Fortunately, as a result of advances in medical imaging many new measurement techniques are available and, because most are non-contact, they are compatible with the hygiene standards required for industrial food processing. Also, in principle non-contact instruments are able to respond immediately to changes in temperature, whereas contact instruments may have a lengthy time delay before they reach thermal equilibrium with the object to be measured. The purpose of this chapter is to describe the minimally- and non-invasive techniques available for temperature measurement, with particular attention to their adaptability to an industrial environment, including those already in use for off- or on-line process monitoring and development research. Where possible, examples are given which compare the measured data with that produced by Computerised Fluid Dynamics (CFD) software, since this serves to cross-validate both the measurement technique and the software.

19.2 Minimal and non-invasive measurement techniques

There are a variety of minimal and non-invasive techniques, shown in Table 19.1, that can measure temperatures in the range useful for food processing, i.e. −50°C to 200°C; each having advantages and disadvantages for particular applications, as described below. In addition, the temperature dependence of electrical impedance (Conway *et al.*, 1992; Paulsen *et al.*, 1996; Gersing, 1999); permittivity (Kimoto and Shida, 2000); ultrasound properties (Richardson and Povey, 1990, Simon *et al.*, 1998); and X-ray absorbance have also been used. However, there have been few reported food applications, probably owing to the difficulty in separating the temperature dependence from structural changes in heterogeneous systems. The practical details of temperature measurement on the human body by most of those techniques can be found elsewhere (Miyakawa and Bolomey, 1996).

19.2.1 Fibre optic thermometry

Traditional bimetallic thermocouples are invasive, as they require direct physical contact with the food. As a consequence, they have to be hygienically encased to ensure safety standards are maintained, which lengthens their

Table 19.1 Comparison of some non-invasive techniques used to measure temperature in foods

	Claimed accuracy (°C)	Claimed response time (s)	Temperature range (°C)	Other comments
Fibre optic thermometry	< 1	0.25* 1.9†	−195–450* −40–350†	Point
Infrared thermometry	1 over any 100°C range	0.02	−40–1500	Pseudo 3D
Microwave radiometry	< 0.5	2.5	See text	Weighted average
Magnetic resonance imaging	1 (uniform) 5 (non-uniform)	From 0.5 (1D) to 52 (3D)	See text	Bulk 1, 2 or 3D

* Fluorometry † Interferometry

response time (Berrie, 2001). In addition, in many cases the temperature measurement is influenced by the presence of the thermocouple, which itself is a good thermal and electrical conductor. Fibre optic thermometry is an affordable alternative technique used to measure temperature in environments where the integrity of metallic thermocouples is compromised, e.g. during radio frequency-, microwave- and ohmic-heating; because it is small (< 1 mm) and has a low thermal conductivity it does not deform existing temperature fields and thereby gives accurate, highly localised measurements. It is typically used for medical applications because the chemical inertness of the sensors allow them to survive sterilisation; that is also an important consideration in their application to food. The basic theory of fibre optic thermometry can be found elsewhere (Michalski *et al.*, 2001) and, although many different types exist, two are commonly available (one based on fluorescence, the other on Fabry-Perot interferometry) that are usable over the temperature range of interest for food applications. The former measures a temperature dependent fluorescent signal decay, and the latter thermal expansion at the tip of the fibre optic probe. The mechanism of the latter is versatile in that the same signal conditioner can be used with different probes to measure, amongst other parameters, pressure; hence care must be taken to ensure that the temperature calibration is pressure independent. Apart from the integrity of the hardware, there are no other physical factors that affect its accuracy as a closed system. Nevertheless, it relies upon contact with the object in question; hence there will be an associated response time, which will increase if the probe is encased for protective and/or hygienic reasons.

Although multi-channel signal conditioners exist, they become prohibitively expensive for more than eight channels for which the total cost including sensors can be ~£10,000. Even eight sensors are not always enough to provide a comprehensive array across the sample, thus making it difficult to detect hot and cold spots associated with certain types of heating, such as microwave, for which the individual sensors are otherwise suited. Nevertheless, instruments are available for research purposes that can measure up to 16 spot temperatures in a domestic microwave oven fitted with a turntable. The most obvious use of fibre optic thermometry in an industrial setting is temperature measurement during a microwave pasteurisation or sterilisation process. The centre of the food mass is the slowest heating point for conventional heating modalities and also more often than not for microwave heating; unless the combination of microwave frequency, sample size and properties leads to a focusing effect, in which case the centre will be the fastest heating point. The sensors can determine the time it takes for the centre of the product to reach a predefined temperature, after which the microwave power can be pulsed to stabilise the temperature thereby preventing overheating.

19.2.2 Infrared thermometry

Infrared thermometry relies on the fact that all materials emit energy in the form of electromagnetic radiation. The wavelength of maximum emittance is given by

Wien's displacement law λ_M (μm) = $2.89 \times 10^3/T$, where T is the absolute temperature; for food-heating applications this occurs between 6 and 13 μm. The spectrum of thermal radiation is determined, beyond the properties of a blackbody, by the temperature of a particular body and by the emissivity of its surfaces. The rate of heat flux (radiation emitted per unit area) is given by $Q = \sigma T^4 \epsilon$, where σ is the Stefan-Boltzmann constant. The emissivity ϵ is the fraction of the radiation emitted by an object compared to that emitted by a perfect radiator (blackbody) at the same temperature; $\epsilon = 0.96$ for water, ice and most organic substances and this value is often used for food applications.

There are many devices that either enable measurement at one point, or can be mechanically scanned to provide an image, albeit at decreased sensitivity; details of that technology can be found elsewhere (Ridley, 2001). Modern thermal imaging cameras, available since the early 1990s, are based on a Focal Plane Array technology giving images of 256 × 256 or 320 × 240 pixel resolution over the 7.5–13 μm spectral range. A matrix of individual sensors is lined up in rows and columns, each creating a pixel of information; the data can be acquired almost continuously at a rate of 60 images per second and recorded on digital media, which allows automated analysis of the images by software.

The accuracy of the measurement is a function of emissivity and environmental factors. Since the emissivity can vary with surface condition, temperature, wavelength and time, use of a constant value may not always be appropriate; as a consequence, some instruments provide predefined emissivity look-up tables, although even these should be treated with caution. Commercial instruments operate either between 3 and 5 μm, or 8 and 13 μm to avoid atmospheric absorption, particularly by water vapour. Even so, care must be taken because steam, dust and smoke, etc. may obstruct the instrument's optics, leading to measurement errors. Two other variables, reflection and transmission, complicate infrared temperature measurement; hence polished materials such as metals should be avoided since they can reflect radiant energy to the scanner. It is also recommended that the temperature measurement should not be taken through glass which has very distinct reflection and transmission properties. Nevertheless, it has been shown that if the transmission properties are taken into account, temperature measurement is possible through certain materials such as hardware cloth, a microwave oven door, or plastic film in food packaging (Goedeken *et al.*, 1991; Mullin and Bows, 1993).

The major disadvantage of infrared thermometry is that since the penetration depth of the radiation is limited to millimetres this technique can only be used to measure surface temperatures. Nevertheless, it is an effective way of mapping non-uniform temperature distributions, particularly in ready-meal trays where, since they are shallow, one can assume that the surface temperature distributions are representative of those within (Bows and Joshi, 1992). As such, it has proved to be an invaluable technique for the development of new microwave configurations (Bows, 1999, 2000; Bows *et al.*, 1999). To develop models that can be used to estimate the internal temperature one can use infrared thermometry in conjunction with thermocouples (Ibarra *et al.*, 1999, 2000). Either way, there is no guarantee that a food cooked using this technology is 100% safe, though a continuous surface temperature measurement is likely to be more useful in process control than an intermittent, manual, central reading. As infrared thermometry can only measure surface temperature, those inside most food processing equipment cannot be measured; however, windows which are transparent to infrared can be used as viewing ports into closed vessels, and fibre optic cables can be used to separate the sensor from the detector in hostile environments as described previously (Section 19.2.1).

19.2.3 Microwave radiometry

Microwave radiometry is similar to infrared thermometry in that it is based on the measurement of the temperature-dependent intensity of electromagnetic thermal radiation; however, the measurements are made in the microwave frequency region. The microwave power radiated depends on the absolute temperature and microwave absorption of the material, as well as on the change in microwave properties at the surface, or 'emissivity'. Since most foods are partially transparent to electromagnetic radiation in the microwave frequency range, which allows the measurement system to couple with radiation up to several centimetres below the surface of an object, this approach has the potential to probe the whole volume of some food products; the dependence of penetration depth on wavelength offers the potential for profiling measurements (in contrast, infrared radiation is readily absorbed, which restricts measurements to a superficial layer; Section 19.2.2). Since, in practice, the microwave properties are mostly dependent on water content, in a composite food the higher water content components will be the main source of the microwave signal. Salt and mineral contents have a modest effect; fats, oils and proteins are relatively and air completely – transparent to microwaves. Microwaves are almost completely unaffected by particulates that block infrared radiation; however, like infrared, it would be impossible to detect radiation from within processing equipment unless it was suitably adapted for the sensor.

The technical details (Foster and Cheever, 1992; Leroy *et al.*, 1998) as well as the industrial and medical applications (Land, 2001) are dealt with in a number of informative reviews. In general, a microwave radiometer consists of a microwave probe which couples with radiation generated within the source object, and a microwave radiometer receiver which measures the equivalent temperature; the signal is very small and therefore requires high gain amplification. The signal can be collected all around the source, from within a cavity, or from the surface in front of an antenna type probe. A cavity has been used in a product developed to measure chilled and frozen food temperatures (from −25 to 20°C) throughout the distribution chain. This is an ideal application because: the signal is immune from external influences; there is no operational dependence; and the temperature distribution will be fairly uniform given that chilled and frozen ready meals are usually packaged in thin trays. Figure 19.1 shows a comparison of the average weighted temperature measured by microwave

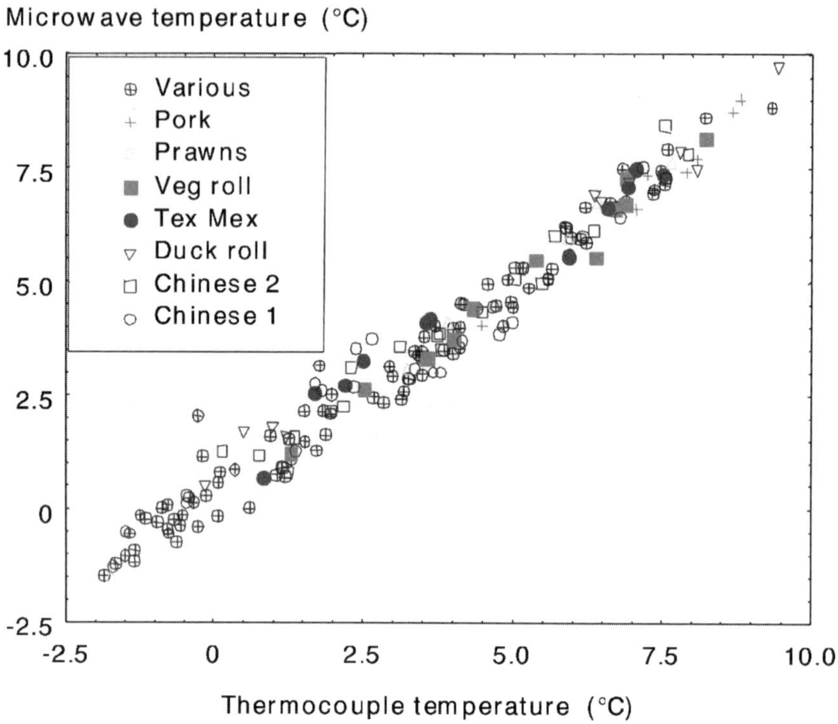

Fig. 19.1 Graph of a 'weighted average' temperature measurement by microwave radiometry vs. point measurements using thermocouples (courtesy of Dr D.V. Land, University of Glasgow, UK).

radiometry against point thermocouple measurements taken in various factories, all of which had been carefully inter-calibrated (data courtesy of Dr D.V. Land, University of Glasgow, UK). It is believed that much of the 0.9°C spread can be attributed to the thermocouple measurement as well as any temperature gradients that remain in the food, and that under ideal conditions it should be no more than 0.1–0.2°C (private communication).

As with most techniques, there has to be a compromise between the time required for the temperature measurement and the temperature resolution achievable. In the above application, this off-line measurement had no minimum time limit related to the speed of the production line. However, an on-line measurement would require use of an antenna for detection of moving objects, therefore has less time available for measurement, as well as exposing it to external influences. Nevertheless, currently it is one of the most promising techniques for on-line food process control, and developments are under way to use the technology at higher temperatures on a production line. Although the upper and lower limits of temperature measurement have not been explored, there are no obvious technical constraints.

19.3 Magnetic resonance imaging: principles, measurements and processing

19.3.1 Basic principles

The nuclei of the protons in water (and fat) in food are effectively magnetic so that when placed in a strong magnetic field their spins align either parallel (low energy) or antiparallel (high energy) to that field (represented by quantum numbers +½ and −½ respectively; Fig. 19.2); the population difference between those energy levels, which is the basis for the NMR signal, is dependent on the magnetic field strength ($\mathbf{B_0}$). The net magnetisation, $\mathbf{M_0}$, lies in the direction of the applied field (z direction). A radio frequency (RF) pulse is transmitted that can interact with the aligned nuclear spins at a particular 'resonance' frequency ν (Hz) given by

$$\nu = \gamma \mathbf{B}_0 / 2\pi \qquad [19.1]$$

where γ is the magnetogyric ratio of the protons; this causes the orientation of the aligned nuclei to invert and the precessions of the individual spins are brought into phase. The combination of RF pulse-duration and -power, $\mathbf{B_1}$, which rotates all the magnetisation into the x,y plane, transverse to the direction of the $\mathbf{B_0}$ field, for detection is called a '90° pulse'; this induces the maximum possible NMR signal, which is proportional to the water (and/or fat) proton content. When the RF pulse is turned off, the overall magnetisation along the z axis recovers with a characteristic time constant T_1; the phase relationship of the nuclear spins in the transverse (x,y) plane decays with a characteristic time constant T_2. Both T_1 and T_2 are measures of water mobility and dependent on food structure and its temperature.

To observe magnetic resonance phenomena of nuclei as a function of their position in real space (i.e. to produce an image) a small magnetic field gradient, $\mathbf{G}$, is applied in addition to $\mathbf{B}_0$ (Fig. 19.3). As a result, the resonant frequency of the nuclear spins is then dependent on the spatial position of the molecules within the sample, and is given by:

$$\nu(\mathbf{r}) = (\gamma \mathbf{B}_0 + \gamma \mathbf{G} \cdot \mathbf{r}) / 2\pi \qquad [19.2]$$

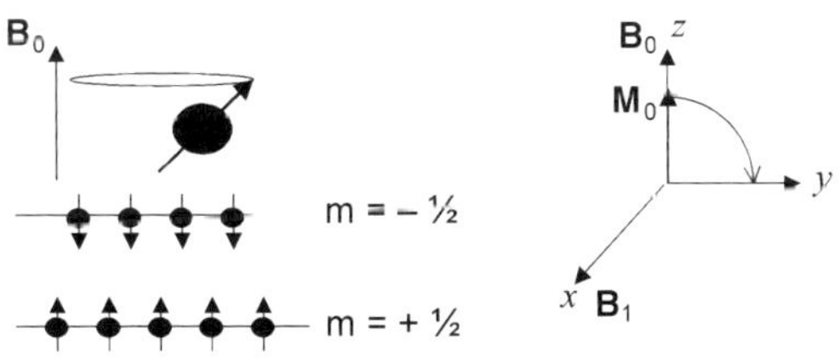

Fig. 19.2 The NMR phenomenon. The application of a strong magnetic field, $\mathbf{B}_0$, aligns the nuclear spins either parallel (m = + ½) or anti-parallel (m = ½) to the field; the energy difference between the two states results in a population difference producing a net magnetisation in one direction, $\mathbf{M_0}$; circle denotes precessional motion of the net magnetisation vector. The combination of RF pulse-duration and power, $\mathbf{B_1}$, which rotates all the magnetisation into the x,y plane which is transverse to the direction of the $\mathbf{B_0}$ field for detection is called a '90° pulse'; this induces the maximum possible NMR signal.

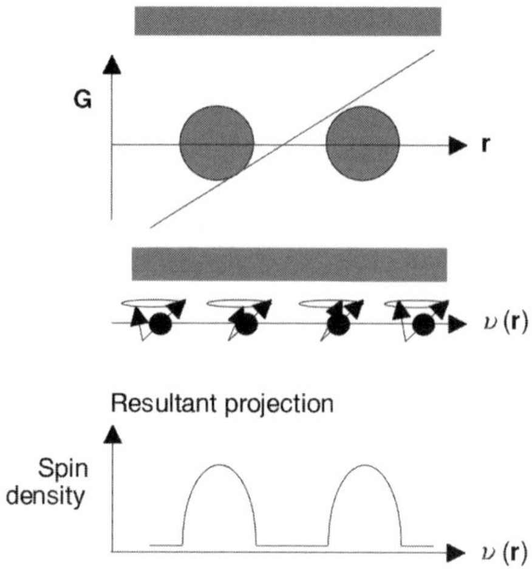

Fig. 19.3 Magnetic Resonance Imaging. The magnetic field gradient (**G**) across a sample space (**r**) acts to produce a similar spatial dependence in the Larmor frequency of the spins $\nu(\mathbf{r})$. Consequently the spin density at various positions can be extracted from the MR signal to produce an image.

where **r** is the position vector of the nuclear spin. MRI uses three orthogonal magnetic field gradients for spatial localisation of an object in three dimensions. For example, one gradient is used to control the position of the 'slice' and the other two the resolution within that slice to give a 2D image; a more comprehensive description of the basics of MRI can be found in Hashemi and Bradley (1997).

Both the magnitude and phase of the MRI signal can be sensitised in the MRI experiment to various properties. The magnitude of the MRI signal is mainly a function of liquid proton density $\mathbf{M_0}$ (related to water content) but also depends on the relaxation times, T_1 and T_2 (related to water mobility). The effect of molecular diffusion in the presence of a magnetic field gradient on magnetic resonance (MR) signals has been described in some of the classical NMR papers (Hahn, 1950; Carr and Purcell, 1954) and summarised by Callaghan (1993). Diffusion causes an attenuation of the overall magnitude of the MR signal due to a loss of phase between the spins of water molecules that move randomly along the direction of a magnetic field gradient, the extent of which depends on the sizes of the diffusion coefficient and that gradient. Thus, the application of a series of pulses of known gradient strength during a Pulsed Gradient Spin Echo (PGSE) experiment can be used to determine the diffusion coefficient (Stejskal and Tanner, 1965).

The return to equilibrium of the MR signal is detected in the receiver coil with respect to the scanner frequency; the phase of that signal is related to the field strength and hence depends on the homogeneity of the static magnetic field. It is also sensitive to chemical shift (see below), and to differences in magnetic susceptibility between substances. Those differences in any object alter the local magnetic field and, as a consequence, the intensity of the MR response, when placed in a uniform magnetic field. Most food materials are diamagnetic; however, large differences in susceptibility, such as at an air/water interface, cause magnetic field gradients which lead to dephasing of spins and, as a consequence, signal loss as described previously. In contrast with the diffusion experiments, the application of a series of known gradient strength pulses sensitises the phase of the MRI signal to flow, thereby enabling measurement of the flow velocity in any chosen direction (Pope and Yao, 1993; Fukushima, 1999). Importantly, MRI is the only method which can quantitate in 3D the flow of opaque fluids and separately measure the flow of water and lipid in mixed fluids.

19.3.2 Hardware

The strong magnetic field which is required for MRI is typically generated by a superconducting solenoid magnet, the coils of which have to be immersed in baths of liquid helium and nitrogen for cooling. Low cost, permanent magnets can be used but there is a limit to their field strength and consequently to the signal-to-noise, that can be achieved; nevertheless, the MRI scanners using permanent magnets which have been developed for clinical applications do provide excellent images.

Within the typically cylindrical magnet bore are other cylindrical formers, which house the shim- and gradient-coils. The shim-coils allow the $\mathbf{B_0}$ magnetic field homogeneity to be optimised, which is essential for optimal signal-to-noise and accurate image reconstruction. The gradient coils provide the magnetic field gradients in the three orthogonal directions used for spatial encoding of the MR information. Situated within the gradient coils is the RF probe which is used both to transmit pulses of RF radiation to manipulate the magnetisation of the nuclei, and also to receive the resultant signal from the sample. The shape of the coils in the probe can be matched to the dimensions of the object and can vary from a basic loop of wire ('surface coil'), to a more complex cylindrical cage ('birdcage coil').

Only containers made of plastic, glass, ceramic, paper and wood can be used; pieces of non-magnetic metals (aluminium, copper) can also be tolerated though they may destroy the local magnetic field homogeneity which is important for accurate image reconstruction, and also shield regions of the sample from the radio frequency radiation. Ferromagnetic metals, especially iron, strongly affect the magnetic field and hence iron objects should not be taken near the magnet for practical and safety reasons.

19.3.3 Temperature, flow and other related measurements

The major advantage of MRI over other tomographic techniques is that the MRI signal can be sensitised, though not always exclusively, to a specified property, e.g. water/fat-content, or -mobility, diffusion, flow, pH, redox or temperature; a more comprehensive description of MRI applications in food science can be found in some dedicated books (McCarthy, 1994; Hills, 1998a). Although temperature is an important variable in thermal processing, its role is often complicated by accompanying structural changes as well as interrelated heat and mass transfer processes. Therefore the unique power of MRI is not only that it can be used to measure in three dimensions the temperature distribution, but also other relevant parameters including thermal convection, moisture loss and redistribution, starch gelatinisation, protein denaturation and phase transitions either before, during or after the heating process.

Several MRI properties of water have been used to measure temperature in foods (Webb and Litchfield, 1999) including the relaxation times T_1 and T_2 (Hulbert *et al.*, 1995, 1997), diffusion coefficient (Le Bihan *et al.*, 1989; Hall *et al.*, 1990; Sun *et al.*, 1993, 1994) and chemical shift (Hindman, 1966; Ishihara *et al.*, 1995; Kantt *et al.*, 1997, 1998); a comparison of the sensitivity and accuracy of each MRI parameter is given in Table 19.2. Chemical shift refers to the variations in resonance frequency caused by different chemical environments, for example as in between water (-OH) and fat (-CH) protons. Each parameter has specific strengths and weaknesses; however, the chemical shift, sensitised in the phase of the MRI signal (phase mapping), appears to be the only parameter that has the potential to be independent of structural changes which accompany cooking since its temperature sensitivity is based on hydrogen bonds with vicinal water rather than the mobility of water in the food matrix. In addition, T_1, T_2 and diffusion coefficient methods are based on the magnitude of the MRI signal and therefore are dependent on moisture content, which is likely to change during thermal processing.

Although phase mapping has advantages in terms of signal-to-noise, spatial- and temporal-resolution, its major disadvantage is the necessity for a room temperature reference image which is spatially co-registered with the heated image. Hence errors in the temperature measurement can be caused by shrinkage (Nott *et al.*, 2003) and local magnetic susceptibility changes associated with movement of pockets of air during cooking. For systems which contain fat, the solution is to use the signal from fat as an internal reference (Kuroda *et al.*, 1995, 1997, 2000) since its resonance frequency and phase do not change significantly with temperature. Fortuitously, the magnetic susceptibility changes in fat are similar to those of water hence the phase difference between water and fat leaves the phase changes due to the temperature dependent MR-frequency of water as the unique dependency. Unfortunately the signal-to-noise and sensitivity of using fat as an internal reference are poor, hence in most cases it is only feasible to provide a bulk temperature measurement by that method (Walton and McCarthy, 1999).

The accuracy of all the MRI methods of temperature measurement is a function of water content, and thus is best for water-rich foods. Currently, the upper and lower limits of measurable temperature are not known; however, those will depend on the application, and will also be strongly dependent on water

Table 19.2 Sensitivity and accuracy of the MRI parameters of water used to measure temperature in real and model food systems. Taken from two-dimensional slice images unless stated otherwise

Sample	Measurement	Temperature Range (°C)	Spatial resolution	Temporal resolution	Sensitivity	Error	Reference
Model Food Gel	Diffusion	18–42	1.5 mm^2	30 s	13%/°C*	< 1.26°C	Sun *et al.* (1993)
Potato	Diffusion	20–50	0.75 mm^2	10 s	13.5%/°C*	0.5°C	Sun *et al.* (1994)
Polyacrylamide Gel	Diffusion	34–44	4 mm^2, 0.8 cm^2 ROI[†]	2.5 min	2.4%/°C	0.5°C	Le Bihan *et al.* (1989)
Muscle	Diffusion	28–38	–	3.6 min	2.5%/°C	–	Hall *et al.* (1990)
	T_1				1.5%/°C		
Carrot	T_1	20–83	Bulk	–	3%/°C	–	Hulbert *et al.* (1995)
Water	Chemical Shift	0–100	Bulk	–	0.01 ppm/°C	–	Hindman (1966)
NaCl solution	Chemical Shift	35–51	0.36 mm^2	15 s	0.0098 ppm/°C	1°C	Ishihara *et al.* (1995)
Agar gel					0.0103 ppm/°C		
Boiled egg white					0.0091 ppm/°C		
Chicken muscle					0.0114 ppm/°C		
Gel	Chemical Shift	20–60	0.88 mm^2	8 s	0.0098 ppm/°C	< 5°C	Kantt *et al.* (1997)
Raw Potato					0.0104 ppm/°C		
Cooked Potato					0.0107 ppm/°C		

* Pseudo Self Diffusion Coefficient [†] Region-of-interest

content. In general signal-to-noise ratio (SNR) of the measurement decreases with increasing temperature. The upper limits of MRI temperature measurement have not been tested in water-based foods as the system must be under pressure to prevent boiling; the challenge is to engineer an MRI compatible chamber capable of withstanding high temperatures and pressures. It is also pertinent to consider measurement of temperatures below ambient. SNR decreases rapidly below 0°C as the food system freezes; nevertheless, in some foods there is non-freezing water that is sufficiently mobile to give an MRI signal.

As well as temperature, the phase of the MRI signal can be sensitised to flow velocities up to 500 cm/s. Flow fields have been measured in a range of geometries related to food processing including those through a baffle system (Derbyshire *et al.*, 1994), an expansion-contraction-expansion chamber (Newling *et al.*, 1997a,b), a manifold (Herchel Smith Laboratory, unpublished), a pipe- and screw-extruder (Wang *et al.*, 1999, 2000; Amin *et al.*, 2003). Also of relevance to thermal processing, velocity measurement has been applied to Rayleigh Benard convection (Gibbs *et al.*, 1993; Weis *et al.*, 1996) and, more recently in combination with temperature mapping for a microscopic study on porous media, the results of which compared favourably with CFD data (Weber and Kimmich, 2002).

19.3.4 Batch processing

One of the most important concepts in batch processing is the 'spatial time-temperature history' since it is that which determines not only the homogeneity of the cooking itself, but also the 'microbiological kill' in the product. Typically, such products are sauces with meat and/or vegetable particulates; hence fluid rheology and the thermal properties of the different materials are relevant, as well as heat transfer from the medium (water, steam, hot air) through the container (glass, plastic, metal).

Many studies have shown that, in principle, MRI can be applied to aspects of batch retort processing since it provides spatially co-registered maps of temperature and structural heterogeneity, e.g. the position of particulates in a jar (Nott and Hall, 1999). Recent studies have extended this approach to microwave cooking of starch sauces in jars since this addresses the possibility of 'minimal processing'; previous studies of in-container microwave heating have been confined to transparent liquids with different viscosities (Prosetya and Datta, 1991, Anantheswaran and Liu, 1994). Figure 19.4 shows corresponding maps of temperature, structure (gelatinisation) and velocity (thermal convection) after microwave heating of a base sauce consisting of a raw 2% modified waxy maize starch, 0.5% xanthan suspension and 2% NaCl (added to modify the dielectric properties) in a jar. Microwave heating of the base sauce produces an internal focusing effect which promotes intense heating across much of the centre of the jar; in contrast, the addition of 2% NaCl promotes surface heating. The T_2 maps of the jars at room temperature show that for the base sauce, high temperatures promote faster cooking and create a convection plume up the

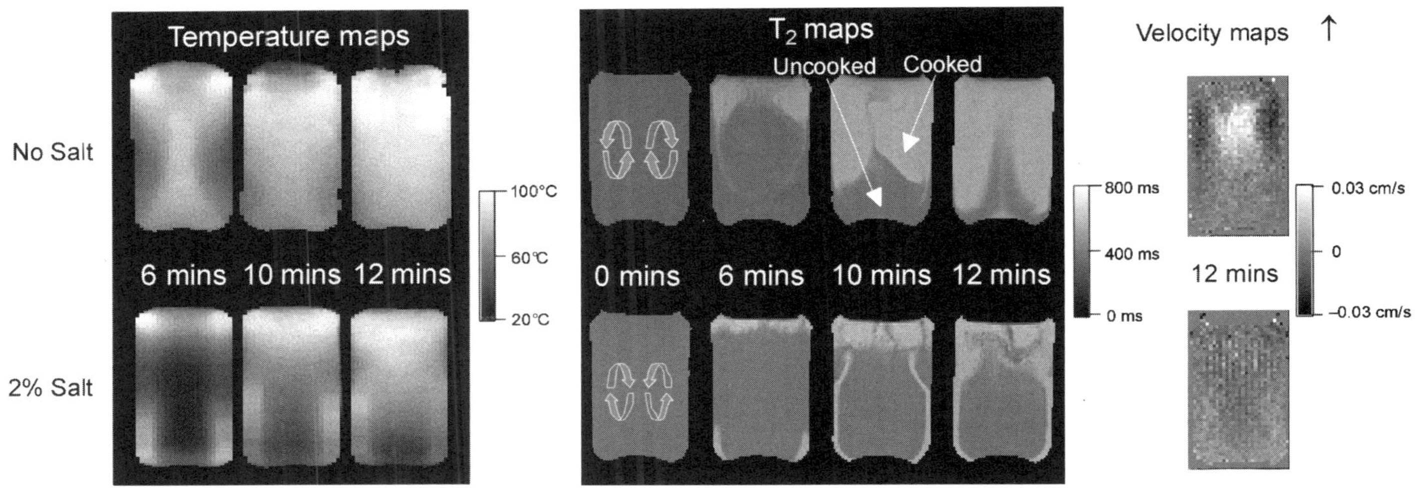

Fig. 19.4 Maps of temperature (from 3D phase mapping), gelatinisation (from 2D T_2 mapping) and thermal convection (from 2D velocity mapping), all measured by Magnetic Resonance Imaging. 330 ml jars containing a 2% waxy maize starch/0.5% xanthan sauce were heated for 6, 10 and 12 minutes in a 800 W domestic microwave oven (Goldstar M1164-TE) at 20% power with turntable rotation. 2% NaCl was used to influence heating uniformity by affecting the dielectric and rheological properties of the suspension. Arrow indicates direction of velocity encoding.

centre of the jar; in contrast, the addition of 2% NaCl results in localised heating near the inner surface of the container such that much of the jar's contents remains uncooked. Those observations are confirmed by spatially co-registered velocity maps. Preferential heating at the top of the jar could be a function of its shape, and/or the field in the microwave cavity.

Although none of the individual sets of data provided by the above protocols is unique, their combination certainly is. And although it is not easy to create full-scale batch retort processing conditions within an MRI magnet, it is certainly feasible to study smaller-scale models and to ensure that the uniformity of heating is optimised and thereby to use those data to optimise the composition and rheology of the product.

19.3.5 Continuous processing

MRI temperature mapping has been used in combination with finite element modelling software to calculate the fluid-to-particle convective heat transfer coefficient during aseptic processing of food particulates (Hulbert *et al.*, 1997; Kantt *et al.*, 1998). Since flow can produce artefacts in an MRI image unless it is coherent (or non-turbulent) either the flow has to be stopped during, or siphoned off for, acquisition; alternatively, the acquisition time can be shortened by use of fast imaging protocols.

Apart from temperature, one of the most important concepts in continuous heat processing is the 'residence time' since it is the length of time that the fluid, and of more concern the particulates, are exposed to temperature which ultimately determines the extent both of cooking and of overall microbiological kill. Although separate measurements by MRI of temperature and of velocity are relatively mature, it is only recently that studies have reported simultaneous measurement of both temperature and velocity by MRI. One such study used T_1 weighted 'tagging' to measure the temperature of the water by its effect on the tag, whilst the flow motion was visualised via the distortion of the tag pattern (Ogawa *et al.*, 2000); the resultant temperature map was compared with a theoretical map obtained by analysing forced convective heat transfer in laminar pipe flow.

Sun *et al.* developed a thermostatted glass tube rheometer (essentially a simple heat exchanger) which was optimised for use in an MRI scanner (Sun *et al.*, 1999). It allowed flow- and temperature-equilibria to be established in the system before it was inserted in the scanner without disruption of the fluid flow. Subsequently, the phase of the MRI signal was used for measurement of both temperature and velocity through the tube (Sun and Hall, 2001). Temperature maps were calculated using the isothermal flowing fluid as the reference; both the temperature and velocity maps compared well with CFD modelling by FIDAP (Fluent Inc., East Hartford, USA).

Since MRI has a proven ability for simultaneous measurement of temperature and velocity on simple systems, there is no reason why it cannot be applied to more complicated foods and geometries. Consequently MRI can provide experimental data which can be used to validate CFD models which would otherwise have to be based on assumed input parameters.

19.3.6 Other thermal technologies

Much of the recent development in MRI temperature mapping has been concerned with the clinical practice of interventional thermal therapy procedures (Lufkin, 1999) which often involve radio frequency or microwave heating of human tissues. Whereas it is possible to use MRI to monitor those procedures, most domestic and industrial food applications involve microwave cavities made of metal which cannot be studied inside a MRI magnet. Although temperatures can be monitored *in situ* using fibre optic thermometry and infrared thermal imaging (Sections 19.2.1 and 19.2.2), those techniques are restricted to either point- or surface-measurements and cannot, therefore, adequately locate the hot- and cold-spots inherent to microwave heating. Clearly the strength of MRI is that it can capture such three-dimensional heating patterns before their dissipation through thermal conduction (Nott *et al.*, 1999, 2000). Although a 'hedgehog' array of thermocouples can be used after heating in a microwave oven, despite the affordability of thermocouples there is inevitably a spatial resolution limitation. MRI has also been used to measure temperature in particulates during ohmic heating (Ruan *et al.*, 1999, 2001; Ye *et al.*, 2003) as described in Chapter 12.

19.4 Future trends

Although traditional thermocouples will continue to be used for most applications, there are many potential uses for non-invasive temperature measurement techniques, particularly for on-line measurements during a heating process. In principle, non-contact techniques can immediately respond to changes in temperature, although they will have an associated acquisition time related to the hardware configuration being used by a particular technique.

Techniques such as fibre-optic and infrared thermometry are affordable and easy-to-use, but lack the spatial resolution necessary for 100% assurance of the microbiological safety of the food. Nevertheless, they can be used to detect deviations in a well-defined system and both have the potential to be integrated within a production line. In contrast, compared to the other techniques mentioned in this review, microwave radiometry exhibits many *a priori* advantages in terms of temperature sensitivity, ease of handling, cost, the potential to probe the whole of a food sample and is currently being developed for on-line food process control. Although it is comparatively an immature field in terms of the development of the technology and its use, it is encouraging to note that this area is being developed in collaboration with end-users and, therefore, it is likely that as a result more products will emerge on the market.

MRI is certainly the most versatile technique for research studies, and there is already a substantial body of literature on its application to chemical engineering (Gladden, 2003), including food processing applications. However, the larger size MRI systems designed for medical scanning are expensive and require considerable expertise to run; it is encouraging that affordable 'user-friendly' systems are being developed for small samples (1–2 cm). In principle, the technology already exists for both off- or on-line process control (McDonald, 1995; Tellier and Mariette, 1995; Hills, 1998b; Schenz *et al.*, 1999; Pykett, 2000; De Los Santos *et al.*, 2001); sadly, however, in practice there currently is a general lack of investment and commitment in the food industry. Most of the commercial advances in on-line MR are associated with the oil and related industries; those have confirmed that the main challenge is to protect the MRI hardware from vibrations, temperature changes and radio frequency interference in an industrial environment, as well as protecting the users from the associated high magnetic fields.

Although much MRI research has focused on traditional heating modalities, it is important to note that MRI can uniquely provide a comprehensive, multi-dimensional set of data, not only of temperature, but also of other food-related parameters. Consequently this multi-dimensional technique can uniquely provide data to validate existing CFD models which would otherwise have to make assumptions, or to aid the development of new models. The current revolution in methods for packaging foods for retail consumption has stimulated a search for more innovative methods of thermal processing, whether by more appropriate use of traditional methods, new thermal technologies or by a combination of both. It is the authors' view that MRI will be a valuable technique for providing the experimental information necessary for many aspects of those innovations.

19.5 Sources of further information and advice

19.5.1 Fibre-optic thermometry

Davidson Instruments, Inc., 4800 Research Forest Drive, The Woodlands, TX 77381, USA
Tel: +1 281–364–6033
Web: www.davidson-instruments.com

Fiso Technologies, Inc., 500 St Jean Baptiste Avenue, Suite 195, Quebec, G2E 5R9, Canada
Tel: +1 418 688 8065
Fax: +1 418 688 8067
Web: www.fiso.com

Ipitek, 2330 Faraday Avenue, Carlsbad, CA 92008, USA
Tel: +1 760 438 1010
Fax: +1 760 438 2412
Web: www.ipiteksensors.com

Luxtron Corporation, 3033 Scott Boulevard, Santa Clara, CA 95054–3316, USA
Tel: + 1 408 727 1600
Fax: +1 408 727 1677
Web: www.luxtron.com

19.5.2 Infrared thermometry

FLIR Systems, 16505 SW 72nd Avenue, Portland, OR 97224, USA
Tel: +1 800 322 3731
Web: www.flirthermography.com

Land Instruments International, Dronfield, Derbyshire, S18 1DJ, UK
Tel: +44 (0) 1246 417691
Fax: +44 (0) 1246 410585
Web: www.landinst.com

19.5.3 Microwave radiometry

Microwave Thermometry Group, Department of Physics and Astronomy, University of Glasgow, Glasgow G12 8QQ, UK
Tel: +44 (0) 141 330 4703
Fax: +44 (0) 141 330 5881
Web: www.physics.gla.ac.uk

Loma Systems, Head Office, Sales and Manufacturing, Southwood, Farnborough, Hampshire GU14 0NY, UK
Tel: +44 (0) 1252 893300
Fax: +44 (0) 1252 513322
Web: www.loma.com

19.5.4 Magnetic Resonance Imaging

Academic contacts that have expertise in food applications of MRI:

Herchel Smith Laboratory for Medicinal Chemistry, University of Cambridge School of Clinical Medicine, University Forvie Site, Robinson Way, Cambridge, CB2 2PZ, UK
Tel: +44 (0) 1223 336805
Fax: +44 (0) 1223 336748
Web: www.hslmc.cam.ac.uk

Department of Food Science & Technology, University of California, Davis, CA 95616, USA
Tel: +1 530 752 8921
Fax: +1 530 752 4759
Web: www-foodsci.ucdavis.edu

Manufacturers of NMR and MRI systems that cater for non-medical applications:

Bruker BioSpin MRI GmbH, Rudolf-Plank-Str. 23, D-76275 Ettlingen, Germany
Tel: +49 (0) 7243 50 45 31
Fax: +49 (0) 7243 50 45 39
Web: www.bruker-biospin.de/MRI/index.html

Resonance Instruments Ltd, 31a Avenue One, Witney, Oxfordshire, OX28 4XZ, UK
Tel: +44 (0) 1993 700442
Fax: +44 (0) 1993 700363
Web: www.resonance.com

Varian, Inc., 3120 Hansen Way, Palo Alto, CA 94304–1030, USA
Tel: +1 800 356 4437
Fax: +1 650 494 7186
Web: www.varianinc.com

Companies and institutions that specialise in industrial magnetic resonance for process monitoring:

Invensys Foxboro, 33 Commercial Street, Foxboro, MA 02035, USA
Tel: +1 508 549 2424
Fax: +1 508 549 4999
Web: www.foxboro.com/nmr

Process Control Technology, 3620 Soderburg Drive, Ft. Collins, CO 80526–4914, USA
Tel: +1 970 377 2763
Fax: +1 970 482 8907
Web: www.pctnmr.com

Process NMR Associates, LLC, 87A Sand Pit Rd, Danbury, CT 06810, USA
Tel: +1 203 744 5905
Fax: +1 203 743 9297
Web: www.process-nmr.com

Progression Systems, 231 Sutton Street, Unit 1D, North Andover, MA 01845, USA
Tel: +1 978 738 9700
Fax: +1 978 738 0092
Web: www.progression-systems.com

Sensor Systems and Nondestructive Evaluation (NDE) Technology Department, Southwest Research Institute, P.O. Drawer 28510, San Antonio, Texas 78228–0510, USA
Tel: +1 210 684 5111

Fax: +1 210 522 3547
Web: www.swri.edu

19.6 References

AMIN, M.H.G., HANLON, A.D., HALL, L.D., MARRIOTT, C., ABLETT, S., WANG, W. and FRITH, W.J. (2003) A versatile single-screw extruder system designed for magnetic resonance imaging measurements *Measurement Science and Technology*, **14**, 1760–1768.

ANANTHESWARAN, R.C. and LIU, L. (1994) Effect of viscosity and salt concentration on microwave-heating of model non-Newtonian liquid foods in a cylindrical container *Journal of Microwave Power and Electromagnetic Energy*, **29**, 119–126.

BERRIE, P.G. (2001) Pressure and temperature measurement in food process control. In *Thermal Technologies in Food Processing* (Ed, Richardson, P.) Woodhead, Cambridge.

BOWS, J. and JOSHI, K. (1992) Infrared imaging feels the heat in microwave ovens *Physics World*, **5**, 21–22.

BOWS, J.R. (1999) Variable frequency microwave heating of food *Journal of Microwave Power and Electromagnetic Energy*, **34**, 227–238.

BOWS, J.R. (2000) A classification system for microwave heating of food *International Journal of Food Science and Technology*, **35**, 417–430.

BOWS, J.R., PATRICK, M.L., JANES, R., METAXAS, A.C.R. and DIBBEN, D.C. (1999) Microwave phase control heating *International Journal of Food Science and Technology*, **34**, 295–304.

CALLAGHAN, P.T. (1993) *Principles of Nuclear Magnetic Resonance Microscopy,* Clarendon Press, Oxford.

CARR, H.Y. and PURCELL, E.M. (1954) Effects of diffusion on free precession in nuclear magnetic resonance experiments *Physical Review*, **94**, 630–638.

CONWAY, J., HAWLEY, M., MANGNALL, Y.F., AMASHA, H. and VAN RHOON, G.C. (1992) Experimental assessment of electrical impedance imaging for hyperthermia monitoring *Clinical Physics and Physiological Measurement*, **13(Sup A)**, 185–189.

DE LOS SANTOS, A., DERWIN KING, J. and STEAD, D.C. (2001) Industrial Magnetic Resonance – Food and Other Industries. In *Magnetic Resonance in Food Science: A View to the Future* (Eds, Webb, G.A., Belton, P.S., Gil, A.M. and Delgadillo, I.) Royal Society of Chemistry, Cambridge, pp. 246–251.

DERBYSHIRE, J.A., GIBBS, S.J., CARPENTER, T.A. and HALL, L.D. (1994) Rapid 3-dimensional velocimetry by Nuclear Magnetic Resonance Imaging *AIChE Journal*, **40**, 1404–1407.

FOSTER, K.R. and CHEEVER, E. (1992) Microwave radiometry in biomedicine: a reappraisal *Bioelectromagnetics*, **13**, 567–579.

FUKUSHIMA, E. (1999) Nuclear Magnetic Resonance as a tool to study flow *Annual Reviews of Fluid Mechanics*, **31**, 95–123.

GERSING, E. (1999) Monitoring temperature-induced changes in tissue during hyperthermia by impedance methods. In *Electrical Bioimpedance Methods: Applications to Medicine and Biotechnology*, Vol. 873, pp. 13–20.

GIBBS, S.J., CARPENTER, T.A. and HALL, L.D. (1993) Magnetic-Resonance-Imaging of Thermal-Convection *Journal of Magnetic Resonance Series A*, **105**, 209–214.

GLADDEN, L.F. (2003) Magnetic resonance: Ongoing and future role in chemical engineering research *AIChE Journal*, **49**, 2–9.

GOEDEKEN, D.L., TONG, C.H. and LENTZ, R.R. (1991) Design and calibration of a continuous temperature measurement system in a microwave cavity by infrared imaging *Journal of Food Processing and Preservation*, **15**, 331–337.

HAHN, E.L. (1950) Spin echoes *Phys. Rev.*, **80**, 580–594.

HALL, A.S., PRIOR, M.V., HAND, J.W., YOUNG, I.R. and DICKINSON, R.J. (1990) Observation by MR Imaging of in vivo temperature changes induced by radio frequency hyperthermia *Journal of Computer Assisted Tomography*, **14**, 430–438.

HASHEMI, R.H. and BRADLEY, W.G. (1997) *MRI: the basics,* Williams and Wilkins, Baltimore.

HILLS B.P. (1998a) *Magnetic Resonance Imaging in Food Science,* John Wiley & Sons, New York.

HILLS B.P. (1998b) On-line MRI for process control and quality assurance. In *Magnetic Resonance Imaging in Food Science,* John Wiley & Sons, New York, pp. 152–159.

HINDMAN, J.C. (1966) Proton resonance shift of water in gas and liquid states *Journal of Chemical Physics*, **44**, 4582–4592.

HULBERT, G.J., LITCHFIELD, J.B. and SCHMIDT, S.J. (1995) Temperature mapping in carrot using T_1 weighted magnetic resonance imaging *Journal of Food Science*, **60**, 780–785.

HULBERT, G.J., LITCHFIELD, J.B. and SCHMIDT, S.J. (1997) Determination of convective heat transfer coefficients using 2D MRI temperature mapping and finite element modeling *Journal of Food Engineering*, **34**, 193–201.

IBARRA, J.G., TAO, Y., WALKER, J. and GRIFFIS, C. (1999) Internal temperature of cooked chicken meat through infrared imaging and time series analysis *Transactions of the ASAE*, **42**, 1383–1390.

IBARRA, J.G., TAO, Y. and XIN, H.W. (2000) Combined IR imaging-neural network method for the estimation of internal temperature in cooked chicken meat *Optical Engineering*, **39**, 3032–3038.

ISHIHARA, Y., CALDERON, A., WATANABE, H., OKAMOTO, K., SUZUKI, Y., KURODA, K. and SUZUKI, Y. (1995) A precise and fast temperature mapping using water proton chemical shift. *Magnetic Resonance in Medicine*, **34**, 814–823.

KANTT, C.A., SCHMIDT, S.J., SIZER, C.E., PALANIAPPAN, S. and LITCHFIELD, J.B. (1998) Temperature mapping of particles during aseptic processing with Magnetic Resonance Imaging *Journal of Food Science*, **63**, 305–311.

KANTT, C.A., WEBB, A.G. and LITCHFIELD, J.B. (1997) Temperature measurement of foods using chemical shift magnetic resonance imaging as compared with T_1-weighted temperature mapping *Journal of Food Science*, **62**, 1011–1016.

KIMOTO A. and SHIDA K. (2000) Imaging of temperature change distribution in the brain phantom by means of capacitance measurement *IEEE Transactions on Instrumentation and Measurement*, **49**, 591–595.

KURODA, K., SUZUKI, Y., ISHIHARA, Y., OKAMOTO, K. and SUZUKI, Y. (1995) Temperature mapping using the water proton chemical shift obtained with 3D-MRSI: Feasibility *in vivo*. *Magnetic Resonance in Medicine*, **35**, 20–29.

KURODA, K., OSHIO, K., CHUNG, A.H., HYNYNEN, K. and JOLESZ, F.A. (1997) Temperature mapping using the water proton chemical shift: A chemical shift selective phase mapping method *Magnetic Resonance in Medicine*, **38**, 845–851.

KURODA, K., MULKERN, R.V., OSHIO, K., PANYCH, L.P., NAKAI, T., MORIYA, T., OKUDA, S., HYNYNEN, K. and JOLESZ, F.A. (2000) Temperature mapping using the water proton chemical shift: Self-referenced method with echo-planar spectroscopic imaging *Magnetic Resonance in Medicine*, **43**, 220–225.

LAND, D.V. (2001) An efficient, accurate and robust radiometer configuration for microwave temperature measurement for industrial and medical applications *Journal of Microwave Power and Electromagnetic Energy*, **36**, 139–153.

LE BIHAN, D., DELANNOY, J. and LEVIN, R.L. (1989) Temperature mapping with MR Imaging of molecular-diffusion – application to hyperthermia *Radiology*, **171**, 853–857.

LEROY, Y., BOCQUET, B. and MAMOUNI, A. (1998) Non-invasive microwave radiometry thermometry *Physiological Measurement*, **19**, 127–148.

LUFKIN, R.B. (Ed.) (1999) *Interventional MRI,* Mosby, London.

MCCARTHY, M.J. (1994) *Magnetic Resonance Imaging in Foods,* Chapman and Hall, New York.

MCDONALD, P.J. (1995) The use of nuclear magnetic resonance for on line process control and quality assurance. In *Food Processing: Recent Developments* (Ed., Gaonker, A.G.) Elsevier Science, Amsterdam, pp. 23–36.

MICHALSKI, L., ECKERSDORF, K., KUCHARSKI, J. and MCGHEE, J. (2001) Fibre optic thermometers. In *Temperature Measurement* John Wiley & Sons, Chichester, England.

MIYAKAWA, M. and BOLOMEY, J. CH. (Ed.) (1996) *Non-Invasive Thermometry of the Human Body,* CRC Press, Boca Raton.

MULLIN, J. and BOWS, J. (1993) Temperature-measurement during microwave cooking *Food Additives and Contaminants*, **10**, 663–672.

NEWLING, B., GIBBS, S.J., DERBYSHIRE, J.A., XING, D., HALL, L.D., HAYCOCK, D.E., FRITH, W.J. and ABLETT, S. (1997a) Comparisons of magnetic resonance imaging velocimetry with computational fluid dynamics *Journal of Fluids Engineering – Transactions of the ASME*, **119**, 103–109.

NEWLING, B., GIBBS, S.J., HALL, L.D., HAYCOCK, D.E., FRITH, W.J. and ABLETT, S. (1997b) Chemically resolved NMR velocimetry *Chemical Engineering Science*, **52**, 2059–2072.

NOTT, K.P. and HALL, L.D. (1999) Advances in temperature validation of foods *Trends in Food Science and Technology*, **10**, 366–374.

NOTT, K.P., HALL, L.D., BOWS, J.R., HALE, M. and PATRICK, M.L. (1999) Three-dimensional MRI mapping of microwave induced heating patterns *International Journal of Food Science and Technology*, **34**, 305–315.

NOTT, K.P., HALL, L.D., BOWS, J.R., HALE, M. and PATRICK, M.L. (2000) MRI phase mapping of temperature distributions induced in food by microwave heating *Magnetic Resonance Imaging*, **18**, 69–79.

NOTT, K.P., SHAARANI, S.MD. and HALL, L.D. (2003) The effect of microwave heating on potato texture studied with magnetic resonance imaging. In *Magnetic Resonance in Food Science: Latest Developments* (Eds, Belton, P.S., Gil, A.M., Webb, G.A. and Rutledge D.) Royal Society of Chemistry, Cambridge, pp. 38–45.

OGAWA, K., TOBO, M., IRIGUCHI, N., HIRAI, S. and OKAZAKI, K. (2000) Simultaneous measurement of temperature and velocity maps by inversion recovery tagging method *Magnetic Resonance Imaging*, **18**, 209–216.

PAULSEN, K.D., MOSKOWITZ, M.J., RYAN, T.P., MITCHELL, S.E. and HOOPES, P.J. (1996) Initial in vivo experience with EIT as a thermal estimator during hyperthermia *International Journal of Hyperthermia*, **12**, 573–591.

POPE, J.M. and YAO, S. (1993) Quantitative NMR Imaging of flow *Concepts in Magnetic Resonance*, **5**, 281–302.

PROSETYA, H. and DATTA, A. (1991) Batch microwave-heating of liquids – an experimental study *Journal of Microwave Power and Electromagnetic Energy*, **26**, 215–226.

PYKETT, I.L. (2000) NMR – A powerful tool for industrial process control and quality assurance *IEEE Transactions On Applied Superconductivity*, **10**, 721–723.

RICHARDSON, P.S. and POVEY, M.J.W. (1990) Ultrasonic temperature measurement and its potential for food processing systems *Food Control*, **1**, 54–57.

RIDLEY, I. (2001) Practical aspects of infra-red remote thermometry. In *Instrumentation and Sensors for the Food Industry* (Eds, Kress-Rogers, E. and Brimelow, C.J.B.) Butterworth-Heineman, Oxford, pp. 355–402.

RUAN, R., CHEN, P., CHANG, K., KIM, H.J. and TAUB, I.A. (1999) Rapid food particle temperature mapping during ohmic heating using FLASH MRI *Journal of Food Science*, **64**, 1024–1026.

RUAN, R., YE, X., CHEN, P. and DOONA, C.J. (2001) Ohmic heating. In *Thermal technologies in food processing* (Ed., Richardson, P.) Woodhead, Cambridge.

SCHENZ, T.W., DAUBER, B., NICHOLLS, C., GARDNER, C., SCOTT, V.A., ROBERTS, S.P. and HENNESY, M.J. (1999) Online MR imaging for detection of spoilage in finished packages. In *Advances in Magnetic Resonance in Food Science* (Eds, Belton, P.S., Hills, B.P. and Webb, G.A.) Royal Society of Chemistry, Cambridge, pp. 264–271.

SIMON, C., VAN BAREN, P. and EBBINI, E.S. (1998) Two-dimensional temperature estimation using diagnostic ultrasound *IEEE Transactions On Ultrasonics Ferroelectrics and Frequency Control*, **45**, 1088–1099.

STEJSKAL, E.O. and TANNER, J.E. (1965) Spin diffusion measurements: spin echoes in the presence of a time-dependent field gradient *Journal of Chemical Physics*, **42**, 288–292.

SUN, L., AMIN, M.H.G., HALL, L.D., WOLF, B., FRITH, W.J. and ABLETT, S. (1999) A versatile thermostatted glass tube MRI rheometer *Measurement Science & Technology*, **10**, 1272–1278.

SUN, L. and HALL, L.D. (2001) An experimental solution of the Graetz problem in heat exchangers *International Communications in Heat and Mass Transfer*, **28**, 461–466.

SUN, X.Z., LITCHFIELD, J.B. and SCHMIDT, S.J. (1993) Temperature mapping in a model food gel using Magnetic Resonance Imaging *Journal of Food Science*, **58**, 168–172, 181.

SUN, X.Z., SCHMIDT, S.J. and LITCHFIELD, J.B. (1994) Temperature mapping in a potato using half Fourier transform MRI of diffusion. *Journal of Food Process Engineering*, **17**, 423–437.

TELLIER, C. and MARIETTE, F. (1995) On-line applications in food science *Annual Reports on NMR Spectroscopy*, **31**, 105–122.

WALTON, J.H. and MCCARTHY, M.J. (1999) New method for determining internal temperature of cooking meat via NMR spectroscopy. *Journal of Food Process Engineering*, **22**, 319–330.

WANG, W., WALTON, J.H. and MCCARTHY, K.L. (1999) Flow profiles of power law fluids in scraped surface heat exchanger geometry using MRI *Journal of Food Process Engineering*, **22**, 11–27.

WANG, W., WALTON, J.H. and MCCARTHY, K.L. (2000) Mixing in scraped surface heat exchanger geometry using MRI *Journal of Food Process Engineering*, **23**, 403–418.

WEBB, A.G. and LITCHFIELD, J.B. (1999) Magnetic resonance temperature mapping. In *Advances in Magnetic Resonance in Food Science* (Eds, Belton, P.S., Hills, B.P. and Webb, G.A.) Royal Society of Chemistry, Cambridge, pp. 241–255.

WEBER, M. and KIMMICH, R. (2002) Rayleigh-Benard percolation transition of thermal convection in porous media: Computational fluid dynamics, NMR velocity mapping, NMR temperature mapping *Physical Review E*, **66**, art. no.-056301.

WEIS, J., KIMMICH, R. and MULLER, H.P. (1996) NMR imaging of thermal convection patterns *Magnetic Resonance Imaging*, **14**, 319–327.

YE, X., RUAN, R., CHEN, P., CHANG, K., NING, K., TAUB, I.A. and DOONA, C. (2003) Accurate and fast temperature mapping during ohmic heating using proton resonance frequency shift MRI thermometry *Journal of Food Engineering*, **59**, 143–150.

Part V

Analysing microbial inactivation in thermal processing

20

Analyzing the effectiveness of microbial inactivation in thermal processing*

M. Peleg, University of Massachusetts at Amherst, USA

20.1 Introduction: microbial heat inactivation

Traditionally, microbial heat inactivation has been considered a process which follows a first order kinetics. Consequently, the heat resistance of microbial cells and spores has been expressed in terms of '*D* values', the time needed to reduce their population by one log cycle (base ten) at a given temperature. The temperature dependence of *D* has been assumed to obey a log linear relationship, which has produced the '*Z* value', the temperature span needed to shorten *D* by one log cycle (base ten). Or alternatively, the temperature dependence of the exponential inactivation rate, *k*, the reciprocal of *D*, has been assumed to obey the Arrhenius equation, originally developed for simple chemical reactions. Either way, the effectiveness of any given thermal preservation process has been evaluated in term of an '*F* value'. It is a measure of the equivalence of the changing temperature integrated lethal effect, at the coldest point in the product, to an isothermal process at a reference temperature. Traditionally, for heat sterilization of low acid foods, it has been 121.1°C (250°F) with the inactivation target being *C. botulinum* spores. There are various numerical and graphical methods to incorporate these kinetic assumptions and considerations in the calculation of the theoretical efficacy of thermal processes. In the commercial sterilization operations, a safety factor, which varies among products, is also added to the calculation. Consequently, in most thermally preserved foods, the actual duration of the treatment exceeds the 'theoretical requirement' by a substantial margin.

* This chapter has been reproduced at the publisher's request.

In light of the impressive safety record of the canning industry, there has been little incentive to reexamine the traditional methods to calculate sterility. Yet, there are a few recognized problems with their theoretical foundations, which may have safety implication in at least some heat treated foods. Or more specifically:

- There is a growing evidence that the semi-logarithmic survival curves of many microorganisms and spores, including pathogens of food safety concern, are not linear as the first order kinetics implies. Thus, forcing a straight line through the curved experimental survival data to produce a '*D* value' is not a permissible option. (In cases where the isothermal semi-logarithmic survival curve of the organism or spore has an upper concavity, or 'tailing', then forcing a straight line may result in under processing and hence, can at least potentially, increase the safety risk. On the other hand, if the semi-logarithmic survival curve has a downward concavity, then the result will be over processing to a level well beyond that which is required to guarantee the product safety.)
- Even if microbial inactivation were a process, which followed a first order kinetics, the Arrhenius equation or log linear model to express the rate constant's temperature dependence would not be a good choice. The reason is that the logarithmic transformation of k or D and the assumption of a linear relationship between log k or log D and $1/T$ or T, respectively, gives an inappropriate weight to the low temperatures of the treatment at the expense of the high temperatures, where most of the inactivation actually takes place. (As previously shown (Campanella and Peleg, 2001, Peleg *et al*., 2002) there is no reason to assume that there is a universal analogy between microbial mortality and simple chemical reactions that the use of the Arrhenius equation entails.)
- The formula to calculate the '*F* value' contains the reference temperature as a term. But since the '*F* value' can be translated into a survival ratio, the latter will be independent of the reference if, and only if, log D vs T is a linear relationship. Also, theoretically at least, according to the current method to calculate sterility, the same process can be safe depending on the number of points used in the regression to determine the 'D' values.

The existence of these three problems is not a trivial matter and a rethinking of the whole issue of microbial survival during heat treatments is clearly warranted. One can also add, that the fundamentals of the current mathematical methods to estimate microbial survival in thermal processes were formulated in a time when the ability to perform complicated calculations was extremely limited. Hence, there was a premium on linear models whose parameters could be determined graphically or with the use of a mechanical desk calculator. With today's software and computation power, mathematical simplicity, although by no means undesirable, need not be the prime consideration as it had been in the past. As will be shown below, we can now formulate survival models that are based on realistic assumptions rather than on idealization and unproven analogies between microbial mortality and certain physical phenomena in the non-living realm.

The following discussion will focus on how microbial heat inactivation can be described, and even predicted, by mathematical models constructed almost exclusively on the basis of the targeted organisms or spores experimentally observed inactivation patterns without the assumption that microbial inactivation is a process that follows a universal first order kinetics or any other preconceived kinetic model.

20.2 Survival curves, the Weibull distribution function and heat resistance

Survival curves, microbial included, depict the fraction of survivors as a function of time. Hence, each can be considered the cumulative form of a distribution of temporal mortality events. This has long been recognized and several authors have suggested a variety of distribution functions to describe them (e.g., Casolari, 1988; Little *et al.*, 1994; Stephens *et al.*, 1994; Anderson *et al.*, 1996; Linton *et al.*, 1996; Augustin *et al.*, 1998). If the thermal resistance of an organism or a spore is expressed as the time needed for its destruction or inactivation, then the survival curve's *slope* has rate units and hence the relationship between the survival curve's shape and the inactivation kinetics (Peleg and Penchina, 2000). Since different organisms and spores, in different media and at different temperatures, can have different distributions of inactivation times, the shape of their isothermal survival curves can vary accordingly. Consequently, many microbial survival curves, when plotted on semi-logarithmic coordinates, have an upward or downward concavity. They can also have a sigmoid shape or exhibit a 'shoulder' or a 'tail'. The interpretation of the different shapes in terms of the mortality pattern has been discussed in a series of recent papers (Peleg and Penchina, 2000; Peleg, 2002, 2003). Suffice it is to say, that upward concavity is a manifestation of the rapid elimination of the weak members of the population leaving progressively sturdier survivors, while downward concavity is an indication that accumulated damage sensitizes the survivors. A true linear semi-logarithmic survival curve would indicate that all the population members have the same probability of being inactivated at any given time (the equivalent of a radioactive decay). The important point here is that any mathematical survival model should be derived from the actual shapes of the experimental isothermal semi-logarithmic survival curves, determined in the pertinent medium, and not from the assumption that all inactivation processes obey a single universal law. One must also take into consideration that the general shape of the isothermal semi-logarithmic survival curves of the same organism need not remain fixed. Thus, a concavity inversion as the treatment's temperature increases or decreases is by no means unusual.

The Weibull distribution function has been a successful model of many unrelated systems, where destruction and survival are involved. One would therefore expect that a ubiquitous mortality pattern would emerge if many organisms and spores had a Weibull, or 'Weibull like', distribution of heat resistances. Consequently, we have chosen this distribution as a model for much of this chapter's discussion. It can be shown that *C. botulinum* spores (in the range of 101 to 121ºC, and those of various bacilli, can indeed be considered as having a 'Weibull type' distribution of heat resistances (Peleg and Penchina, 2000). Similarily, the heat inactivation of several food borne pathogens, most notably that of *Salmonella*, can also be described as being governed by an underlying Weibull distribution of resistances (Peleg and Cole, 1998; Mattick *et al.*, 2001; Van Broekel, 2002). The proposed methodology to deal with such organisms, as it will become evident, can be extended to other survival patterns, which include 'tailing', 'shoulders' and concavity inversion, i.e. sigmoidal isothermal semi-logarithmic survival curves (Peleg and Penchina 2000; Peleg, 2002, 2003). Needless to say, the same method also applies to organisms or spores whose semi-logarithmic survival curves are linear, which is just a special simple case of the Weibull distribution where the shape factor is an equal one (see below).

20.2.1 The model

An isothermal survival pattern of a population governed by heat resistances having a Weibull distribution (Fig. 20.1) is characterized by:

$$\log_{10} S = -b(T)t^{n(T)} \tag{20.1}$$

where S is the momentary survival ratio, $S = N(t)/N_0$ where $N(t)$ and N_0 are the momentary and initial number of viable cells or spores, respectively, and $b(T)$ and $n(T)$ temperature dependent coefficients representing the distribution's scale and shape factor respectively. Consider the following assumptions (Peleg and Penchina, 2000):

- The microbial or sporal population is sufficiently large that Eq. 20.1 can be used to describe the changes in its size at all temperatures and times in the pertinent range.
- The temperature dependence of the survival parameters, in our case, Eq. 20.1's coefficients $b(T)$ and $n(T)$, can be described algebraically.
- The treatment's non-isothermal temperature 'profile', $T(t)$, can also be expressed algebraically.
- The momentary semi-logarithmic survival (or inactivation) rate, $d\log_{10}S/dt$, is the isothermal rate at the momentary temperature, $T(t)$, at a time, t^*, which corresponds to the momentary survival ratio, (Fig. 20.2).

According to Eq. 20.1:

$$t^* = \left[\frac{-\log_{10} S(t)}{b(T)}\right]^{1/n(T)} \tag{20.2}$$

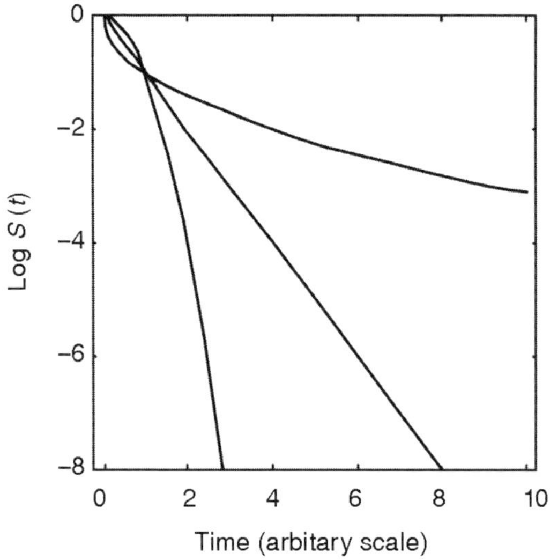

Fig. 20.1 Schematic view of the semi-logarithmic survival curves of microorganisms or spores whose heat resistances has a Weibull distribution with a shape factor, $n(T)$ in Eq. 20.1, smaller, equal and bigger than one.

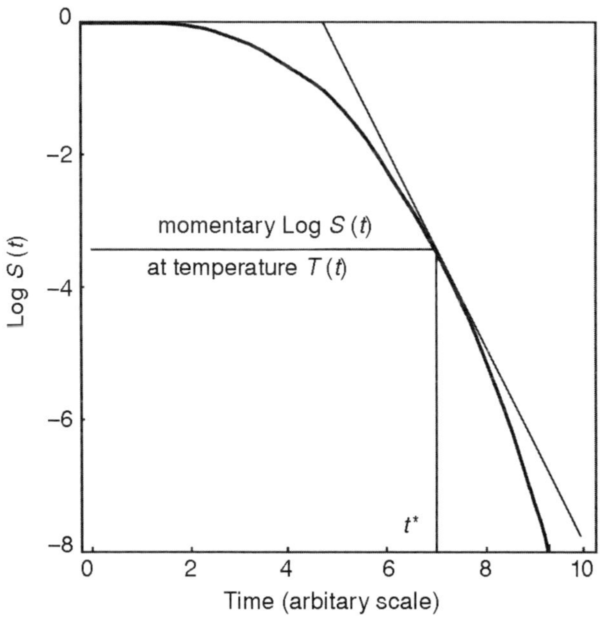

Fig. 20.2 Schematic view of the construction of the differential equation that describes a non-isothermal semi-logarithmic survival curve.

and

$$\frac{d\log_{10}S(t)}{dt} = -b(T)n(T)t^{*n(T)-1} \tag{20.3}$$

Combining Eqs. 20.2 and 20.3 yields the differential equation (Peleg and Penchina, 2000):

$$\frac{d\log_{10}S(t)}{dt} = -b[T(t)]n[T(t)]\left\{\frac{-\log_{10}S}{b[T(t)]}\right\}^{\frac{n[T(t)]-1}{n[T(t)]}} \tag{20.4}$$

whose solution, $\log_{10}S(t)$, is the non-isothermal survival curve, which corresponds to the particular temperature profile, $T(t)$. In programs like Mathematica® (Wolfram Research, Champaign, IL), the one used to produce the plots shown in this chapter, defining the terms $b[T(t)]$ and $n(t) = n[T(t)]$ is a trivial matter. As will be seen below, the differential equation can be solved numerically to produce the survival curve, $\log_{10}S(t)$, even for very elaborate temperature profiles, which include heating and cooling, sudden interruptions and regular, irregular and even random temperature fluctuations.

20.3 Analyzing the survival ratio dependence on temperature

20.3.1 Expressing *b*(*T*)

It is a well-known fact that microbial thermal inactivation starts in earnest only at a certain temperature. The same also applies to bacterial spores. Well below this temperature, there is hardly any mortality or inactivation (there might even be growth) and above it, the process accelerates. These features are implemented in the empirical log logistic model (Campanella and Peleg, 2001; Peleg *et al.*, 2002):

$$b(T) = \log_e\{1 + \exp[k(T - T_c)]\} \tag{20.5}$$

where k and T_c are constants. The reader will notice that at $T \leq Tc, b(T) \approx 0$ and at $T \geq Tc, b(T) = k(T - Tc)$, i.e., $b(T)$ increases linearly with temperature as shown in Fig. 20.3. The almost perfect fit of Eq. 20.5 to the $b(T)$ vs. t relationships of *C. botulinum* spores and *Salmonella* and *Listeria* cells has been reported by Peleg *et al.* (2002). If the observed increase of $b(T)$ is clearly non linear, Eq. 20.5 can be amended by adding an exponent to the right side, i.e.,

$$b(T) = \{\log_e[1 + \exp\,[k(T - T_c)]]\}^m \tag{20.6}$$

where m is a constant $(m > 1)$. Although neither Eq. 20.5 nor 20.6 can be used for extrapolation, one can assume that the increase in $b(T)$ will be at least linear at temperatures slightly higher than those for which experimental data are available. Nevertheless, a scenario where there can be a certain degree of survival, irrespective of temperature, cannot be ruled out a priori. Hence extreme caution is needed even if Eq. 20.5 is only used to estimate the borderline case by extrapolation.

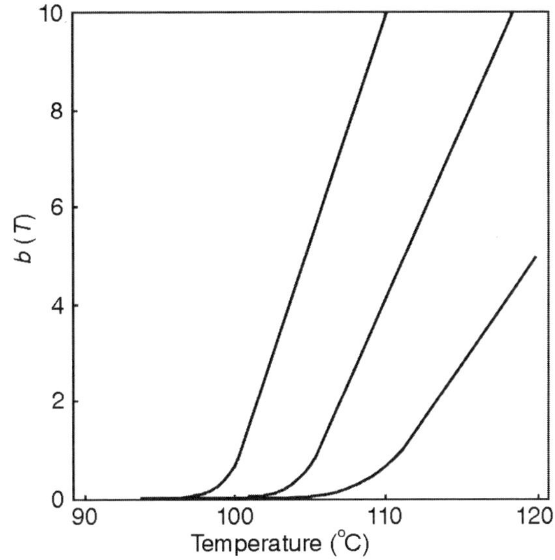

Fig. 20.3 The log logistic expression of the survival parameter $b(T)$ in Eq. 20.1 (from left to right organisms with increasing heat resistance, higher T_c and lower k). (Note that the same model would apply to $k(T)$ if an organism's semi-logarithmic survival curves happen to be linear.)

20.3.2 Expressing $n(T)$

While $b(T)$ in Eq. 20.1 primarily accounts for the overall steepness of the survival ratio drop, n(T) primarily expresses the semi-logarithmic survival curve's concavity direction and its degree of curvature. A power smaller than one, $n(T) < 1$, represents upward concavity ('tailing'), and bigger than once, $n(T) > 1$, downward concavity. When $n(T) = 1$, as already stated, the semi-logarithmic survival curve is linear. In contrast with $b(T)$, it is difficult to know, a priori, how $n(T)$ changes with temperature. At least in some cases, it can be assumed to be constant or practically constant at the pertinent lethal temperature range (e.g., Campanella and Peleg, 2001; Mattick *et al.*, 2001). This need not be a coincidence but a consequence of the general nature of failure phenomena (van Broekel, 2002).

20.4 Simulating heating and cooling curves

Theoretically, the temperature at the coldest point in a processed can, a pouch or a glass container, can be calculated or be closely approximated, *provided* that the heat transfer mechanism and the thermal properties of the treated contents are known. The same applies to the mean temperature and residence times' distribution of a fluid product passing through a heat exchanger, including a holding time tube, if the flow regime and the fluid's thermal and flow properties

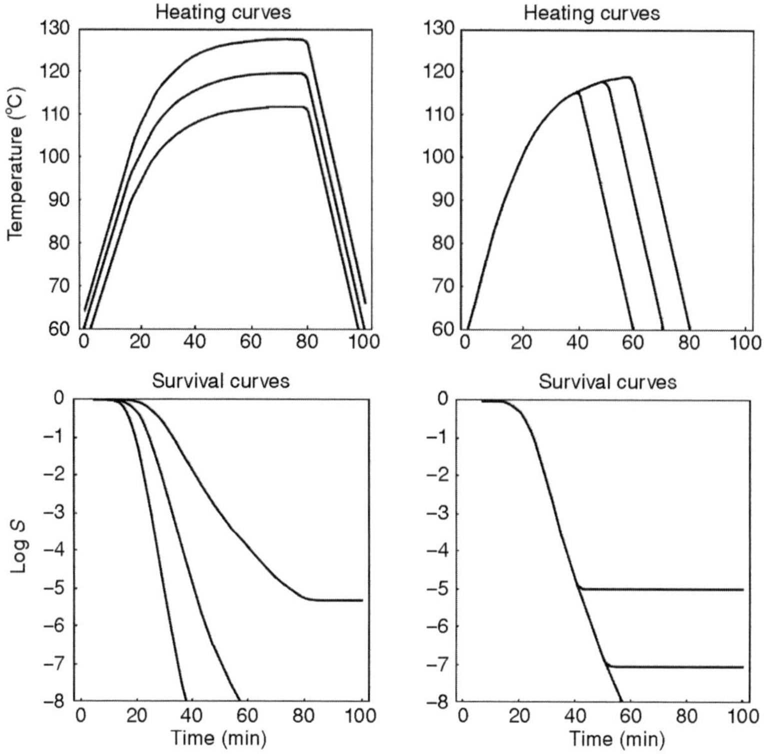

Fig. 20.4 Simulated sporal inactivation patterns in heat sterilization processes having different target temperatures (left) and durations (right).

are known. In reality, most notably in foods treated in a sealed container, accurate calculation of the temperature profile can be complicated by inhomogeneties at various levels, within and outside the container. Consequently, the temperature profile at the coldest point is monitored at various locations in a retort, for example, using a set of thermocouples. Usually the heating and cooling curve has the characteristic shape shown in Fig. 20.4 (Toledo, 1999). Such curves can be described by the empirical model (Campanella and Peleg, 2001):

$$T(t) = \frac{T_{asymp} - \log_e\{1 + \exp[k_1(t - t_{c1})]\}}{1 + \exp[k_2(t_{c2} - t)]} \tag{20.7}$$

or a similar model, where T_{asymp} is the target or asymptotic temperature, 120°C for example, and k_1, k_2, t_{c1} and t_{c2} are constants. The realistic temperature profiles shown in Figs 20.4 and 20.5 were generated with Eq. 20.7 as a model. It is a combination of the logistic and log logistic equations and hence $T(t)$, when expressed in this way, is an ordinary algebraic continuous function. Thus, despite the discontinuous appearance of the temperature profile curve at the onset of cooling (Figs 20.4 and 20.5), the term $T(t)$ so defined can be used to express the changes in the survival parameters, $b[T(t)]$ and $n[T(t)]$ during the

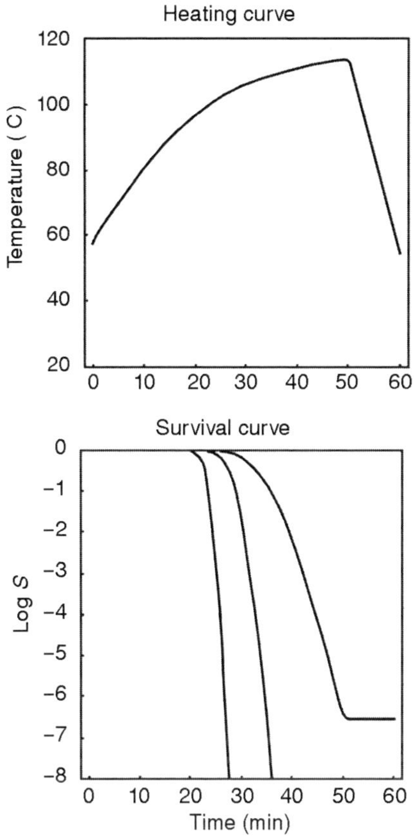

Fig. 20.5 A simulated comparison between the non-isothermal semi-logarithmic survival curves of three hypothetical spores having the different heat resistances shown in Fig. 20.3 with $n(T) = 0.6$.

whole process cycle. These two terms, in turn, can be incorporated into the survival curve's differential equation (Eq. 20.4), which will produce the corresponding survival curve. Despite its cumbersome mathematical structure, Eq. 20.7 is a very convenient mathematical model for temperature profiles simulations. Not only does it fit well (which would be expected from a model having five adjustable parameters) but it also enables easy control of the simulated temperature profile's shape. Or more specifically:

Parameter	*Increases*	*Decrease*
T_{asymp}	elevates the whole curve	lowers the whole curve
k_1	increases the cooling rate	decreases the cooling rate
t_{c1}	postpones the cooling	advances the cooling
k_2	increases the heating rate	decreases the heating rate
t_{c2}	increases the initial temperature (shifts the curve to the right)	decreases the initial temperature (shifts the curve to the left)

If only the heating part of the profile is of interest to the analysis, then one can use the logistic component only, i.e.,

$$T(t) = \frac{T_{asymp}}{1 + \exp[k_2(t_{c2} - t)]} \tag{20.8}$$

Either way, in at least some cases, t_{c2} can be dropped (i.e., $t_{c2} = 0$). This reduces the number of adjustable parameters without much effect on the fit. The heating part of the curve can also be described by the three parameter model:

$$T(t) = T_0 + \frac{t}{k_i' + \dfrac{t}{T_{asymp} - T_0}} \tag{20.9}$$

where T_0 is the initial temperature and k_1' a constant, which can serve as a 'skeleton' for more elaborate profiles which contain oscillations.

'Regular' and 'irregular' temperature fluctuations can be simulated with a variety of models (Peleg, 2002), for example:

$$T(t) = T_0 + \frac{[1 + c_i \Sigma x_i \sin(\omega_i t)]t}{c_2 + \dfrac{t}{T_{asymp} - T_0}} \tag{20.10}$$

and

$$T(t) = T_0 + \Sigma \frac{c_3(c_4 + R_{ni})}{1 + \exp[(R_{nj} - t)(c_5 + R_{ni})]} \tag{20.11}$$

respectively. When Eq. 20.10 is used as a model the oscillations' amplitude is controlled by the c_i and x_i and the frequencies by the ω_i's. In Eq. 20.11, they are controlled not only by fixed parameters c_3, c_4 and c_5 but also by two sets of random numbers; $R_{ni}(0 < R_{ni} < 1)$, which primarily controls the temperature's random 'increments' and R_{nj}, which primarily controls their frequency. (Each R_{nj} is approximately the process duration divided by i, the number of random temperature 'increments'.) Although these models are far from being 'handy' they can still be used to examine the potential qualitative effect of process instabilities on the survival ratio – see below.

Steam interruption(s) and subsequent intermediate cooling can be simulated using models containing 'If statements' to separate the heating and cooling regimes before, during and after the steam supply had been interrupted and restored. A 'simplified' example is the model:

$$T(t) = \text{if}[t \le t_{\text{interp}}, T(t),\ \text{if}[t \le t_{\text{interp}} + \Delta t, T(t_{\text{interp}})\ \text{Decayterm}(t - t_{\text{interp}}), T(t - t_{\text{end of cooling}})]] \tag{20.12}$$

where $T(t)$ is the temperature rise equation of a non-interrupted cycle, e.g., Eq. 20.7, 20.8, or 20.9, t_{interrpt} the time when the steam interruption started, Δt the interruption duration, Decayterm ($t - t_{\text{interup}}$) a decay function describing the temperature drop during the steam stoppage and $T_{\text{end of cooling}}$ and $t_{\text{end of cooling}}$ are the temperature and time when the steam supply was restored, respectively. The model is called 'simplified' because rarely, if ever, will the steam interruption and restoration be even approximately instantaneous, and because the heating curves before and after the interruption need not follow the same model. Nevertheless, such a model can capture the qualitative aspects of a steam stoppage and show how it can affect the process efficacy. At least in principle, the simulations that it produces can be used to determine to what extent the safety of the product has been compromised and what kind of a 'remedial' treatment would be necessary. One should keep in mind, though, that this sort of model would not account for possible sporal germination or microbial growth, which might be accompanied by toxin production, if the steam interruption is sufficiently long.

20.5 Applications of survival patterns in food processing

The theoretical effect of the target temperature and the initiation of the cooling on the inactivation patterns of spores are in Fig. 20.4. They were generated using survival parameters similar to those of *C. botulinum* spores (Campanella and Peleg, 2001) derived from the published results of Anderson *et al.* (1996). These were obtained in a buffered solution rather than in a real food. One could therefore expect that in any given real food, the survival parameters, especially, $b(T)$, would be somewhat different. For the sake of this discussion, we will assume that the hypothetical food in question offers a certain degree of protection to the spores, which would be expressed by a simultaneous elevation of T_c and the lowering of k in Eq. 20.5. We will therefore assign $T_c = 105°C$ instead of 102°C and $k = 0.25°C^{-1}$ instead of about $0.3°C^{-1}$. We shall also assume, for the sake of simplicity, that $n(T)$ is practically unchanged and remains constant, e.g., $n(T) \approx 0.3$. We can now generate survival curves with the new survival parameters and check whether any contemplated heating-cooling cycle produces a preset level of inactivation or not. Or similarly, one can generate a series of heating-cooling cycles in order to identify the conditions which will produce or surpass any given survival ratio deemed critical for microbial safety. Examples of such evaluations are shown in Fig. 20.4. In the simulations shown in the figure, and all subsequent simulations, the lowest marked survival ratio is 10^{-8}. Eight orders of magnitude reduction seems to be the highest level that can be quantitatively determined by routine procedures in food microbiology laboratories. Hence, and unlike the traditional '12D reduction', it can be determined directly, without extrapolation. It is the author's opinion that a safety factor should be based on an actually observed survival ratio rather than on the extrapolation of the survival curve to survival levels where no experimental data are available.

20.5.1 Heat effects on different microorganisms

The emergence of new food-borne pathogens and discovery of heat-resistant strains of old ones has recently become a food safety issue. The question that arises is whether existing thermal, or other preservation processes are sufficient for their inactivation and if not, how should the process be adjusted to cope with the new or potential danger. If survival parameters of a newly discovered pathogen can be expressed in a form suitable for the model, then one can assess the efficacy of a present or planned process by running a simulation with the corresponding temperature profile. This is demonstrated in Fig. 20.5 where simulated survival curves of a hypothetical heat sensitive and of a resistant strain are compared. Obviously, and like in any computer simulation of survival patterns, the result should be confirmed with experimental data. But while these will provide the ultimate criterion of the product's safety, the simulation would still be a useful tool. They will enable rapid identification of processes, which are potentially safe or risky, thus narrowing the conditions range for the more time consuming and expensive experimental validation. Similar procedures can also be used to assess the potential effect of factors like pH, salt contents, etc. If their effect can be expressed in terms of the survival parameters, e.g., k and T_c in Eq. 20.5 and the magnitude of $n(T)$ in Eq. 20.1, then simulations could be used to identify the range of potential treatment conditions that will produce a safe product which still satisfies organoleptic and other quality requirements.

20.5.2 Unstable processes

Let us examine a hypothetical erratic process where the heating was almost totally out of control. This could be a result of 'periodic' or 'random' variations in the steam supply. The temperature profiles, $T(t)$, in such processes can be described by models with fixed or random coefficients, of the kind given in Eqs. 20.10 and 20.11 for example (Peleg, 2002). Two exaggerated examples are shown in Fig. 20.6. The main purpose of the figure is to demonstrate that the complexity of the differential equation when such temperature profiles are introduced is not a hindrance to its numerical solution by Mathematica®. Therefore, at least in principle, survival curves of this kind can be generated and analyzed in order to evaluate the risk stemming from an irregular steam supply and determine if an additional treatment would be needed if it happens. The same procedure could also be used to determine the duration of the additional treatment once the need has been established.

20.5.3 Steam stoppage

The effect of an accidental steam shut-down in mid-process on the theoretical survival ratio can also be examined by the described procedure, at least in principle. This is demonstrated in Fig. 20.7, where the temperature profiles were generated using Eq. 20.12 as the 'simplified' model. As before, the purpose is not to account for a specific event, but to show that the method works even when the temperature profile expression contains 'If statements'. Still, it is clearly evident that interrupted heating of same duration at different stages of the process can produce dramatically different effects on the residual survival ratio,

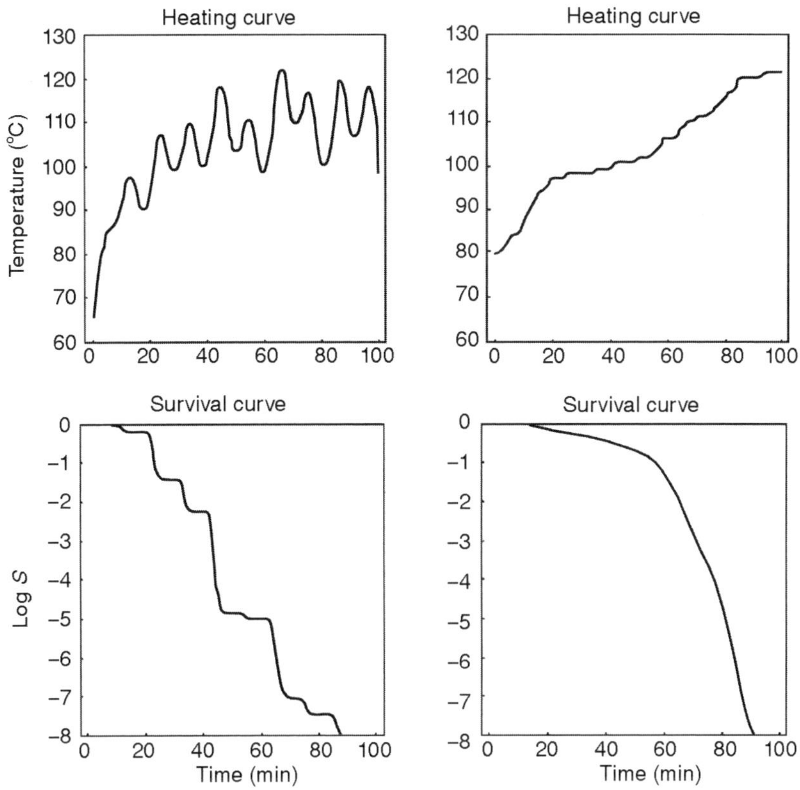

Fig. 20.6 Simulated sporal inactivation patterns during two uncontrolled heating processes, with smooth (right) and random (left) temperature oscillations.

and hence on the process safety. Again, a procedure of the kind that generated these demonstrations can be used to study the effect of steam interruptions on the final survival ratio in real food products and to determine how much additional treatment would be needed in order to guarantee their microbial safety.

20.6 Conclusion

The two main objectives of the chapter were to present an alternative interpretation of microbial inactivation kinetics and to demonstrate the capabilities of an unconventional mathematical procedure to assess the safety of thermal processes. No effort has been made to accurately match any particular thermal process or product in detail, or to obtain the accurate survival parameters of any particular organism, or spores, in a relevant food medium. At the time this chapter is being written, the described method has been verified for only a limited number of profiles involving *Salmonella* and *Listeria*.

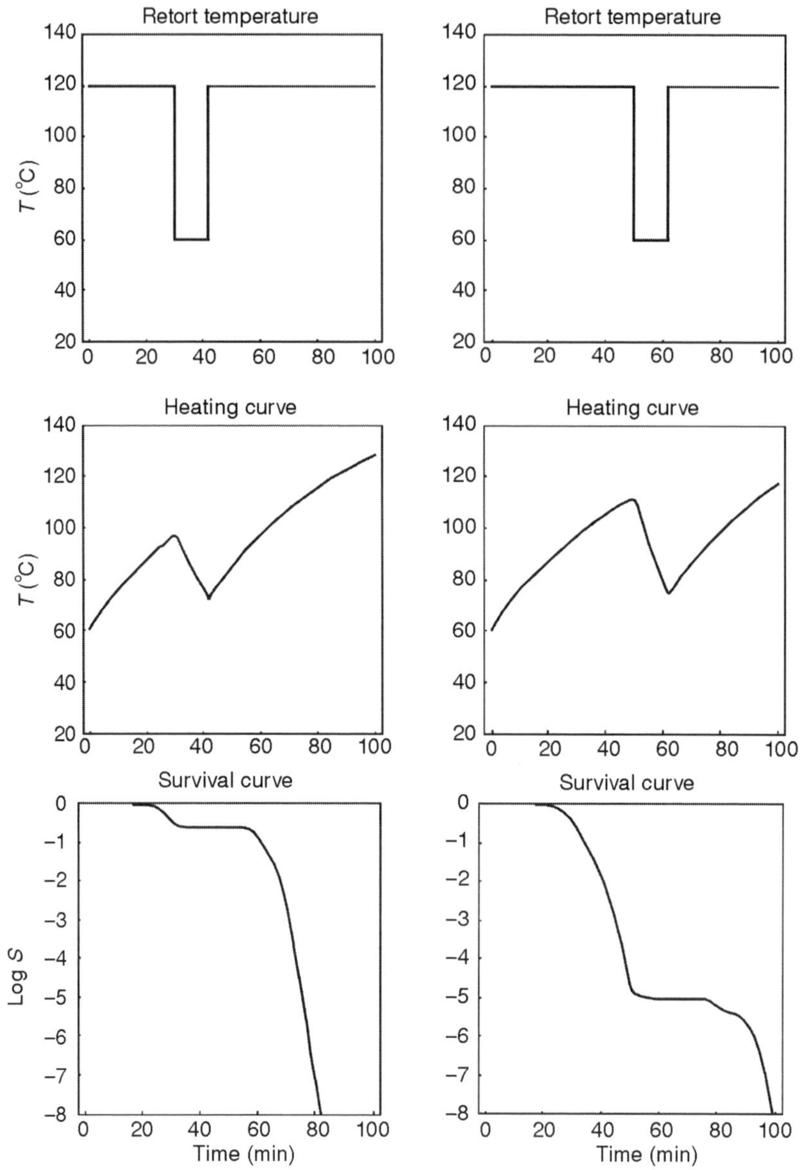

Fig. 20.7 Simulated steam interruptions of the same duration and their expected effect on the sporal inactivation pattern.

Nevertheless, the method's development will continue and, hopefully, its usefulness could be demonstrated in a larger number of organisms and spores of food safety concern. In light of the growing evidence and recognition that microbial inactivation *need not be* and in many cases is not a process which follows a first order kinetics, governmental regulating agencies will have to face this reality and revise the procedures to assess the safety of thermal preservation processes. Moreover, the inadequacy of the current official methods to calculate sterility, which are based on forcing straight lines through curved data sets has become all too obvious. Therefore, there will be an increasing pressure from the food industry and other quarters to develop alternative methods to calculate a process's safety based on more solid foundations than those of the ones now in use. With the computation power available today, such methods can be devised and easily implemented. As has been demonstrated in this chapter, and the publications on which it is based, the development of these methods will require a new critical view of the kinetics of microbial mortality and the abandonment of traditional concepts that have been around for many years. Still unresolved is the pertinent medium issue. Currently, in order to obtain the 'isothermal' survival data, a microbial culture is heated and cooled in a capillary or narrow metal tube in order to shorten the come-up and cooling times as much as possible. Such experiments are very difficult if not impossible to perform with highly viscous foods or foods which contain particulates. Consequently, the 'isothermal' survival data, in many cases, are obtained by experiments performed with a surrogate medium and not with the actual food. The problem would be eliminated if the survival parameters could be determined directly by experiments performed under non-isothermal conditions, where good mixing of the treated sample is the only requirement. It has been recently demonstrated that the calculation of the survival parameters from non-isothermal survival curves is theoretically possible. This has been demonstrated with *Salmonella* whose inactivation pattern in one non-isothermal heat treatment could be predicted from patterns observed in different non-isothermal treatments (Peleg and Normand, 2004; Peleg *et al.*, 2003). The method however still requires a refinement and validation with other organisms and spores before it could be considered for general use. The described method should also be adapted for treatments where the temperature increase and decrease are accompanied by changes in other factors (e.g., moisture contents, anti-microbial concentration, etc.), which can affect the inactivation rate. Although this has not yet been done, it seems that the mathematical tools already available will be sufficient for such application as well.

20.7 References

ANDERSON, W.F., MCCLURE, P.J., BAIRD-PERKER, A.C. and COLE M.G. (1996). The application of a log-logistic model to describe the thermal inactivation of *C. Botulinum* 213b at temperatures below 121.1°C. *Applied Bacteriol.* **80**, 283–290.

AUGUSTIN, J.C., CARLIER, V. and ROZJIER, J. (1998) Mathematical modeling of the heat resistance of *L. monocytogenes. Applied Microbiol.* **84**, 185–191.

CAMPANELLA, O.H. and PELEG, M. (2001). Theoretical comparison of a new and the traditional method to calculate *C. botulinum* survival during thermal inactivation. *J. Sci. Foods Agric.* **81**, 1069–1076.

CASOLARI, A. (1988) Microbial death. In M.J. Bazin and J. I. Prosser, *Physiological Models in microbiology* (pp. 1–44) Boca Raton, FL: CRC Press.

LINTON, R.H., CARTER, W.H., PIERSON, M.D., HACKEY, C.R. and EIFERT, J.D. (1996). Use of a modified Gompertz equation to model non-linear survival curves of *Listeria* monocytogenes. Scott, A. *J. Food Protec.* **59**, 16–23.

LITTLE, C.L., ADAMS, M.R., ANDERSON, W.A. and COLE, M.B. (1994). Application of the Log-logistic model to describe the survival of Yersinis enterocolitica at sub-optimal pH and temperature. *Intnl. J. Food Microbiol.* **22**, 63–71.

MATTICK, K.L., LEGAN, J.D., HUMPHREY, T.J. and PELEG, M. (2001). Calculating *Salmonella* inactivation in non-isothermal heat treatments from non-linear isothermal survival curves. *J. Food Protec.* **64**, 606–613.

PELEG, M. (2000). Microbial survival curves – The reality of flat shoulders and absolute thermal death times. *Food Res. Intnl.* **33**, 531–538.

PELEG, M. (2002). A model of survival curves having an 'activation shoulder'. *J. Food Sci.* **67**, 2438–2443.

PELEG, M. (2003). Calculation of the non-isothermal inactivation patterns of microbes having sigmoidal isothermal semi-logarithmic survival curves. *Crit. Rev. Food Sci. Nutr.* **43**, 645–658.

PELEG, M. and COLE, M.B. (1998). Reinterpretation of microbial survival curves. *Crit. Rev. Food Sci.* **38**, 353–380.

PELEG, M. and NORMAND, M.D. (2004). Predicting microbial survival curves from non-isothermal inactivation data. *Crit. Rev. Food Sci. Nutr.* (in press).

PELEG, M., NORMAND, M.D. and CAMPANELLA, O.H. (2003). Estimating microbial survival parameters from survival curves obtained under varying conditions – The linear case. *Bull. Mathaemat. Biol.* **65**, 219–234.

PELEG, M. and PENCHINA, C.M. (2000). Modeling microbial survival during exposure to a lethal agent with varying intensity. *Crit. Rev. Food Sci. Nutr.* **40**, 159–172.

PELEG, M., ENGEL, R., GONZALEZ-MARTINEZ, C. and CORRADINI, M.G. (2002). Non Arrhenius and non WLF kinetics in food systems. *J. Sci. Food Agric.* **82**, 1346–1355.

PELEG, M., PENCHINA, C.M. and COLE, M.B. (2001). Estimation of the survival curve of *Listeria monocytogenes* during non-isothermal heat treatments. *Food Res. Intnl.* **34**, 383–388.

STEPHENS, P.J., COLE, M.B. and JONES, M.V. (1994). Effect of heating rate on the thermal inactivation of *Listeria monocytogenes. J. Appl. Bacteriol.* **77**, 702–710.

TOLEDO, R. (1999). *Fundamentals of Food Process Engineering* (2nd ed.). Aspen. Gaithersburg, MD. pp. 315–397.

VAN BROEKEL, M.A.J.S. (2002). On the use of the Weibull model to describe thermal inactivation of microbial vegetative cells. *Intnl. J. Food Microbiol.* **74**, 139–159.

21

Evaluating microbial inactivation models for thermal processing

A. H. Geeraerd, V. P. Valdramidis, K. Bernaerts and J. F. Van Impe, Katholieke Universiteit Leuven, Belgium

21.1 Introduction

Heat treatment, in various forms, is one of the most widely used food preservation techniques. This chapter deals with the question of how to describe, in an accurate and at the same time generic way, the link between a heat treatment and the resulting microbial load on/in a selected food product. More specifically, modelling approaches within the discipline predictive microbiology dealing with the development of mathematical models able to describe the microbial inactivation during a heat treatment step are under consideration.

This chapter has three main sections. Firstly, in Sections 21.2–21.4 an overview of existing model types and their features for describing microbial inactivation as a function of time (primary models) is given. Features like the ability to describe the so-called shoulder and/or tailing in a microbial survival curve and the dynamic character of the model, are succinctly summarised. Hereafter, the attention in Section 21.5 is turned to secondary models, which describe primary model parameters (typically the specific inactivation rate) as a function of environmental factors. These factors can be both related with the heat treatment, e.g., temperature, and the food product, e.g., pH. A general application-driven modelling recipe, highlighting the different steps necessary to provide an accurate prediction of microbial death, is introduced in Section 21.6. The three covered steps, dealt with in parallel, are: (i) modelling the thermal history of the food product, (ii) formulating some backstage microbial considerations, and (iii) selection of a suitable microbial kinetic model. In the latter step, suitable models, selected based on the results presented in Sections

21.2–21.5 are incorporated. The different steps considered are exemplified making predictions of the *Escherichia coli* K12 inactivation by a thermal profile characterising an air impingement system as a case-study. Some future trends are highlighted in Section 21.7.

21.2 Description of primary models of inactivation

In this section, some considerations concerning a general framework for modelling the microbial evolution are presented. The framework and associated remarks are at the basis of model features along which the reviewed inactivation models can be catalogued, as summarised in Section 21.4.4.

A general expression for the evolution (incorporating growth, inactivation, and/or survival) of a (mixed) microbial population N consisting of i species in a homogeneous food product as a function of time, consists of the following set of n differential equations (Vereecken *et al.*, 2000; Bernaerts *et al.*, 2003).

$$\frac{dN_i}{dt} = \mu_i(N_i, < N_j >_{j \neq i}, < S >, < P >, < env >, < phys >) \cdot N_i \qquad [21.1]$$

In this so-called *primary model*, μ_i represents the specific evolution rate of a microbial species i, depending on one or more of the following variables:

1. N_i, the cell density of the species itself [cfu/mL],
2. $< N_j >$ cell densities of other species [cfu/mL],
3. $< S >$ concentrations of available substrates, like glucose, lactose, ... [amount/mL],
4. $< P >$ concentrations of microbial metabolites, like lactic acid, protons, ... [amount/mL],
5. $< env >$ actual environmental conditions as temperature, high pressure, salt concentration, fatty acid concentration, ...
6. $<phys>$ physiological state of the species, for instance, as influenced by the temperature history.

To complete the description, differential equations for time-varying state variables (e.g., dP/dt) are to be formulated, while the influence of $<env >$ is incorporated through secondary model types (see Section 21.5). The total set of equations incorporates all factors which may influence the microbial evolution in a sound way. To date, most models focus either on growth or inactivation. For growth, where $\mu_i > 0$, all six variables as mentioned above are important, as all internal processes related with cell growth and division have to take place. On the contrary, for inactivation, with $\mu_i < 0$, the main determining factors are $<env>$ and $<phys>$. The former factor includes temperature, pH, ... but also protective components like fatty acids. The latter refers to all possible cell components determining the cells' sensitivity to inactivation (e.g., membrane proteins).

Three remarks have to be formulated with respect to expressions [21.1]. Firstly, the variable *time* is not explicitly appearing in the right-hand side of the equation, making these differential equations *autonomous* ones. Obviously, time has an influence on the microbial evolution, but this is indirectly, for instance, due to the accumulation of a microbial metabolite. Secondly, the expressions are valid for local values of, e.g., temperature. If the food product is not homogeneous, the microbial evolution is depended on space as well. Thirdly, measurements like the initial value of the microbial population *N(0)* are not appearing in the right-hand side of expressions [21.1]: obviously, *N(0)* comes *only* into play when solving the set of differential equations to obtain a simulation or prediction of the microbial population N as a function of time t.

A large number of models (as will be presented in Section 21.4) are available in literature which can *not* be framed in this general expression. These so-called *static models* can be represented as $N = f(t, N(0))$. In comparison with the general expression [21.1], the following simplifications have been made:

1. The relation is an explicit expression of N as function of time and explicitly incorporates *N(0)*. As pointed out in Van Impe *et al.* (1992), explicit functions are valid only under constant environmental conditions. In other words, the use of models of this type in realistic time-varying conditions can only be exploited by approximating the time-varying conditions (e.g., temperature) with piece-wise constant temperature regions and by resetting certain variables at the start of each time interval, for instance, *N(0)*. Needless to say, this is an approximation *by hand* of the numerical solution of the underlying differential equation, which is a rather cumbersome and artificial way of working. The use of dynamic models is highly preferential when dealing with time-varying conditions (this will be illustrated in Section 21.6).
2. $< N_j >$, $< S >$, $< P >$, $< env >$ and $<phys>$ are not considered as time-varying variables but rather as constant (initial) conditions.

Turning the attention now to the thermal inactivation of vegetative micro-organisms, there are eight [see, e.g., Xiong *et al.* (1999) and Devlieghere *et al.* (2004)] commonly observed types of inactivation or survivor curves (i.e., when plotting *log(N)* as a function of time t): linear curves, curves with a shoulder, linear curves with tailing, sigmoidal-like curves, biphasic curves, biphasic curves with a shoulder, concave and convex curves (Fig. 21.1). Note that in this subdivision, tailing indicates a resistant population (no significant further inactivation achievable), whereas a biphasic pattern indicates two microbial populations with a difference in heat sensitivity.

In the following paragraphs different existing inactivation models, tested for vegetative micro-organisms, are presented. In the next section, dynamic models, which can be framed in Expression [21.1] are presented, while Section 21.4 deals with static model formulations. A reference table according to modelling features is presented in Section 21.4.4.

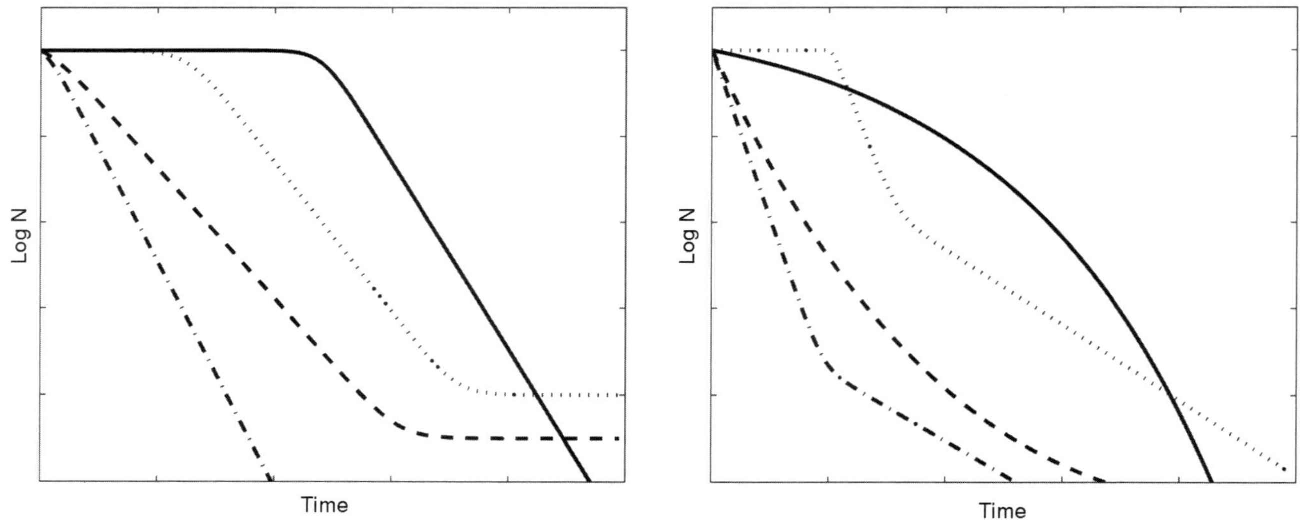

Fig. 21.1 Commonly observed types of inactivation curves. Left plot: linear curves (-·-·), linear curves with tailing (- -), sigmoidal-like curves (· · ·), curves with a shoulder (—). Right plot: biphasic curves (-·-·), concave curves (- -), biphasic curves with a shoulder (· · ·), and convex curves (—).

21.3 Dynamic inactivation models

In this section, three dynamic models are discussed: (i) the classical loglinear approach, (ii) a biphasic model, and (iii) a sigmoidal-like model including shoulder and tailing effects.

21.3.1 Loglinear model

Microbial inactivation or survival processes are traditionally described as loglinear (Bigelow and Esty, 1920; Anonymous, 2000).

$$\frac{dN}{dt} = -k \cdot N \quad \text{with} \quad k = \frac{\ln(10)}{D} \qquad [21.2]$$

or, in its more familiar format valid under static conditions:

$$\log\left[\frac{N(t)}{N(0)}\right] = -(1/D) \cdot t \qquad [21.3]$$

Herein, N represents the microbial cell density [cfu/mL], $N(0)$ the initial microbial cell density [cfu/mL], k [1/min] the first-order inactivation constant and D [min] the decimal reduction time. In this model it is assumed that the rate of destruction is proportional to the number of organisms present and that the death of an individual is dependent upon the random chance that one key molecule or 'target' within it receives sufficient heat (see, for example, Casolari, 1988). Observe how this equation is the most basic approximation for microbial inactivation modelling and is the most simple version of expression [21.1] (with μ_i denoted with -k). Despite the world-wide use of this model, especially in the canning industry for the so-called '12D process' focusing on the proteolytic strains (Group I) of *Clostridium botulinum* (see, for example, ICMSF, 1996), a lot of deviations have been observed (particularly at lower temperatures and for vegetative cells) indicating that inactivation kinetics are not always following first-order loglinear relationships (see, for example, references in Whiting, 1995, or Anonymous, 2000). Describing non-loglinear inactivation kinetics wrongly with expression [21.2] or [21.3] yields (highly) erroneous values for the inactivation parameter D.

As indicated above, eight types of survivor curves can be distinguished (Fig. 21.1). Explanations for these deviations have been searched for and can be categorised as the mechanistic and the vitalistic conception (Cerf, 1977). According to the *vitalistic conception*, on the one hand, individuals in a population are not identical, e.g., due to phenotypic variation between cells (Humpherson *et al.*, 1998). This means, in other words, that underlying the eventual microbial death there is a mechanism at the molecular level, such as the inactivation of certain vital enzymes or DNA (Van Boekel, 2002), which may vary between individuals. Variation of this mechanism from cell to cell leads to a non-identical response to the implemented heat stress, and as such, to deviations from loglinear inactivation kinetics at population level.

On the other hand, considering the *mechanistic theory*, two explanations arise: (i) deviations are a normal feature, i.e., some micro-organisms are inaccessible for the heat, or acquire heat resistance during the treatment, or (ii) deviations are experimental artefacts (Cerf, 1977). This may include phenomena like clumping of micro-organisms or factors more related to the experimental protocol. For the second one, some examples include heterogeneity of treatment, differences on counting techniques, on the protocol, the medium, the physiological stage of the inoculum, the rate of increase in the temperature, the environmental conditions prior to the heat treatment, etc. (Mackey and Derrick, 1987; Bréand *et al.*, 1998; Huang and Juneja, 2001).

21.3.2 Biphasic model

Cerf (1977) proposed a two-fraction model, which can be formulated as follows (see also Xiong *et al.*, 1999):

$$\frac{dN}{dt} = -\left(\frac{\ln(10)}{D_1} \cdot f + \frac{\ln(10)}{D_2} \cdot (1 - f)\right) \cdot N \qquad [21.4]$$

Herein, f is the fraction of the initial population in a major subpopulation (with $f \cdot N(0) = N_1(0)$) and $(1 - f)$ is the fraction of the initial population in a minor subpopulation (with $(1 - f) \cdot N(0) = N_2(0)$), the latter fraction being more heat resistant that the former one. D_1 and D_2 [min] are the decimal reduction times for the two fractions, respectively. The model describes biphasic curves, which are generally considered to represent a mix of two species or strains having different heat resistances. The dynamic version of the model (as the sum of two first order inactivation kinetics) makes it applicable under time-varying environments. Again, this model can be framed in expression [21.1], taking $i = 2$ and $N = N_1 + N_2$.

Biphasic thermal inactivation kinetics were also studied without the use of a fraction factor (Humpherson *et al.*, 1998). This makes the transition to the second (more resistant) population less smooth.

21.3.3 Sigmoidal-like model

A third dynamic model, consisting of two coupled differential equations, is the Geeraerd *et al.* (2000) model constructed for microbial inactivation by mild heating.

$$\frac{dN}{dt} = -k_{\max} \cdot \left(\frac{1}{1 + C_c}\right) \cdot \left(1 - \frac{N_{res}}{N}\right) \cdot N \qquad [21.5]$$

$$\frac{dC_c}{dt} = -k_{\max} \cdot C_c \qquad [21.6]$$

Herein, C_c (*critical component*) is related to the physiological state of the cells [-], k_{max} is the specific inactivation rate [1/min], and N_{res} is the residual population density [cfu/mL]. The model has four degrees of freedom: two states $N(0)$ and $C_c(0)$, and two parameter values k_{max} and N_{res}. The first factor at the right-hand side of equation [21.5] models the loglinear part of the inactivation curve and is equivalent to the classical first-order inactivation kinetics, as in equation [21.2]. The second factor describes the shoulder effect and is based on the hypothesis of the presence of a pool of protective components around or in each cell (Mossel *et al.*, 1995). Gradually, this pool is destroyed. In case of a shoulder, $1/(1 + C_c(0))$ takes on a small (positive) value. Towards the end of the shoulder region $1/(1 + C_c(t))$ becomes (approximately) equal to one, due to the component C_c undergoing heat inactivation following a first-order relationship (see equation [21.6]). Finally, the last factor of equation [21.5] implies the existence of a more resistant subpopulation N_{res} which can be framed within the mechanistic or vitalistic concepts as described in Cerf (1977). This model can exhibit a loglinear behaviour with and without shoulder and/or tailing revealing a smooth transition between each phase. It is important to remark that, for this model, tailing is considered for a population not undergoing any significant inactivation anymore, in contrast with, for instance, the Whiting model (Whiting, 1993) or the aforementioned biphasic model.

Although the model is derived in a dynamic formulation, an explicit solution, valid for constant environmental conditions, can be obtained.

$$N(t) = (N(0) - N_{res}) \cdot e^{-k_{max}t} \cdot \left(\frac{1 + C_c(0)}{1 + C_c(0) \cdot \mathrm{e}^{-k_{max}t}} \right) + N_{res} \qquad [21.7]$$

Another formulation of the static model is as follows:

$$N(t) = (N(0) - N_{res}) \cdot e^{-k_{max}t} \cdot \left(\frac{\mathrm{e}^{k_{max}S_l}}{1 + (\mathrm{e}^{k_{max}S_l} - 1) \cdot \mathrm{e}^{-k_{max}t}} \right) + N_{res} \qquad [21.8]$$

In this equation $C_c(0)$ is replaced by $\mathrm{e}^{k_{max}S_l} - 1$ as derived in Geeraerd *et al.* (2000). S_l [min] is a parameter that stands for the length of the shoulder. Observe that in this formulation, all parameters have a clear biological/visual meaning, which is interesting with respect to the parameter estimation procedure for this non-linear model (initial values are easy to obtain) and because they can be interpreted afterwards.

The model structure has been successfully applied to survival data of different micro-organisms and different treatments, such as *Listeria monocytogenes* and *Lactobacillus sakei* during a mild thermal inactivation (Geeraerd *et al.*, 2000), *Monilinia fructigena* and *Botrytis cinerea* during a pulsed white light treatment (Marquenie *et al.*, 2003), the Acid Tolerance Response (ATR) of *Salmonella enterica* and *L. monocytogenes* (Greenacre *et al.*, 2003), and the inactivation of *L. monocytogenes* in a pH-modified chicken salad during cold storage (Guentert *et al.*, 2003).

GInaFiT (**G**eeraerd et al. **Ina**ctivation Model **Fi**tting **T**ool), a freeware Add-in for Microsoft® Excel, based on the explicit form [21.8] of the inactivation model, can be downloaded via the KULeuven/BioTeC-homepage at http://

www.kuleuven.ac.be/cit/biotec/index.htm, topic "Downloads". The tool provides parameter estimates and Asymptotic Standard Errors, as well as the calculation of some statistical measures like the Sum of Squared Errors (with the error defined as the difference between the measured and the predicted *log(N)* values), and the (Adjusted) R^2.

21.4 Static inactivation models

In this section, only a selection of existing static inactivation models is presented. For a review of two models developed by Casolari (1988), the model of Whiting (1993) as derived from the logistic based model of Kamau *et al.* (1990), the model of Chiruta *et al.* (1997), the model of Daughtry *et al.* (1997), and the model of Xiong *et al.* (1999), reference is made to Geeraerd *et al.* (2000). Model features will, for all models considered, be listed in Table 21.1.

21.4.1 Sigmoidal models

Cole *et al.* (1993) have developed a model, for studying the thermal inactivation of *Listeria monocytogenes*, based on a distribution of heat sensitivity within the population of heated cells. The model expression with its four parameters is as follows:

$$\log N(t) = \alpha + \frac{\omega - \alpha}{1 + e^{4\sigma(\tau - \log t)/(\omega - \alpha)}} \qquad [21.9]$$

Herein, α is the upper asymptote [-] ($\sim N(0)$ in [21.7]), ω is the lower asymptote [-] ($\sim N_{res}$ in [21.7]), τ is the position of maximum slope [-] and σ the value of this maximum slope [1/min] ($\sim k_{max}$ in [21.7]). Although the model can simulate a loglinear curve with and without shoulder and tail, it has a symmetric log-logistic nature that cannot be motivated based on experimental data (Geeraerd *et al.*, 2000).

Bhaduri *et al.* (1991) and Linton *et al.* (1995) used a modified Gompertz equation to describe non-linear survival curves for *Listeria monocytogenes* Scott A. The latter group of authors used the following formulation:

$$\log \frac{N(t)}{N(0)} = Ce^{-e^{BM}} - Ce^{-e^{-B(t-M)}} \qquad [21.10]$$

Herein, *C, B* and *M* represent regression coefficients. The equation is capable of describing both loglinear and sigmoidal survival curves. Nevertheless, differentiating this equation for retrieving its dynamic form yields an equation that explicitly depends on the initial population, limiting its applicability.

The model of Buchanan *et al.* (1993) describes a shoulder effect followed by exponential decay and was subsequently extended by Bréand (1998). After this extension, it is a three phase linear model in which shoulder and tailing effects are included:

Table 21.1 Model features (proposed for vegetative micro-organisms) with respect to the eight possible inactivation curves shapes (Fig. 21.1) and their design characteristics. Partly originating from Xiong *et al.* (1999), Geeraerd *et al.* (2000), Devlieghere *et al.* (2004), Valdramidis *et al.* (2004a). '+': the model can describe this kind of curvature; '-': the model cannot describe this kind of curvature

	Loglinear	Loglinear − shoulder	Loglinear + tail	Loglinear + tail	Biphasic	Biphasic +shoulder	Multiphasic (Concave)	Multiphasic (Convex)
Models suitable for use in both constant and changing environmental conditions								
First order	+	−	−	−	−	−	−	−
Cerf	+	−	+	−	+	−	−	−
Geeraerd	+	+	+	+	−	−	−	−
Models only to be used under constant environmental conditions or by approximating a time-varying environment as successive static conditions (i.e., the model is inherently static, or inconsistent at time zero, or inconsistent at changing conditions)								
Casolari I	+$	−	−	−	−	−	+$	−
Casolari II	+$	+$	+$	+$	−	−	+$	+$
Kamau	+*	+	+	−	+	−	−	−
Whiting	+*	+	+	+	+	+	−	−
Cole	+	+	+	+	−	−	−	−
Modified Gompertz	+#	+#	+#	+#	−	−	−	−
Daughtry	+	−	−	−	+	−	+	−
Membré	−	−	−	−	−	−	−	+
Chiruta	+	−	−	−	+	−	+	−
Extended Buchanan	+	+	+	+	−	−	−	−
Weibull	+	−	−	−	−	−	+	+
Xiong	+	+	+	+	+	+	−	−

+$: the model misses a degree of freedom, implying a strong correlation between the slopes of the survival curve at distinct moments in time, for example, a long shoulder is automatically followed by a slow inactivation phase (Geeraerd *et al.*, 2000).
+* the model only approximates loglinear behaviour.
+# the slope of the loglinear part cannot be directly related to one (or a combination of) parameter value(s), see, e.g., Membré *et al.* (1999).

$$\log N = \log N(0) - k_0 t \quad 0 < t \leq t_{\text{lag}} \tag{21.11}$$

$$\log N = \log N(0) - k_0 t - k(t - t_{\text{lag}}) \quad t_{\text{lag}} < t \leq t_{\max} \tag{21.12}$$

$$\log N = \log N(0) - k_0 t - k(t_{\max} - t_{\text{lag}}) \quad t_{\max} \leq t \tag{21.13}$$

Herein, k_0 [1/min] is the initial inactivation rate, t_{lag} [min] is an initial lag phase during which the population decreases slowly, $t_{\max}$ [min] is the time at which a lower value is reached where, again, the population decrases slowly. This model can have a loglinear curvature with and without shoulder and tailing effects but without smooth transition zones. The model is inherently non-differentiable (as it consists of three separate phases) which limits its applicability to constant environmental conditions.

21.4.2 Convex model

Membré *et al.* (1997) developed an exponential primary model to describe convex inactivation curves of *Salmonella typhimurium*.

$$\log N(t) = (1 + \log N(0)) - e^{kt} \tag{21.14}$$

Herein, k corresponds to the shape of the destruction curve [1/min] and was used as an indicator of the rate of bacterial destruction. The minimalistic feature (only two parameters) of this empirical model was a main motivation for the authors to use this model. A dynamic version of this model cannot be extracted in a way to eliminate time and $N(0)$ in the right-hand side of the equation.

21.4.3 Convex-concave curves

The Weibull model is based on the concept that a survival curve is the cumulative form of a distribution of varying inactivation times within a microbial population. This approach can be considered close to the vitalistic theory as proposed by Cerf (1977). It has been tested for several micro-organisms (Peleg and Cole, 1998; Fernández *et al.*, 1999; Huang and Juneja, 2001; Van Boekel, 2002) and applied for recalculating the efficiency of a sterilisation treatment by Mafart *et al.* (2002).

$$\frac{N(t)}{N(0)} = \exp\left(-\left(\frac{t}{\alpha}\right)^{\beta}\right) \tag{21.15}$$

Herein, α [min] is a scale parameter, and β [-] is a shape parameter. It can describe the non-linearity of convex ($\beta > 1$) and concave ($\beta < 1$) survivor curves, and the classical first-order approach ($\beta = 1$). The different shapes of the inactivation curves can all be explained in terms of statistical properties of different underlying distributions of resistances or sensitivities, rather than being a manifestation of lethality kinetics of different orders.

Van Boekel (2002) and Mafart *et al.* (2002) observed a strong correlation between parameters α and β. The dependency of the parameters is due to the model structure (e.g., an error in α will be balanced by an error in β). This drawback can be circumvented by fixing the value of β (Peleg and Penchina, 2000; Mafart *et al.*, 2002; Fernández *et al.*, 2002) retaining two model parameters ($N(0)$ and α). It should be observed that by doing so the curvature of the model is restricted to being loglinear, concave or convex.

Derivation of the Weibull model with respect to time results in a dynamic version in which time cannot be eliminated.

21.4.4 Summarising primary inactivation models

The presented inactivation models are summarised in Table 21.1. The subdivision is based on their dynamic or static character and on their capabilities to describe one or more of the eight feasible survivor curve shapes, as presented in Fig. 21.1.

21.5 Description of secondary models of inactivation

Secondary models aim at describing the effect of environmental or processing factors, $< \mathit{env} >$ in expression [21.1], on the survival/inactivation kinetics of micro-organisms. Traditionally, when studying environmental factors at inactivation conditions, it is assumed that organisms within a population are identical, having the same constant value for $< \mathit{phys} >$, and that the first-order (log-linear) primary model in its static form [21.3] can describe their inactivation. Consequently, most secondary models focus on the description of k, or, alternatively, D, as a function of a number of environmental conditions. It should be noted that most studies for secondary models are based on spore-forming bacteria rather than on non-spore formers and that most studies are at several distinct but constant environmental conditions, enabling the use of [21.3].

21.5.1 Effect of temperature

In most conditions, temperature is the key factor controlling the survival/inactivation of bacteria during a thermal treatment. The two basic modelling approaches for relating temperature and inactivation rate are (see, e.g., Anonymous, 2000).

1. The Arrhenius model, which reads as follows:

$$k = A \exp\left(-\frac{E_a}{RT}\right) \qquad [21.16]$$

with k the specific reaction – in this case, inactivation – rate [1/min], E_a the so-called activation energy of the reaction system [J/mol], T the absolute temperature [K], R the universal gas constant [J/(mol K)] and A the so-called collision factor [1/min].

2. The Bigelow model (Bigelow, 1921), which is the traditional approach in the canning industry for studying the effect of temperature on the microbial thermoresistance:

$$D = D_{ref} 10^{-(T - T_{ref})/z} \quad [21.17]$$

Herein, D and D_{ref} [1/min] denote the decimal reduction time at temperature T [°C] and at a reference temperature T_{ref} [°C], respectively, and z [°C] represents the thermal resistance constant, i.e., the number of degrees change of temperature required to achieve a tenfold change in D-value. The model describes in a linear way the effect of temperature on the logarithm of the decimal reduction time.

When comparing the Arrhenius and the Bigelow model, it is clear that a temperature dependent relationship between E_a and z exists. However, the magnitude of the two coefficients is significantly different and any influence of temperature is negligible as long as the temperature reference is within the range used for data collection (Anonymous, 2000).

21.5.2 Effect of multiple factors

Turning now to the combined influence of different environmental conditions like *pH* and salt concentration on *k* or *D,* the most generally used models can be catalogued as (i) Modified (or linear) Arrhenius, where *ln(k)* is linearly related with $1/T, pH, pH^2$ … (ii) Bigelow-type, and (iii) polynomial or response surface models. Features and some key references are succinctly collected in Table 21.2.

Generally speaking, and as indicated in this table, response surface models include interaction terms between environmental factors (referring to synergistic/antagonistic effects), while Modified Arrhenius and Bigelow type models do not (infering additive effects). Evidence for an interacting effect between temperature and pH is presented by the studies of López *et al.* (1996) and Fernández *et al.* (1996), for the spores of *Bacillus stearothermophilus* and *Clostridium sporogenes*. To illustrate this point, consider the (additive) model of Mafart and Leguérinel (1998) which reads as follows

$$\log D = \log D^* - \frac{(T - T^*)}{z_T} - \frac{(pH - pH^*)^2}{z_{pH}^2} \quad [21.18]$$

Herein, T^* is the reference temperature (for example, 121.1°C), pH^* is the pH of maximum heat resistance (generally pH 7 for spores), z_T is the commonly used thermal z-value, z_{pH} is the distance of pH from pH^* which leads to a tenfold reduction of the decimal reduction time, D^* is the D-value at T^* and pH^*. The model was modified by Gaillard *et al.* (1998a) by adding an interaction term:

$$\log D = \log D^* + C_1(T - T^*) + C_2(pH - pH^*)^2 + C_3(T - T^*)(pH - pH^*)^2 \quad [21.19]$$

with C_1, C_2, C_3 polynomial-type regression coefficients. (Observe that this

Table 21.2 Model features and key references of commonly used secondary inactivation modelling approaches

Model features for modelling k_{max} or D	Model Type		
	I. Modified Arrhenius	II. Bigelow	III. Polynomial (response surface)
Biological interpretability of parameters	None	Yes	None
Number of parameters as function of d factors			
$d = 1$ (e.g., T)	2 (Davey *et al.*, 1993)	2 (Bigelow, 1921)	Not used
$d = 2$ (e.g., T and pH)	4 (Davey *et al.*, 1993)	3 (Mafart and Leguérinel,1998)	6 (Fernández *et al.*, 1996), 3 (Gaillard *et al.*, 1998a)
$d = 3$ (e.g., T, pH, and a_w)	5 (Cerf *et al.*, 1996)	4 (Gaillard *et al.*, 1998b)	10 (Blackburn *et al.*, 1997)
$d > 3$	unknown	unknown	15 (Juneja *et al.*, 1999)
Factors are considered:	Additive	Additive	Additive or Interacting
Correlation between parameters	Low	Low	Considerable
Extendibility towards additional factors	Possible, limited to additive effects	Possible, limited to additive effects	Possible
General applicability	Universally tested	Universally tested	Universally tested
Quality of fit	Acceptable	Acceptable	Acceptable

model is no longer to be considered as a Bigelow type model as it cannot be rephrased using *z* values.) Adding this interaction term, the obtained R^2 value improved (slightly) when describing *Bacillus cereus* inactivation. However, the necessity of incorporating such interaction terms is still debated in literature. Despite the (small) increasing quality of fit when including interaction terms (Gaillard *et al.*, 1998a), the authors stress the biological significance of the parameter values for the original additive version and the minor increase in quality of fit for the extended version. This view is clearly not shared by research groups developing polynomial models, who prefer to incorporate interaction effects between different environmental factors like, for example, temperature, pH, % salt and % sodium pyrophosphate (Juneja *et al.*, 1999).

Based on the model features presented in Table 21.2, it can be concluded that Bigelow-type models seem preferable over Arrhenius type models (because of the biological significance of the parameter values) for case studies where factors act additively, whereas polynomial relationship or equations like [21.19] are to be used when a prominent interaction effect is present.

Moreover, as for all modelling approaches, one must be cautious when translating these secondary models based on laboratory data to situations in real food products. As mentioned in Section 21.2, other relevant factors include the physiological state of the cells (exponential/stationary), fat content, carbohydrates, ... For example, heat resistance data for various pathogens as listed in ICMSF (1996) indicate the strong influence of the type of substrate on the microbial resistance. These factors are often not modelled, making the obtained *D* and z values species specific and product specific.

21.6 Modelling the interaction between micro-organisms, food and heat treatment

In this section, a general application-driven modelling recipe, highlighting the different steps necessary to provide an accurate prediction of microbial death, is presented. Hereto, an overview of the process of analysing the interaction between (i) micro-organisms, (ii) a specific food product, and (iii) a heat treatment is illustrated. The different modelling steps as presented in Fig. 21.2, are exemplified by simulating the surface decontamination of *Escherichia coli* K12 in a laboratory set-up of an air impingement system. In the following subsections, the three main parts of the recipe are presented.

21.6.1 Modelling the target heat process

The left part of the recipe (Fig. 21.2) represents in summary the (outcome of) modelling approaches related with the heat transfer modelling. Different modelling techniques exist in order to relate product (e.g., heat capacity, moisture content) and process characteristics (e.g., air temperature and velocity in a forced convection oven), as abundantly illustrated in other chapters of this

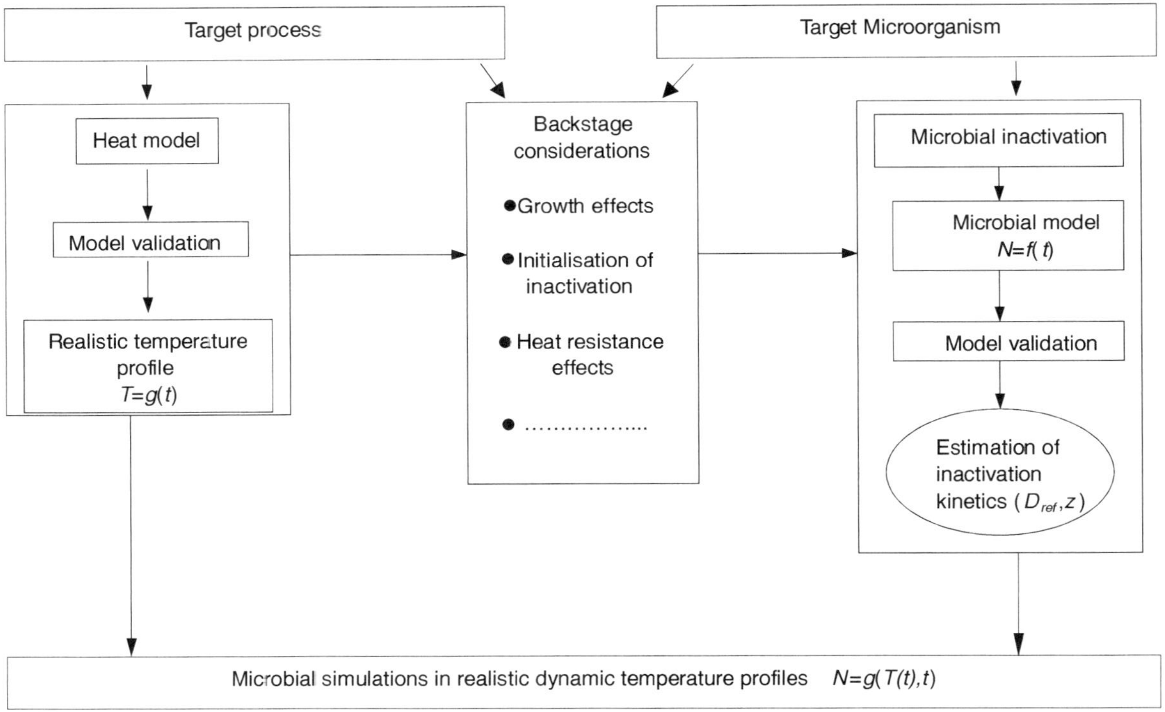

Fig. 21.2 Application-driven general modelling scheme relating the heat treatment, food properties and microbial inactivation phenomena.

book (for example, Chapters 4, 5, 13 and 15). Having constructed a heat transfer model, an important step is the validation of this model: how well does the model (with identified parameter values) compare to actual temperature measurements, possibly heterogeneously distributed over the food surface and the food interior? This heat model validation step is the subject of Chapter 16.

The final outcome of this part of the modelling recipe is a realistic (local) temperature profile, characterising both the chosen process and food product. Note that this outcome is indicated as $T = g(t)$ with g referring to the numerical solution of, for example, a set of coupled (partial) differential and algebraic equations making up the heat transfer model.

For the case-study used in this section, modelling approaches and validation studies to describe the temperature evolution in a laboratory set-up of an air impingement system using Teflon slices as model system, are presented in Kondjoyan and Havet (2003) and Valdramidis *et al.* (2004b). Calculated temperature profiles of this validation procedure, generated as part of the European project QLK1–CT-2001–01415 'Predicting microbial death during heat treatments on foods' (BUGDEATH), Co-ordinator: University of Bristol, UK, will be presented in Section 21.6.4.

21.6.2 Modelling the target micro-organism: microbial inactivation

Heat inactivation kinetics are classically determined in liquid media at various constant temperature levels. For this case-study, it was decided to assume equivalence of this type of experiments with surface decontamination kinetics, necessitating the need to assess this hypothesis during the validation phase (see Section 21.6.4).

E. coli K12 MG1655 (Jensen, 1993) has been chosen as a surrogate for the food-borne pathogen *E. coli* O157:H7. More importantly, coliforming bacteria are relevant for faecal surface contamination of food products. After preparation of the inoculum in Brain Heart Infusion broth (see Valdramidis *et al.*, 2004b, for technical details), isothermal inactivation experiments took place in sterile glass capillary tubes (Hirschmann) immersed in a constant temperature circulating water bath (GR150-S12, Grant) at temperatures of 54, 54.6, 56.6, 58.6 and 60.6°C. The come-up time for the temperature in the capillaries has been estimated as being 30 s in the worst case (Verboven P., personal communication) and data up to that time for all the isothermal experiments were omitted (as the study focuses on isothermal experiments) regarding them as part of the microbial prehistory. Calculations for the come-up time have been done based on heat conduction modelling by the use of the Fourier equation (Nicolaï *et al.*, 2001). As illustrated by the data points in Fig. 21.3, the microbial data at hand, when expressed as $\log(N)$ as function of time t, exhibit a shoulder followed by a linear behaviour.

The dynamic non-linear model of Geeraerd *et al.* (2000), constructed for microbial inactivation by mild heating, and presented in equations [21.5] and [21.6], has been selected as the model of use. The choice of this model was based on the possibility to describe loglinear inactivation curves preceded by a

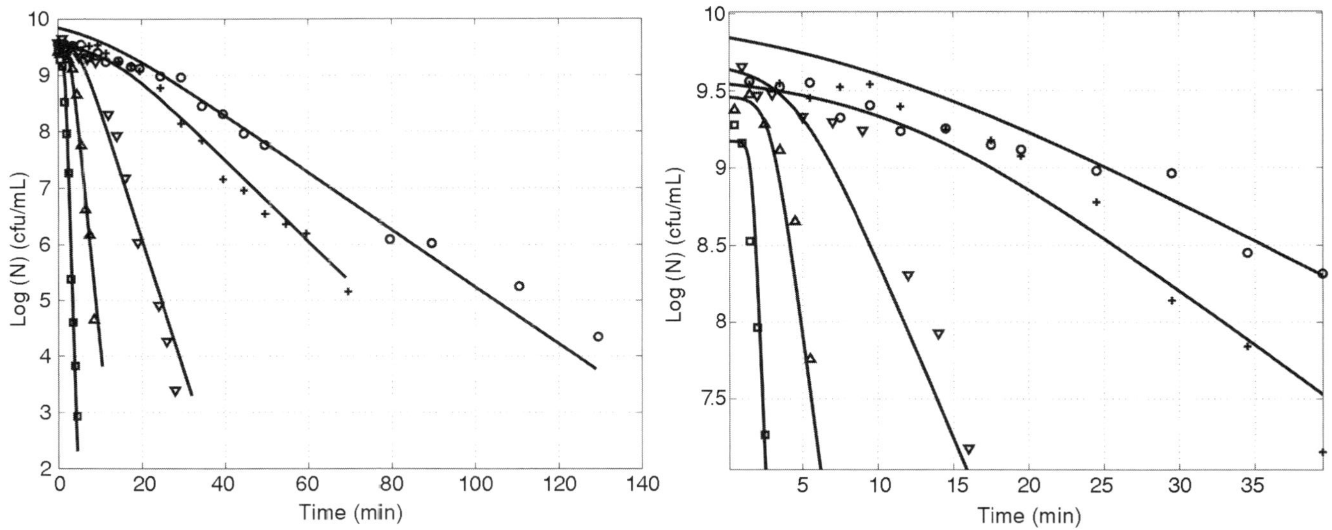

Fig. 21.3 One step regression of experimental data of *E. coli* K12 MG1655 at 54 ('o'), 54.6 ('+'), 56.6 ('∇'), 58.6 ('Δ') and 60.6ºC ('□') using equation [21.22] incorporated in equation [21.21]. The right plot is a detailed part of the left plot, focusing on the shoulder region.

shoulder, as observed in our data, and the ability to predict microbial inactivation under dynamic environmental conditions. As no tailing effect is observed in the present data (Fig. 21.3), equation [21.5] was reduced by omitting this factor (or setting $N_{res} = 0$).

$$\frac{dN}{dt} = -k_{\max} \cdot \left(\frac{1}{1 + C_c}\right) \cdot N \qquad [21.20]$$

The model can also be used in its explicit form as in equation [21.7] when constant temperature conditions are examined, as is the case for these isothermal experiments. If tailing is not present, equation [21.7] can be simplified to

$$N(t) = N(0)\mathrm{e}^{-k_{\max} t} \left(\frac{1 + C_c(0)}{1 + C_c(0)\mathrm{e}^{-k_{\max} t}}\right) \qquad [21.21]$$

Observe that here preference is given to the static version with the $C_c(0)$ parameter instead of the shoulder length S_l, as this shoulder length does not exist under time-varying conditions, which will be necessary to implement in Section 21.6.4. For the secondary model, the Bigelow model (equation [21.17]) was chosen for relating the specific inactivation rate with temperature.

$$k_{\max}(T) = \frac{\ln 10}{D_{ref}} \cdot \exp\left(\frac{\ln 10}{z} \cdot (T - T_{ref})\right) \qquad [21.22]$$

For parameter identification based on the classical Sum of Squared Errors criterion, a *one step* or *global* regression is applied in order to prevent the accumulation of fitting errors (see, e.g., Fernández *et al.*, 2002; Valdramidis *et al.*, 2002). This approach estimates the D_{ref} and z-values in a single step by the incorporation of the (secondary) Bigelow model for k_{max} (equation [21.22]) in the thermal inactivation (primary) model (equation [21.21]). The degrees of freedom for the parameter identification step are $N(0)$ and $C_c(0)$, for each inactivation curve separately, and one D_{ref} and z-value. In order to ensure that all these degrees of freedom have the same order of magnitude and, moreover, in order to normalise the variance on $N(t)$, a logarithmic transformation of $N(0)$ and $C_c(0)$ (denoted as $n(0)$ and $c_c(0)$) is performed prior to the parameter identification step. Observe that this step is denoted as $N = f(t)$ in Fig. 21.2, with f indicating an analytical expression, in contrast with $T = g(t)$ obtained in Section 21.6.1.

The result of this modelling step is shown in Fig. 21.3. Estimated inactivation kinetic parameters and the corresponding standard deviations are $D_{57°C} = 3.41 \pm 0.09$ min and $z = 3.94 \pm 0.07$°C. The initial average values of the parameters $n(0)$ and $c_c(0)$ (namely, at 9.48 and 1.98) are used as nominal values for performing the dynamic simulations, which is the subject of Section 21.6.4.

21.6.3 Modelling the target micro-organism: backstage microbial considerations

The third (and central) step in this modelling recipe includes aspects of both the target process and (mainly) the target micro-organisms part. Describing accurately the microbial inactivation in a dynamic temperature environment necessitates the formulation of a number of *backstage* microbiological considerations. They are denoted as backstage considerations because they are not incorporated within the mathematical model(s).

The major microbiological considerations can be formulated as follows:

1. Is there any microbial growth possible during the time to reach the desired inactivation temperature (i.e., come-up time)?
2. What is the (lowest) temperature at which inactivation starts?
3. Is there any increase in the heat resistance of the micro-organism due to the gradual increase of the temperature?

Answers to the questions – based on literature and/or experimental observations – lead to the formulation of a number of facts, which may reasonably differ for different target micro-organisms and target thermal treatments. It should be realised that even when these issues are neglected, some hypotheses are inherently and in a hidden way used during the dynamic simulations (as will be presented in Section 21.6.4) and, as such, may induce unrealistic kinetic predictions. Therefore, it is imperative to unravel the backstage considerations.

For the case-study of *E. coli* the following answers can be formulated:

1. Cardinal temperature values of *E. coli* O157:H7 are (i) the minimum growth temperature, approximately 8°C, (ii) the optimum growth temperature, approximately 37°C, and (iii) the maximum growth temperature, approximately 44–45°C. Additionally, the highest growth rate of *E. coli* 2 ETEC is not exceeding 3 logs [cfu/mL]/3.5 h when the temperature is between 32.8 and 40°C (ICMSF, 1996). This means that for heating rates in which inactivating temperatures are reached within minutes, no relevant growth of *E. coli* is expected.
2. Thermal inactivation experiments of *E.coli* K12 MG1655 at 49.5°C (results not shown), following the same protocol for the data generation as the one presented, show a shoulder length S_l of approximately 180 min. Therefore, it is assumed that inactivation of *E. coli* starts at temperatures above 49.5°C.
3. To the best of our knowledge, there are no literature data available (reporting counts of viable microbial population) determining quantitatively the effect of rising temperatures on the heat resistance of *E. coli*. However, studies on *L. monocytogenes* have shown a maximum thermotolerance induced at heating rates below (or equal to) 0.7°C/min (Stephens *et al.*, 1994). It can be assumed that in case of fast heating treatments the time-frame is too short to induce an increase in the heat resistance of *E. coli* K12 MG1655.

21.6.4 Microbial simulations in realistic dynamic temperature profiles

Joining all elements presented in the previous sections, realistic simulations of the microbial behaviour under a dynamic temperature profile $T = g(t)$, becomes feasible. At this point, the importance of choosing a 'dynamic' model for describing the microbial inactivation becomes clear. By making use of equations [21.6] and [21.21], with the Bigelow expression [21.17] incorporated, the following dynamic equation is obtained for use in time-varying temperature conditions;

$$\frac{dN}{dT} = -\frac{\ln 10}{D_{ref}} \cdot \exp\left(\frac{\ln 10}{z} \cdot (T - T_{ref})\right) \cdot \left(\frac{1}{1 + C_c}\right) \cdot N \qquad [21.23]$$

$$\frac{dC_c}{dT} = -\frac{\ln 10}{D_{ref}} \cdot \exp\left(\frac{\ln 10}{z} \cdot (T - T_{ref})\right) \cdot C_c \qquad [21.24]$$

Herein, z, D_{ref}, $n(0)$ and $c_c(0)$ are fixed at their values as identified in Section 21.6.2. Note however that parameter estimates should be used with caution for temperatures outside the experimental region (i.e., 49.5–54°C, >60.6°C) where extrapolation is made. By taking the following subsequent steps: (i) coupling equations [21.23] and [21.24] with the temperature profile $T = g(t)$ of Section 21.6.1, (ii) taking into account the above mentioned microbiological considerations, and (iii) solving (numerically) for $N = \mathrm{g}(T(t), t)$, the final aim, namely microbial simulations in realistic dynamic temperature profiles, is reached. Results are presented in Fig. 21.4. In the left part of this figure, a temperature profile with a slow heating regime aiming at reaching the final temperature in 1200 seconds is shown. The thermal profile characterising the dry air impingement treatment results in a complete inactivation of the examined micro-organism (observe that the simulated micro-organisms concentration decreases further in the period between 600 and 1200 seconds). On the contrary, the temperature profile with a very rapid come-up time (only 30 seconds are needed to reach the final temperature) appears not sufficient for inactivating the micro-organism (Fig. 21.4, right). Only a really small microbial decrease is achieved.

Some observations are the following. As parameter estimates come from broth experiments, care has to be applied when using these model predictions for cases of a contaminated food product (Murphy *et al.*, 1999; Mossel *et al.*, 1995). Moreover, it is obvious that the complete validation of these predictions and the underlying backstage considerations and model choices can only be addressed by performing microbial experiments at some of the observed dynamic temperature conditions and by using a real, selected food product for surface decontamination studies. Also, it should be observed that, before real-life application of such a recipe would be feasible, important additional targets related with the quality of the food product (for example, structure, texture, enzymatic deterioration, vitamin content, microbial spoilage, ...) should be focused on. For part of these targets, a conflict with microbial safety aspects is apparent. Product quality is dealt with in more detail in Chapter 1.

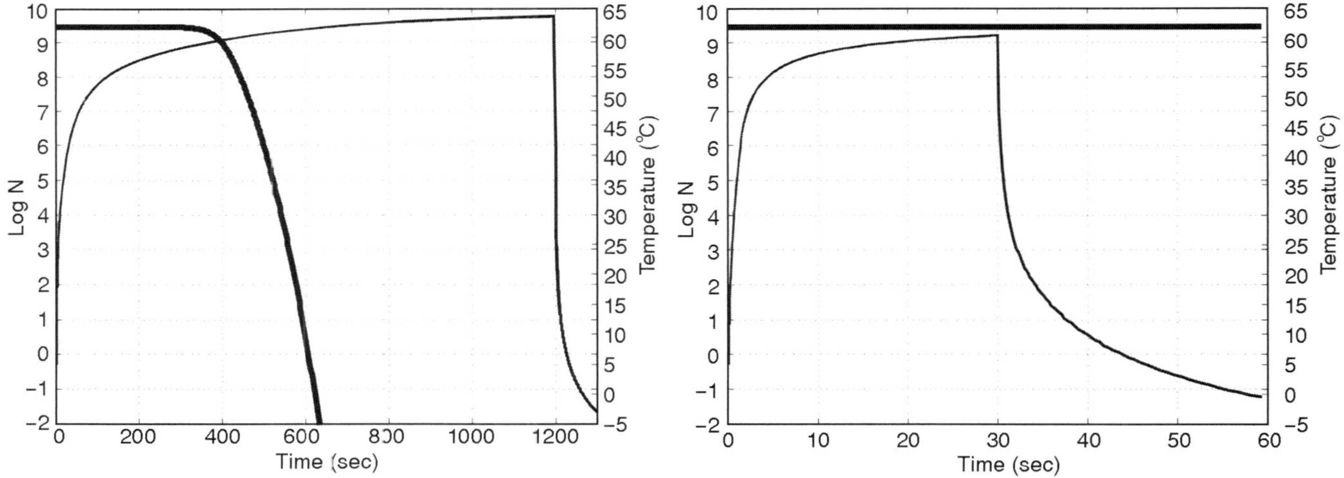

Fig. 21.4 Microbial inactivation simulations (thick line) of *E. coli* K12 MG1655 with the use of temperatures profiles of dry air (thin line) for a fully wetted product. Left plot: air temperature = 350°C and air velocity 1.0 m/s for the first 1200 s, thereafter: air temperature = −20°C and air velocity 50 m/s. Right plot: air temperature = 350°C and air velocity 20 m/s for the first 30 s, thereafter: air temperature 20°C and air velocity 20 m/s.

21.7 Future trends

In this section, a number of future trends are listed with respect to microbial inactivation modelling. Characterising microbial inactivation parameters like D_{ref} and z based on a set of isothermal inactivation curves (as was illustrated earlier in this chapter), is the common approach. Recently, developments are directed towards the direct use of non-isothermal survival curves. For instance, Peleg *et al.* (2003) illustrate how to use survival data obtained under a time-varying temperature profile to identify secondary model parameters, assuming first-order inactivation under isothermal conditions and a certain analytical expression for the heating profile. A similar approach would be possible using equations [21.23] and [21.24] for *identifying* D_{ref} and z on non-isothermal data, assuming certain backstage microbiological considerations as in Section 21.6.3. One can even go one step further by optimising the dynamic temperature profile with respect to its information content, i.e., maximising the information content in the experimental data in order to minimise the Standard Error on the subsequently obtained kinetic parameters (see, Versyck *et al.* (1999) and Valdramidis *et al.* (2002) in the context of inactivation modelling).

The search for non-loglinear inactivation curves is still continued – both with respect to the development of alternative model equations as with respect to the mechanisms underlying this non-loglinear behaviour. For example, the Institute of Food Technologists (IFT) organised a Summit Conference entitled 'Kinetics of Inactivation of Microbial Populations – Emphasizing Models for Non-Log-Linear Microbial Curves' in January 2003, Orlando. This conference follows previous work of IFT (Anonymous, 2000), a monumental review concerning the kinetics of microbial inactivation due to alternative processing techniques like, for example, ohmic and inductive heating, high pressure processing, pulsed light treatment, etc. Techniques which have already found application in the food industry as well as techniques which are (still) at research level are extensively considered. Generally speaking, this report summarises evidence of various types of deviation from expression [21.2] (or [21.3]), for example, for a pulsed electric field treatment, the widespread model of Hülsheger *et al.* (1981) indicates a log-log relationship between N and time t. More recently, several contributions on the Fourth International Conference on Predictive Modelling in Foods, held in Quimper in June 2003, stressed the importance of non-loglinear inactivation model developments, for instance of the Weibull model type (Albert and Mafart, 2003; Hassani *et al.*, 2003; Álvarez *et al.*, 2003; Gómez *et al.*, 2003); the biphasic model type (Geeraerd *et al.*, 2003) or based on a sigmoidal-type model (Guentert *et al.*, 2003).

It should also be taken into account that when using models in real food products or when asking questions about what happens to the microbial population after the heat treatment, one is inevitably confronted with extrapolation of the developed modelling approaches. This necessitates the need for developments focusing on unravelling the mechanisms at the basis of the microbial behaviour, as illustrated by Smelt *et al.* (2002) and (2003) for microbial inactivation, and, more generally in McMeekin *et al.* (2002) for predictive modelling approaches in general dealing with microbial growth, inactivation and survival.

21.8 Acknowledgements

This research is supported by the European Commission as part of project EU QLK1–CT-2001–01415, the Research Council of the Katholieke Universiteit Leuven as part of projects OT/99/24 and IDO/00/008 and DB/03/071 (B2467), the Fund for Scientific Research – Flanders (Belgium) (F.W.O.-Vlaanderen) as part of project G.0213.02 and for the Postdoctoral Fellowship of AG and KB, the Belgian Program on Interuniversity Poles of Attraction and the Second Multi-annual Scientific Support Plan for a Sustainable Development Policy, initiated by the Belgian Federal Science Policy Office. The scientific responsibility is assumed by its authors.

21.9 References

ALBERT I and MAFART P (2003), 'A modified Weibull model for bacterial inactivation', in Van Impe J F, Geeraerd A H, Leguérinel I and Mafart P, *Predictive Modelling in Foods – Conference Proceedings*, KULeuven/BioTeC, Belgium (ISBN 90-5682-400-2), 158–160.

ÁLVAREZ I, VIRTO R, PAGÁN R, RASO J and CONDÓN S (2003), 'Kinetics of inactivation of *Saccharomyces cerevisiae* by pulsed electric fields', in Van Impe J F, Geeraerd A H, Leguérinel I and Mafart P, *Predictive Modelling in Foods – Conference Proceedings*, KULeuven/BioTeC, Belgium (ISBN 90-5682-400-2), 182–184.

ANONYMOUS (2000) 'Kinetics of microbial inactivation for alternative food processing technologies – IFT's response to task order 1, US Food and Drug Administration: How to quantify the destruction kinetics of alternative processing technologies', *J Food Sci*, Suppl, 4–108.

BERNAERTS K, DENS E, VEREECKEN K, GEERAERD A, DEVLIEGHERE F, DEBEVERE J and VAN IMPE J F (2003), 'Modeling microbial dynamics under time-varying conditions', in McKellar R and Lu X, *Modeling Microbial Responses in Foods*, CRC Press, Boca Raton (ISBN 0-8493-1237-X), 243–261.

BHADURI S, SMITH P W, PALUMBO S A, TURNER-JONES C O, SMITH J L, MARMER B S, BUCHANAN R L, ZAIKA L L and WILLIAMS A C (1991), 'Thermal destruction of *Listeria monocytogenes* in liver sausage slurry', *Food Microbiol*, 8, 75–78.

BIGELOW W D and ESTY J R (1920), 'The thermal death point in relation to typical thermophylic organisms', *J Infect Dis*, 27, 602.

BIGELOW W D (1921), 'The logarithmic nature of thermal death time curves', *J Infect Dis*, 29, 528–536.

BLACKBURN C W DE, CURTIS L M, HUMPHESON L, BILLON C and MCCLURE P J (1997), 'Development of thermal inactivation models for *Salmonella enteritidis* and *Escherichia coli* O157:H7 with temperature, pH and NaCl as controlling factors', *Int J Food Microbiol*, 38, 31–44.

BRÉAND S (1998), *Etude biométrique de la résponse d'une population bacteriénne à une variation défavorable de température ou de pH, Applications en microbiologie prévisionelle alimentaire,* PhD Thesis, Université Claude Bernard – Lyon I, France.

BRÉAND S, FARDEL G, FLANDROIS J P, ROSSO L and TOMASSONE R (1998), 'Model of the influence of time and mild temperature on *Listeria monocytogenes* nonlinear survival curves', *Int J Food Microbiol*, 40, 185–195.

BUCHANAN R L, GOLDEN M H and WHITING R C (1993), 'Differentiation of the effects of pH and lactic or acetic acid concentration on the kinetics of *Listeria monocytogenes*', *J Food Prot*, 56, 474–478.

CASOLARI A (1988), 'Microbial death', in Bazin M J and Prosser J I, *Physiological models in microbiology* 2, CRC Press, Boca Raton, Florida, 1–44.

CERF O (1977), 'A review. Tailing of survival curves of bacterial spores', *J Appl Microbiol*, 42, 1–19.

CERF O, DAVEY K R and SADOUDI A K (1996), 'Thermal inactivation of bacteria – a new predictive model for the combined effect of three environmental factors: temperature, pH and water activity', *Food Res Int*, 29, 219–226.

CHIRUTA J, DAVEY K R and THOMAS C J (1997), 'Combined effect of temperature and pH on microbial death in continuous pasteurisation of liquids', in Jowitt R, *Engineering and Food at ICEF7*, Sheffield Academic Press, Sheffield, A109–A112.

COLE M B, DAVIES K W, MUNRO G, HOLYOAK C D and KILSBY D C (1993), 'A vitalistic model to describe the thermal inactivation of *Listeria monocytogenes*', *J Ind Microbiol*, 12, 232–239.

DAUGHTRY B J, DAVEY K R, THOMAS C J and VERBYLA A P (1997), 'Food Processing – a new model for the thermal destruction of contaminating bacteria', in Jowitt R, *Engineering and Food at ICEF7*, Sheffield Academic Press, Sheffield, A113–A116.

DAVEY K R (1993), 'Extension of the generalized chart for combined temperature and pH', *Lebensm-Wiss u-Technol*, 26, 476–479

DEVLIEGHERE F, FRANCOIS K, VEREECKEN K M, GEERAERD A H, VAN IMPE J F and DEBEVERE J (2004), 'Effect of chemicals on the microbial evolution in foods', *J Food Prot* (accepted).

FERNÁNDEZ P S, OCIO M J, RODRIGO F, RODRIGO M and MARTÍNEZ A (1996), 'Mathematical model for the combined effect of temperature and pH on the thermal resistance of *Bacillus stearothermophilus* and *Clostridium sporogenes* spores', *Int J Food Microbiol*, 32, 225–233.

FERNÁNDEZ A, SALMERÓN C, FERNÁNDEZ P S and MARTÍNEZ A (1999), 'Application of a frequency distribution model to describe the thermal inactivation of two strains of *Bacillus cereus', Trends Food Sci Technol*, 10, 158–162.

FERNÁNDEZ A, COLLADO J, CUNHA L M, OCIO M J and MARTÍNEZ A (2002), 'Empirical model building based on Weibull distribution to describe the joint effect of pH and temperature on the thermal resistance of *Bacillus cereus* in vegetative substrate', *Int. J Food Microbiol,* 77, 147–153.

GAILLARD S, LEGUERINEL I and MAFART P (1998a), 'Modelling combined effects of temperature and pH on the heat resistance of spores of *Bacillus cereus*', *Food Microbiol*, 15, 625–630.

GAILLARD S, LEGUERINEL I and MAFART P (1998b), 'Model for combined effects of temperature, pH and water activity on thermal inactivation of *Bacillus cereus* spores', *J Food Sci,* 63, 887–889.

GEERAERD A H, HERREMANS C H and VAN IMPE J F (2000), 'Structural model requirements to describe microbial inactivation during a mild heat treatment', *Int J Food Microbiol*, 59, 185–209.

GEERAERD A H, WOUTERS P C and VAN IMPE J F (2003), 'A combined primary and secondary model for describing pulsed electric field microbial inactivation', in Van Impe J F, Geeraerd A H, Leguérinel I and Mafart P, *Predictive Modelling in Foods – Conference Proceedings*, KULeuven/BioTeC, Belgium (ISBN 90-5682-400-2), 173–175.

GÓMEZ N, GARCÍA D, ÁLVAREZ I, CONDÓN S and RASO J (2003), 'Modelling the inactivation of *Listeria monocytogenes* by pulsed electric fields in media of different pH', in Van Impe J F, Geeraerd A H, Leguérinel I and Mafart P, *Predictive Modelling in Foods – Conference Proceedings*, KULeuven/BioTeC, Belgium (ISBN 90-5682-400-2), 185–187.

GREENACRE E J, BROCKLEHURST T F, WASPE C R, WILSON D R and WILSON P D G (2003), '*Salmonella enterica* Serovar Typhimurium and *Listeria monocytogenes* Acid Tolerance Response Induced by Organic Acids at 20ºC: Optimization and Modeling', *Appl Environ Microbiol*, 69(7), 3945–3951.

GUENTERT A M, LINTON R H, MOHTAR R H, TAMPLIN M L, LUCHANSKY J B and COUSIN M A (2003), 'Growth and inactivation of *Listeria monocytogenes* in pH-modified chicken salad during cold storage', in Van Impe J F, Geeraerd A H, Leguérinel I and Mafart P, *Predictive Modelling in Foods – Conference Proceedings*, KULeuven/BioTeC, Belgium (ISBN 90-5682-400-2), 167–169.

HASSANI M, ÁLVAREZ I, RASO J, CONDÓN S and PAGÁN R (2003), 'Comparing predicting models for heat inactivation of *Listeria monocytogenes* and *Pseudomonas aeruginosa* at different pH', in Van Impe J F, Geeraerd A H, Leguérinel I and Mafart P, *Predictive Modelling in Foods – Conference Proceedings*, KULeuven/BioTeC, Belgium (ISBN 90-5682-400-2), 164–166.

HUANG L and JUNEJA V K (2001), 'A new kinetic model for thermal inactivation of microorganisms: Development and Validation using *Escherichia coli* O157:H7 as a test organism', *J Food Prot*, 64, 2078–2082.

HÜLSHEGER H, POTEL J and NIEMANN E-G (1981), 'Killing of bacteria with electric pulses of high field strength', *Radiat Environ Biophys*, 20, 53–65.

HUMPHERSON L, ADAMS M R, ANDERSON W A and COLE M B (1998), 'Biphasic Inactivation kinetics in *Salmonella enteriditis* PT4', *Appl Environ Microbiol*, 64, 459–464.

ICMSF (1996) *Microorganisms in Foods 5. Characteristics of Microbial Pathogens*, Blackie Academic and Professional, London.

JENSEN K F (1993), 'The *Escherichia coli* K-12 'Wild Types' W3110 and MG1655 have an *rph* frameshift mutation that leads to pyrimidine starvation due to low *pyre* expression levels', *J Bact,* 175, 3401–3407.

JUNEJA V K and EBLEN B S (1999), 'Predictive thermal inactivation model for *Listeria monocytogenes* with temperature, pH, NaCl, and sodium pyrophosphate as controlling factors', *J Food Prot*, 62, 986–993.

KAMAU D N, DOORES S and PRUITT K M (1990), 'Enhanced thermal destruction of *Listeria monocytogenes* and *Staphylococcus aureus* by the lactoperoxidase system', *Appl Environ Microbiol*, 56, 2711–2716.

KONDJOYAN A and HAVET M (2003), 'Modeling heat transfer in a jet of hot air to decontaminate meat products', *Sci Aliments,* 23, 157–161.

LINTON R H, CARTER W H, PIERSON M D and HACKNEY C R (1995), 'Use of a modified Gomprertz equation to model nonlinear survival curves for *Listeria monocytogenes*

Scott A', *J Food Prot*, 58, 946–954.

LÓPEZ M, GONZÁLEZ I, CONDÓN S and BERNARDO A (1996), 'Effect of pH heating medium on the thermal resistance of *Bacillus stearothermophilus* spores', *Food Microbiol*, 28, 405–410.

MACKEY B M and DERRICK C M (1987), 'Changes in heat resistance of *Salmonella typhimurium* during heating at rising temperatures', *Lett Appl Microbiol*, 4, 13–16.

MAFART P and LEGUÉRINEL I (1998), 'Modeling combined effects of temperature and pH on heat resistance of spores by a linear-Bigelow equation', *J Food Sci*, 63, 6–8.

MAFART P, COUVERT O, GAILLARD S and LEGUÉRINEL I (2002), 'On calculating sterility in thermal preservation methods: application of the Weibull frequency distribution model', *Int J Food Microbiol*, 72, 107–113.

MARQUENIE D, GEERAERD A H, LAMMERTYN J, SOONTJENS C, VAN IMPE J F, MICHIELS C W and NICOLAÏ B M (2003), 'Combinations of pulsed white light and UV-C or mild heat treatment to inactivate conidia of *Botrytis cinerea* and *Monilia fructigena*' , *Int J Food Microbiol*, 85, 186–196.

MCMEEKIN T A, OLLEY J, RATKOWSKY D A and ROSS T (2002), 'Predictive microbiology: towards the interface and beyond', *Int J Food Microbiol*, 73, 395–407.

MEMBRÉ J-M, MAJCHRZAK V and JOLLY I (1997), 'Effects of temperature, pH, glucose and citric acid on the inactivation of *Salmonella typhimurium* in reduced calorie mayonnaise', *J Food Prot*, 60, 1497–1501.

MEMBRÉ J-M, ROSS T and MCMEEKIN T (1999), 'Behaviour of *Listeria monocytogenes* under combined chilling processes', *Lett Appl Microbiol*, 28, 216–220.

MOSSEL D A A, CORRY J E L, STRUIJK C B and BAIRD R M (1995), *Essentials of the microbiology of foods*, John Wiley & Sons, Inc., West Sussex.

MURPHY R Y, MARKS B P, JOHNSON E R and JOHNSON M G (1999), 'Inactivation of *Salmonella* and *Listeria* in ground chicken breast meat during thermal processing', *J Food Prot*, 62, 980–985.

NICOLAÏ B M, VERBOVEN P and SCHEERLINCK N (2001), 'The modelling of heat and mass transfer', in Tijskens L M M, Hertog L A T M and Nicolaï B M, *Food Process Modelling*, Woodhead Publishing Limited and CRC Press LLC, Cambridge, UK and Boca Raton, FL, USA, 60–86.

PELEG M and COLE M B (1998), 'Reinterpretation of microbial survival curves', *Crit Rev Food Sci Nutr*, 38, 353–380.

PELEG M and PENCHINA C M (2000), 'Modeling microbial survival during exposure to a lethal agent with varying intensity', *Crit Rev Food Sci Nutr,* 40, 159–172.

PELEG M, NORMAND M D and CAMPANELLA O H (2003), 'Estimating microbial inactivation parameters from survival curves obtained under varying conditions – the linear case', *Bull Math Biol*, 65, 219–234.

SMELT J P P M, HELLEMONS J C, WOUTERS P C and VAN GERWEN S J C (2002), 'Physiological and mathematical aspects in setting criteria for decontamination of foods by physical means', *Int J Food Microbiol*, 78, 57–77.

SMELT J P P M, SCURRA K and BULL M (2003), 'The effect of clumping on the kinetics of microbial inactivation. A proposal for an experimental approach to validate a quantitative model', in Van Impe J F, Geeraerd A H, Leguérinel I and Mafart P, *Predictive Modelling in Foods – Conference Proceedings*, KULeuven/BioTeC, Belgium (ISBN 90-5682-400-2), 161–163.

STEPHENS P J, COLE M B and JONES M V (1994), 'Effect of heating rate on the thermal inactivation of *Listeria monocytogenes*', *J Appl Bact*, 77, 702–708.

VALDRAMIDIS V P, BERNAERTS K, GEERAERD A H and VAN IMPE J F (2002), 'Application of a novel procedure to quantify thermal inactivation kinetics (PART II)', in Axelsson L, Tronrud E S and Merok K J, *Proceedings and abstracts of the 18th international ICFMH Symposium*, Lillehammer, Norway, 343.

VALDRAMIDIS V P, GEERAERD A H, BERNAERTS K, DEVLIEGHERE F, DEBEVERE J and VAN IMPE J F (2004a), 'Accurate modelling of non-loglinear survivor curves', *Int Dairy Fed Bull* (selected for the Special Issue following the Workshop on Revisiting Heat Resistance of Microorganisms in Milk, Kiel (Germany), May 5–8, 2003).

VALDRAMIDIS V P, BELAUBRE N, ZUNIGA R, FOSTER A M, HAVET M, GEERAERD A H, SWAIN M J, BERNAERTS K, VAN IMPE J F and KONDJOYAN A (2004b), 'Development of predictive modelling approaches for surface temperature and associated microbiological inactivation during hot dry air decontamination', *Int J Food Microbiol* (selected for the Special Issue following Fourth International Conference on Predictive Modelling in Foods, Quimper (France), June 15–19, 2003).

VAN BOEKEL M A J S (2002), 'On the use of Weibull model to describe thermal inactivation of microbial vegetative cells', *Int J Food Microbiol*, 74, 139–159.

VAN IMPE J F, NICOLAÏ B M, MARTENS T, DE BAERDEMAEKER J and VANDEWALLE J (1992), 'Dynamic mathematical model to predict microbial growth and inactivation during food processing', *Appl Environ Microbiol*, 58, 2901–2909.

VEREECKEN K, GEERAERD A, BERNAERTS K, DENS E, POSCHET F and VAN IMPE J (2000), 'Predicting microbial evolution in foods: general aspects of modelling approaches and practical implementation', *Journal A,* Special issue on modelling and control in bioprocesses, 41, 45–55.

VERSYCK K J, BERNAERTS K, GEERAERD A H and VAN IMPE J F (1999), 'Introducing optimal experimental design in predictive modeling: A motivating example', *Int J Food Microbiol*, 51, 39–51.

WHITING R C (1995), 'Microbial modeling in foods', *Crit Rev Food Sci Nut*, 35, 467–494.

WHITING R C and BUCHANAN R L (1993), 'A classification of models in predictive microbiology – a reply to K R Davey', *Food Microbiol*, 10, 175–177.

XIONG R, XIE G, EDMONDSON A E and SHEARD M A (1999), 'A mathematical model for bacterial inactivation', *Int J Food Microbiol*, 46, 45–55.

22

Identifying and dealing with heat-resistant bacteria

J. T. Rosnes, Norconserv, Norway

22.1 Introduction: the problem of heat-resistant bacteria

Heat treatment is used to produce safe and shelf stable foods and to eliminate pathogenic microorganisms. This is an ultimate objective but a more accurate definition is to reduce the probability of survival and/or growth of the microorganisms in a particular food to an acceptable low level. Basic knowledge of combined technologies can then give the necessary shelf-life. In the last decade, a large industry has developed within making ready meals. In 2000 the ready-meals market in the UK, France, Germany, Spain and Italy was estimated to be €5.7 billion. Compared to 1996, this means an increase of 19%. This growth of ready meals will be at the expense of canned and ambient meal products, which will also suffer from the increase in frozen ready-meal consumption. As refrigerated processed foods with extended durability (REPFED) are becoming increasingly popular, new challenges for food safety are introduced. The demand for nutritious and healthy foods has encouraged the development of low-heat treatments and new methods for heat treatments. This subsequently leads to survival of many microorganisms, and there is a possibility for germination and growth to high numbers that may be a health risk for the consumer.

22.2 Heat-resistant bacteria and their growth potential

Temperature is one of the most important environmental factors influencing the growth and survival of organisms. It can affect living organisms in either of two opposing ways. As temperature rises, chemical and enzymatic reactions in the cell proceed at more rapid rates and growth becomes faster. However, above a certain temperature, proteins, nuclear acids, and other cellular components are sensitive to high temperatures and may be irreversibly denaturated. Thus we find that for every organism there is a minimum temperature below which growth no longer occurs, an optimum temperature at which growth is most rapid, and a maximum temperature above which growth is not possible. Organisms that grow at temperatures above 45 to 50°C are called thermophiles, and are adapted to tolerate high temperatures without cell damage. Temperatures as high as these are found in nature only in certain restricted areas, e.g. soils to full sunshine, fermenting materials, volcanic areas and hydrothermal vents and hot springs. Within the food processing industry, many local high temperature areas are found, ideal for thermophilic growth. In dealing with heat-resistant bacteria, survival of a short heat treatment, with subsequent growth at chilled or ambient temperature, is often a greater challenge for quality and safety than growth of vegetative thermophilic bacteria from raw materials.

Thermal processing is the primary method for adding value and ensuring microbial safety of food products. Although numerous technologies, e.g. irradiation, ultra high pressure, pulsed electric fields, use of bacteriocins, are new promising technologies for the food industry, the application of heat will certainly continue as the dominant means to impact desirable characteristics, add economic value and ensure product safety. Additionally, major shifts in consumer demand and regulatory burden have increased the importance of thermal processing of these areas. Today, the consumer spends more money on food than any other item, and thermal processing has allowed him to be both geographically and climatically independent for his nutritional needs.

During the development of any new heat treated product it is essential to assess the combined effects of the total system such as heat process, preservatives, packaging and storage conditions in order to ensure that the product is of a good microbiological standard and does not present any food safety hazard. In order to produce such high quality products there is a trend towards using fewer preservatives and minimal heat treatment. These developments have serious implications with regard to the time and temperature of the heat process and hence the lethal effect on any potential spoilage or food poisoning bacteria. One of the drawbacks with conventionally produced cook-chill products has been the relatively short shelf-life of the products, for example, in the catering sector a storage period of 5 days has been recommended. REPFEDs are defined by the Codex Committee on Food Hygiene as products with shelf-life of more than 5 days and, based on the processing and packaging conditions used, the shelf-life may range from about 10 days to more than 5 weeks.

21.2.1 Survival of heat treatments

The different microorganisms themselves have inherently different resistance to high temperatures. Vegetative cells and yeasts are generally the most susceptible

Table 22.1 Spore-forming bacteria causing problems in different foods

Food	Bacteria	References
Milk	*Bacillus cereus* *Bacillus stearothermophilus* *B.subtilis*	Griffiths and Phillips 1990; Guinebretiere *et al.* 2003; Larsen and Jorgensen 1999; Lin *et al.* 1998; Shehata *et al.* 1983
Fruit products, juice	*B.subtiles, B.coagulans* *Byssochlamys spp.*	Anderson 1984; Bayne and Michener 1979; Rodriguez *et al.* 1992; Roland and Beuchat 1984
Soy protein formulae	*Desulfotomaculum nigrificans*	
Vegetables	*B.subtiles, B.cereus* *B.stearothermophilus* *C.thermosaccharolyticum* *C.botulinum spp.*	Austin *et al.* 1998; Carlin *et al.* 2000; Dodds 1989; Valero *et al.* 2002
Fish	Non-prot. *C.botulinum*, type E and B	Cann *et al.* 1966; Eklund 1982; Hyytia *et al.* 1998; Korkeala *et al.* 1998
Meat	Non-prot. *C.botulinum*, type B *C.perfringens*	De Lacy *et al.* 1998; Gill 1979; Taormina *et al.* 2003
Poultry	*Clostridium* spp.	Kalinowski and Tompkin 1999
Honey	Prot. *C.botulinum*, type A and B	Hauschild *et al.* 1988; Nevas *et al.* 2002

while endospores are much more resistant. The type of foodstuff to be heat treated will often have associated microorganisms with a high thermal resistance which it is important to inactivate to ensure sterility or control of that product (Table 22.1).

Almost the only procedure for killing spores which is both effective and tolerable in foods, has been heating. But even here, the most heat-resistant spores require so severe a treatment that commonly has an undesirable effect on the food. This enforces processes that use as little heating as possible, and lead to three broad categories of treatment:

1. In food whose quality is not required, by high degree of heating, a 'nearly sterile' condition can be achieved (i.e. a predominant proportion of units – e.g. cans or bottles – contain no viable spores). Even here, quality is usually better when heating is minimised. Foods in this class will be devoid of enzyme activity and, when packed in hermetic containers needed to maintain sterility, are called 'conserves'.
2. In foods which tolerate only a rather lower degree of heating, some of the more resistant spores will remain alive though they may not develop. If the spores remain dormant, they cause no change in the food. It is as if the

spores were not present, and the food is consequently described as 'commercially sterile'.

3. In foods sensitive to heating, it is only possible to destroy vegetative cells and the more sensitive bacterial spores, and some auxiliary factors are needed to ensure preservation even for limited periods. This type of treatment is commonly called 'pasteurisation'. Today there are many processing methods, specially designed for ready meals, which are used in the catering and retail sector, e.g. sous vide, boil in bag, cook and chill, cap cold. One must expect survivors in all of these processes, and safety must be strictly monitored to avoid food poisoning.

Foods that are heat treated lack the initial normal flora of the raw materials and surviving bacteria have little competition from other species. In addition, packaging conditions often favour growth of anaerobic or facultative anaerobic bacteria, e.g. members of heat-resistant spore-forming *Bacillus* and *Clostridium* spp. Foods are complex ecosystems composed of intrinsic factors inherent to the food, e.g. pH, water activity and nutrients, and extrinsic factors external to it, e.g. temperature, gaseous environments and the presence of other bacteria. Both intrinsic and extrinsic factors can be manipulated to preserve food by designing conditions in the food that limit bacterial growth. Foods can also be heterogeneous on a micrometer scale, and this heterogeneity and its associated gradients of pH, oxygen, nutrients, etc. are key ecological factors in foods (Boddy and Wimpenny, 1992). Several distinct microenvironments are found in the food. This is well illustrated by the food poisoning outbreaks in 'aerobic' foods caused by 'obligate' anaerobe *Clostridium botulinum*. Growth of *C.botulinum* in potatoes, sautéed onions, and cole slaw exposed to air has caused botulism outbreaks (Lund, 1992). The oxygen in these foods is driven out during cooking and diffuses back in so slowly that the bulk of the products remain anaerobic.

22.2.2 Spore-formers as food spoilage organisms

The spores are biochemical inert, and therefore not themselves responsible for spoilage. Spoilage is actually effected by vegetative cells, and one of the points to be remembered is that the various environmental factors may have distinctly different effects on spores and vegetative cells. The capacity of the almost inert spores to spoil food depends on their ability to germinate, outgrowth, and achieve extensive vegetative multiplication in the food. Interruption of this chain of events at any point will prevent spoilage. Earlier it was claimed that most of the spore-formers were mesophilic and only an important problem under warm conditions, and with little problem for chilled foods (Ingram, 1969). Over the last few years several psychotropic spore-formers have been characterised, along with serious pathogens of *Bacillus* species.

22.3 Types of heat-resistant microorganisms

22.3.1 Spore-forming bacteria

Several reviews have focused on spore resistance mechanisms (Gerhardt, 1988; Gerhardt and Marquis, 1989; Gombas, 1983; Setlow and Johnson, 1997). Spores formed by the genera *Bacillus*, *Clostridium, Desulfotomaculum* and *Sporolactobacillus* present practical problems in food microbiology. *Bacillus* and *Clostridium* spp. and some closely related genera respond to slowed growth or starvation by initiating the process of sporulation. The different stages of the genus *Bacillus* sporulation has been extensively studied (Setlow, 1993, 1994), including the molecular biology of sporulation (Losick and Straiger, 1992; Setlow and Johnson, 1997) and spore resistance (Beaman and Gerhart, 1986; Farkas, 1994, Kiss *et al.*, 1978; Marquis *et al.*, 1994; Rutherford *et al.*, 2000). An unequal cell division is the first notable morphological event in sporulation. This creates the smaller forespore compartment and the larger mother cell compartment. As sporulation proceeds, the mother engulfs the forespore, resulting in a cell (the forespore) within a cell (the mother cell), each with a complete genome (Setlow, 1993). This is called an endospore because it is formed within the mother cell. There is a defined pattern of gene expression throughout sporulation which is ordered not only temporally but also spatially, as some genes are expressed only in the mother cell or forespore (Errington, 1993). The gene expression pattern is controlled by the ordered synthesis and activation of new sigma (σ; specificity) factors for RNA polymerase. At least four of these are synthesised specifically for modulation of gene expression during sporulation. A number of DNA binding proteins, both repressors and activators, also regulate gene expression during sporulation (Errington, 1993; Stragier, 1994).

The developing forespore synthesises a large amount of small acid soluble proteins (SASP), some of which coat the spore chromosome and protect the DNA from damage. In addition, the spore becomes both metabolically dormant and extremely resistant to a variety of harsh treatments including heat, radiation, and chemicals. As a consequence of these physiological changes, spores can survive extremely long periods in the absence of exogenous nutrients.

22.3.2 Other heat-resistant organisms

Most studies on sporulation, spores and spore germination have been carried out with species of either aerobic bacilli or the anaerobic clostridia. However, other genera form similar spores; among these are the genera *Sporosarcina, Sporolactobacillus* and *Thermoactinomyces* (Slepecky and Leadbetter, 1994). Studies using rRNA sequence analysis have shown that the genera *Bacillus, Sporosarcina, Sporolactobacillus* and *Thermoactinomyces* are quite closely related. *Bacillus and Clostridium* are more distantly related but clearly derived from a common ancestor, most likely a sporeformer.

Viruses

Outbreaks of viral food poisoning have occasionally been reported from inadequately processed foods, and filter feeding mussels from contaminated waters are in this respect vulnerable. Assuming contamination, viruses in these foods cannot multiply, and they can be heat inactivated before someone eats the food. Hepatitis A viruses may be transferred by water and certain foods. Thermal processing is generally effective, although some further attention to the adequacy of milk pasteurisation to inactivate hepatitis A virus is needed (Parry and Mortimer, 1984). In the United States 1,293 cases of hepatites A from shellfish were reported from the period 1961 to 1981 (Guzewich and Morse, 1986). Also parvovirus is shown to survive when processed at milk pasteurisation times (Fassolitis *et al.*, 1985). The British Government recommends heating molluscs to 85–90°C for at least 90 seconds to destroy viruses (IFST, 1996).

Fungi

The most heat-resistant fungal spores are much less resistant than most bacterial spores, but in some the heat resistance is high enough to cause major problems in processed acidic foods. Spores of most mesophilic species are killed after a few minutes' exposure to temperatures between 65 and 79°C. The ascospores of filamentous fungi are more heat resistant than their conidia (Pitt and Christian, 1970), while the ascospores and vegetative cells of yeasts appear to possess heat resistance broadly similar to that of fungal conidia (Put *et al.*, 1976). One of the significant fungal spores in heat processed foods are those of *Byssoclamys* spp., with D-values in the range 1–12 minutes at 90°C (Bayne and Michener, 1979). Ascospores of *Neosartorya fischeri* are almost as heat resistant (Splittstoesser and Splittstoesser, 1977). There are reports of moderate heat resistance in mesophilic species belonging to the genera *Streptomyces, Nocardia* and *Micromonospora* (Roberts and Hitchins, 1969). Characteristics of capillary and test tube procedures for thermal inactivation kinetic analysis of microbial cells have been studied for mould spores (Engel and Teuber, 1991; Fujikawa *et al.*, 2000)

Prions

Prions consist of proteins and are strictly not live microorganisms. Their relation to foodborne diseases and tolerance to heat treatment, however, give prions a position among the heat-resistant 'pathogens'. The Transmittable Spongiform Encephalopathies (TSE) are the best known prion diseases. The TSE are fatal neurodegenerative disorders, affecting several animals and humans. Other diseases from prions are 'mad cow disease' – Bovine Spongiform Encephalopathies (BSE) – and the human variant Creutzfeldt-Jakob Disease (CJD). The knowledge of heat inactivation kinetics of scrapie and CJD prions is incomplete and fragmentary. This occurs also because the inactivation kinetics of prions is not exponential. By analysis of available data, it is concluded that usual food processing parameters are largely unable to kill prions. Official authorities have suggested treatments of infected material at 134–136°C for 18–60 min (American Neurological Association, 1986; Dept. of Health and Social Security, 1984). Casolari (1998) suggests that the required prions lethality could be reached by changing the food processing technology, and that the aseptic process is the most promising alternative.

22.4 Thermal inactivation kinetics of bacterial spores

22.4.1 Early studies of heat inactivation

Heat treatment is probably one of the oldest methods to preserve foods. The preservative effect is dependent on a sealed package which surrounds the food to avoid reinfection. Nicolas Appert was the acknowledged inventor of canned food products. In 1810 he demonstrated that heating foods in hermetically sealed containers could render them shelf stable for long periods at ambient temperatures (Appert, 1810). Appert carried out practical experiments without the knowledge of heat stable spore-forming bacteria and it took almost a century before Bigelow *et al.* (1920) made the first mathematical evaluations of heat sterilisation applied to canned foods. Then Ball (1923) published 'Methods of calculation for determining the time necessary to process canned foods'. Even today Ball's method is widely employed in many countries. Early studies of the heat resistance of bacterial spores were made in order to develop processes for production of canned foods. The concept of the thermal death point was determined by sealing a suspension containing a known number of spores into tubes, immersing the tubes in a water bath at the required temperature for given periods of time after which the tubes were cooled rapidly and tested for sterility (Bigelow and Esty, 1920).

22.4.2 Heat resistance of spores

Dehydration, mineralisation, and thermal adaptation are three physiological determinants that affect the protoplast in thermo resistance (Gerhardt and Marquis, 1989). It is generally thought that the mechanisms of heat resistance involve restricting the mobility of heat-labile components of the spore core, e.g., proteins and genetic material, thereby preventing denaturation (Gombas, 1983). Dehydration of the protoplast is assumed to be the only determinant necessary for the heat resistance of spores (Gerhardt and Marquis, 1989). Once protoplast dehydration is obtained, its maintenance requires an intact cortex, but not a coat or exosporium. Mineralisation of the protoplast, usually by calcium, affects heat resistance. Heat resistance of *Bacillus* spores was positively related to calcium content and inversely to magnesium content (Murrell and Warth, 1965). *C.botulinum* spores have been shown to be more susceptible to thermal inactivation with increased iron content, and added Fe or Cu also appeared less able to repair heat-induced injuries than spores with added manganese or zinc (Kihm *et al.*, 1990). Mineralisation of spores is accomplished by a reduction in the water content of the protoplast, and at least part of the increased heat resistance associated with mineralisation is due to dehydration (Beaman and Gerhart, 1986). Although often associated with minerals, DPA is not thought to be necessary for attaining heat resistance, but may have a role in retaining the resistance (Gerhardt and Marquis, 1989). The spores of thermophilic species are inherently more resistant than those of mesophilic or psychrophilic species (Warth, 1978). However, increasing the sporulation temperature can produce additional heat resistance (Heinen and Lauwers, 1981). In spores, both the inherent and imposed components of thermal adaptation are reflected in the protoplast water content over much, but not all, of the temperature range (Beaman and Gerhart, 1986).

22.4.3 Heat inactivation of bacteria

Pathogen inactivation with heat treatment is measured with regard to individual effects of pH, salt, temperature, oxygen level, etc. in multiple factional experiments. The effects and interactions of these parameters are then assessed in lowering the heat resistance of foodborne pathogens. Thermal death models and inactivation kinetics are developed which predict the target pathogens survival within a specific range of food formulation variables. These models can be used by food processors and regulatory agencies to ensure critical food safety margins, by predicting the combined effects of multiple food formulation variables.

Early studies of the heat resistance of bacterial spores were made in order to develop processes for the production of canned foods. Esty and Meyer (1922) concluded that for *C.botulinum* spores, that showed the maximum heat resistance, heat treatments were necessary to sterilise suspensions containing billions of spores in phosphate buffer at pH 7 as follows: 4 min at 120ºC. They also showed that at 100ºC there was a logarithmic decrease of cells; that germination of heated spores could occur after very long delays; and that in juices from some canned foods, the heat resistance of the spores was greater than in the reference phosphate buffer.

When a homogeneous population of viable spores is exposed to a lethal temperature, the number of spores that remain viable is observed to decrease exponentially (or logarithmically) with time, and the rate of this exponential decrease varies with temperature within the lethal range, $N = N_0 e^{-kt}$, where N is the actual number of viable cells, N_0 is the initial number of viable cells, k is the rate constant (time^{-1}) of heat destruction at constant temperature T, and t is the heat processing time. Mathematically, this reaction can be described by the general rate equation for a first-order reaction in terms of a rate constant at a reference temperature and activation energy for use of the Arrhenius equation to describe the temperature dependency of the rate constant (Table 22.2).

The z-value model is useful because a base 10 logarithm is used. Because of the limited parameters in this model, the D and z values, it is not useful for non-linear inactivation curves which have shoulder and/or tail characteristics. The

Table 22.2 Models used to describe inactivation rates of microorganisms

Models[1]	Mathematical description
Arrhenius model[2]	$K = A \exp[(-E_a/R)]$
D-value[3]	$D = D_r 10^{-(T-Tr)Z}$
z-value	$Z = 2.303RTT_r/E_a$

[1] Fujikawa and Itoh (1998).
[2] A is the rate constant at infinite temperature, E_a is the activation energy of inactivation, R is the gas-law constant, T in degrees Kelvin.
[3] D_r = D-value at the reference temperature T_r.

Arrhenius model uses the natural logarithm and can predict various non-linear inactivation curves with combinations of first-order reactions (Fujikawa and Itoh, 1996a, 1996b). The logarithms in the models are different and the relationships in these models are not mathematically equivalent. It is, however, possible for the predictions of the models to be similar, and comparisons have been done on spores (Johnson *et al.*, 1977) and on vegetative cells (Fujikawa and Itoh, 1998). Many approaches to model bacterial heat resistance have since been proposed taking into account effects of environmental factors (Chea *et al.*, 2000; Couvert *et al.*, 1999; Gaillard *et al.*, 1998a,b; Juneja *et al.*, 1995; Leguerinel *et al.*, 2000; Mafart *et al.*, 2001, 2002; Periago *et al.*, 1998).

It has been assumed that inactivation models can be applied to a real thermal process by dividing the process into a series of small time steps, and then computing the cumulative inactivation by stepping through these time steps and recalculating the survivors after each interval. Different strategies aimed toward this type of process modelling have been presented (Schellekens *et al.*, 1994; Van Impe *et al.*, 1995; Zwietering and Hasting, 1997). Van Impe *et al.* (1995), in particular, developed a system theory approach for predictive microbiology in a dynamic environment. Although their model was a significant contribution, temperature was the only transient variable included. They noted that there is still a need to include the influence of water activity, which also varies during dynamic thermal processes.There are now available excellent programs using tertiary inactivation models (simple to use, user friendly, Window-based), e.g. the USDA ARS Pathogen Modelling Program (PMP, v 7.0, http://www.arserr.gov/mfs/PATHOGEN.HTM), there they are inferring that this tool could be used to relate cooking parameters to pathogen lethality. Another example is the AMI Process Lethality Spreadsheet (American Meat Institute, http://www.amif.org/processlethalityinstr.htm). Both tools have a number of limitations relevant to the real thermal process, and consequently, there is still a need to further extend the methods of quantitative microbiology by coupling the inactivation models with validated process (heat and mass transfer) models to evaluate the lethality of commercial cooking systems.

22.4.4 Thermal destruction curves

Heat resistance is commonly expressed as the decimal reduction time, or *D*-value. The *D*-value is the time required to inactivate 90% of the population at the specified temperature. *D*-values vary considerably among bacterial species and strains, even within the same group or type. Thermal death time or TDT studies are conducted in a laboratory to determine the heat resistance of microorganisms. Several factors which may affect the heat resistance of the organism are considered, and the theoretical straight-line semi-logarithmic survivor curve is, in practice, not always obtained, and different deviations like concave curves, curves with shoulders, sigmoid-type curves and tailing are observed. For example, the addition of starch, bicarbonate, and other supplements and of lysozyme to a medium can greatly increase the estimated number of spores of some *Clostridium* species that survive heat treatment (Lund and Peck, 1994).

In TDT studies, inoculated menstruum in a comminuted form (water, buffer solution, culture medium, or food material) is distributed in small diameter test tubes which are subsequently sealed. The sealed tubes of inoculated product are generally heated in a thermostatically controlled bath of mineral oil, propylene glycol or some other suitable medium and heated for varying lengths of time at a series of different temperatures. The goal of the TDT tests is to find the breaking point or time between destruction and survival of the organism at each of the different temperatures and to calculate survivor curves. There are different measuring devices like sealed glass tubes, open glass tubes, metal tubes, can, flask and thermoresistor, excellently reviewed by Pflug and Gould (2000). The chief advantages of the TDT tube methods are: it employs simple, inexpensive equipment; bacterial growth in media and spoilage changes in some food products may be observed visually without opening the tubes; tubes may be opened for subculture with little danger of contamination; and space required for incubation of unopened TDT tubes is small. The methods also suffer from disadvantages, including: time consuming operations; appreciable heating and cooling lags; splashes of content; flocculation; and high labour costs.

Due to differences in techniques, pre-treatment of strains, and medium, some published results are difficult to interpret. For example, the published literature on the heat resistance of *Listeria monocytogenes* has been conflicting in that some results indicate survival (Beams and Girard, 1958) and some not at pasteurisation temperatures (Bradshaw *et al.*, 1985). Donnelly *et al.* (1987) have compared the methods used to determine heat resistance of this bacterium.

22.4.5 Heat resistance of spores to moist heat

In the laboratory, water activity (a_w) can be measured, but it is difficult to subject spores and vegetative microorganisms to ideal heating conditions. The methods used are designed to minimise the time for temperature equilibration, or 'come-up time' to ensure that the whole sporal or cell population experiences the same heating conditions. When short heating times are used, e.g. less than 3 min, the substrate volume must be very small. The vehicle in which the heating is carried out is designed in such a way that the sample will reach the test temperature in a fraction of a minute (Pflug and Gould, 2000). Laboratory conditions therefore often have different exposure time at a_w equilibrium, compared to real food productions. Death rate is a function of water activity (a_w) or equilibrium relative humidity (ERH) and in general spores are killed quickly and at comparatively low temperatures in a steam autoclave compared with a dry-heat oven. Moist heat is a condition where the a_w is close to 1.0, but maximum heat resistance is found at a_w values between 0.2 and 0.4. However, some experiments have conflicting conclusions on the effect of higher a_w values than 0.4 on heat resistance. Mazas *et al.* (1999) concluded that it was impossible to predict the behaviour for a given microorganism when it is heated in media with an intermediate a_w.

Water activity also influences the thermal resistance to vegetative pathogens, even though few studies have quantified the effects of moisture content or a_w on microbial inactivation.

Blankenship (1978) and Goodfellow and Brown (1978) reported *Salmonella* survivors on the surfaces of fully cooked, dry roasted beef and suggested that thermal resistance was enhanced by the reduction in water activity (a_w) near the meat surface. Kirby and Davies (1990) worked with dehydrated cultures of *S.thyphimurium* and reported an increased thermal resistance, but they were heating pure cultures rather than food products (Hsieh *et al.*, 1976). The inactivation rate (k) for *Salmonella anatum* NF4 and *Staphylococcus aureus* 196 E in pasta was shown to be a function of a_w, however, this relationship was not fundamentally modelled. In each of the above studies, the authors reported that thermal inactivation was highest at high a_w (>0.95), decreased with decreasing a_w until it reached a minimum between 0.6 and 0.8 a_w, and subsequently increased again as a_w approached 0.

22.4.6 Spore activation, germination and outgrowth

There are three stages that a spore undergoes successively in forming a vegetative cell after a heat treatment: activation, germination and outgrowth (Keeney and Evenchick, 1969). Inactivated spores will not germinate spontaneously until activation occurs. Activation is usually a reversible process and the activated spore retains nearly all the spore properties, whereas germination is irreversible and involves loss of all spore properties, especially heat resistance. Activation can be brought about in different ways, but mainly from the following: sub-lethal heat treatment; exposure to chemicals, particularly calcium dipicolinate, low pH (1 to 1.5), thiol compounds, and strong oxidising agents; and ageing of the spore population, by storing spores at a low temperature for a period of time. When using sub-lethal heat to activate spores, the optimum temperature and exposure time for a maximum colony count depend on the species and strain of organism and may also vary from batch to batch. Other factors are known to be the composition of the medium in which the pores were grown, the medium in which activation occurs, the age of the spore suspension, and the temperature used for sporulation.

One of the most accurate methods for determining spore germination utilises the loss in heat resistance as a quantitative measure of the number of germinated spores. Briefly, the method consists of heating a spore suspension at a time-temperature relationship which is lethal to the germinating spores but not to ungerminated spores and plating the survivors. A control suspension of ungerminated spores is treated similarly and the percentage germination calculated. Another technique often employed is to heat the spores at the beginning of the germination experiment, followed by a second heating after incubation in a suitable medium. The difference in spore counts before and after incubation is used for calculating the percentage germination. Several precautions must be taken when using this method. Despite the spore's extreme dormancy, if given the appropriate stimulus (often nutrient such as sugar or amino acids) the spore can rapidly return to life via spore germination. Within minutes of exposure to a germinant, spores lose most of their unique characteristics, including loss of DPA by excretion and loss of the cortex SASP by degradation. The spore's resistance is also lost in the first minutes of germination, when active metabolism of both endogenous and exogenous compounds begins and macromolecular synthesis is initiated. Eventually the germinated spore is converted back to a growing vegetative cell through the process of outgrowth.

An example of a spore-forming pathogenic organism that can rapidly be activated and germinate in cook-in vacuum-packaged products is *Clostridium perfringens*. With a generation time of 7.1 min at 41°C *C.perfringens* is one of the fastest growing bacteria. The heat-resistant spores can survive at 100°C for 1 hour and for products with a low heat treatment it is therefore important to chill the product to temperatures below the minimal growth temperature to avoid germination. A general guide for chilling of cook and chill foods, in order to preserve the appearance, texture, flavour, nutritional quality and safety of the heated food, is that chilling should commence as soon as possible after heating. In any event within no more than 30 min after leaving a cooker, the food should be chilled to between 0°C and 3°C within a further 90 min (ACMSF, 1992). US Department of Agriculture (USDA) stabilisation guidelines require that the chilling process not allow more than 1 log total growth of *C.perfringens* in finished products (US Department of Agriculture-Food Safety and Inspection Service, 1999). To insure that chilling guidelines are met, meat products should be chilled from 54.4 to 26.6°C in 90 min and from 26.6 to 4.4°C in 5 hours or less (Danler *et al.*, 2003). Customised stabilisation processes are permitted provided the procedure has been confirmed to meet the USDA guidelines.

22.4.7 Safety from a public health standpoint

Esty and Meyer (1922) carried out pioneer work when they introduced the approach of F_0 values for *C.botulinum*. This work was later corrected for heating and cooling (Townsend, 1938). They used 10^{11} of the most resistant pathogenic

C.botulinum spores that they could grow per unit tested, and used as their endpoint no growth in about 10 replicates, which was a survival level of no more than 10^{-1}, so the log reduction for their experiments was 12. Still, after more than 80 years since publication, these data are quoted as the basis for selecting the minimum F_o value for preserving low-acid foods against botulism. In order to compare the relative capacities of heat processes, the unit of food pasteurisation and sterilisation processes is chosen arbitrarily to be 1 min at 121ºC (250ºF) and the criterion of process adequacy is the extent to which the bacterial population is reduced. The primary aim is to obtain an adequate reduction of organisms of public health concern. The minimum process should be such as to reduce any population of the most-resistant pathogenic botulinum spores to 10^{-12}. Proteolytic *C.botulinum*, type A, is so far known to have the most-resistant spores, which have a D_{121} value of about 0.21. The minimum process requirement for low-acid canned foods (pH > 4.5) in terms of its equivalent in minutes at 121ºC are then $F_o = 0.21 \times 12 = 2.52$ or 2.5. Thus, the 'botulinum cook' reduces the population of the most resistant pathogenic *C.botulinum* by an arbitrarily established factor of 12 decimal values (12D concept), and it provides a substantial margin of safety in low-acid canned foods.

Several guidelines and codes of practice are made with respect to safe production of REPFEDs (ACMSF, 1992, 1995; Betts, 1996; ECFF, 1996). Most of these are targeted at preventing growth and toxin production by non-proteolytic *C.botulinum*. In these guidelines several product extremes are recommended, e.g. storage temperature, heat treatment and shelf-life. A general recommendation is that the heat treatments or combination processes utilised should reduce the number of viable spores of non-proteolytic *C.botulinum* by a factor of 10^6 (6D). Accordingly, a minimum heat treatment of 90ºC for 10 min or equivalent lethality in the slowest heating point of the product has been recommended by ACMSF (1992, 1995). The recommendations from ACMSF are an important step towards harmonising food manufacturing practice with governmental concern, but there are still improvements to be done. For example the minimum temperature at which growth and toxin production occurs is often reported to be 3.3ºC (included in the ACMSF recommendations). Graham *et al.* (1997) showed, however, that toxin production could occur at 3.0ºC with prolonged storage.

22.5 New thermal inactivation processes: microwaves, radio frequency and high-pressure processing

With heat processing of food both desirable and undesirable changes occur. Useful changes are texture softening, starch swelling, protein coagulation and formation of taste and aroma components. Loss of vitamins and minerals, formation of thermal reaction components of biopolymers and loss of fresh appearance, flavour and texture are among the undesirable changes (Ohlsson, 2002). During the last years, new thermal methods are tested in order to overcome or at least minimise undesirable quality changes. Examples of such methods are HTST (high temperature short time), ohmic heating, infra red heating, high frequency or radio frequency heating and microwave heating. Microwave and radio frequency heating refers to the use of electromagnetic waves of certain frequencies to generate heat. Microbial cells in food may be differently heated compared with the food matrix, which could result in specific inactivation. This has been examined for microwaves and radio frequency waves.

22.5.1 Microwave heating

Microwave heating has many advantageous over other methods, such as gamma radiation and ultraviolet radiation. Microwave ovens do not need a large space, do not heat the air, can be used continuously, save time and energy, produce well-distributed heat conduction, and preserve more of the nutrients in the food (Wang *et al.*, 2003). Many studies have reported non-thermal effects of microwave energy on microorganisms in foods, and bacteria reported to be inactivated by microwave heating include *Bacillus cereus, Campylobacter jejuni, Clostridium perfringens,* pathogenic *Escherishia coli, Enterococcus, Listeria monocytogenes, Staphylococcus aureus* and *Salmonella* (Chipley, 1980; Heddleson and Doors, 1994; Knutson *et al.*, 1987; Rosenberg and Bögel, 1987). Kozempel *et al.* (1998) summarised the theories that have been advanced to explain non-thermal effects of electromagnetic energy and investigated the effect of microwave energy on yeast and bacteria at $\leq$ 45°C. They concluded that there were no non-thermal effects. Welt *et al.* (1994) also concluded that the effect of microwave energy on viability of *Clostridium sporogenes* was indistinguishable from the effect of conventional heating.

22.5.2 Radio frequency energy

The existence of non-thermal effect of radio frequency (RF) energy on microorganisms has been debated for more than 50 years. This would be attractive; because fresh-like qualities and nutrients would be maintained in foods if RF energy could cause non-thermally inactivate microorganisms. So far there are no reports that RF technique is being practised today by the food industry. No non-thermal effects of RF energy were detected on *Escherichia coli* K-12, *Listeria innocua,* or yeast in apple cider, beer, deionised water, liquid whole egg, and tomato juice; nor were there any synergistic effects of RF energy with heat (Geveke *et al.*, 2002)

22.5.3 High pressure

Different microorganisms have different levels of resistance to high hydrostatic pressure processing (HPP). While spores are relatively stable against pressure alone, the combination of elevated temperatures and pressures of, e.g. 400 MPa is effective against spores.

The inactivation of bacterial spores through HPP, unlike the vegetative bacteria, occurs in two steps: (1) high pressure causes spore germination, and (2) high temperature inactivates the germinated spores. The resistance of bacterial spores to physiochemical agents is genetically determined, but can be influenced by many environmental factors, such as sporulation temperature, pH, and water activity. The effect of sporulation temperature is one of the strongest produced by any environmental factors.

22.6 Identifying heat-resistant bacteria

22.6.1 Reference organisms

Establishment of traditional thermal processes for foods has been based on two factors: (1) knowledge of the thermal inactivation kinetics of the most heat-resistant pathogen of concern for each specific food product, or other microbes potentially causing spoilage, discoloration, rancidity, off-flavours, etc. in food products; and (2) determination of the nature of heat transfer properties of the food system. The established process is thereafter often confirmed using an inoculated test pack study tested under actual production conditions. In cases where it is not practical to use the reference organism, surrogate mircro-organisms that can mimic the pathogen are used as biological indicators. The surrogate or reference microorganisms must be carefully selected. For each specific food product one must choose between the bacteria best known, or their relatives, to be responsible for causing foodborne diseases: *Aeromonas hydrophila, Bacillus cereus, Campylobacter jejuni, Clostridium botulinum, C.perfringens,* pathogenic *Escherichia coli, Listeria monocytogenes, Salmonella* serovars, *Shigella* spp. *Vibrio* spp. and *Yersinia enterocolotica.*

Non-proteolytic *C.botulinum*, type A is used as a reference organism for canned, shelf-stable foods. It is widely accepted that in the case of mild heat treatment below 90°C only vegetative bacteria can be chosen for this purpose, because spores are not killed during pasteurisation. Non-proteolytic *C.botulinum* can multiply and form toxins at 3.0–3.3°C and is the principal microbiological safety concern in REPFEPs. Pathogens other than non-proteolytic *C.botulinum* may pose a hazard to safety of REPFEDs under certain conditions. Proteolytic *C.botulinum* and *C.perfringens* are of concern when the temperature exceeds 10°C for prolonged periods. Several isolates of *Bacillus cereus* have been found able to grow slowly under 10°C, but high numbers are needed to pose a genuine safety hazard. An adequate heat treatment destroying critical microorganisms in the product should indicate that all harmful microorganisms (including viruses) are reduced to below harmful levels (Incze *et al.*, 1999).

22.6.2 Identification of spore-formers in foods

Both the quality and microbiological safety of pasteurised foods with extended shelf-lives, require good control and monitoring of critical process parameters throughout the entire manufacture and distribution process (Betts and Gaze, 1995). To control microbiological risks associated with each step of the processing, microbial risk assessment has been used as a means of selecting hazards for consideration in a HACCP process (Hazard Analysis Critical Control Points). Food safety tools, risk assessments and users' guides for HACCP have been described by many authors (Jouve *et al.*, 1999; Mayes, 1992; Serra *et al.*, 1999; Voysey, 2000). Microbial sampling of the end products is usually a part of the risk assessment. Buchanan (2000), however, points out that if such sampling is to be used effectively, it is necessary to have a clear understanding of the purpose of such testing and the limitation of the methods employed. Thus, the testing methods and the number of tests may differ greatly for different processing methods and different kinds of food. Some bacteria that can be determined in the products may survive, but will never be able to grow and produce toxins in the food. Reliable methods used in laboratory experiments and industrial production that can reveal both potential for germination and growth, not only survival, will therefore be very valuable.

Relatively few data appear to have been published on the survival and growth of spore-forming bacteria in commercial products treated as sous vide or other mild heat treatments, however, some reports are available (Alberto *et al.*, 2003; Carlin *et al.*, 2000; Lin *et al.*, 1998; Mulak *et al.*, 1995; Sarrías *et al.*, 2002; Taormina *et al.*, 2003; Valero *et al.*, 2002). Nissen *et al.* (2002) analysed 2,168 samples representing 24 different products of ready-made meals which were analysed for total viable counts and spore-forming bacteria, directly after production and after 3–5 weeks of storage. A total of 350 isolates of *Bacillus* spp. were collected, but no *Clostridium* strains were detected. Only 11 of the 113 tested strains were able to grow at 7°C in broth, and none of the psychotropic strains were able to produce substantial amounts of toxins to cause food poisoning.

Although *Clostridia* are regarded as the most serious microorganisms, the aerobic spore-forming bacilli are among the most frequently isolated and important groups of microorganisms occurring in foods. Due to their versatile metabolism, strong saccharolytic and proteolytic activity as well as the high resistance of their spores, they are among the main spoilage organisms in foods (Granum and Baired-Parker, 2000; Ingram, 1969). Of the aerobic spore-formers, special testing methods have been developed for *B.cereus* to control its potential role in food-poisoning (ICMSF, 1978). The species identity of other bacilli remains unknown in the microbiological investigation of foods as usually carried out. Information of this kind, however, would be of primary importance in monitoring good manufacturing practice, as there are wide differences in the properties of *Bacillus* species, e.g. their heat resistance.

22.6.3 Serological and DNA identification methods

To accomplish identification of spore-forming bacteria is beyond the capacity of most laboratories as the number of tests to be performed is high. For detection of spore-formers several international methods are available, e.g. the Bacteriological Analytical Manual (FDA). However, these methods may take several days and are unable to detect low numbers of target cells in food. Thus, conventional spore-formers are detected by their ability to grow on selective plating media, and the isolates are identified by morphological, cultural and biochemical characteristics. Even a simplified identification requires as many as 14 tests (Norris *et al.*, 1981). Other methods require twice this number and sometimes more (Deák and Timár, 1988, Leonard *et al.*, 1997).

Advances made in detection methods have led to the use of more precise methods like use of antibodies and use of polymerase chain reaction (PCR). Both monoclonal, polyclonal and recombinant antibodies are used to identify spore-forming bacteria (Koo *et al.*, 1998a,b). PCR protocols have been developed for the detection of spore-formers in pure cultures systems and food samples using specific primers (Helps *et al.*, 1999; Klijn *et al.*, 1994; Manzano *et al.*, 2003; Miwa *et al.*, 1998; Tsen *et al.*, 2000).

Phylogenetic analysis, based on 16S ribosomal DNA (rDNA), has shown that *Bacillus cereus, Bacillus mycoides* and *Bacillus thuringiensis* are closely related. A method to differentiate between this strain was developed by (Manzano *et al.*, 2003) using the polymerase chain reaction combined with a restriction endonuclease (PCR-RE) technique. Amplifications were specific for the *B.cereus* group. Specific amplifications and good differentiations were obtained using pure strains, suggesting the possibility of using the method described to identify the *B.cereus* group directly in food samples.

22.7 Sources of further information and advice

Ohlsson, T. (2002). Chap. 2. Minimal processing of foods with thermal methods. In: Eds: Ohlsson, T. and Bengtsson, N. *Minimal processing technologies in the food industry*. pp. 4–33.Woodhead Publishing Ltd. Cambridge, England.

Ohlsson, T. (2002). Chap. 3. Minimal processing of foods with non-thermal methods. In: Eds: Ohlsson, T. and Bengtsson, N. *Minimal processing technologies in the food industry*. pp. 34–60. Woodhead Publishing Ltd. Cambridge, England.

Juneja, V. K. (2002). Chap. 2. Thermal inactivation of microorganisms. In: Eds: Juneja, V.K. and Sofos, J.N. *Control of foodborne microorganisms*. pp 13–53. Marcel Dekker, Inc., New York.

Lewis, M. and Heppell, N. (2000). Chap. 2. Kinetics for microorganisms death and changes in biochemical components. In: Eds: Lewis, M. and Heppell, N. *Continuous thermal processing of foods*. pp 33–56. *Pasteurization and UHT sterilisation*. Aspen Publishers, Maryland.

22.8 References

ACMSF Advisory committee on the microbiological safety of food. Report on vacuum packaging and associated processes. (1992). London, UK, HMSO.

ACMSF. Annual report (1995). London, UK, Her Majesty's Stationery Office.

ALBERTO, F., BROUSSOLLE, V., MASON, D.R., CARLIN, F. and PECK, M.W. (2003) Variability in spore germination response by strains of proteolytic *Clostridium botulinum* types A, B and F. *Lett. Appl. Microbiol.* **36**, 41–45.

AMERICAN NEUROLOGICAL ASSOCIATION. (1986) Precautions in handling tissues, fluids and other contaminated materials from patients with documented or suspected Creutzfelt-Jacob disease. *Ann. Neurol.* **19**, 75–77.

ANDERSON, R.E. (1984) Growth and corresponding elevation of tomato juice pH by *Bacillus coagulans*. *J. Food Science* **49**, 647–649.

APPERT, N. (1810) L'art de conserver pendant plusiers annes toutes les substances animales et vegetables. (Translated by B.G. Bitting, Chicago, IL, 1920). In (1961) *Introduction to thermal processing of foods* (Eds. Goldblitch, S. A., Joslyn, M. A., and Nickerson, J. T. R.) AVI, Westport, CT.

AUSTIN, J.W., DODDS, K.L., BLANCHFIELD, B. and FARBER, J.M. (1998) Growth and toxin production by *Clostridium botulinum* on inoculated fresh-cut packaged vegetables. *J. Food Prot.* **61**, 324–328.

BALL, C.O. (1923) Determining, by methods of calculation, the time necessary to process canned foods. *Bull. National Research Council* **37**, 9–76.

BAYNE, H.G. and MICHENER, H.D. (1979) Heat resistance of *Byssochlamys* ascospores. *Appl. Environ. Microbiol.* **37**, 449–453.

BEAMAN, T.C. and GERHART, P. (1986) Heat resistance of bacterial spores correlated with protoplast dehydration, mineralization, and thermal adaptation. *Appl. Environ. Microbiol.* **52**, 1242–1246.

BEAMS, R.E. and GIRARD, K.F. (1958) The effect of pasteurization on *Listeria monocytogenes*. *Can. J. Microbiol.* **4**, 55–61.

BETTS, G.D. (1996) Code of practice for the manufacture of vacuum and modified atmosphere packaged chilled foods with particular regard to the risk of botulism. Chipping Campden, UK, Campden and Chorleywood Food Research Association.

BETTS, G.D. and GAZE, J.E. (1995) Growth and heat resistance of psychrotropic *Clostridium botulinum* in relation to 'sous vide' products. *Food Control* **6**, 57–63.

BIGELOW, W.D., BOHART, G.S., RICHARDSON, A.C. and BALL, C.A. (1920) Heat penetration in processing canned foods. *Bull. Natl. Canners Assoc.,* **16–L**, 128.

BIGELOW, W.D. and ESTY, J.R. (1920) The thermal death point in relation to time of typical thermophilic organisms. *J. Infect. Dis.* **27**, 602–617.

BLANKENSHIP, L.C. (1978) Survival of a *Salmonella typhimurium* experimental contaminant during cooking of beef roasts. *Appl. Environ. Microbiol.* **35**, 1160–1165.

BODDY, L. and WIMPENNY, J.W. (1992) Ecological concepts in food microbiology. *Soc. Appl. Bacteriol. Symp. Ser.* **21**, 23S–38S.

BRADSHAW, J.G., PEELER, J.T., CORWIN, J.J., HUNT, J.M., TEINEY, J.T. and LARKIN, E.P. (1985) Thermal resistance of *Listeria monocytogenes* in milk. *J. Food Prot.* **48**, 743–745.

BUCHANAN, R.L. (2000) Acquisition of microbiological data to enhance food safety. *J. Food Prot.* **63**, 832–838.

CANN, D.C., WILSON, B.B., SHEWAN, J.M. and HOBBS, G. (1966) Incidence of *Clostridium botulinum* type E in fish products in the United Kingdom. *Nature* **211**, 205–206.

CARLIN, F., GUINEBRETIERE, M.-H., CHOMA, C., PASQUALINI, R., BRACONNIER, A. and NGUYEN-THE, C. (2000) Spore-forming bacteria in commercial cooled, pasteurised and chilled vegetable purées. *Food Microbiology* **17**, 153–165.

CASOLARI, A. (1998) Heat resistance of prions and food processing. *Food Microbiology* **15**, 59–63.

CHEA, F.P., CHEN, Y., MONTVILLE, T.J. and SCHAFFNER, D.W. (2000) Modeling the germination kinetics of *Clostridium botulinum* 56A spores as affected by temperature, pH, and sodium chloride. *J. Food Prot.* **63**, 1071–1079.

CHIPLEY, J.R. (1980) Effects of microwave irradiation on microorganisms. pp. 129–145. London, Academic Press Inc.

COUVERT, O., LEGUERINEL, I. and MAFART, P. (1999) Modelling the overall effect of pH on the apparent heat resistance of *Bacillus cereus* spores. *Int. J. Food Microbiol.* **49**, 57–62.

DANLER, R.J., BOYLE, E.A., KASTNER, C.L., THIPPAREDDI, H., FUNG, D.Y. and PHEBUS, R.K. (2003) Effects of chilling rate on outgrowth of *Clostridium perfringens* spores in vacuum-packaged cooked beef and pork. *J. Food Prot.* **66**, 501–503.

DE LACY, K.M., BRODA, D.M. and BELL, R.G. (1998) In vitro assessment of psychrotrophic *Clostridium* spp. as possible causative agents of bone-taint in beef. *Food Microbiology* **15**, 583–589.

DEÁK, T. and TIMÁR, É. (1988) Simplified identification of aerobic spore-formers in the investigation of foods. *Int. J. Food Microbiol.* **6**, 115–125.

DEPARTMENT.OF HEALTH AND SOCIAL SECURITY, (1984) Management of patients with spongiform encephalopathy (CJD). 84 (16). DHSS Circular DA.

DODDS, K.L. (1989) Combined effect of water activity and pH on inhibition of toxin production by *Clostridium botulinum* in cooked, vacuum-packed potatoes [published erratum appears in *Appl Environ Microbiol* 1989 Jul; 55(7): 1855]. *Appl. Environ. Microbiol.* **55**, 656–660.

DONNELLY, C.W., BRIGGS, E.H. and DONNELLY, L.S. (1987) Comparison of the heat resistance of *Listera monocytogenes* in milk as determined by two methods. *J. Food Prot.* **50**, 14–17.

ECFF (1996) ECFF Guideline for the hygienic manufacture of chilled foods. Helsinki, Finland. The European Chilled Food Federation.

EKLUND, M.W. (1982) Significance of *Clostridium botulinum* in fishery products preserved short of sterilization. *Food Technology* **36**, 107–112, 115.

ENGEL, G. and TEUBER, M. (1991) Heat resistance of ascospores of *Byssochlamys nivea* in milk and cream. *Int. J. Food Microbiol.* **12**, 225–233.

ERRINGTON, J. (1993) *Bacillus subtilis* sporolation: regulation of gene expression and control of morphogenesis. *Microbiol. Rev.* **57**, 1–33.

ESTY, J.R., AND MEYER, K.F. (1922) The heat resistance of spores of *Bacillus botulinus* and allied anaerobes. XI. *J. Infect. Dis.* **31**, 650–663.

FARKAS, J. (1994) Tolerance of spores to ionizing radiation: mechanisms of inactivation, injury and repair. *J. Appl. Bacteriol.* **76**, 81–90.

FASSOLITIS, A.C., PEELER, J.T., JONES, V.I. and LARKIN, E.P. (1985) Thermal resistance of three parvoviruses: a possible human isolate, the minute virus of mice, and the latent rat virus. *J. Food Prot.* **48**, 4–6.

FUJIKAWA, H. and ITOH, T. (1996a) Characteristics of a multicomponent first-order model for thermal inactivation of microorganisms and enzymes. *Int. J. Food Microbiol.* **31**, 263–271.

FUJIKAWA, H. and ITOH, T. (1996b) Tailing off of thermal inactivation curve of *Aspergillus niger* spores. *Appl. Environ. Microbiol.* **62**, 3745–3749.

FUJIKAWA, H. and ITOH, T. (1998) Thermal inactivation analysis of mesophiles using the Arrhenius and z-value models. *J. Food Prot.* **61**, 910–913.

FUJIKAWA, H., MOROZUMI, S., SMERAGE, G.H. and TEIXEIRA, A.A. (2000) Comparison of capillary and test tube procedures for analysis of thermal inactivation kinetics of mold spores. *J. Food Prot.* **63**, 1404–1409.

GAILLARD, S., LEGUERINEL, I. and MAFART, P. (1998a) Model for combined effects of temperature, pH and water activity on thermal inactivation of *Bacillus cereus* spores. *J. Food Science* **63**, 887–889.

GAILLARD, S., LEGUERINEL, I. and MAFART, P. (1998b) Modelling combined effects of temperature and pH on the heat resistance of spores of *Bacillus cereus*. *Food Microbiology* **15**, 625–630.

GERHARDT, P. (1988) The refractory homeostatsis of bacterial spores. In *Homeostatic mechanisms in microorganisms*, Wittenbury, R., Banks, J.G., Gould, G.W. and Board, R.B. pp. 41. Bath, United Kingdom: Bath University Press.

GERHARDT, P. and MARQUIS, R.E. (1989) Spore thermoresistance mechanisms. In *Regulation of procaryotic development* ed. Smith, I., Slepecky, R.A. and Setlow, P. p. 43. Washington DC: American Society for Microbiology.

GEVEKE, D.J., KOZEMPEL, M., SCULLEN, O.J. and BRUNKHORST, C. (2002) Radio frequency energy effects on microorganisms in foods. *Innovative Food Science & Technologies* **3**, 133–138.

GILL, C.O. (1979) Intrinsic bacteria in meat – a review. *J. Appl. Bacteriol.* **47**, 367.

GOMBAS, D.E. (1983) Bacterial spore resistance to heat. *Food Technology* **37**, 105.

GOODFELLOW, S.J. and BROWN, W.L. (1978) Fate of *Salmonella* inoculated into beef for cooking. *J. Food Prot.* **51**, 598–605.

GRAHAM, A.F., MASON, D.R., MAXWELL, F.J. and PECK, M.W. (1997) Effect of pH and NaCl on growth from spores of non-proteolytic *Clostridium botulinum* at chill temperature. *Lett. Appl. Microbiol.* **24**, 95–100.

GRANUM, P. E. and BAIRD-PARKER, T. E. (2000) *Bacillus* spp. In *The microbiological safety and quality of food*, Lund, B.M., Baird-Parker, T.E. and Gould, G.W. pp. 1029–1039. USA: Aspen Publishers.

GRIFFITHS, M.W. and PHILLIPS, J.D. (1990) Incidence, source and some properties of psychrotrophic *Bacillus* spp. found in raw and pasteurized milk. *J. Soc. Dairy Technology* **43**, 62–70.

GUINEBRETIERE, M.H., GIRARDIN, H., DARGAIGNARATZ, C., CARLIN, F. and NGUYEN-THE, C. (2003) Contamination flows of *Bacillus cereus* and spore-forming aerobic bacteria in a cooked, pasteurized and chilled zucchini puree processing line. *Int. J Food Microbiol.* **82**, 223–232.

GUZEWICH, J.J. and MORSE, D.L. (1986) Sources of shellfish in outbreaks of probable viral gastroenteritis: implecations for control. *J. Food Prot.* **49**, 389–394.

HAUSCHILD, A.H., HILSHEIMER, R., WEISS, K.F. and BURKE, R.B. (1988) *Clostridium botulinum* in honey, syrups and dry infant cereals. *J. Food Prot.* **51**, 892–894.

HEDDLESON, R.A. and DOORS, S. (1994) Factors affecting microwave heating of foods and microweve induced destruction of foodborne pathogens – a review. *J.Food Prot.* **57**, 1025–1037.

HEINEN, W. and LAUWERS, A.M. (1981) Growth of bacteria at 100°C and beyond. *Arch. Microbiology* **129**, 127.

HELPS, C.R., HARBOUR, D.A. and CORRY, J.E.L. (1999) PCR-based 16S ribosomal DNA detection technique for *Clostridium estertheticum* causing spoilage in vacuum-packaged chill-stored beef. *Int. J. Food Microbiol.* **52**, 57–65.

HSIEH, F., ACOTT, K. and LABUZA, T.P. (1976) Death kinetics of pathogens in a pasta product. *J. Food Science* **41**, 516–519.

HYYTIA, E., HIELM, S. and KORKEALA, H. (1998) Prevalence of *Clostridium botulinum* type E in Finnish fish and fishery products. *Epidemiol. Infect.* **120**, 245–250.

ICMSF. Micro-Organisms in foods, Vol. 1. 274 (1978). Toronto, Toronto Press. Their significance and methods for enumeration.

IFST. IFST position statement on foodborne viral infection. (1996). London, UK, Institute of Food Science and Technology.

INCZE, K., KÖRMENDY, L., KÖRMENDY, I. and ZSARNÓCZAY, G. (1999) Consideration of critical microorganisms and indicator enzymes in connection with the pasteurization of meat products. *Meat Science* **51**, 115–121.

INGRAM, M. (1969) Spore-formers as food spoilage organisms. In *The bacterial spore*, Gould, G.W. and Hurst, A. pp. 349–610. London: Academic Press.

JOHNSON, U., SNYGG, B.G., HARNULV, B.G. and ZACHRISSON, T. (1977) Testing two models for the temperature dependence of the heat inactivation rate of *Bacillus stearothermophilus*. *J. Food Science* **42**, 1251–1252.

JOUVE, J.L., STRINGER, M.F. and BAIRED PARKER, A.C. (1999) Food safety management tools. *Food Sci. Tech. Today* **13**, 82–91.

JUNEJA, V.K., MARMER, B.S., PHILLIPS, J.G. and MILLER, A.J. (1995) Influence of the intrinsic properties of food on thermal inactivation of spores of non-proteolytic *Clostridium botulinum*: development of a predictive model. *J. Food Safety* **15**, 349–364.

KALINOWSKI, R.M. and TOMPKIN, R.B. (1999) Psychrotrophic clostridia causing spoilage in cooked meat and poultry products. *J. Food Prot.* **62**, 766–772.

KEENEY, M. and EVENCHICK, Z. (1969) Heat activation. In *The bacterial spore*, Gould, G.W. and Hurst, A. pp. 359–396. London: Academic Press.

KIHM, D.J., HUTTON, M.T., HANLIN, J.H. and JOHNSON, E.A. (1990) Influence of transition metals added during sporulation on heat resistance of *Clostridium botulinum* 113B spores. *Appl. Environ. Microbiol.* **56**, 681–685.

KIRBY, R.M. and DAVIES, R. (1990) Survival of dehydrated cells of *Salmonella typhimurium* LT2 at high temperatures. *J. App. Bact.* 241–246.

KISS, I., RHEE, C.O., GRECZ, N., ROBERTS, T.A. and FARKAS, J. (1978) Relation between radiation resistance and salt sensitivity of spores of five strains of *Clostridium botulinum* types A, B, and E. *Appl. Environ. Microbiol.* **35**, 533–539.

KLIJN, N., BOVIE, C., DOMMES, J., HOOLWERF, J.D., VAN DER WAALS, C.B., WEERKAMP, A.H. and NIEUWENHOF, F.F.J. (1994) Identification of *Clostridium tyrobutyricum* and related species using sugar fermentation, organic acid formation and DNA probes based on specific 16S rRNA sequences. *Syst. Appl. Microbiol.* **17**, 249–256.

KNUTSON, K.M., MARTH, E.H. and WAGNER, M.K. (1987) Microwave heating of food. *Lebensm. Wiss. Technol.* **20**, 101–110.

KOO, K., FOEGEDING, P.M. and SWAISGOOD, H.E. (1998a) Construction and expression of a bifunctional single-chain antibody against *Bacillus cereus* spores. *Appl. Environ. Microbiol.* 2496.

KOO, K., FOEDING, P. M. and SWAISGOOD, H. E. (1998b) Development of a streptavidin-conjugated single-chain antibody that binds *Bacillus cereus*. *Appl. Environ. Microbiol.* **64**(7) 2479–2502.

KORKEALA, H., STENGEL, G., HYYTIA, E., VOGELSANG, B., BOHL, A., WIHLMAN, H., PAKKALA, P. and HIELM, S. (1998) Type E botulism associated with vacuum-packaged hot-smoked whitefish. *Int. J. Food Microbiol.* **43**, 1–5.

KOZEMPEL, M.F., ANNOUS, B.A., COOK, R.D., SCULLEN, O.J. and WHITING, R.C. (1998) Inactivation of microorganisms with microwaves at reduced temperatures. *J. Food Prot.* **61**, 582–585.

LARSEN, H.D. and JORGENSEN, K. (1999) Growth of *Bacillus cereus* in pasteurized milk products. *Int. J. Food Microbiol.* **46**, 173–176.

LEGUERINEL, I., COUVERT, O. and MAFART, P. (2000) Relationship between the apparent heat resistance of *Bacillus cereus* spores and the pH and NaCl concentration of the recovery medium. *Int. J. Food Microbiol.* **55**, 223–227.

LEONARD, C., CHEN, Y. and MAHILLON, J. (1997) Diversity and differential distribution of IS231, IS232, and IS240 among *Bacillus cereus*, *Bacillus thuringiensis* and *Bacillus mycoides*. *Microbiology* **143**, 2537–2547.

LIN, S., SCHRAFT, H., ODUMERU, J.A. and GRIFFITHS, M.W. (1998) Identification of contamination sources of *Bacillus cereus* in pasteurized milk. *Int. J. Food Microbiol.* **43**, 159–171.

LOSICK, R. and STRAIGER, P. (1992) Crisscross regulation of cell-type specific gene expression during development in *Bacillus subtilis*. *Nature* (London) **355**, 601–604.

LUND, B.M. (1992) Ecosystems in vegetable foods. *Soc. Appl. Bacteriol. Symp. Ser.* **21**, 115–126.

LUND, B.M. and PECK, M.W. (1994) Heat resistance and recovery of spores of non-proteolytic *Clostridium botulinum* in relation to refrigerated, processed foods with an extended shelf-life. *Soc. Appl. Bacteriol. Symp. Ser.* **23**, 115–128.

MAFART, P., COUVERT, O. and LEGUERINEL, I. (2001) Effect of pH on the heat resistance of spores comparison of two models. *Int. J. Food Microbiol.* **63**, 51–56.

MAFART, P., COUVERT, O., GAILLARD, S. and LEGUERINEL, I. (2002) On calculating sterility in thermal preservation methods: application of the Weibull frequency distribution model. *Int. J. Food Microbiol.* **72**, 107–113.

MANZANO, M., COCULIN, L., CANTONI, C. and COMI, G. (2003) *Bacillus cereus, Bacillus thuringiensis*, and *Bacillus mycoides* differentiation using PCR-RE technique. *Int. J. Food Microbiol.* **81**, 249–254.

MARQUIS, R.E., SIM, J. and SHIN, S.Y. (1994) Molecular mechanisms of resistance to heat and oxidative damage. *J. Appl. Bacteriol.* **70**, 40–48.

MAYES, T. (1992) Simple users guide to the hazard analysis critical control point concept for the control of food microbiological safety. *Food control.* **3**.14–19.

MAZAS, M., MARTINEZ, S., LOPEZ, M., ALVAREZ, A.B. and MARTIN, R. (1999) Thermal inactivation of *Bacillus cereus* spores affected by the solutes used to control water activity of the heating medium. *Int. J. Food Microbiol.* **53**, 61–67.

MIWA, N., NISHINA, T., KUBO, S., ATSUMI, M. and HONDA, H. (1998) Amount of enterotoxigenic *Clostridium perfringens* in meat detected by nested PCR. *Int. J. Food Microbiol.* **42**, 195–200.

MULAK, V., TAILLIEZ, R., EB, P. and BECEL, P. (1995) Heat resistance of bacteria isolated from preparations based on seafood products. *J. Food Prot.* **58**, 49–53.

MURRELL, W. G. and WARTH, A. D. (1965) Composition and heat resistance of bacterial spores. In *Spores III* ed. Campbell, L.L. and Halvorsen, H.O. p. 1. Ann Arbor, MI: American Society for Microbiology.

NEVAS, M., HIELM, S., LINDSTRÖM, M., HORN, H., KOIVULETHO, K. and KORKEALA, H. (2002) High prevalence of *Clostridium botulinum* types A and B in honey samples detected by polymerase chain reaction. *Int. J. Food Microbiol* **72**, 45–52.

NISSEN, H., ROSNES, J.T., BRENDEHAUG, J. and KLEIBERG, G.H. (2002) Safety evaluation of sous vide-processed ready meals. *Lett. Appl. Microbiol.* **35**, 433–438.

NORRIS, J. R., BERKELEY, R. C. W., LOGAN, N. A. and O'DONELL, A. G. (1981) The genera *Bacillus* and *Sporolactobacills*. In *The procaryotes* ed. Starr, M.P. pp. 1709–1742.

OHLSSON, T. (2002) Minimal processing of foods with thermal methods. In *Minimal processing technologies in the food industry*, Ohlsson, T. and Bengtsson, N. pp. 4–33. Cambridge, England: Woodhead Publ. Ltd.

PARRY, J.V. and MORTIMER, P.P. (1984) The heat sensitivity of hepatites A virus determined by a simple tissue culture method. *J. Med. Virol.* **14**, 277–283.

PERIAGO, P.M., FERNANDEZ, P.S., SALMERON, M.C. and MARTINEZ, A. (1998) Predictive model to describe the combined effect of pH and NaCl on apparent heat resistance of *Bacillus stearothermophilus*. *Int. J. Food Microbiol.* **44**, 21–30.

PFLUG, I. J. and GOULD, G. W. (2000) Heat treatment. In *The microbiological safety and quality of food*, Lund, B.M., Baird-Parker, A.C. and Gould, G.W. pp. 36–64. Gaithersburg, Maryland: Aspen Publishers, Inc.

PITT, J.I. and CHRISTIAN, J.H. (1970) Heat resistance of xerophilic fungi based on microscopical assessment of spore survival. *Appl. Microbiol.* **20**, 682–686.

PUT, H.M.C., DE JONG, J. and SAND, F.E.M.J. (1976) Heat resistance studies on yeast spp. causing spoilage in soft drinks. *J. Appl. Bact.* **40**, 135–152.

ROBERTS, T. A. and HITCHINS, A. D. (1969) Resistance of spores. In *The bacterial spore* ed. Gould, G.W. and Hurst, A. pp. 611–670. London: Academic Press.

RODRIGUEZ, J.H., COUSIN, M.A. and NELSON, P.E. (1992) Oxygen requirements of *Bacillus licheniformis* and *Bacillus subtilis* in tomato juice; ability to grow in aseptic packages. *J. Food Science* **57**, 937–976.

ROLAND, J.O. and BEUCHAT, L.R. (1984) Influence of temperature and water activity on growth and patulin production by *Byssochlamys nivea* in apple juice. *Appl. Environ. Microbiol.* **47**, 205–207.

ROSENBERG, U. and BÖGEL, W. (1987) Microvawe pasteurization, sterilization, blanching, and pest control in the food industry. *Food Technology* **41**, 92–99.

RUTHERFORD, G.C., REIDMILLER, J.S. and MARQUIS, R.E. (2000) Method to sensitize bacterial spores to subsequent killing by dry heat or ultraviolet irradiation. *J. Microbiol. Methods* **42**, 281–290.

SARRÍAS, J.A., VALERO, M. and SALMERÓN, M.C. (2002) Enumeration, isolation and characterization of *Bacilllus cereus* strains from Spanish raw rice. *Food Microbiology* **19**, 589–595.

SCHELLEKENS, M., MARTENS, T., ROBERTS, T.A., MACKEY, B.M., NICOLAI, B.M., VAN IMPE, J.F. and DE BAERDEMAEKER, J. (1994) Computer aided microbial safety design of food processes. *Int. J. Food Microbiol.* **24**, 1–9.

SERRA, J.A., DOMENECH, E., ESCRICHE, I. and MARTORELL, S. (1999) Risk assessment and critical points from the production perspective. *Int. J. Food Microbiol.* **46**, 9–26.

SETLOW, P. (1993) DNA structure, spore formation and spore properties. In *Regulation of bacterial differentiation*, Piggot, P.J., Youngman, P. and Moran, C.P. pp. 181–194. Washington DC: American Society for Microbiology.

SETLOW, P. (1994) Mechanisms which contribute to the long-term survival of spores of *Bacillus species*. *J. Appl. Bacteriol.* **17**, 49–60.

SETLOW, P. and JOHNSON, E.A. (1997) Spores and their significance. In *Food Microbiology. fundamentals and frontiers*, Doyle, M.P., Beuchat, L.R. and Montville, T.J. pp. 30–65. Washington D.C.: ASM Press.

SHEHATA, A.E., MAGDOUB, M.N.I., SULTAN, N.E. and EL-SAMRAGY, Y.A. (1983) Aerobic mesophilic and psychrotrophic spore-forming bacteria in buffalo milk. *J. Dairy*

Sci. **66**, 1228–1231.

SLEPECKY, R. A. and LEADBETTER, E. R. (1994) Ecology and relationships of endospore-forming bacteria: changing prspectives. In *Regulation of bacterial differentiation*, Piggot, P.J., Moran, C.P. and Youngman, P. pp. 195–206. Washington, DC: American Society for Microbiology.

SPLITTSTOESSER, D.F. and SPLITTSTOESSER, C.M. (1977) Ascospores of *Byssochlamys fulva* compared with those of a heat resistsnt *Aspergillus*. *Journal of Food Science* **42**, 685–688.

STRAGIER, P. (1994) A few good genes. In *Regulation of bacterial differentiation*, Piggot, P.J., Moran, C.P. and Youngman, P. pp. 207–246. Washington, DC: American Society for Microbiology.

TAORMINA, P.J., BARTHOLOMEW, G.W. and DORSA, W.J. (2003) Incidence of *Clostridium perfringens* in commercially produced cured raw meat product mixtures and behavior in cooked products during chilling and refrigerated storage. *J. Food Prot.* **66**, 72–81.

TOWNSEND, C. T., ESTY, J. R., AND BASELT, F. C. (1938) Heat resistance studies on spores of putrefactive enaerobes in relation to determination of safe processes for canned foods. *Food Res.* **3**, 323–246.

TSEN, H.-Y., CHEN, M.-L., HSIEH, Y.-M., SHEU, S.-J. and CHEN, Y.-L. (2000) *Bacillus cereus* group strains, their hemolysin BL activity, and their detection in foods using a 16 S RNA and hemolysin BL gene-targed multiplex polymerase chain reaction system. *J. Food Prot.* **63**, 1496–1502.

US DEPARTMENT OF AGRICULTURE-FOOD SAFETY AND INSPECTION SERVICE. Performance standards for certain meat and poultry products: technical paper. 64, 732–749. 1999. US Department of Agriculture.

VALERO, M., HERNÁNDEZ-HERRERO, L.A., FERNÁNDEZ, P.S. and SALMERÓN, M.C. (2002) Characterization of *Bacillus cereus* isolates from fresh vegetables and refrigerated minimally processed foods by biochemical and physiological tests. *Food Microbiology* **19**, 491–499.

VAN IMPE, J.F., NICOLAI, B.M., SCHELLEKENS, M., MARTENS, T. and DE BAERDEMAEKER, J. (1995) Predictive microbiology in a dynamic environment: a system theory approach. *Int. J. Food Microbiol.* **25**, 227–249.

VOYSEY, P. An introduction to the practice of microbiological risk assessment for the food industry applications. Guideline No. 28, 1–104. 2000. Campden and Chorleywood Food Research Association Group.

WANG, J.C., HU, S.H. and LIN, C.Y. (2003) Lethal effect of microwaves on spores of *Bacillus* spp. *J. Food Prot.* **66**, 604–609.

WARTH, A.D. (1978) Relationship between the heat resistance of spores and the optimum and maximum growth temperatures of *Bacillus* species. *J. Bacteriol.* **134**, 699.

WELT, B.A., TONG, C.H., ROSSEN, J.L. and LUND, D.B. (1994) Effect of microwave radiation on inactivation of *Clostridium sporogenes* (PA 3679) spores. *Appl. Environ. Microbiol.* **60**, 482–488.

ZWIETERING, M.H. and HASTING, A.P.M. (1997) Modelling the hygienic processing of foods – a global process overview. *Food & Bioproducts Processing* **75**, 159–167.

Part VI

Appendix

23

Optimising the thermal processing of liquids containing solid particulates

N. Heppell, Oxford Brookes University, UK

23.1 Introduction: problems with heating liquids containing solid particulates

The aseptic processing of liquid foods which contain solid particulates is, in essence, a simple system in which, as with liquid UHT systems, the foodstuff is pumped in continuous flow through a series of appropriate heat exchangers to a high temperature (generally 130–150°C, held in a holding tube until sterile, then cooled and packed aseptically. UHT (or aseptic processing) systems for sterilisation of pure liquids have been commonplace for many decades, but the introduction of solid particulates into a liquid complicates the process in several ways:

- The equipment must be chosen and sized so that the particulates can physically pass through the whole process without damage to themselves or blocking the equipment. This is accomplished by applying the rule-of-thumb that the smallest passage or gap through which the foodstuff must flow should not be less than three times the size of the largest solid particulate. This also has the effect of increasing the diameter of pipes and equipment generally over that normally used and therefore reducing the flow velocity in these.
- The foodstuff must be pumped at a relatively low, steady flowrate to about 4 bar pressure (to prevent boiling) without damage to the solid particulates. After the heating, holding and cooling sections, the pressure must then be reduced back to atmospheric pressure under aseptic conditions, before packaging. The latter is more difficult as the particulates are cooked and therefore more fragile and more susceptible to damage. See Lewis and Heppell (2000) for further details on pumping and back-pressure arrangements.

- Maintaining aseptic conditions is more difficult, especially during pressure release after processing, pre-sterilisation of equipment surfaces before processing commences and in filling and sealing of aseptic packages.
- Ensuring sterility of the foodstuff. This is the major difficulty in processing liquids containing solid particulates. Each phase must have received a sufficient thermal treatment to ensure sterility of the foodstuff as a whole and, if the solid particulates vary in size, shape or material (e.g. meat cubes, potato cubes, green beans or peas), it may be necessary to ensure each of these solid phases has received an adequate heat treatment as well. For sterility, the temperature history and residence time distribution of each phase must be known The data are relatively easy to determine for the liquid phase, but much more difficult to measure for the solid phase during transportation by the liquid. Both of these aspects are complicated by the need to keep the particulates suspended in the liquid and not settle out during horizontal or vertical flow. This is normally achieved by increasing the viscosity of the liquid phase using starches, gums, pectin or other edible polymers. Increased viscosity, nearly always non-Newtonian in nature, usually ensures flow is laminar rather than turbulent, especially in pipeline sections, with its concomitant velocity profile across the pipe. The increased viscosity of the liquid phase can also decrease the rate of heat transfer between the liquid and solid phases due to the reduction of the difference between the velocities of liquid and solid, increasing the boundary layer and thus slowing the rate of heating of the solid particulate centre. A significantly longer holding time would then be necessary for sterility of the particulates but resulting in over-processing of the liquid phase and an associated loss in organoleptic quality.

These problem areas are considered in depth in this chapter. In addition, the reader is directed to the chapters on verification of the process (Chapters 16 and 18) describing techniques to verify the whole process, as well as alternative heating methods, e.g. the use of Ohmic heating covered in Chapter 12.

23.2 Residence time distribution of solid particulates and liquid phase

23.2.1 Transport of the particulate solids through the process plant

The physical aspects of the flow of liquids containing solid particulates is extremely complex and little is known about such systems, especially in viscous, non-Newtonian fluids containing large, solid particulates with approximately the same density as the liquid. To date, it is not possible to predict the flow, or residence time distribution, of either the solid or liquid phases with any confidence, despite the many studies recently carried out in this area. Much of this work has concentrated on either single particles or low concentrations of multiple particles, which does not entirely represent the situation present in aseptic processing of such foods, and little is understood on particulate-particulate or particulate-wall interactions. Single particle studies have highlighted difficulties in establishing the effect of particulate size, shape and density and fluid density and viscosity and flow channel size on particulate behaviour, or even whether the particulate rotates or not. The extremely large number of interacting factors to be studied mean that it is unlikely that a full understanding of particulate behaviour will be achieved quickly.

When considering forces on the particulates by a transported liquid, the following effects have been found to be important:

- Drag by the liquid on the particulate – which depends upon the density difference between liquid and solid, inertia of the particulate and viscosity of the liquid
- Lift – caused by the Magus effect due to a rotating particulate and the Saffman force which causes particulates to migrate even if they are not rotating.
- Virtual mass – where liquid is dragged along with the particulate as it moves and has the effect of making the particulate seem larger than it actually is.

In addition to the residence time distribution of the particulates it is also important to understand the residence time distribution of the liquid, as this is generally shorter; it is also important to ensure this phase is sterilised correctly. The liquid residence time distribution is affected by the particulate behaviour and again needs to be measured rather than predicted.

The majority of studies on the residence time distribution of solid particulates concentrate on detecting a particle as it enters a section of equipment and determining the time spent within that section. One factor of importance is the minimum particulate residence time in different sections of the process plant, as this may determine the sterility of the foodstuff as a whole. Some work has reported this as a ratio to the average residence time for the whole foodstuff (based on volume flowrate and cross-sectional area of the flow channel).

23.2.2 Measurement methods for residence time distribution of particulates

Most studies on residence time distribution have focused on visualising or detecting flow of particulates are either within or at the inlet and outlet of process equipment. There are a variety of methods which have been used for measurement of particulate residence times. These include the following:

- *Magnetic particle methods*. The movement of a magnet near an electrical coil induces a voltage in that coil, depending on the velocity and orientation of the magnet, among other physical factors. This voltage can be used to trigger timers and measure the time taken between coils. By embedding a small magnet into a food particle and positioning coils at different points in the aseptic processing plant, the residence time of the particulate in the different plant sections can be measured (e.g. Segner *et al*., 1989). A similar technique is the Hall effect: the method again relies on the movement of a magnet, but the detector in this case is a semiconductor, whose electrical resistance is altered as the magnet passes, and can again be used to trigger a timer to measure the time for the particulate to travel between sensors (Tucker and Withers, 1994). The advantage of this method over the magnet/coil method is that the magnet can be detected through a metal pipe wall and the use of several sensors at a single point may help determine the radial position of the particulate.

- Particulate velocity can be measured directly using an ultrasonic Doppler effect transducer which uses the change in frequency of an ultrasonic wave reflected from a particulate to determine its velocity. This technique will only give the velocity of particulates at a given point but not information on movement of particulates across the pipe to different radii later in the equipment or the residence time within a section of process equipment. It is therefore useful as a research tool but not in process design.
- *Optical methods*. By tracking of particulates using a video camera (static or moving, with or without mirrors), a large amount of information can be gathered about particulate behaviour in aseptic processing systems. Residence time, velocity, radial position (using mirrors at right angles) and migration, particle rotation, leapfrogging, etc. can all be observed directly and measured. By incorporating small polystyrene particulates (Balasubramaniam and Sastry, 1996), it is possible to determine fluid velocity at the same time as particulate velocity. The disadvantage of this method is that both the equipment and liquid must be transparent, limiting its use to experimental systems only.
- Other methods of value include magnetic resonance imaging and radioactive tracers, especially positron emission particle tracking (PEPT).

The difficulties with all these methods is to ensure that the residence time distribution is not affected by the measurement methods and is representative of foodstuffs as a whole. Implanting magnets into particulates may change their density, and even selecting non-food particulates of the relevant (and variable) density to use in experimental systems requires care. Real foodstuffs are inevitably more complex.

There are also methods of measuring the residence time distribution of the liquid phase, which is altogether easier. Selection of a tracer is the most important aspect, as it must not affect the flow behaviour of the liquid and be readily detected, preferably by an in-line detection system. Salt, sugar and dyes are the most common tracer materials used, detected respectively by conductivity, refractive index and light absorption.

23.3 Liquid-particulate heat transfer

23.3.1 Heating of the solid particulates

The temperature history at the centre of the particulates, the slowest heating point, must be known to ensure that adequate heat treatment has been received by the foodstuff overall. In all conventional processes, the liquid is heated first and heat must then transfer from the liquid to the cooler solid through a boundary layer around the solid particulate, which can itself be a large resistance to heat transfer, then through the solid to its thermal centre. The rate of heating at the centre of a particulate can be predicted mathematically from a knowledge of the size, shape, and thermal properties of the solid and the rate of heat transfer across the boundary layer. An extra additional problem is that, in the holding tube as the particulate solids heat up, conservation of heat means that the liquid temperature falls, which can affect the rate of heat transfer.

The rate of heating of a solid body can be calculated using analytical solutions to the Fourier equation, which have been solved by many workers, and solutions are available for standard shapes of bodies (flat plates, spheres and infinite cylinders), constant thermal and physical properties, constant initial uniform temperature distribution and an instantaneous rise in liquid temperature. Solutions are presented graphically (Gurney and Lurie, 1923; Heisler, 1947; Schneider, 1963) or may be calculated from analytical solutions to the Fourier equation (e.g. Rohsenow and Choi, 1961). Mathematical models have also been developed and have many advantages over the analytical solutions in that they are not constrained by the conditions listed above. Models have been published for spheres (Lewis and Heppell, 2000; Sastry and Cornelius, 2002) and short cylinders (Teixeira *et al.*, 1969). Finite element or Finite volume methods are also commonplace.

Prediction of liquid-particulate heat transfer coefficient may be made using published correlations, though the accuracy of these is not always high. Correlations are usually in the form

$$\mathrm{Nu}_p = \mathrm{K}(\mathrm{Re}_p)^n(\mathrm{Pr})^m \tag{23.1}$$

where Nu_p Re_p and Pr are the particle Nusselt number, particle Reynolds number and Prandtl numbers respectively.

The particle Reynolds number contains the term $(U_L - U_p)$ the relative liquid-particulate velocity and the higher this value, the higher the heat transfer coefficient predicted. It should be noted that for non-Newtonian Power Law fluids, the generalised Reynolds and Prandtl numbers should be used. This is likely to further increase the inaccuracy of the heat transfer coefficient value returned from the correlation, as very few correlations have been derived for non-Newtonian fluids and unsteady state heat transfer.

For all shapes of particulates, there is a minimum heat transfer which corresponds to the conduction solution of the cold particulate in an infinite hot medium but, for a sphere, it can be shown that this corresponds to a Nusselt Number value of 2.0, for steady state heat transfer only. The correlation of Ranz and Marshall (1952), was derived for spheres:

$$\mathrm{Nu}_p = 2.0 + 0.6(\mathrm{Re}_p)^{0.5}(\mathrm{Pr})^{0.33} \tag{23.2}$$

This correlation has been modified by Whitaker (1972) to give a two-term correlation for the Reynolds number.

Actual measurement of the temperature at the centre of a particulate is much more difficult, however, when the solid is actually being transported by the liquid. Some form of mobile temperature indicator is required that will not affect the way the solid particulate behaves, allowing the temperature somewhere within the particulate to be read from outside the equipment, or possibly after processing and recovery from the outlet.

23.3.2 Methods for measuring liquid-particulate heat transfer coefficients

There have been a variety of interesting and inventive methods used to determine liquid-particulate heat transfer coefficients:

- *Static particulate methods*. Many early attempts to measure liquid-solid heat transfer used a particle containing a thermocouple which was held statically and heated liquids were pumped past, or the particle was held in a stirred tank of liquid (e,g, Chang and Toledo, 1989; Chandarana *et al.*, 1989; Clement *et al.*, 1997). The validity of the results from this type of work for use on aseptic processing systems must be in doubt as the particulate is not subject to the same conditions as it would be in a flowing liquid. In particular, the solid particulate does not move or rotate and the liquid velocity profile will not be the same at the same liquid-particulate relative velocity. However, the technique is relatively cheap and easy and will give an indication of the order of heat transfer coefficient likely to be present. Nevertheless, confirmation of the data will still be required in a continuous-flow thermal processing system.
- *Moving thermocouple.* The above method has been adapted to allow movement of the particulate down a tube. Using a fine-wire thermocouple attached to a transducer particle, the thermocouple wire is withdrawn from the downstream end of a holding tube at the same speed as the particle would travel at, having been determined beforehand (Sastry *et al.*, 1989, 1990). Although an improvement on the static particle apparatus, the presence of the thermocouple wires will probably modify the particle behaviour compared to a free-moving particle. Also, measurements cannot be taken in some elements of the aseptic process, e.g. scraped-surface heat exchangers or coiled tubes. Further experimental details and results are summarised in Sastry and Cornelius (2002).
- *Marker chemicals or microorganisms.* These have been developed and used by several workers. In general, a marker which degrades or changes with known, measured thermal kinetics is used, such as spores of *Bacillus stearothermophilus* or *Clostridium sporogenes*, both of which have high heat resistance, or an enzyme with appropriate z-value, such as α-amylase. These markers are incorporated (or immobilised) into a solid particulate and subjected to a thermal treatment, then removed or dissolved and the marker recovered. The immobilisation technique of Dallyn *et al.* (1977) for entrapping microorganisms into and releasing them from calcium alginate particles is extremely useful in this respect. The technique, as used for

measuring liquid-particulate heat transfer coefficients, is to pass the marker immobilised in a particle through the system of interest, collect it and evaluate the number of surviving spores (or concentration of enzyme surviving). This is then compared to the results from mathematical calculation of surviving spores expected for different liquid-particulate heat transfer coefficient values using a prediction of the temperature profile in the body from the physical properties of that material together with the thermal inactivation kinetics measured for the marker in the particle material (Heppell, 1985) The method using microorganisms is very labour-intensive, however, but easier with enzymes.

- *Liquid Crystal Indicators.* The use of thermochromic paints in food processing has been reviewed by Balasubramaniam and Sastry (1995) and developed for use in measuring the liquid-particulate heat transfer coefficient by these authors in a series of publications. The basis of the technique is the use of thermochromic paints – which change colour depending on their temperature – coated onto a test particulate and injected into a test system. The colour of the paint is monitored, as a function of time, by video camera, and the colour translated into the relevant temperature. The time-temperature relationship can be used to evaluate the liquid-particle heat transfer coefficient from a mathematical model. As the colour of the particle must be measured, the whole system must be transparent and its use is therefore limited to model systems. However, the technique is a powerful one. Further experimental details and results are summarised in Sastry and Cornelius (2002). Mwangi *et al.* (1993) used a similar technique but with a melting point indicator which changes colour at a set temperature, immobilised in polymethylmethacrylate.
- *Calorimetry of particulates or liquid.* Calorimetry is based on the loss of heat from the liquid to the solid to achieve an equilibrium. Either a sample of particulates and liquid combined is removed, or a particulate alone is quickly removed from a suitable point in the system and left to equilibrate under insulated conditions. From a heat balance, the average temperature of the particulates can be determined and therefore the liquid-particulate heat transfer coefficient deduced from a mathematical model. Temperature measurement is critical for this method.

23.4 Measurement of the overall thermal treatment received: time-temperature integrators (TTIs)

It is useful to have a method for determining the overall thermal treatment actually received by a particulate, rather than predicting it by residence time and heat transfer data. The best method to date is the use of a time-temperature integrator (TTI) in which a thermolabile marker is immobilised within a solid particulate. The marker compound may already be present in the foodstuff (intrinsic) or added to the foodstuff (extrinsic). Markers which have been used are microorganisms or spores enzymes or any chemical or physical marker with a first-order degradation kinetics. Intrinsic markers are less useful as there must be a reasonable concentration which survives the process to ensure a good value for the sterilisation efficiency is obtained (rather than just a minimum value). It would be extremely useful to be able to translate the overall thermal treatment received to an equivalent F_0 value to enable this to be compared to target values required for Public Safety and for commercial sterility, which requires the thermal kinetics of the change in marker to have a z-value of about 10°C. Time-temperature integrators have been extensively reviewed by Maesmans *et al.* (1993).

The use of TTIs requires the marker to be immobilised within the particulate at this lowest heating point in the foodstuff rather than being evenly distributed throughout the particulate. To obtain a realistic process validation, all physical and thermal properties of the particulate are very important and it must simulate the foodstuff in all respects, not only size and density but also thermal conductivity and specific heat to ensure the particle behaves physically and thermally as the foodstuff. It is not necessary to have microorganisms or spores as a marker which are representative of the natural microbial population or with the same thermal death rate; but it is important that the rate of degradation change with temperature (z value) is typical for microorganisms or spores Common markers are *Bacillus stearothermophilus*, *Clostridium sporogenes* and *Bacillus coagulans*, and enzymic markers such as α-amylase from *Bacillus subtilis* (van Loey *et al.*, 1996) or *Bacillus licheniformis* (De Cordt *et al.*, 1992) have been used to validate aseptic processes.

TTIs are important in developing aseptic processes and especially so in measuring improvements in thermal processing efficiency. A large number of TTI particulates need to be used, however, but fortunately, this is possible as a large number of particulates can be processed at the same time.

23.5 Optimising heat transfer

The aseptic processing of liquids containing solid particulates has many constraints which it is not possible to model in terms of the high viscosity of fluid required, the particle sizes and composition of particles for transport. There have been many research studies investigating the physical aspects of the system and attempts to correlate the various factors involved continue to be carried out.

The intention of this chapter is to propose methods which might improve the quality of aseptically processed particulate foods. The key method for improving any aseptic product is to maximise the heating rate of the whole of the food, hold for the minimum length required, and to cool as quickly as possible.

One option that must be mentioned here is to avoid the problem altogether and use an alternative system to process particulate liquids. One system allows the solid particulates and liquids to be processed separately, using an agitated vessel such as the double-cone Jupiter system (APV Co UK) heated by steam injection to sterilise the solids and a conventional UHT process plant for the liquid. Each phase may then be given the appropriate optimum thermal treatment, before they are combined and aseptically packed. Another technique is the use of an ohmic heating system. Otherwise, when heat treating both phases as a whole, the broad aim is to increase heat transfer to the solid phase as much as possible, and/or to increase the residence time of the particulates compared to the liquid to allow them to receive a sufficient thermal treatment. Many studies (and the basic theory) have shown that the rate of heating when the liquid is agitated, i.e. it has a high Reynolds number, is much more rapid than would occur in tube flow. The effect of agitation on the residence time of the particles must be investigated, though, to ensure that the minimum particulate residence time is not decreased.

It is easy to show that improvements to the upstream part of the process, i.e. the heating section, have a much greater effect on the overall thermal treatment obtained. Figure 23.1 demonstrates the temperature history and associated F_0 value at the centre of a 10 mm sphere of beef during heating in a typical aseptic process. Three conditions have been used: (1) infinite heat transfer coefficient throughout; (2) minimum heat transfer coefficient throughout (Nu = 2); and (3) an infinite heat transfer coefficient during the heating phase only, the minimum heat transfer in the holding tube. No effect of residence time distribution has been used; it has been assumed the same for each phase.

It is also possible to show, using the same mathematical model, that increasing the temperature of the feed to the process will reach the target F_0 value required for the food in a shorter residence time, e.g. starting at a feed

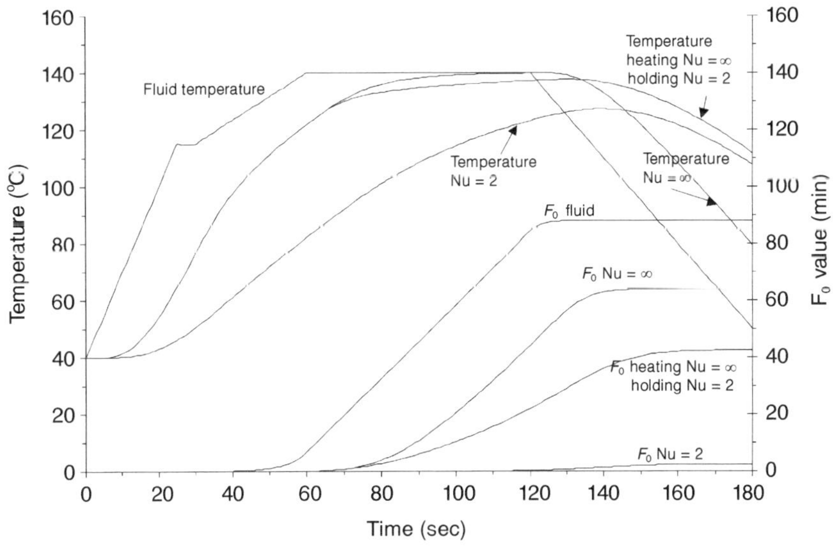

Fig. 23.1 Temperature history and F_0 value at the centre of a 10 mm sphere of beef under different process conditions during a typical aseptic process.

temperature of 80°C has a significant effect over a temperature of 40°C (Lewis and Heppell, 2000). This may happen anyway, as it is likely that, for a significant number of foods, the solid particulates will have received some cooking (fried, boiled, etc.) before being aseptically processed.

Methods of increasing liquid agitation include:

- Use of scraped surface heat exchangers rather than tubular heaters, in the aseptic process. The agitation inherent in scraped surface heat exchangers would be expected to, and indeed has been shown to, increase the liquid-particulate heat transfer coefficient, depending on mutator speed (Balasubramaniam and Sastry, 1996).
- Changes in the shape and direction of the flow channel, and other methods of inducing turbulence in the liquid need to be investigated. The use of contorted tubes (e.g. a bank of U-shaped tubes), inclusion of impellers, mixers or static mixers or other turbulence promoters (similar in principle to those used in Ultrafiltration systems) into the flow channel or use of spiral-wound heat exchanger tubes as a holding tube, all have potential. The use of ultrasound to help disrupt the boundary layer and improve heat transfer may also be useful as would the use of pulsed, oscillatory flow, rather than steady flow, which may be induced by incorporating a reciprocating cylinder after the feed pump but before the holding tube. Several of these techniques have been investigated in the past for other applications and found to increase heat transfer rates in heat exchangers.

One device which has been shown to improve the rate of heating of particulates is the Stork Rota-Hold in which particulate solids are held back on slowly rotating blades which allow the fluid to pass through, and therefore give the particulate a longer holding time to ensure sterility. The gap between the blades can be adjusted to trap different sizes of particles and the rate of rotation of the blades can be adjusted to give the appropriate residence time. Additionally, the flow of liquid over the particles is increased, thus improving heat transfer.

Methods of improving liquid-particulate heat transfer coefficient can be evaluated for effectiveness using either TTI particulates in a commercial system, or laboratory-based techniques such as liquid crystal indicators.

23.6 Conclusions and future trends

A large body of research work to date has, quite rightly, concentrated on increasing understanding of the physical principles behind the behaviour of solid particulates during transportation by a fluid, but the complexity of the situation means that much more work is still required before this is achieved. However, a more practical approach into measuring and, importantly, manipulating residence time distribution and liquid-particulate heat transfer to improve and optimise the product quality would be of immediate benefit to process designers and food processors and could be achieved without substantial research input.

23.7 Sources of further information and advice

LEWIS, M.J. and HEPPELL, N.J. (2000). *Continuous thermal processing of foods.* Gaithersberg, Maryland, USA, Aspen Publishers Inc.

SASTRY, S.K. and CORNELIUS, W.D. (2002). *Aseptic processing of foods containing solid particulates.* New York, John Wiley & Sons Inc.

23.8 References

BALASUBRAMANIAM, V.M. and SASTRY, S.K. (1995). Use of liquid crystals as temperature sensors in food processing research. *Journal of Food Engineering* 26: 219–230.

BALASUBRAMANIAM, V.M. and SASTRY, S.K. (1996). Fluid-to-particle convective heat transfer coefficient in a horizontal scraped surface heat exchanger determined from relative velocity measurement. *Journal of Food Process Engineering* 19: 75–95.

CHANDARANA, D.I., GAVIN, A.I. and WHEATON, F.W. (1989). Simulation of parameters for modelling aseptic processing of foods containing particulates. *Food Technology* 43(3): 137–143.

CHANG, S.Y. and TOLEDO, R.T. (1989). Heat transfer and simulated sterilisation of particulate solids in a continuously flowing system. *Journal of Food Science* 54(4): 1017–1023, 1030.

CLEMENT, I., DUQUENOY, A., JUNG, A. and MORISSET, V. (1997). Measuring the coefficient of heat transfer between a liquid and suspended particles. *Engineering & Food at ICEF-7.* Jowitt, R., Sheffield Academic Press.

DALLYN, H., FALLOON, W.C. and BEAN, P.G. (1977). Method for the immobilization of bacterial spores in alginate gel. *Laboratory Practice* 26 (10), 773–775.

DE CORDT, S., HENDRICKX, M., MAESMANS, G. and TOBBACK, P. (1992). Immobilized Alpha-Amylase from Bacillus-Licheniformis – a Potential Enzymatic Time Temperature Integrator for Thermal-Processing. *International Journal of Food Science and Technology* 27(6): 661–673.

GURNEY, H.P. and LURIE, J. (1923). Charts for estimating temperature distributions in heating or cooling of solid shapes. *Industrial and Engineering Chemistry* 15 (11), 1170–1172.

HEISLER, M.P. (1947). Temperature charts for induction and constant temperature heating. *Transactions of the American Society of Mechanical Engineers* 69 (4), 227–236.

HEPPELL, N.J. (1985). Measurement of the liquid-solid heat transfer coefficient during continuous sterilisation of foodstuffs containing particles. In: *Symposium on Aseptic Processing and Packaging of Foods*, Tylosand, Sweden, 9–12 September 1985. Lund, Sweden: Institute of Technology, pp. 108–114.

LEWIS, M.J. and HEPPELL, N.J. (2000). *Continuous thermal processing of foods.* Gaithersberg, Maryland, USA, Aspen Publishers Inc.

MAESMANS, G., HENDRICKX, M., DECORDT, S. and TOBBACK, P. (1993). Theoretical Considerations on Design of Multicomponent Time- Temperature Integrators in Evaluation of Thermal-Processes. *Journal of Food Processing and Preservation* 17(5): 369–389.

MWANGI, J.M., RIZVI, S.S.H. and DATTA, A.K. (1993). Heat transfer to particles in shear flow: application in aseptic processing. *Journal of Food Engineering* 19(1): 55–74.

RANZ, W.E. and MARSHALL, W.K.JR. (1952). Evaporation from drops. *Chemical Engineering Progress* **48**, 141–146.

ROHSENOW, W.M. and CHOI, H. (1961). *Heat, Mass and Momentum Transfer*. New Jersey, USA: Prentice-Hall Inc.

SASTRY, S.K., HESKITT, B.F. and BLAISDELL, J.L. (1989). Experimental and modelling studies on convective heat transfer at the particle-liquid interface in aseptic processing systems. *Food Technology* 43 (3), 132–136, 143.

SASTRY, S.K., LIMA, M., BRIM, J., BRUNN, T. and HESKITT, B.F. (1990). Liquid-to-particle heat transfer during continuous tube flow: influence of flow rate and particle to tube diameter ratio. *Journal of Food Process Engineering* 13 (3), 239–253.

SASTRY, S.K. and CORNELIUS, W.D. (2002). *Aseptic processing of foods containing solid particulates*. New York, John Wiley & Sons Inc.

SCHNEIDER, P.J. (1963). *Temperature Response Charts*. New York, USA: John Wiley & Sons Inc.

SEGNER, W.P., RAGUSA, T.J., MARCUS, C.L. & SOUTTER, E.A. (1989). Biological evaluation of a heat transfer simulation for sterilising low-acid large particulate foods for aseptic packaging. *Journal of Food Processing and Preservation* 13, 257–274.

TEIXEIRA, A.A., DIXON, J.R., ZAHRADNIK, J.W. and ZINSMEISTER, G.E. (1969). Computer determination of spore survival distributions in thermally processed conduction heated foods. *Food Technology* 23, 352–354.

TUCKER, G.S & WITHERS, P.M. (1994) Determination of residence time distribution of non-settling food particles in viscous carrier fluids using Hall effect sensors. *Journal of Food Process Engineering*. 17, 401–422.

VAN LOEY, A., HENDRICKX, M., DE CORDT, S., HAENTJENS, T. and TOBBACK, P. (1996). Quantitative evaluation of thermal processes using time- temperature integrators. *Trends in Food Science & Technology* 7(1): 16–26.

WHITAKER, S. (1972). Forced convection heat transfer for flow in pipes, past flat plates, single cylinders, single spheres and for flow in packed beds and tube bundles. *American Institute of Chemical Engineers Journal* 18 (2), 361–371.

Index